TEXTBOOK
OF
DAIRY CHEMISTRY

Textbook
of
DAIRY CHEMISTRY

M.P. MATHUR

D. DATTA ROY

P. DINAKAR

National Dairy Research Institute
Karnal 132 001

भाकृअनुप
ICAR

Published by

Directorate of Knowledge Management in Agriculture
Indian Council of Agricultural Research
Krishi Anusandhan Bhavan, Pusa
New Delhi 110 012

PRINTED : DECEMBER 1999
REPRINTED : JULY 2005
REPRINTED : DECEMBER 2008
REPRINTED : JULY 2019
REPRINTED : APRIL 2024

Project Director (DKMA) : Dr R Roy Burman
Incharge (English Editorial) : Ravindra Verma
Incharge (Production Unit) : Punit Bhasin

ISBN : 978-81-7164-046-X

Price: ₹550

Published by Dr R Roy Burman, Project Director (DKMA), Directorate of Knowledge Management in Agriculture, Indian Council of Agricultural Research, Krishi Anusandhan Bhavan I, Pusa, New Delhi 110 012 and printed at M/s Dolphin Printo Graphics, 1E/18, Jhandewalan Extension, New Delhi 110 055.

PREFACE

PROGRESS and advancement of civilization have direct links with documentation in black and white. Man's each stride on moon, was a march forward or a leap for the humanity. Now in the last decade of the 20th century, man has crossed the national boundaries with regards to the advancement and progress of knowledge which after 1912 has accumulated like an immense wealth. From this immense wealth of knowledge dairy science developed in leaps and bounds since the year 1917 and today we have come to the conclusion that milk is a delicately balanced biochemical fluid.

This textbook on Dairy Chemistry is an attempt to cross the national boundaries but simultaneously it is intended to provide the basic fundamentals of Dairy Chemistry to students graduating in dairy science. It is a fact that books releated to Dairy Chemistry by Indian authors are very few.

With the introduction of the new, revised common syllabus and development of a common co-ordinated teaching programme in dairy education in India, it was realized that it is high time that a comprehensive textbook on Dairy Chemistry at undergraduate level is written.

This textbook is a sincere venture and effort to provide a compact, varied and vivid picture of Dairy Chemistry to B.Tech (Dairy Technology) students in India. Clarity and presentation of each topic have been maintained so as to cover the syllabus in Dairy Chemistry for B.Sc. (Agriculture) and B.Sc. (Food Technology) of the numerous Agricultural Universities in India. The subject matter has been presented lucidly through tables and flow diagrams thereby enabling us to provide the vast accumulated data of research findings in Dairy Chemistry with clarity and simplicity.

Though written with an International approach this textbook is meant to be a coveted and prized collection and document for B.Tech (Dairy Technology) students all over India and adjoining nations.

Apart from providing basic information on milk and its composition, properties, proteins, carbohydrates, lipids, minerals, minor constitutents, legal standards, nutritional importance and chemistry of Indian and occidental products, the topics themselves have a good buffering capacity thereby maintaining the standard throughout the textbook.

For the easy comprehension of the students and for providing a firm base, additional information has been provided in chapters on comparative aspects, physico-chemical, bio-chemical and biological approach to common dairy processes and analytical aspects of Dairy Chemistry. Some topics of common interest are added in current topics chapter.

For the benefit of students some review questions have been given at the end of the chapters. A list of research laboratories having interest in Dairy Chemistry and standards of various milk and milk products in India have been provided in the Appendix. Attempts have been made to highlight research findings of this Institute in Dairy Chemistry.

We are indebted to the authors, editors and publishers of various scientific books, periodicals and pamphlets from which some of the information presented in the text has been gathered. In particular, we have freely consulted the following :

1. Jenness, R. and Patton, S. : Principles of Dairy Chemistry
2. Rangappa, K., and Achaya, K.T. : Indian Dairy Products.
3. De, S. : Outlines of Dairy Technology.
4. Webb, B.H., Johnson, A. and Alford, J.A. : Fundamentals of Dairy Chemistry.
5. Indian Standard Specification
6. Indian Dairyman (1980–1992).
7. Dairy Guide (1986–1992).
8. Dairy Hand Book (Processing), NDRI, Karnal (1980, 1986).

We are thankful to fellow scientist, Dr B.N. Mathur for his sincere efforts in revising the manuscript. Mr. P. Dinakar left for USA for higher studies. In his absence, the entire burden of editing, drawing of structures, checking of spellings and rewriting of a couple of chapters was shouldered entirely by Mr. P.K. Gupta, Technical Officer, Dairy Chemistry Division, National Dairy Research Institute, Karnal. We acknowledge with gratitude the dedicated help rendered by Mr. P.K. Gupta for the last couple of years. Without his endeavours, this manuscript could not have been as it is at present.

We express our deep gratitude to Dr N.C. Ganguli for revising the contents of the book and offering various suggestions and to the Indian Council of Agricultural Research, New Delhi for approving the project under the scheme, "University Level Books – 1991–92". All our efforts to write this manuscript would have been in vain but for the strong, devoted, dedicated and timely action by our Director, Dr O.S. Tomer. It is only because of his sincere efforts that this manuscript will see the light of the day. Our gratitude knows no bound for him.

M.P. Mathur
D. Datta Roy
P. Dinakar

INTRODUCTION

MILK is a lacteal secretion of the mammary glands of mammals primarily to provide an almost complete food to their off-springs. Such secretions occur normally after parturition under the influence of inherent lactogenic hormones which when injected into the venous blood of pregnant mammals may cause secretion of milk prior to parturition. However, such a process is temporary and cannot be practically utilised primarily because of the hormonal imbalance it may produce and the milk so secreted cannot be defined as milk. It is only after normal parturition, the secretion of the mammary gland is considered first as colostrum and later as milk. Colostrum is a repository of immunoglobulins and thereby provides natural immunity to the new born off-springs of mammals. A marvel of nature is colostrum and so indeed is milk.

Milk is rich in proteins, lipids, carbohydrates, enzymes, minerals, vitamins and other minor constituents. Essential, semi-essential and non-essential amino acids are provided by milk proteins to the mammalian off-spring.

The biological value of milk proteins is very high and when supplemented with cereals, the biological value gets enhanced. This enhancement of the biological value is known as the "supplementary value" of milk proteins. The protein content of milk of any species is related to the days required to double the weight of any species. For example, the protein content of rabbit's milk is highest and the time required to double the weight of the off-springs is the shortest. The percentage of protein in human milk is lowest, and hence human babies take the longest time to double their weights. The stability of casein fractions has been attributed to their non-possession of tertiary structures, and hence casein micelles are more stable than many other proteins. However, whey proteins are less heat stable as α-lactalbumin. ß-lactoglobulin and immunoglobulins are globular proteins possessing secondary and tertiary structure and undergo three dimensional conformation changes when heated. The role of α-lactalbumin in lactose biosynthesis has been well-defined. Its effect is allosteric i.e. it acts as a modifier protein in lactose synthetase reactions. The role of ß-lactoglobulin in augmenting human eyesight has been well documented. Immunoglobins provide immunity to mammalian off-springs while lactoferrin, the iron-containing whey protein plays a bacteriostatic role and also provides iron to the mammalian off-springs. Lactoferrin is present in rich amounts in colostrum but decreases disconcertingly as colostrum slowly transforms to milk, thereby making milk deficient in iron.

Milk lipids exist in milk as fat in water emulsion in the shape of globules which are stabilized by a membrane primarily of lipo-protein complex. Milk lipids provide the necessary calories for the maintenance and growth of mammalian off-springs.

Triesters of glycerol and fatty acids are the main lipids of milk. The lower chain fatty acids and part of the medium chained fatty acids are biosynthesized in the mammary glands of the mammals. The higher fatty acids rise from depot fats of the body of mammals, chylomicra and low density lipo-proteins.

Milk lactose, the principal sugar of milk, exists in true solution. Apart from providing the necessary calories for the maintenance and growth of the off-springs, lactose is the substrate for microbes in producing many cultured dairy products from bovine milk. Milk also contains 0.3 to 0.6% oligosaccharides. Lactose exists in milk as the α- monohydrate and ß-anhydrous of which the alpha form is less soluble and hence has a tendency to crystallize out thereby causing problems of sandiness in the ice-cream, sweetened, condensed and evaporated milk industry. However, this problem is not evident in spray-dried powders.

Milk contains an abundance of minerals which exist in true solution as well as in a colloidal state. Trace minerals have been also found in milk. Calcium exists in milk as a colloidal, tertiary calcium phosphate and in true solution also. The tertiary calcium phosphate exists in milk in the hydroxy-appetite form. Milk contains more than thirty enzymes but their exact role is yet to be well-defined.

Fat soluble and water soluble vitamins are also present in milk. The fat soluble vitamins are A, D, E and K. The water soluble vitamins, i.e. the B- complex from thiamine to *p*-amino benzoic acid are all present in milk. Vitamin C (ascorbic acid) is also present.

Milk is a perishable food item and from times immemorial it has been converted to products in which heat, acid and rennet have been used. Most of the Indian Dairy products are heat desiccated and in some, acid is also used. Heat desiccated and acid treated products have a longer keeping quality than milk. Some of the heat/acid coagulated products are *Paneer* (cheese) and *Chhana.* From the Vedic times, heat desiccated milk products are a part and parcel of the Indian Dairy Products e.g. *khoa, rabri, ghee,* etc.

Conversion of milk to milk products is based on two aspects e.g., it reduces the bulk and enhances the keeping quality. One of the western dairy products owes its birth due to the above two criteria. This product commonly known as cheese was discovered by the nomads accidentally.

To ensure quality milk and milk products flow, out of Dairy Plant premises and reach the consumer, a practical and systematic quality control programme has to be carried out in the Dairy Plants. To check any adulteration and prevent pathogens from entering the finished product, strict legal standards have to be enforced and adhered to. Here, the analytical aspects of Dairy Chemistry play a vital role. The compliance with legal standards bear a direct relationship with a practically oriented genuine quality control programme. A textbook in Dairy Chemistry is not complete when Dairy Physics or physical properties of milk are not taken care of. Choicest of the occidental dairy products leave the Dairy Plant premises marked with the indelible stamp of very high keeping, safety, and aesthetic quality. This is due to the strict adherence to legal standards through an intelligent and farsighted quality control programme. However, apart from the western and some Indian Dairy products which squeeze out of the Dairy Plant premises of organized sectors, most of the Indian Dairy Products are prepared by "Halwais" who have an apathy and scant regard for quality and have, for the most part, an urge for more profit and least concern for aesthetic, safety and keeping quality of the finished product. Hence, particularly strong emphasis has been given to the analytical aspects of Dairy Chemistry, legal standards and quality control in the present text book.

This preamble lends a glance to the fabulous landscape and the gorgeous depths in literature of Dairy Chemistry one can plunge and probe into.

CONTENTS

1

COMPOSITION OF MILK

MAMMALS are characterized by the presence of the mammary glands capable of producing milk required for nourishment of their off-spring. Milk is the sole source of energy for the young ones till they are weaned and are capable of fending for themselves. Thus, the milk of each species of mammals is tailor-made to suit the requirements of its own offspring. When man changed from a nomadic hunter to a settler and adopted farming, one of the turning points was the domestication of cattle. He found that he could rear cattle quite easily and benefit from the copious flow of milk which is remarkably compatible to the needs of humans also. Thus, man is unique in exploiting the milk produced by species other than his own.

The new born infant is not equipped to ingest and assimilate nutrients from any food source other than milk. Consequently, milk has to provide all the ingredients required for early growth, in an easily acceptable form. In short, milk is a complete food in itself. It provides fat, carbohydrates, proteins, enzymes, minerals, vitamins, etc. in the right proportions and in a tasty, acceptable form. Milk is, therefore a preferred and indispensable item in the diets of convalescents and adults. However, milk is deficient in iron, vitamin C and D. For human infants mother's milk is an ideal food. However, cow, buffalo and goat milks may be fed to infants as suitable replacement for mother's milk. Even the diet of human adults can very well be supplemented by milk of these species.

The Gross Composition – The Main Ingredients

Milk is the liquid food secreted by the mammary gland for the nourishment of the newly born, containing water, fat, protein, lactose and minerals. Milk contains on an average 87 per cent water, 3.9 per cent fat, 4.9 per cent lactose, 3.5 per cent protein and 0.7 per cent minerals, vitamins and other minor constituents. Milk of ruminants like cow, buffalo and goat is ideally suited for human consumption and meet the basic dietary requirement of human beings.

Milk may be defined as the lacteal secretion, practically free from colostrum, obtained by the complete milking of one or more healthy animals. According to PFA Rules (1976) cow milk should contain not less than 8.5 per cent of milk solids-not-fat and not less than 3.5 per cent of milk fat while buffalo milk should contain not less than 9.00 per cent of milk as solids-not-fat and not less than 6.00 per cent of milk fat.

The advent of parturition is concomittant and simultaneous with the biosynthesis and secretion of milk, a biological phenomenon triggered by lactogenic hormones. For the first few days after parturition mammary glands secrete, a fluid known as colostrum which has a strong odour, a bitter taste, a slight reddish yellow colour and contains a high percentage of immunoglobulins. It is a rich source of all milk constitutents except lactose, potassium and pantothenic acid. The Table 1.1 shows the gradual

Table 1.1. Transition from colostrum to normal milk

Time after calving (Hr.)	Total protein (%)	Casein (%)	Albumin (%)	Fat (%)	Lactose (%)	Ash (%)	Total solids (%)
0	17.5	5.08	11.34	5.10	2.19	1.01	26.20
6	10.0	3.51	6.30	6.85	2.71	0.90	20.46
12	6.05	3.00	2.96	3.80	3.71	0.89	14.53
24	4.52	2.76	1.48	3.40	3.98	0.86	12.77
30	4.01	2.56	1.20	4.90	4.27	0.83	13.63
36	3.98	2.77	1.03	3.55	3.97	0.84	12.22
48	3.74	2.63	0.99	2.80	3.97	0.83	11.46
72	3.86	2.70	0.97	3.10	4.37	0.84	11.86
96	3.76	2.68	0.82	2.80	4.72	0.83	11.85
120	3.86	2.68	0.87	3.75	4.76	0.85	12.67
168	3.31	2.42	0.69	3.45	4.96	0.84	12.13

Source : Fundamentals of Dairy Chemistry Webb., Johnson and Alford.

change from colostrum to normal milk. After about 5 days after parturition, the secretion of udder is considered as the normal milk.

Milk Fat: Milk fat is the most variable constitutent in comparison with other nutrients of milk. Triglycerides, diglycerides and monoglycerides comprise about 98–99 per cent of milk fat. The triglyceride mixture is very soft and brings smoothness and taste to milk and its products. Apart from triglycerides, milk fat consists of phospholipids, sterols, carotenoids, fat soluble vitamins (A, D, E and K) and traces of fatty-acids. These contribute 1–2 per cent of the total fat in milk. Milk fat provides approximately one-half of the total calories.

Milk Proteins: Casein comprises eighty per cent of the total milk protein, α, ß and K are the three main fractions of milk casein. α-casein is a proteolytic breakdown product of ß-casein when milk is stored at refrigeration temperature. The rest, twenty per cent of milk protein consists of lactoglobulin and lactalbumin commonly known as whey protein. Lactalbumin comprises three fractions, viz., α-lactalbumin, ß-lactoglobulin and serum albumin. All the fractions of casein, α-lactalbumin and ß-lactoglobulin exhibit polymorphism. Lactoglobulin is made up of euglobulin and pseudoglobulin. In addition to this, other minor fractions like protease-peptone, lactoferrin and other proteins associated with enzymes form the part of milk proteins. A complex of proteins and enzymes absorbed on the surface of the fat globules of milk is usually designated as the membrane proteins of the fat globules. Milk proteins contain essential amino acids which are needed by the body. About 90–95 per cent protein can be digested in the body when ingested in the form of milk and its products. Milk proteins are very important because these form a part of the dietary sources of protein which forms proteinaceous materials and structure of the body particularly of the infants and even adults.

Milk Carbohydrates: The principal sugar of milk is lactose. It is a disaccharide consisting of a glucose and a galactose unit. Its systematic name is 4-0-ß-D galactopyranosyl-D-glucopyranose. It is a reducing sugar and the reducing group lies at carbon number 1 of the glucose moiety. Milk also contains the sialic acid derivative such as N-acetyl Neuraminic acid. Human milk is a rich source of sialic acid derivatives and lactose containing oligosaccharides. In human milk its percentage varies from 0.3–0.6 percent. Amongst the lactose containing milk oligosaccharides, those containing nitrogen have specific importance in infant nutrition as a result of their association with "Bifidus factor" a growth factor for the growth of *B. bifidium* found predominantly in the intestines of the breast-fed babies. Growth of bifidus organisms in the intestines of new born infants produce unfavourable condition for the survival and growth of enteropathogenic and viral organisms in the intestines of human babies. Also, growth of *B. bifidium* and *B. bifidium* var pensylvanicus helps in the abosrption of minerals by infants and promotes biosynthesis of vitamin B complex.

Milk Minerals: The amount of non combustible matter in milk is designated as the ash content of milk. In normal milk the value remains constant at 0.7 percent. A much higher value than this is indicative of abnormal conditions in the secreting mammary glands. The important minerals present in milk ash are calcium, phosphorus, magnesium and sulphur. Calcium and phosphorus are the pre-requisites for bone formation in the body.

Milk Vitamins: All the recognised vitamins are found in milk. Some are not present in significant quantities but for others, milk is a very rich source and is indeed one of the best available source. Milk is a rich source of vitamin A, B_{12}, other B group vitamins. Vitamin D, E and K are also found in milk. The B vitamins act as coenzyme in many enzyme catalyzed reactions. Hence the metabolic activities in human and animal bodies are controlled by the levels of B vitamins. A vitamin-deficient diet is known to cause serious diseases.

Table 1.2 depicts the important constitutents of milk in brief. While Table 1.3 deals with percentage of some constitutents present in milk of different species.

Table 1.2. Important constituents present in milk

- Milk
  - Water
  - Total solids
    - Fat
      - Triglycerides
      - Phospholipids
      - Sterols
      - Fat soluble vitamins A,D,E,K
    - Solids not fat
      - Nitrogenous component
        - Protein
          - Casein: α-casein, β-casein, K-casein
          - Non casein: α-lactalbumin, β-lactoglobulin, Immunoglobulin
        - Non protein nitrogen: Ammonia, amino acids, urea, uric acid, creatinine
      - Carbohydrate
        - Lactose
        - Other sugar, Oligosaaccharides, Sialic acid
      - Minerals: Calcium, Magnesium, Phosphorus, Sodium, Potassium, Citrate, Chloride, Sulphate
      - Enzymes: Protease, Lipase, Catalase, Peroxidase, Amylase, Carbonic Anhydrase, Xanthine-oxidase, Aldolase, Phosphatase
      - Vitamins: (Water soluble) Vit.B Complex, Vit. C.

Table 1.3. Constituents of cow, buffalo, goat, sheep, human and camel milk (per cent)

Constituents	Cow Milk	Buffalo Milk	Goat Milk	Sheep Milk	Human Milk	Camel Milk
Moisture	87.6	81.0	85.2	80.71	87.4	87.61
Protein	3.3/3.4	4.3/3.9	3.7/3.5	4.8	1.6	2.98
Fat	3.6/4.9	8.8/6.6	5.6/4.5	6.0	3.75	5.98
Carbohydrates	4.5/4.1	5.0/5.2	4.7/4.6	4.9	6.98	3.26
Mineral matters (ash)	0.7	0.8/0.7	0.8	0.8	0.21	0.70
Calcium	0.12	0.21	0.17	–	–	–
Phosphorus	0.09	0.13	0.12	–	–	–
Iron	0.2	0.2	0.3	–	–	–
Vitamin A (IU/100g)	180	162	182	189.7	189.9	–
Vitamin B_{12} (mg/100g)	0.05	0.04	0.04	–	–	–
Riboflavin (mg/100 g)	0.2	0.19	0.15	–	–	–
Vitamin C (mg/100g)	1.6	1.5	1.5	5.5	4.0	–
Nicotinic acid (mg/100g)	0.1	0.1	0.1	–	–	
Calories	67	117	72	–	–	–
Total solids	13.7	17.02	13.5	16.3	12.57	–
Solids-not-fat	8.8	9.2	9.0	10.3	8.80	–

Factors Affecting Composition

Milk contains the same gross constitutents, but differs widely in proportion. This is due to the availability of different species of mammal. In order to determine an average gross composition, various factors such as number of samples taken from different herd, place, breed, various parts of country, different periods of lactation are taken into consideration. These factors are conveniently grouped into (a) animal factors and (b) environmental and others. However, for human consumption, cow, buffalo, goat and sheep are generally considered suitable and studied in detail.

Animal Factor

Species: The variation in milk composition in different species depend on so many factors that it is difficult to determine accurately the relationship betwen a specific factor and its effect on composition. However, the composition of milk of each species is designed to provide adequate nutrition for natural rate of growth which is an inherited trait of new born mammals for that species. The faster the rate of growth, the more concentrated are the milk components needed for this growth (Table 1.4).

Table 1.4. Growth rate of young mammals in relation to protein content of their mother milk

Species	Days required to double birth weight	Protein percentage in milk
Man	180	1.6
Cow	47	3.5
Sheep	15	4.80
Pig	14	5.21
Cat	9.5	7.00
Rabbit	6	10.38

Breed: The composition of milk is strongly influenced by the breed. The variation in composition due to the breed is most manifest in fat, though other solid constituents are also affected but to a much lesser degree. Generally, milk containing a higher percentage of fat is also rich in solids-not-fat and vice versa (when milk of one breed is compared with that of another). The variation in composition of milk are evident from the following Tables 1.5, 1.6 and 1.7.

Table 1.5. Variation in composition of milk in different breeds (per cent)

Breeds	Water	Fat	Protein	Lactose	Ash	SNF
Jersey	85.27	5.14	3.80	5.04	0.75	9.59
Guernsey	85.47	4.96	3.84	4.98	0.75	9.57
Ayrahire	87.10	3.85	3.34	5.02	0.69	9.05
Short horn	87.43	3.03	3.32	4.89	0.73	8.94
Friesian	88.01	3.45	3.15	4.65	0.68	8.48

Source : Dairy Chemistry and Animal Nutrition by M.M. Rai

Table 1.6. Variation of composition of milk due to breed of cow

Composition in per cent	Red Sindhi	Gir	Tharparkar	Sahiwal	Crossbred
Lactometer (20°C)	29.41	29.03	29.36	29.88	29.27
Acidity	0.15	0.16	0.16	0.16	0.16
Fat	4.90	4.73	4.55	4.55	4.50
Crude proteins	3.42	3.32	3.36	3.33	3.37
Lactose	4.91	4.85	4.83	5.04	4.92
Ash	0.70	0.66	0.68	0.66	0.67
SNF	8.76	8.63	8.70	8.82	8.63
Total solids	13.66	13.30	13.25	13.37	13.13

Source : Indian Dairy Products, Rangappa and Achaya

Table 1.7. Variation of composition of milk according to the breed of buffalo

Breed	Fat	SNF
Murrah	6.8	10.1
Jaffarabadi	7.3	10.1
Surti	8.4	10.3

Source : Indian Dairy Products, Rangappa and Achaya

Different quarters of the udder: The four quarters of the udder are separate units physiologically as well as anatomically.

The quarters milked first yield milk having the highest fat content; milk from the last quarters milked has the lowest fat content. In general, the gross composition of milk from different quarters may show differences, however, the yield from right hand quarter averaged slightly more than the rest. When a milking machine is used and all four quarters are milked simultaneously, the milk from the different quarters will be very similar in fat percentage (Table 1.8).

Table 1.8. Fat content of milk from different parts of udder (Fat per cent).

Animal	L.Fore	R.Fore	Average L.side	L.Hind	R.Hind	Average R. side
1.	4.36	4.33	4.39	4.42	4.20	4.29
2.	4.42	4.68	4.37	4.31	4.38	4.52
3.	4.19	4.12	4.28	4.32	4.27	4.19

Source : Indian Dairy Products, Rangappa and Achaya

Lactation Period: The fat content of milk varies widely during the lactation period, but the general trend of the fat percentage is similar to the trend of the solids-not-fat. Fat content is high after parturition, decreases to a low during the first or second month of lactation, and then gradually increases during the remainder of the lactation period. In addition to this composition of fat is also changed by lactation. The only change in lactose percentage attributable to stage of lactation is a slight decrease towards the end.

The chloride content increases with the advancement of lactation. Calcium decreases to a minimum concentration and then increases, whereas total phosphorus declines throughout. Natural acidity commences from a high level and decreases to the normal value during the first three months and remains fairly steady until the last month when it declines sharply.

Feeds and Nutritional Level: Feeding more than the normal amount of feed, or feeding additional concentrates, has been reported to increase the solids-not-fat slightly. Feeding excess protein does not appreciably alter the protein content of milk, but the non-protein nitrogen content and at time fat content may be enhanced. When the animals are allowed fresh pasturage the solids-not-fat increases. The lactose-content is not changed.

It has been found that feeding of oils such as palm oils, butter fat, lard and coconut oils increases the fat percentage while cod liver oil or herring oil produce opposite effect. The results are variable not only in the individuals but in the same animal at different times. The unsaturated fats usually have depressing action and the saturated fat increase the fat percentage. The specific effects of cod liver oil is largely removed either by hydrogenation which renders the fat less saturated or by feeding the fat in a form in which it is protected from degradation and biohydrogenation in rumen. Food fats modify the composition of milk fat to a limited extent. However, the technology to prevent rumen degradation and biohydrogenation of fat has enabled to modify the fatty acid profile tailored to our requirements. In several studies the content of essential fatty acids has increased by 35 per cent or so.

An increase in the proportion of certain groups of fatty acids in the food fat effect a similar but slightly modified action on the composition of the milk fat. This is particularly true of fatty acids of the oleic type which produce soft fat. Similarly, grass feeding increased the unsaturation of milk fat.

Disease: Pronounced changes in the composition of milk are associated with an inflammed condition in the udder. The composition of milk from an animal affected by mastitis disease shows a low lactose and casein percentage, increase in chloride content, increase in percentage of soluble nitrogen and a consequent reduction in the casein number, reduction in natural acidity. Osmotic pressure remains normal and there is no change in freezing point but non-fatty soilds are low.

At the onset of the foot and mouth disease, there is a marked reduction in the volume of milk secreted, whether the udder is inflammed or not. If the udder is not inflammed, moderate increase in the percentages of fat, protein, and ash occurs, with a decrease in the percentage of lactose. If the quarters are inflammed, the changes are much greater and resemble those of severe mastitis.

Age of the Animal: The average fat percentage of milk decreases slightly with the age of the cow. A gradual though irregular, decrease in solids-not fat content is also noted with age. The decline in solids-not-fat percentage with age is approximately twice the magnitude of the decline in fat percentage. Lactose and casein are the solids-not-fat components, that are most affected by the ageing of the cow. Decline in solids-not-fat ranging from 0.21 to 0.45 percent during the first 7 years of lactation have been reported. Maximum milk yields may occur from the fifth to the ninth lactation. Age factor is most potent in advance lactation.

Heat or Oestrum: The effect of oestrum on milk composition is real but inconsistent. There is usually much variation in the fat test, together with a slight decrease in yield during this period. These changes have been attributed to increased excitability and nervousness causing the cow to either hold up some of the milk or to secrete less milk.

Gestation: Gestation can affect the composition of milk in an indirect manner by hastening the end of lactation which caused marked changes in the composition of milk. It is also accepted that gestation does cause a direct change in the composition of milk. An increase in milk solids, especially solids-not-fat, starts at about the fourth month of pregnancy and continues to increase till the end of the lactation. Cows which are open show no rise or negligible rise in solids content during a comparable period in their lactation.

Hormones: Prolactin, one of the hormones of the pituitary gland is responsible for the maintenance of lactation. It is a heat stable hormone. When prolactin is given to cows past peak production an increase in the yield and fat percentages have been observed. Sometimes lactose concentration increases following treatment with pituitary extracts. Thyroxin of thyroid glands control the metabolic rate of the body. When the activity of the body processes increases, there is an increase in the fat percentage also. Excess feeding of hormone has a depressing effect on production. Oestrogen of ovaries and the female sex organs has two fold effect on lactation that is stimulating and depressing. The stimulating effect is characterised by increase in fat and non-fatty solids with decrease in yield. The percentage of protein is also raised but casein number remains normal. The depressing effect is due to increased doses of compound.

Environmental and Others Factors

Milking Intervals: The composition of milk is not markedly affected by different intervals between milkings except for the percentage of fat. When milking is done at longer intervals, the yield is also more with a corresponding smaller percentage of fat, whereas milk drawn after short intervals yield smaller quantities with higher amount of fat. A definite variation is generally observed in the fat content of both the individual and herd milk between the morning and evening bulk milk samples (Table 1.9).

Table 1.9. Variations in fat content of milk due to the time of milking (Fat per cent).

Time of milking	Red Sindhi Cow	Gir cow	Buffalo
Morning	6.0	3.2	7.1
Evening	6.3	6.2	7.9

Source : Indian Dairy Products, Rangappa and Achaya

Milking Efficiency: The percentage of fat increases markedly during milking; hence the udder must be effectively stripped if the highest percentage of possible fat is desired in the milk. If high, fat milk is left in the udder at one milking, it may be obtained by stripping at a subsequent milking. No appreciable difference other than in fat have been found between fore milk and strippings. The fore-drawn milk contains about 1 to 2 percent fat and as milking proceeds the fat percentage increases to about 6 percent or more in strippings. However, in general it was observed that the first portion of the milk drawn is richest in solids-not-fat but poorest in fat and vice-versa. In other words the fat content increases and the S.N.F. decreases with the progress of milking (Table 1.10).

Table 1.10. Composition of different portions of a milking

Portions	Fat per cent	S.N.F. per cent
1	1.04	8.64
2	1.92	8.63
3	3.02	8.51
4	4.40	8.37
5	5.32	8.15
6	7.63	7.77

Source : Indian Dairy Products, Rangappa and Achaya

Season: Both fat and solids-not-fat shows a slight but well defined variation during the course of the year. A variation due to season in the fat percentage shows maximum values in May and minimum in November, while solids-not-fat shows highest value in October for typical Red Sindhi cow breed (Table 1.11). The lower values for solid-not-fat occur in July and September. So far as milk production is concerned, season includes a number of separate factors, such as temperature, hours of sunshine, length of the day, exercise, rainfall, drought, stall feeding or pasture feeding, influx of newly calved cows, thunderstorms, etc.

Table 1.11. Effect of season on yield and composition of Red Sindhi cow milk in particular region

Month	Season	Yield (Litre)	Total Solid (%)	Fat (%)	S.N.F. (%)	Lactose (%)	Ash (%)	Protein (%)
January	Winter	5.12	13.830	5.14	8.652	4.656	0.784	3.349
February		4.92	13.439	4.86	8.573	4.691	0.746	3.143
March	Spring	4.34	13.574	5.03	8.541	4.668	0.757	3.186
April		4.82	13.312	4.82	8.453	4.738	0.743	3.176
April		4.82	13.312	4.82	8.453	4.738	0.743	3.176
May	Summer	4.01	13.545	5.19	8.360	4.517	0.748	3.254
June		4.00	13.237	4.84	8.355	4.480	0.751	3.139
July		3.34	13.358	5.11	8.240	4.232	0.749	3.243
August	Rains	4.80	13.614	5.18	8.438	4.449	0.756	3.205
September		5.38	13.141	4.85	8.291	4.602	0.731	3.085
October		6.16	13.443	4.82	8.683	4.609	0.774	3.237
November	Winter	6.01	13.332	4.72	8.601	4.700	0.747	3.231
December		5.01	13.459	4.84	8.618	4.575	0.753	3.265

Source : Indian Dairy Products, Rangappa and Achaya

However, the following points are clear :

a. An increase in temperature favours slight decrease in fat percentage and vice-versa.

b. The amount of vitamin D is directly related to the exposure of the cows to sunshine.

c. Carotene and riboflavin reach maximum level when green feeds are available in plenty.

d. Summer drought depresses the percentage of solid-not-fat but has no effect on fat percentage.

Lactose is the least variable of the gross components on a percentage basis, but because of its osmotic relationship in milk, it increases slightly as the other solids decrease and decreases as the other solids increase.

Weather: In dry months the total yields of milk tends to decrease, together with a decrease in the solid-not-fat and an increase in fat percentage. In wet months there is a tendency for both the solid-not-fat and the fat percentage to decrease, with or without an increase in the yield of milk, depending on the level of nutrition. Controlled humidity, as opposed to natural weather conditions, does not appear to influence milk composition when the temperature is kept below 24°C.

Exercise: Dairy cows are benefited by milk exercise. During a period of exercise feed consumption is increased, milk production is maintained, and the percentage of fat is increased over a corresponding period of rest.

Excitement: During periods of sexual excitement, storms, worrying by dogs or insects, etc. both yield and composition are affected but no generalisation can be made. Some cows are more sensitive to unusual conditions than others.

INTER-RELATIONSHIP BETWEEN THE MILK CONSTITUENTS

On the whole, the major constituents of milk are independent of each other.The concentration of the constituents are influenced by a large variety of factors described earlier . Physiologically, milk has to maintain isotonic equilibrium with blood. Consequently, it is seen that constituents of milk which contribute towards the osmotic pressure, namely the lactose, chloride, sodium, etc., bear a relationship with each other. Fat is completely independent of any other constituents and is influenced only by the hereditary and environnmental factors like breed, type of feed, interval of milking, stage of lactation and so on.

Fat and Solids-not-fat

Generally it is seen that an increase in the fat content is concomitant with an increase in the solid-not-fat content. This relation is only empirical and not quantitated in specific terms. The protein and fat content are positively correlated and since the lactose content is fairly constant, the above relation between fat and solids not fat is understandable. This correlation is significant only from the point of view of fixing legal standards and pricing policy of milk. This correlation also gives a broad indication of adulterated milk with water particularly in India where attempts to dilute buffalo milk are quite common.

Vieths Ratio

In normal milk, the ratio of percentage of lactose:protein:ash bears a definite relationship. Vieth estimated this as 13:9:2. For Indian milks this ratio may not hold good and has been estimated as 12.4:8.6:2 for cow milk and 12.4:9.7:2 for buffalo milk. This ratio varies for abnormal milks, when the udder is infected, etc. If milk is diluted with water all the three constituents decrease proportionately and the ratio is still maintained. But the freezing point would change confirming adulteration with water.

Lactose–Chloride Number

The two factor contributing most towards maintaining the osmotic pressure of milk, are lactose and chloride. Hence, a decrease or increase in one of these is matched by a proportionate increase or decrease in the other.

This is particularly true in the case of mastitis infected animals, where it is seen that the lactose concentration decreases, but a simultaneous proportional increase in the chloride content is observed. This relationship has been quantified as the chloride-lactose number or the Koestler number (100 × per cent chloride)/per cent lactose. This number varies between 1.5 and 3.0 for normal milks. The relationship between chloride and lactose arises from the need to maintain isotonic equilibrium with the blood.

REVIEW QUESTIONS

1. Define Milk.
2. What are the different constituents present in milk ?
3. Give the approximate percentage of major constituents, present in the cow and buffalo milk.
4. Which is the major protein of milk
5. What are the isoelectric points of different milk proteins ?
6. Define a fat, what are different constitutents present in the fat ?
7. What is sialic acid ?
8. List the factors which affect composition of milk.
9. Which milk has the highest protein content ?
10. What is colostrum ?
11. What is vieths ratio ?
12. What is the relationship between lactose and chloride content of milk ?
13. What is gestation period ?

REFERENCES

Jenness, R. and Patton, S. 1959. *Principles of Dairy Chemistry.* John Wiley & Sons, Inc. New York, Chapter 1.

Rai, M.M. 1964. *Dairy Chemistry and Animal Nutrition.* Ram Prasad & Sons, Agra-3, Chapter 3.

Rangappa, K.S. and Achaya, K.T. 1974. *Indian Dairy Products.* Asia Publishing House, Bombay. Chapter 2 and 4.

Webb, B.H., Johnson, A. and Alford, J.A. 1978. *Fundamentals of Dairy Chemistry.* The AVI Publ. Co. Inc. West Port, Connecticut 59, 785, Chapter 1.

2

PHYSICAL PROPERTIES OF MILK

THIS chapter briefly elaborates and describes the physical properties of milk. The composition of milk with its inherent constitutents directly influences all the physical properties of milk.

Milk as Stabilized Colloidal Suspension

Milk contains more water than any other constituent and, therefore, the properties of milk are primarily those of an aqueous system. It consists essentially of an oil-water emulsion, the continuous phase being the aqueous one. The oil phase consists of fat in the form of emulsion. The aqueous phase contains proteins along with some mineral matter present in a colloidal state. In addtion to this, some mineral matter (about one third of total mineral matter) and lactose appears in true solution. In normal milk there remains a state of equilibrium between the colloidal state and true solution. Milk can be described as a delicately balanced biochemical fluid. The density of fat globules (at ~ 20°) (g/ml), casein micelles, globular proteins and lipoprotein particles are, 0.92, 1.11, 1.34 and 1.10, respectively. The isoelectric pH of the aforesaid constitutents are ~ 3.8, ~4.6, 4–5.5, ~ 4.1, respectively. Fat globules can be floculated by "agglutinin" in the cold, caseins by acid or by rennet, and globular protein by heat. Fat globules may be filtered through fine-sintered glass, casein micelles by chamber land filter and globular proteins and lipoprotein particles through cellophane. Fat globules are visible through microscope, whereas casein micelles via ultra microscope. The globular proteins and lipoprotein particles are visible with electron microscope. Fat globules, casein micelles, globular proteins and lipoprotein particles have a particle diameter of 0.1–10 μm, 10–300 nm, 3-6 nm, 10 nm, respectively (1mm = 10^3μm = 10^6nm = 10^7A).

Boiling Point

Any liquid boils at the temperature at which its vapour pressure is equal to that of the atmosphere and since the vapour pressure of a solution is always less than the vapour pressure of the pure solvent it follows that the boiling point of a solution will always be higher than that of the pure solvent. Pure water boils at 100°C. Milk constitutents are, therefore, responsible for elevation of the boiling point of milk about 100°C. Milk generally boils at 100.15°C. This is due to alteration of the normal ionic $\rightleftharpoons$ molecular $\rightleftharpoons$ colloidal equilibria as a result of heating. Addition of water to milk lowers the concentration of the dissolved substances responsible for elevating the boiling point and as a result, milk adulterated with water boils at a lower temperature than that of normal milk. Boiling point determination may help in the detection of adulteration, however, there are practical difficulties in determining boiling point of milk and the method is not easily applicable.

Freezing Point

The freezing point of a liquid is the temperature at which the solid, liquid and vapour phases exist in equilibrium with each other. A solution freezes at a lower temperature than the pure solvent, and the depression of the freezing point depends upon the concentration of the dissolved substances. The freezing point depression is a function of osmotic pressure of the solution. Milk is believed to have some osmotic pressure as that of blood on physiological grounds and has a fairly constant value.

The freezing point of milk follows from the total concentration of dissolved substances and is the most constant physical property (mean value as –0.545°C) and addition of water will raise the value (below –0.53°) and this helps in the detection of its adulteration. Hortvet apparatus is generally used to determine the freezing point of milk.

The average freezing point of cow milk is generally taken as –0.555°C and buffalo as –0.560°C. The freezing point of goat and sheep milk are taken as –0.575°C and –0.588°C . For the purpose of deducing the percentage of added water, it is reasonable to adopt small difference of 0.01°C corresponding to less than 2 per cent of added water.

The major components affecting the freezing point of milk are : (i) lactose, (ii) soluble salts. Lactose and chlorides represent 75 to 85% of entire depression. Fat, protein, colloidal calcium phosphate, fat globules, casein colloids, have negligible effect on freezing point because of high molecular weight. The freezing point is a property which depends on the number of molecules rather than the nature of molecules. Materials of high molecular weight contribute to a negligible amount because they contain few molecules/gm. Souring of milk, fortification with lactose, or non fat milk solids or addition of sugar further depresses the freezing point.

The freezing point of milk is raised due to more water in it. Freezing point depression is inversely proportional to amount of water added. Colostrum has slightly lower freezing point (–0.605°C) than normal milk. The freezing point of milk is constant because the blood and milk are in osmotic equilibrium and this equilibrium is maintained by the dissolved substances like sugar and salts. Any depression in lactose (during mastitis) is compensated by an increase in the concentration of sodium chloride and thus, the freezing point of such milks do not alter.

Electrical Conductivity

The ability of electrolyte to conduct a current is measured in terms of specific resistance, which is defined as the resistance in ohms of a column of solution 1 cm. long and 1 sq. cm in cross-section. This measurement is made with an ordinary wheatstone bridge-the electrolyte being contained in a cell between platinum electrodes. The specific conductance is the reciprocal of specific resistance. It is expressed in terms of mhos. The specific conductance is computed from the measured resistance of the cell filled with solution.

Specific conductance = K/R

where K = "cell constant", i.e., a proportionality factor previously determined for the given cell with solution of known specific conductance and R = measured resistance in ohms.

Milk contains various kinds of ions, and so it can conduct an electric current. The principal ions responsible for the conductivity of milk are the sodium, potassium, and chloride ions since they are present in the free state and are in the greatest concentration. A solution of sodium chloride having the same chloride content as milk - about 0.1% has a conductivity of about 0.003 mhos. A range of 0.003 to 0.010 mhos has been observed for milk. Higher values generally represent milk from mastitic udders wherein the concentrationn of sodium and chloride is abnormally higher. The specific conductance of whole milk is lower than that of skim milk.

There is a relationship between percentage chloride and conductivity of milk. With the increase of electrical conductivity, the chloride percentage also increases.

Factors Affecting the Conductivity

Temperature: The electrical conductivity increases due to an increase in the dissociation of electrolytes and decrease in viscosity with the increase in temperature.

Fat: Electrical conductivity decreases with the increase in fat in milk because the fat globules impede the mobility of ions and the fat phase constitute an inert dilutant. Skim milk has higher electrical conductivity than the whole milk.

Acidity: Electrical conductivity increases with an increase in acidity because the colloidal minerals are brought into the soluble state.

Dilution: Electrical conductivity decreases with dilution of milk due to decrease in electrolyte concentration but slightly compensated by increased dissociation of electrolytes.

Concentration of milk: Electrical conductivity increases up to 30 percent concentration of solids, this is due to the increased concentration of electrolytes. Beyond 30 per cent concentration, electrical conductivity decreases due to increase in viscosity.

Neutralization of milk: Addition of neutralizers to milk increases the electrolyte concentration and hence increases the electrical conductivity.

Viscosity

Viscosity is an important physical property of food systems. It is related to the flow properties of fluids which in turn control manufacturing operations in many cases. The viscosity of the fluid is a measure of the resistance to flow. The rate of flow is decreased by internal frictional forces.

Theories of Viscosity

There are three main theories of viscosity which help to visualise the viscosity phenomena at molecular level and with molecular forces.

Molecular Theory: The molecules in a liquid are coupled to each other by intermolecular attraction. Therefore, if two adjacent layers of fluid move with different velocities, each will tend to drag the other due to this intermolecular attraction. This drag will dissipate the velocity gradient within the liquid. As a result of this drag, molecular structure of fluid becomes deformed. The configuration of the molecules surrounding a given molecule affected by such drag is more ellipsoidal than the usual radial distribution. Principle axis of this ellipsoidal distribution determined by the local velocity gradient and the degree of deformation of the molecular structure determines the coefficient of viscosity.

The Rate-Process Theory: The key aspect of this theory is that a molecule has to pass over a potential barrier in order to flow. Two molecules form a pair instantaneously upon collision. This double molecule passes through its neighbours by rotation when there is free space in the adjacent layer. The free space in the adjacent layer is created by local density difference due to temperature fluctuation. When no external stress is applied, forward and backward motion would be equally frequent. However, when stressing force is applied the potential barrier will be deformed so as to favour the 'forward' move. In other words, a molecule in order to move into an unoccupied site, passes through a potential energy bottleneck which is lowered when shear force is imposed.

Diffusion Theory: It suggests that under application of external shear force, the self-diffusion of molecules increases in the direction of shear field, and naturally the displacement of a molecule decreases in the opposite direction of the shear. If in self diffusion, the molecules are considered to have overcome the activation energy for diffusion as in general, the diffusion theory model in essence is similar to the rate process theory model.

Definitions

Dynamic Viscosity (n): This is defined as the tangential force on unit area of one of two parallel planes at unit distance apart with the space between the planes filled with fluid and one of the planes moving relative to the other with unit velocity in its place. The unit of n is the poise (centipoise).

Kinematic Viscosity (r): This is defined by the equation r = n/Q where Q is the density of fluid

Kinematic viscosity = Absolute Viscosity/Density

The unit of kinematic viscosity is the stoke (or centistoke) corresponding to the absolute unit of viscosity, the poise (centipoise).

Poise

Poise is defined as the force in dynes/cm required to maintain a relative velocity of 1 cm per second between two parallel planes 1 cm apart. Centipoise is one hundredth of a poise.

Viscosity of milk is mainly due to the milk proteins and fat present as colloidal system and to a minor extent due to the lactose and salts in solution with water. In case of proteins in addition to their concentration, the number of their particles, size and degree of hydration also affect the viscosity (in centipoise) of milk. In the case of fat, the globular size and the extent of clustering determines their influence on the viscosity of milk. The relative viscosity of cow milk, buffalo milk, skim milk, whey is 2.000 cp, 1.800 cp, 1.500 cp and 1.200 cp, respectively at 20°C and it decreases with an increase of temperature. Viscosity of milk of the Indian animals appears to be in the order : buffalo, cow, goat and ass. The absolute viscosity of water at 20°C is equal to 1.005 centipoise. Viscosity of milk is influenced by state and concentration of protein, fat, temperature and age of the milk. Viscosity of milk increases upon heating such as in super heated condensed milk. Homogenisation increases viscosity. During homogenisation fat globules are greatly sub-divided and this contributes to a higher viscosity of milk. Increase of temperature causes a marked reduction of viscosity; for example at 20°C, milk is about half as viscous as at 0° and at 40°C, is approximately one third of the value at 0°C. The viscosity of pasteurised milk, both homogenised and unhomgenised, increases with age. Viscosity may be measured in three ways.

a) The time of flow under a fixed pressure such as with a Ostwald Viscometer when a liquid flows through a capillary tube, the viscosity of the liquid is directly proportional to the time taken for a certain volume of liquid to flow out under a certain pressure gradient, and also is proportional to the density of the liquid. A comparison of the viscosity of milk is done with that of a standard liquid, distilled water, according to the above principles, and absolute viscosity of milk can be obtained knowing the viscosity of water.

(b) By measuring the force required to move two layers of liquid past each other as with Mac Michael viscometer.

c) By measuring the fall of a ball through a column of liquid as it is accomplished with the falling ball viscometer.

Viscosity of milk increases with the homogenisation of milk, it is due to the increase in the amount of fat surface and the amount of protein bound by fat globules. The extent of clustering and clumping of fat globules are also the contributing factors of homogenisation in increasing the viscosity. In homogenised milk, the amount of casein adsorbed to the fat globules is more than the unhomogenised milk which cause higher viscosity. Apparent viscosity and plasticity are the two terms used in addition to the viscosity in milk and its products. Apparent viscosity is ordinarily met in concentrated fluid milk products, particularly in ice cream mix. It refers to a thickened condition of the product which can be dispelled by agitation. It results from the formation of gel structure in the medium. Plasticity is ordinarily differentiated from viscosity on the basis of the force necessary to cause flow. Liquid of substantial fluidity starts to flow by the driving force of their own weight while the non fluid substance needs external pressure, and are known to exhibit a plastic flow.

Density and Specific Gravity

The density of a substance may be defined as its mass per unit volume at a specific temperature. The specific gravity of a substance is the ratio of its mass at a known temperature to the mass of an equal volume of water at 4°C. The unit of volume normally used is the millilitre and the unit of weight is gram. As 1 millilitre of water at 4°C (3.98°C) weighs 1 gram, the specific gravity of water at 4°C is exactly 1; at temperature higher and lower than 4°C, water expands and, therefore, has a specific gravity, relative to 4°C, of less than 1.

Density = Weight/Volume

Specific gravity = Density of substance/density of water at the same temperature

Specific gravity of milk is the resultant of the values of specific gravities of all its constituents present in milk.

Milk constituents	Sp. gravity at 15°C
Fat	0.930
Protein	1.346
Lactose	1.666
Salts	4.120
Ash	5.500
SNF	1.616
Water	1.000

The average values of specific gravity of normal bovine milk is in the range from 1.029–1.032 and it is slightly more than water since all the constitutents except the fat have more specific gravities, than water. The specific gravities of skim milk varies from 1.034 to 1.036 and of cream is around 0.93.

A milk low in fat will have the lower specific gravity and conversely one rich in fat will have higher specific gravity. This is because in normal milk as the fat percentage increases the solid-not-fat also increases and their combined specific gravities more than offset the effect of the fat.

The specific gravity of milk decreases significantly with increase in temperature between 10 and 60°C. The specific gravity is significantly correlated to the total solid concentration but not to the two individual components of fat and solids-not-fat. Determination of specific gravity is used for detecting adulteration of milk.

The density and specific gravity of milk may be determined by either measuring the weight of a known volume or the volume of known weight. The weight can be determined by Pycnometer, a vessel that can be filled with liquid at a given temperature and weighed simply or with hydrostatic balance. The density of milk can also be measured by means of the lactometer which is a special hydrometer calibrated over the range 1.025–1.035 (25.0–35.0 as lactometer degree) or 0.000–0.040 (0–40 as lactometer degree). There are three types of lactometer used in India (i) Quevenne lactometer (ii) ISI lactometer (iii) Zeal lactometer. Lactometers are calibrated at different temperatures. After taking the lactometer reading at the calibrated temperature of lactometer or corrected lactometer reading at the temperature of milk after applying correction factor, the specific gravity can be calculated as follows :

$$\text{Sp. gravity} = 1 + \text{CLR}/1000$$

Homogenisation has no traceable influence upon the specific gravity of milk, however, Short (1956) reported a small increase of the order of

0.000065 in density at 20°C as a result of homogenising the whole milk but not skim milk. The specific gravity of pasteurised milk is significantly higher than that of raw milk due to some evaporation during pasteurisation.

Oxidation-Reduction Potential

Oxidation can be defined as the loss of electrons and reduction as the gain in electrons. Every oxidation is accompanied by a reduction, and vice versa. Compounds which lose electrons or which gain electrons are characterized as oxidising agents or reducing agents, respectively.

The relative tendency of a substance in solution to yield or take on electrons can be measured if a standard substance that will donate or accept electrons can be inserted in the system to establish a difference in potential. In practice, the potential difference created on a platinum electrode in a solution of an oxidant or reductant is measured by completing a circuit through a calomel half-cell and a potentiometer. The voltage measured under these conditions reflects the oxidising or reducing capacity of the solution. This potential is called the oxidation-reduction potential or the redox potential. It is designated by the symbol "Eh". Positive potential which involves loss of electrons from the platinum electrode, is indicative of oxidising properties, whereas negative potential, which involves gain of electrons at the platinum electrode, reveals reducing capacity. In addition to potential, the oxidation-reduction system also has a quality known as poise. This term which is closely analogous to buffer capacity in pH and acidity equilibria, refers to the resistance that a system offers to change in oxidation-reduction potential when treated with oxidizing or reducing agents.

There are two methods for measuring the Eh of a system. These are electrometric and colorimetric. Of these two, the electrometric is much more reliable. The Eh of milk has usually been determined with the aid of a platinum electrode, calomel half-cell, and a sensitive potentiometer. In order to enhance the sensitivity in this type of circuit, which is connected by means of KCl-agar bridges, various amplifiers and galvanometers have been frequently used. This type of instrumentation yields data on potential but it tells nothing regarding the poise of the system. For these reasons, potentiometric titration with either an oxidizing or reducing agent is often recommended.

In milk, the oxidation-reduction potential is controlled by the factors such as dissolved oxygen, ascorbic acid, riboflavin lactose, cystine-cysteine transformation and pH value. In addition to these, heat treatment, bacterial activity and contamination with trace metals, particularly copper, influence the Eh of milk.

Fresh milk has +0.20 to 0.30 volt O-R potential mainly due to disssolved oxygen content. Heating removes oxygen from milk and also produce –SH groups from ß-lactoglobulin which decreases the O-R potential of milk. *Streptococcus lactis* or Lactobacilli also remove free oxygen from milk causing change in O-R potential in the negative direction, thus O-R potential measurement can be used to know the degree of microbial contamination of milk and its shelf-life and its suitability for storage and oxidative deterioration and other processing operations. Within the normal range of pH of milk, Eh varied to the extent of +0.03V for every unit decrease in pH of milk. Fresh zebu cow milk at 30°C has Eh = +0.258 V. Fresh buffalo milk at 30°C has Eh = +0.312 V. Eh of milk drawn anaerobically = +0.13 V quickly decolourises methylene blue. Reduction of methylene blue occurs at Eh = +0.12 to 0.03 V and that of resazurin occurs at Eh = +0.18 to 0.09 V.

pH and Buffering Capacity

pH is expressed in number which is equal to the negative logarithm of hydrogen ion concentration of the system. Any solution is said to be netural when it shows a pH 7.0, but it is acidic or alkaline when its pH is lower or higher than seven, respectively. Since pH is a logarithmic function, change in one unit of pH indicates a ten fold difference in the H^+ ion concentration of the solution. For example, a solution of 5.0 pH possesses ten times more acidity than of 6.0 pH.

In milk, the pH and not the acidity is the principal criteria that controls the activity like enzymes, colour to indicator, processes of coagulation, growth of bacteria and cultures. Hence, measurement of milk pH is more rewarding than that of milk acidity in various processing aspects and technology of milk.

pH can be measured either colorimetrically using indicators and buffer solutions or electrometrically with a pH meter and a pair of electrodes. The use of pH meter is more suitable for measuring pH of milk.

The pH of normal buffalo milk is 6.7 while that of cow milk is 6.6. In general pH depends on the factors like dilution and concentration of milk, concentration of indicator used and heat treatment. In case of buffalo milk an average pH of 6.79 at 10°C decreased to 6.26 at 70°C. pH of buffalo milk is significantly but negatively correlated to fat content of the milk. But neither total solids nor proteins of milk significantly influence the milk pH.

The Buffering of Milk

Fresh milk acts as a complex buffer because of its content of carbondioxide, protein, phosphate, citrate, and a number of minor constitutents. Bacterial action introduces lactate and other organic anions as additional buffers. Titration methods have been widely used to assess the increase in acidity due to the action of bacteria. Milk behaves as buffer over the pH range from 4.8 (curdling point) to pH 8.3. Titration of fresh milk from pH 6.6 to pH 8.3 requires 13 to 20 ml of 0.1N NaOH per 100 ml (1.3 to 2.0 meq. per 100 ml). Most fresh milk samples fall in the range of 1.5 to 1.8 meq. per 100 ml.

Surface Tension

It is defined as the force (in dynes) acting upon a line of 1 cm. length on the surface of the liquid. There are several methods of measuring surface tension (a) the force required to pull a metal ring from the surface of a liquid, (b) the number of drops formed when a given amount of liquid is allowed to fall from a pipette; and (c) the pressure required to force a bubble through a nozzle immersed in the liquid.

The surface tension of milk is determined by an apparatus called Torsion balance, based on the method of force required to pull a metal ring from the surface of a liquid. It is influenced by a number of factors. An average value ranges between 40 to 60 dynes/cm at 20°C. Salts and sugar do not tend to accumulate in the surface because they do not lower the surface tension. Proteins and lipids depress the surface tension of milk. Proteins of whey are effective depressants of surface tension. Fat globule membrane released from the fat globules are also responsible for low surface tension. Surface tension is also influenced by addition of water to milk and lipolysis of milk.

During lipolysis the surface active substances are produced like fatty acid, mono and diglycerides which lower the surface tension. Homogenisation of raw milk stimulate lipolysis and thus leads to a decrease in surface tension but if the product has been pasteurized prior to homogenisation the effect is just the reverse. Homogenisation of raw milk removes the most surface active proteose peptone fraction present in low concentration from the plasma protein by absorption on the increased fat surface which raises the surface tension. The heat treatment has little effect on surface tension except on sterilised milk which shows higher surface tension, due to denaturation of the protein.

Surface tension of cow and buffalo milk are reported in the Table 2.1.

Table 2.1. Surface tension of cow and buffalo milk

Animal	Surface tension (dynes/cm) 20°C	27°C
Cow	51.84–51.93	51.09–51.19
Buffalo	50.40	49.64

Source : Indian Dairy Products, Rangappa and Achaya

Refractive Index

When light passes at an oblique angle from a less dense medium such as air into a more dense medium such as water, it is bent or refracted. The magnitude of the bending, expressed as a ratio of the sines of the angles of incidence (i) and refraction of the light (r), is the refractive index. It is designated by the letter n.

$$n = \sin i / \sin r$$

The refractive index of a pure liquid is a characteristic constant for that liquid, under specified conditions of temperature and wave length of light. The refractive index of water at 20°C with the D line of the sodium spectrum (589.3 mμ) is n^{20}_{D} = 1.33299 while for milk it ranges as n^{20}_{D} = 1.3440 to 1.3480. The refractive index of milk, is increased over that of water to an extent determined by the number and kind of molecules of dissolved material. Addition of water lowers the refractive index of milk but as considerable variation is observed in the values of genuine milk, it is not possible to use this property as the sole criterion for the genuineness of milk samples.

Milk has various solid constitutents, out of these, protein contributes the maximum towards the refractive index of milk serum while lactose ranks next. Since the fat and calcium caseinate hinder in the passing of light through milk, they are removed from milk by precipitation before refractive index of milk is determined in the older methods of measurement of refractive index. Refractive index of milk itself is somewhat difficult to determine because of its opacity, but by using a refractometer, such as the Abbe refractometer, which employs a thin layer it is possible to make satisfactory measurement of milk.

The values of refractive index (R.I.) at 20°C is 1.4600 for milk. The values of refractive index of milk at 40°C are 1.3460 (1.3450 to 1.3472) for cow milk and 1.3477 (1.3460 to 1.3490) for buffalo milk. Fat in milk does not contribute to the R.I of milk because fractionation occurs at the interface of air and the continuous phase (milk serum/plasma). The individual constitutents make the following contribution to the refractive index at 20°C with the D line of sodium spectrum of milk.

Casein	=	0.0049–0.0060
Serum protein	=	0.0021–0.0035
Lactose	=	0.0063–0.0067
Miscellaneous constitutents	=	0.0013–0.0022
Fat	=	nothing

Butyrorefractometer Reading (BR)

Fat in milk does not contribute to refractive index but ghee has a constant R.I and it depends upon the composition of its constitutent fatty acids. B.R. measures the index of refraction between air and liquid fat, and varies with the nature of the fat. Since ghee often exists as semi-solid or solid at the ordinary temperature, the reading is usually taken at 40°C. Ordinary day light is used, by a compensating arrangement built into the refractometer, the value is obtained corresponding to the D line wavelength of sodium (589 mμ). The instrument used is sometimes calibrated in butyrorefractometer degrees instead of the usual absolute refractive indices, the two values being interconvertible as

B.R. in degrees = 42.0 + Factor (observed refractive index –1.4538)

Refractive index = 1.4538 + Factor (observed B.R. –42.0)

Refractive Index and B.R. vary inversely with temperature correction from B.R. reading = 0.54 BR degrees/°C

correction for refractive index = 0.00038/°C

In the homologous series of saturated fatty acids from butyric to stearic the refractive index rises steeply among the lower numbers, and flattens out at higher chain lengths. A double bond elevates the refractive index : stearic acid has a lower refractive index than oleic acid, which in turn has a lower value than linoleic acid. Ghee exhibits a B.R. at 40°C of about 43.0 corresponding to a refractive index of 1.4545; most oils and fats have a B.R. value of about 75, i.e. a refractive index of 1.4754. An increase of B.R. in ghee will be caused by a decrease in the content of lower fatty acids, or by an increase either in higher saturated or unsaturated fatty acids, particularly the latter. An increase in B.R. shows the adulteration of vegetable oil/animal body fat in ghee. PFA has given B.R. standard for ghee at 40°C which is 40.0–43.0°C in all areas of India except for the cotton tract areas of Maharashtra, Madhya Pradesh, territories of erstwhile states of Saurashtra and Kutch where it is 41.5–45.0°C.

Specific Heat

The specific heat is defined as the amount of heat required to raise the temperature of one gram of a substance through 1°C. It may be measured at a constant volume or at a constant pressure and though the difference in the two values is negligible in case of solids and liquids, it is appreciable in case of gases where the ratio of the two values gives us valuable information about the atomicity of a gas molecule.

Milk of normal constituent has a specific heat of 0.93. This value is also approximately the same as observed for warm milk in which fat is in liquid condition. The different constitutents of milk that is fat, lactose, protein, ash and water has the specific heat 0.5, 0.3, 0.5, 0.7 and 1.0, respectively.

Bound water

Milk contains water to the extent of about 87.5 per cent. Some of the water is firmly bound by the milk protein, phospholipids, by substances adsorbed on the fat globules. Out of 3.18 gms of bound water in 100 gms of milk, 50 percent is held by casein, 30 percent by albumin and globulin and about 15 percent by the fat globule membrane.

REVIEW QUESTIONS

1. Define colloids.
2. What is the average boiling and freezing point of milk and water ?
3. Name the factors responsible for conductivity of milk.
4. What is viscosity ? Name the factors which affect viscosity of milk.
5. Define oxidation-reduction potential, 'poise'.
6. What is pH? What is the normal pH of milk ?
7. Define specific refractive heat index.

REFERENCES

Jennes, R. and Patton, S. 1959. *Principles of Dairy Chemistry.* John Wiley & Sons. Inc., New York, Chapter-8.

Webb, B.H., Johnson, A. and Alford, J.A. 1979. *Fundamentals of Dairy Chemistry.* The AVI Publ. Co. Inc. West Port Connecticut. Chapter-8.

Rai, M.M., 1964. *Dairy Chemistry and Animal Nutrition.* Ram Prasad and Sons, Agra-3. Chapter-2.

Ling, E.R. 1956. *A Text Book of Dairy Chemistry.* Vol. 1 (Third edition revised). Chapman & Hall Ltd. Lond. Chapter-2.

Walstra Pieter and Jennes Robert 1983. *Dairy Chemistry and Physics.* A Wiley - Inter Science Publication, John Wiley & Sons, New York.

Dairy Hand Book (Processing) 1980. National Dairy Research Institute, Karnal.

Dairy Hand Book (Processing) 1986. National Dairy Research Institute, Karnal.

Rangappa, K.S. and Achaya, K.T. *Indian Dairy Products.* Asia Publishing House. Chapter-14.

3

PROTEINS AND ENZYMES OF MILK

Milk Proteins

NEARLY 2/3 of all cells consist of proteins and thereby they are essential constitutents of all cells. The major structural components of all animal tissues are proteins whereas the structural components of plants are cellulose.

Proteins are giant linear polymers of L-α amino acids with the exception of proline which is an amino acid. The amino acids are linked to each other by amide linkage which in protein chemistry is known as peptide bonding. All proteins contain carbon, hydrogen, nitrogen, oxygen, sulphur, phosphorus and traces of other elements.

Milk, the natural secretion and food of young mammals, contains a number of proteins that fulfill their nutritive requirements. Discussion on the proteins of milk is only relevant when we distinguish between the proteins in milk and those obtained from milk by the various chemical and physical fractionation procedures.

Milk proteins are broadly classified into two groups viz., casein and whey proteins. Casein is dispersed in milk in the colloidal state while whey proteins are present in solution phase. Most proteins are less soluble at their isoelectric point but casein coagulates at its isoelectric point (pH 4.6) by acid. The colloidal casein is present in milk as tertiary calcium phosphate caseinate complex or micelle. Each submicelle of calcium-phospho-caseinate complex is bridged together by tertiary calcium phosphate by an ester linkage with the serine hydroxyl groups (Figs. 1, 2).

Casein complex (in milk) i.e. Phosphate, calcium etc. free from	HCl $\rightarrow$ pH 4.6	Casein (ppt) isolated form. Hydrogen caseinate (chemical name)	+	Salts (contained in whey proteins or serum proteins)

The nitrogen content of milk on an average is 0.5%. When protein percentage basing on the above observation is calculated it comes to 3.19 (0.5 × 6.38 (factor) = 3.19%). The average percentage of protein nitrogen in milk is 90% whereas the non-protein nitrogen is 10%. Of the 90% of protein nitrogen casein contributes 73.5% and the rest 16.5% is due to albumin, globulin, and proteose-peptone.

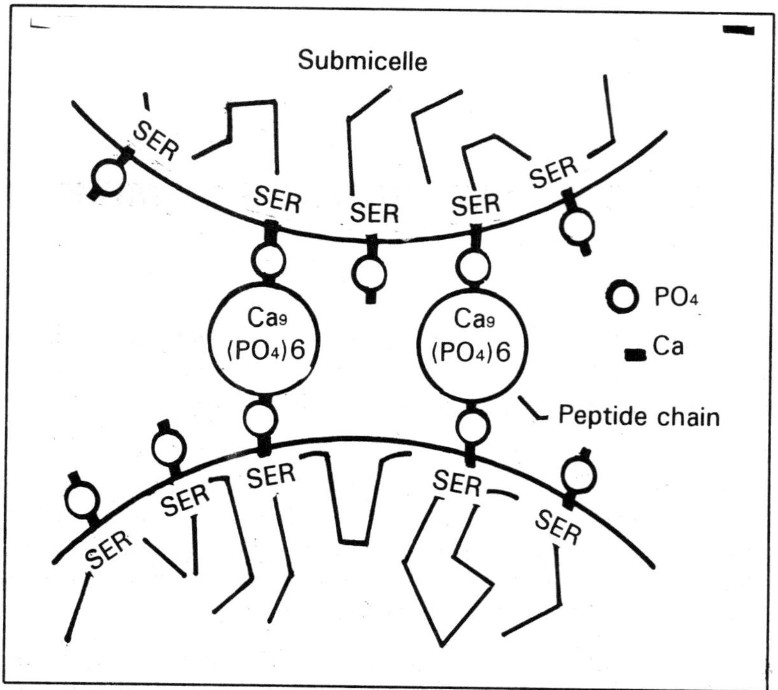

Fig. 1. Schematic diagram of the Binding of two submicelles via $Ca_9(PO_4)_6$ cluster. Source: Development in Dairy Chemistry-I, Proteins by. P.F. Fox

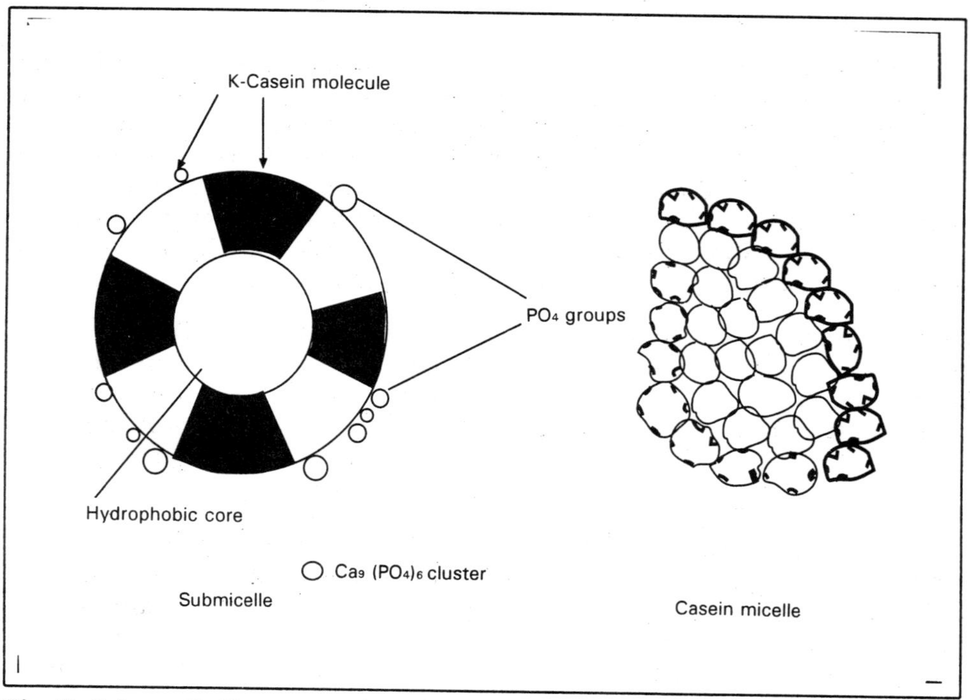

Fig. 2. Schematic Representation of Submicelle and Casein micelle

Caseinate-phosphate System

Casein is the principal milk protein. The casein micelle is an importanı and characteristic macromolecular assembly of mammalian biology, occurring in all milks that have been examined in sufficient detail. Its functions are to form a coagulum in the stomach of the nurshing, allowing the slow release of nutrients down the digestive tract, and to act as a means of transporting calcium and phosphate in a readily assimiable form from mother to young. Also caseins are a source of amino acids. Enzymatic cleavage of casein polypeptide chains can produce various types of biologically active peptides such as the casomorphims, though it is not known if any are physiologically important. Another recent speculation is, that the caseins have a role in protecting the mammary gland against ectopic calcification, a hazard that is common to all tissues in contact with supersaturated calcium solutions. In this regard, the flexibility and relative lack of secondary and tertiary structure in the caseins can be seen as essential features contributing to their biological function.

Unlike any other proteins, milk proteins are very heat stable. A view of the section of casein micelle is given in Figure 3.

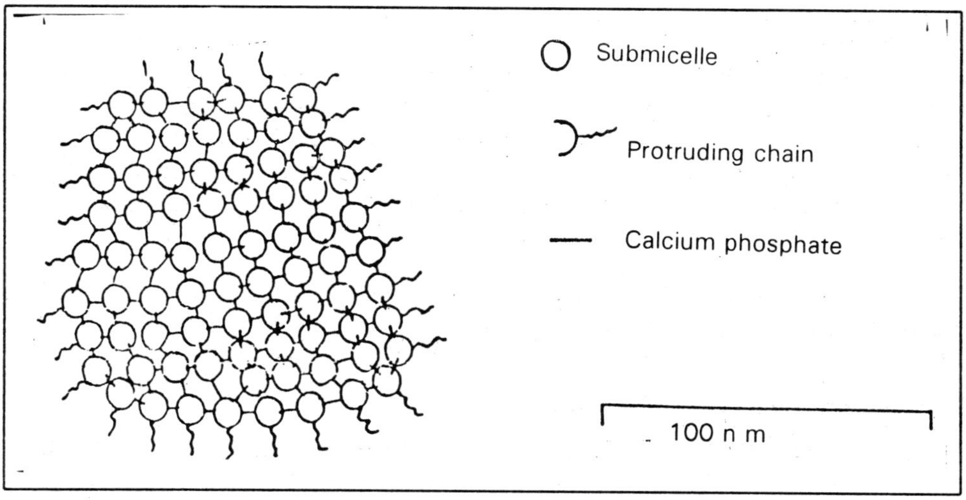

Fig. 3. Section through a casein micelle

It is built of submicelles, roughly spherical aggregates of several casein molecules held together by hydrophobic bonds and salt bridges. Regions of amorphous calcium phosphate link the submicelles to each other, the ester phosphate groups form part of this colloidal phosphate. In this way, almost all regions of casein molecules are severely restricted in mobility. The C-terminal part of the K-casein is, however, predominantly present as flexible "hairs" located at the outside of the micelles. There are essentially two types of submicelles, with or without (much) K-casein. The casein micelles greatly change in properties upon lowering the pH, mostly due to dissolution of colloidal phosphate, at a still lower pH, increased formation of salt bridges predominates. Temeprature also has pronounced effects. Upon lowering it, the micelles become more voluminous, presumably due to protruding hairs of (mainly) ß-casein. Also at high temperature (>70°C), parts of the casein molecules become more flexible.

Casein micelles are very stable. If conditions are changed they may disintegrate or aggregate. Aggregation mainly leads to formation of a gel. The theory of fractal floc formation to gelation serves to explain, the very strong dependence of gelation times on volume fraction of aggregating particles.

The stability against aggregation is primarily due to stearic repulsion, caused by the hairs of K-casein and at low temperature presumably also β-casein. The hairs on different casein micelles may, however, touch, and may lead to lasting contact of the micelles, i.e., aggregation. The bonds formed, can be salt bridges or at high temperature, covalent bonds (chemical crosslinks). Hydrophobic bonds are probably not involved. The probability that casein molecules in different micelles may touch each other for a sufficient time for bonds to be formed appears to depend on electrostatic as well as stearic repulsion which thereby affect aggregation rate. Caseins were once defined as phospho-proteins that precipitate from milk at pH 4.6 but are now defined by their amino acid sequences. The four polypeptide chains of bovine casein are the $\alpha\ s^{-1}$, $\alpha\ s^{-2}$, ß and K-casein. Of these, two αs casein and K-casein are each phosphorylated to different extents and K-casein also exists in a number of glycosylated forms.

The flow sheet (Flow Sheet 3.1) describes a simplified schematic summary of developments in casein nomenclature (Holt, 1993).

Casein Concentrations in Milk

Ion-exchange chromatography in buffers containing dissociating agents has been the technique of choice for the quantification of casein composition. It appears that chromatographic separation on DEAE-cellulose is one of the best techniques. Gel electrophoresis patterns together with the effect on the patterns of chymosin treatment, allowed all fractions apart from 2b, to be identified readily (Fig. 4). Fraction a consists almost entirely of r_2-casein, B-CN (f 106-209), and fraction 1b contains largely r_3-casein, B-CN (f 108-209), with possibly some r_1-casein, B-CN (f 29-209). Fractions 2a, 3, 4a and 4b were, respectively the K-, B-, $\alpha\ s_2^-$, and $\alpha\ s_1^-$ caseins. The identification of the contents of fraction 2b presented some difficulty and its true nature was not established, however, it does not normally amount to much more than 5% of the total casein. Only about 80% of the total K-casein was recovered in fraction 2a, the remainder being present in fraction

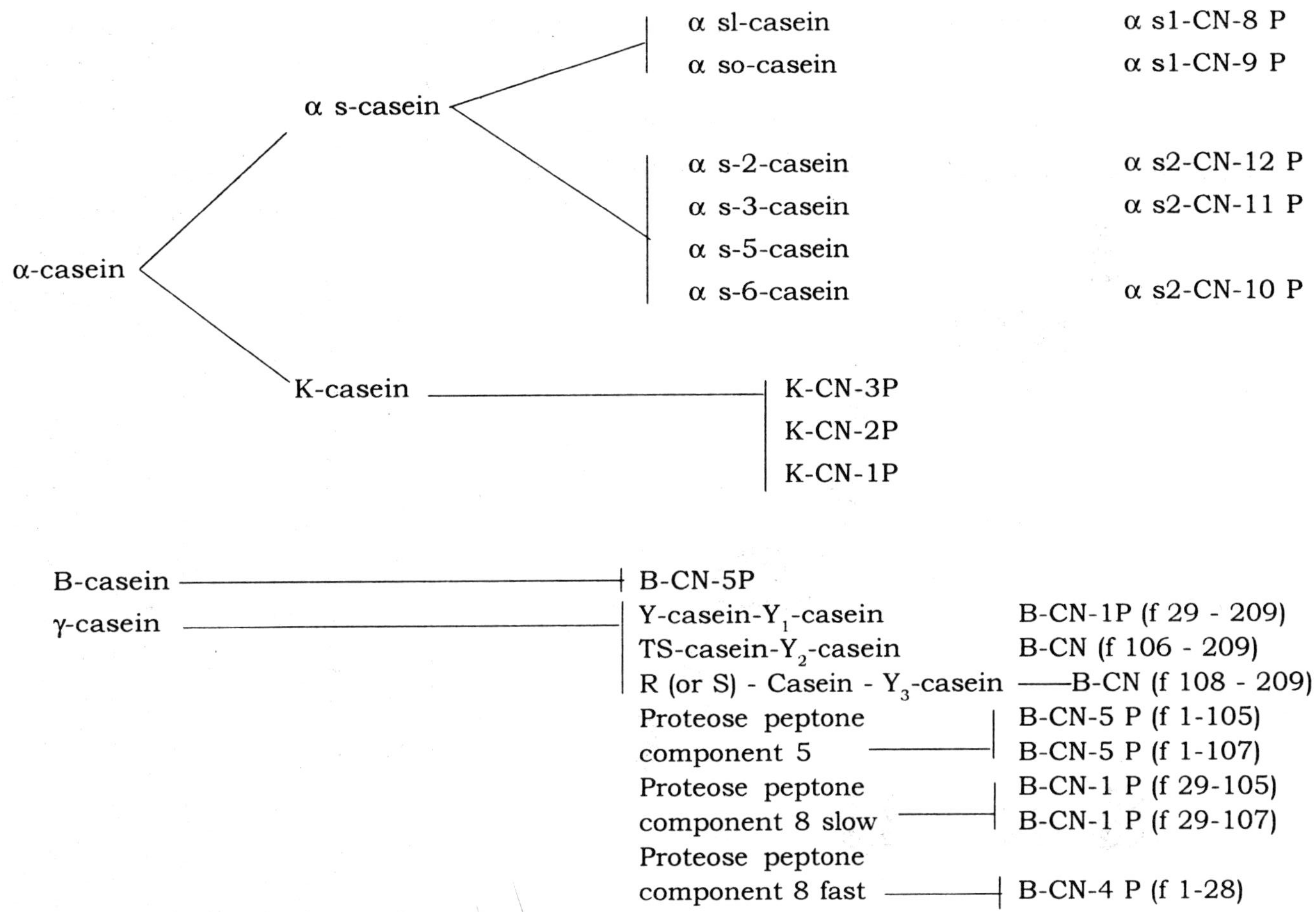

Flow Sheet-3.1: Schematic summary of developments in casein nomenclature

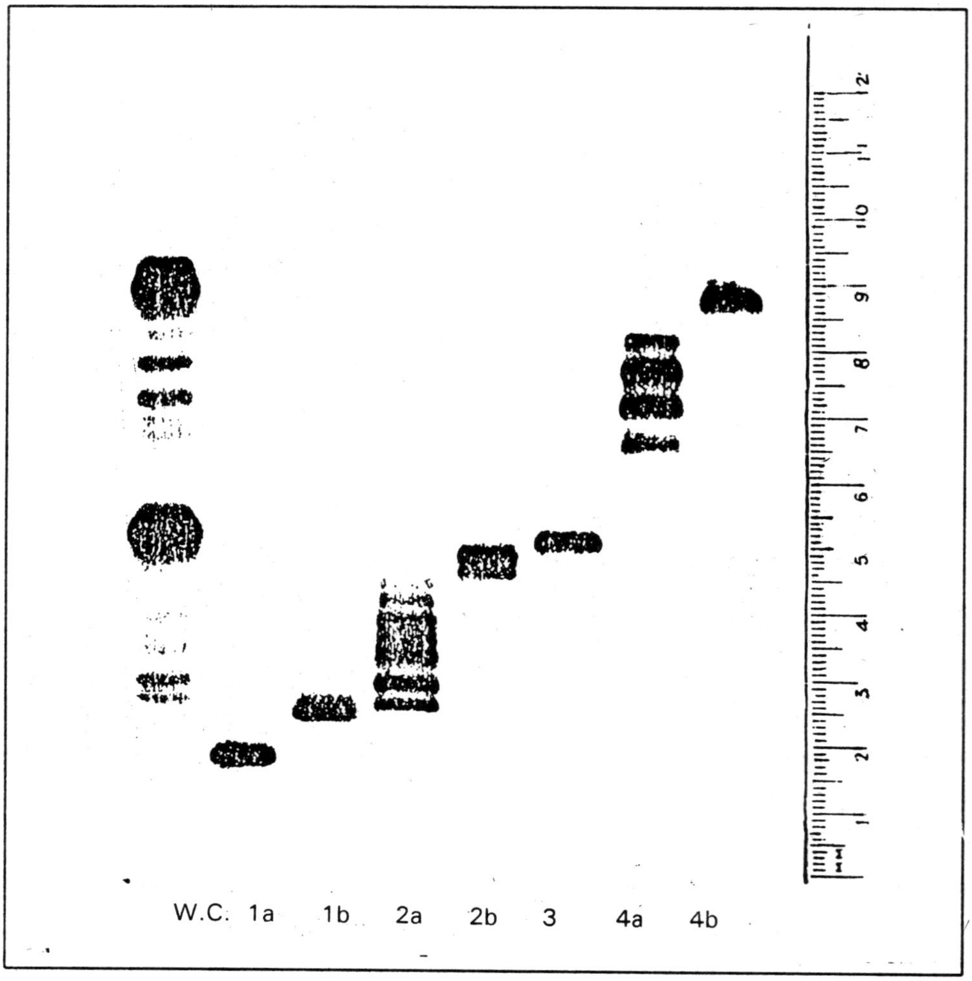

Fig. 4. Electrophoretic pattern of casein and chymo treated casein.

casein. In later reports K-casein was determined independently by gel filtration method, and a comparison of the results indicated that the non K-casein content of fraction 2b amounted to about 2% of total casein.

In Table 3.1, the average weight per cent and mole per cent of the four caseins and the γ-casein fragments are given for bulk milks from creameries in southwest Scotland . The mole fraction of B-casein plus γ-casein equals approximately the mole fraction of αs^{1-} casein, and these are about four times larger than the mole fraction of αs^{2-} casein. Even if about 2% is subtracted from the weight fraction of K-casein, its mole fraction still appears to be a little larger than that of $\alpha\ s^{2-}$ casein. The figures in the table are typical of many of the milk samples from healthy cows in midlactation and each is probably accurate to within 1-2 per cent of whole casein, as indicated by the precision of the method.

Table 3.1. Average casein composition in bulk milk samples

Casein type	Weight (%)	Mole (%)	Casein/ skim milk (g/litre)	Casein/ skim milk (m mol/litre)	Casein/ skim milk (m mol/litre)
α s[1]	38.1	36.4	10.25	0.434	3.52
α s[2]	10.2	9.0	2.74	0.108	1.25
β	35.7	33.6	9.60	0.400	2.00
K	12.8	14.8	3.45	0.176	0.19
r	3.2	6.2	0.88	0.075	0
Total	100.0	100.0	26.92	–	6.96

The mole fractions do not appear to have a general significance, because, in other species, different proportions are found. For example, in human milk the main casein fraction is a β-casein homologue.

The introduction of fast liquid chromatographic procedures for the separation of proteins has, so far, not given as good resolution as traditional methods of all of the individual caseins, particularly the αs caseins. However, alkylation with cystemine and cystamine increases the charge difference between αs_1 and αs_2 caseins and leads to quantitatively satisfactory agreement with traditional methods for most samples. This method was applied to study the variation of casein composition with stage of lactation and it was found that for the greater part of lactation, the proportions of the individual caseins were effectively constant. In the beginning of lactation and at the end, however, there are significant changes, particularly in the β-, r- and K-casein fractions, where the proportions of K- and r-caseins are negatively correlated with the proportion of β-casein.

Major Proteins

Casein precipitates at its isoelectric pH of 4.6 (24-28 g per litre of milk). By salting out the whey left, two distinct fractions are obtained viz. α-lactalbumin (0.6g–0.7g per litre) and ß-lactoglobulin (2.4 g per litre).

Dialysis of concentrated solution of lactalbumin exhaustively at pH 5.2 to remove the salt results in the crystallization of ß-lactoglobulin which is about 50 to 60 percent of the total lactalbumin fraction. This globulin is insoluble in water and needs little amount of salt to bring it into solution. The prefix "ß" used in the name identifies the protein with a protein designated "ß" by Pederson in an ultracentrifugal study. In this study whey proteins resolved into three fractions.

(a) ß-fraction identical with Palmer's ß-lactoglobulin, (b) α-lactalbumin, and albumin of low molecular weight and (c) r-fraction which was the conventional globulin. Skim milk was first decalcified by Pederson. In the sedimentation pattern the peaks were identified by Greek letters in the order of increasing sedimentation constants. The smallest designated as α-lactalbumin, the next larger ß-lactoglobulin and the largest r-lactoglobulin. The r consists of two fractions. The mammary secretory protein α-lactalbumin apart from being a simple constituent of milk serves a major role as a regulatory or modifier protein in the biosynthesis of lactose.

Minor Proteins

1. Proteose-peptone : It is obtained after the removal of major proteins of the milk whey of boiled milk at pH 4.6.

2. Milk contains a protein identical to bovine blood serum albumin (serum albumin, blood, milk).

3. Antibody protein : They are obtained from antigens and the plasma cells of mammary tissue synthesize them. These proteins also known as r-globulins, precipiate with 50 per cent ammonium sulphate and have slow mobility on electrophoresis. They are more commonly known as Immuneglobulins or Immunoglobulins.

4. Iron-containing proteins : They are identical to the transferrin of blood. Milk, saliva, semen and tears contain these proteins called red proteins. They are distinguished as - (i) Lactoferrin-red coloured iron containing protein, specific to milk. (ii) Apolactoferrin-iron free form of lactferrin. (iii) Serum transferrin-red coloured iron containing protein of serum present both in blood and milk.

5. Fat globule membrane protein (FGMP) : Fat globules are stabilised by an adsorbed complex of proteins and phospholipids.

6. Lactollin : Associated with the red protein in small amount, lactollin is isolated from bovine milk.

7. Glycoprotein-α : It can be isolated from milk, assosciated with whey proteins and Immunoglobulins. It is also known as an acid glycoprotein or orsomucoid.

8. Kininogen (cow milk) : It is a protein which when incubated with trypsin or snake venom, releases a material with kininin like ability to contract smooth muscles, associated with whey protein fraction.

9. Other proteins : These include less known proteins that are present in milk. M-I Glycoprotein (Human milk), M_z Glycoprotein, ß-microglobulin, ceruloplasmin, trypsin inhibitor, Folate Binding Protein (FBP) and vitamin B_{12} binding protein.

10. Milk enzymes : More than 30 indigenous milk enzymes are present in milk and they definitely contribute to the protein percentage of milk proteins.

Fractionation of Milk Proteins

The process of fractionation of individual milk proteins determines the concentration, characteristics and other behaviour of individual milk proteins. The process of fractionation is carried out by chemical or other means. Rowland (1939) developed a classical scheme for fractionation and this is depicted hereunder (Flow Sheet 3.2).

Aschaffenburg and Drewry (1959) revised the Rowland scheme. The outline of scheme is given in Flow Sheet 3.3.

Mellander (1939) using free boundary electrophoresis identified casein as α, ß, γ (60%, 17.6%, 2.4% of total casein which is 80% of the total protein), whey protein, Euglobulin, Pseudoglobulin, component III, α-lactalbumin, ß-lactoglobulin and blood serum albumin.

Based on bovine milk, the current knowledge of primary structure suggests that caseins which account for 76-86 percent of the total protein can be represented by four gene products: α_{s1}-casein, α_{s2}-casein, ß-casein, and K-caseins. The others are derived by post-translational processing such as phosphorylation glycosylation or limited proteolysis. The major whey proteins are ß-lactoglobulins, α-lactalbumin, immunoglobulins and serum albumin, in order of decreasing amounts. The proteose-peptones also result from limited proteolysis of other gene products. The heterogenity of milk proteins is further complicated by the fact that genetic variants are recognised to be seven of ß-lactoglobulin, two of α-lactalbumin, four of α_{s1} and α_{s2} casein, two of K-casein (Flow Sheet 3.4).

Preparation of Casein

There are innumerable ways to prepare casein from milk. The pH of diluted skim milk is brought down to 4.6, the isolectric pH of casein with the addition of 10 per cent hydrochloric acid or 1N acetic acid or acetate buffer pH 4.6. The precipitated casein is removed by centrifugation or filtration. It is followed by washing with water, acetone and ultimately dried with solvent ether. Casein may also be prepared by replacement of 10 percent hydrochloric acid with 10 per cent acetic acid.

Another method to obtain casein is by means of rennet coagulation. Skim milk is first warmed to 35°C and rennet is added at the rate of approximately 115g per 460 litres of milk to coagulate casein in 15-20 minutes. The curd is cut, heated to 65°C and the whey drained. Further washings of the curd takes place at 30°C and the curd is dried at 40–50°C. Rennet casein is also known as para casein.

Occasionally, saturated sodium chloride is used to coagulate casein from milk. However, this may precipitate whey protein also in minute amounts sometimes. Ammonium sulphate (264g/lite) is also used to precipitate casein. The coagulated casein is removed by centrifugation at 9500 rpm for 30 minutes.

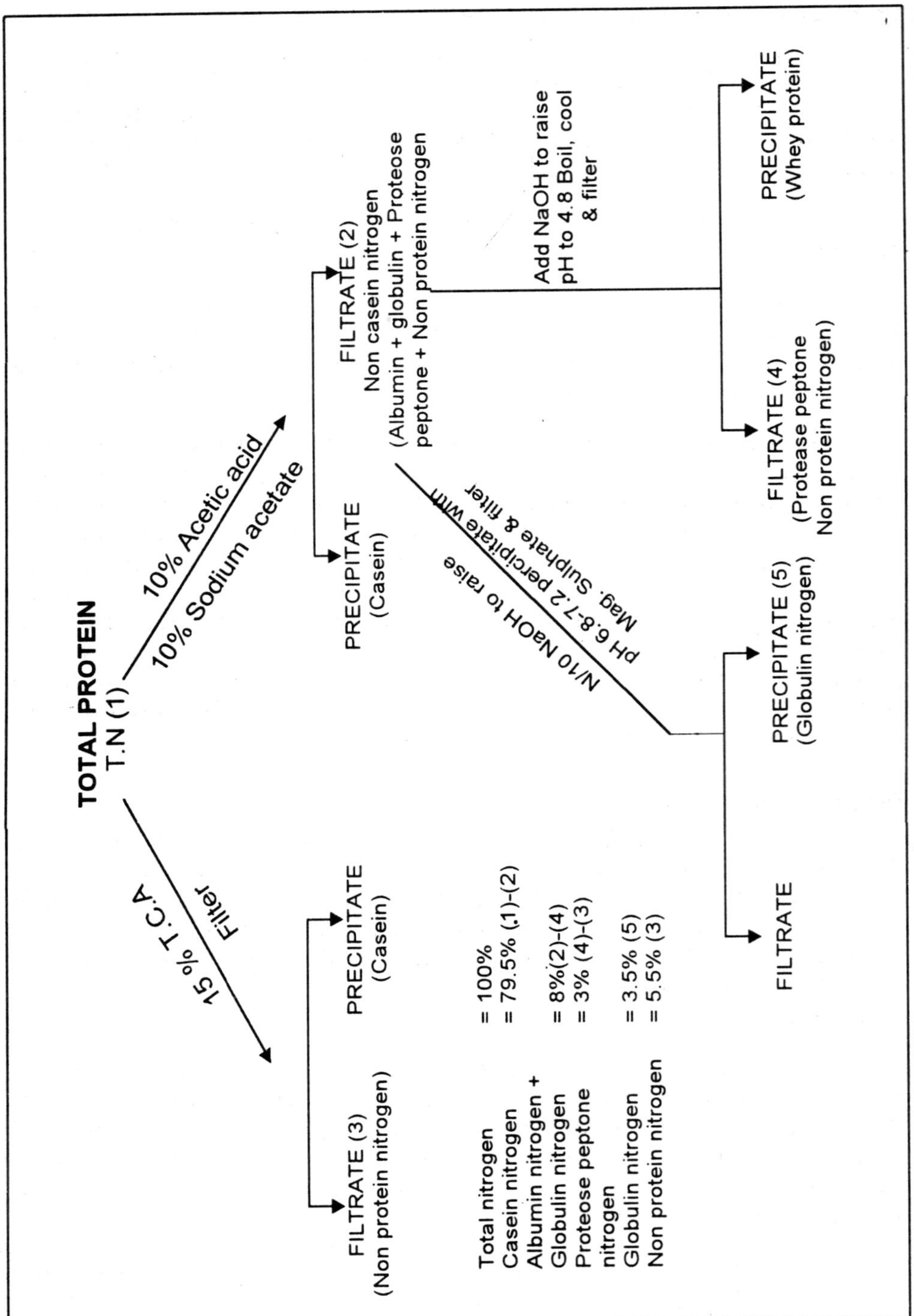

Flow Sheet-3.2: Fractionation of Milk Proteins by Row Land (1939)

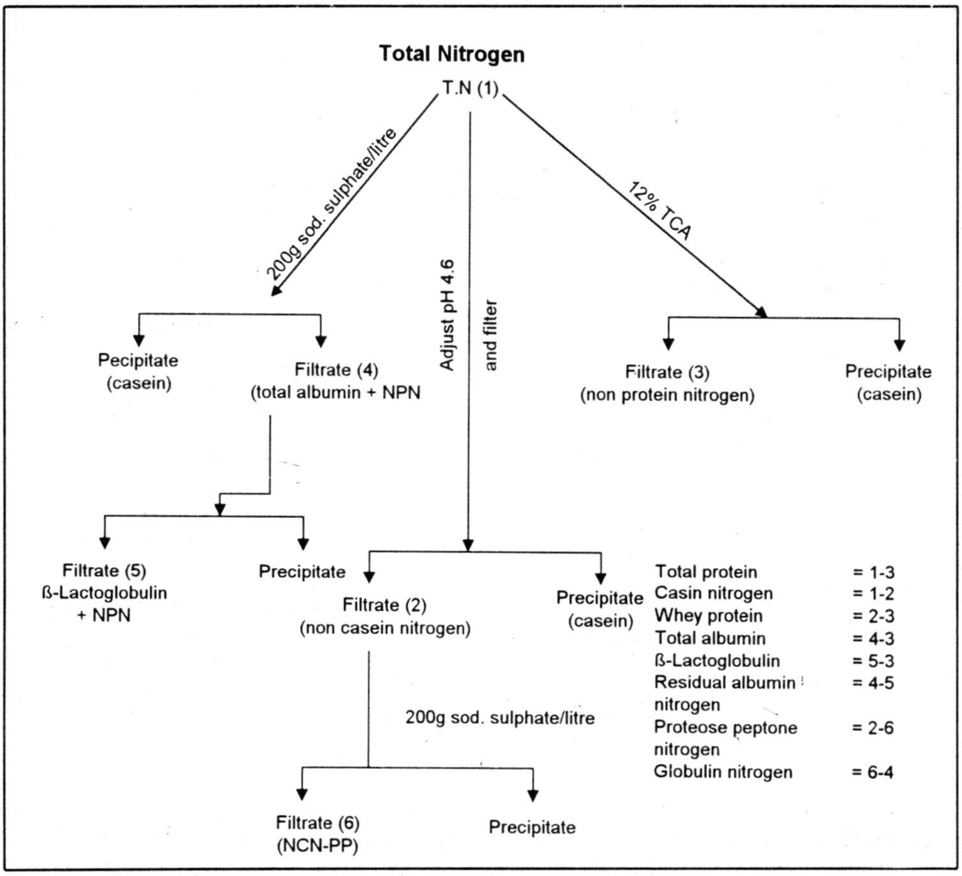

Flow Sheet-3.3: Protein fractionation by Aschafenburg and Drewry (1959)

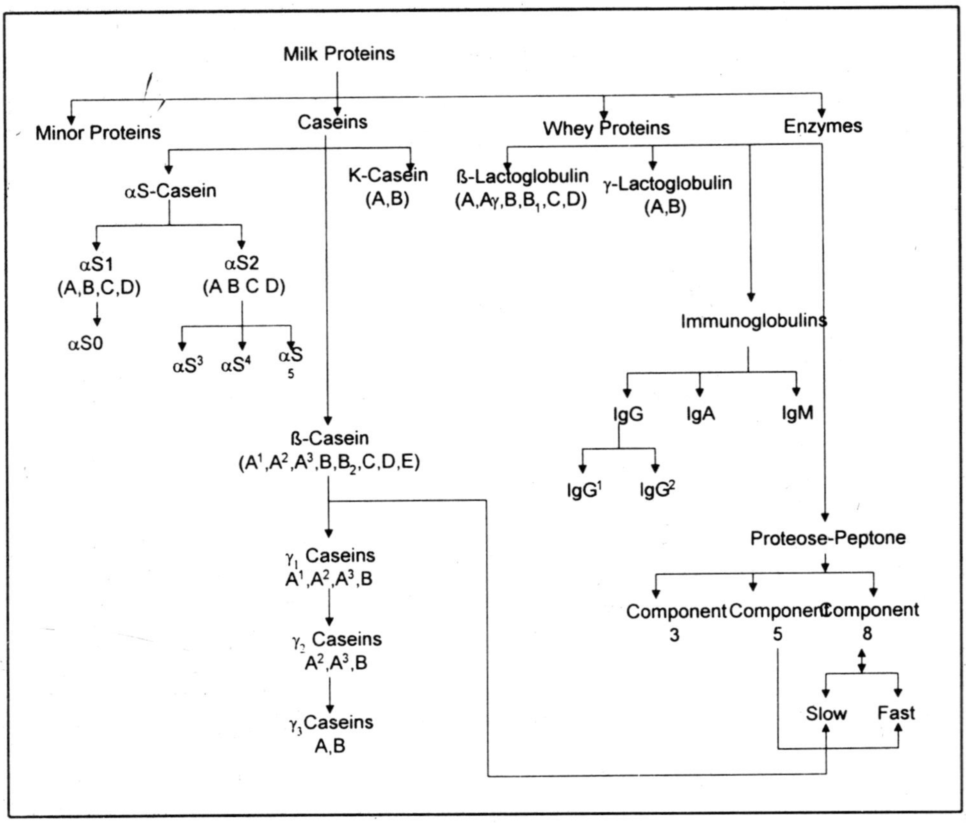

Flow Sheet-3.4: Distribution of fractions and proteins in bovine milk ((Source: Development in Dairy Chemistry-I Proteins by P.F. Fox)

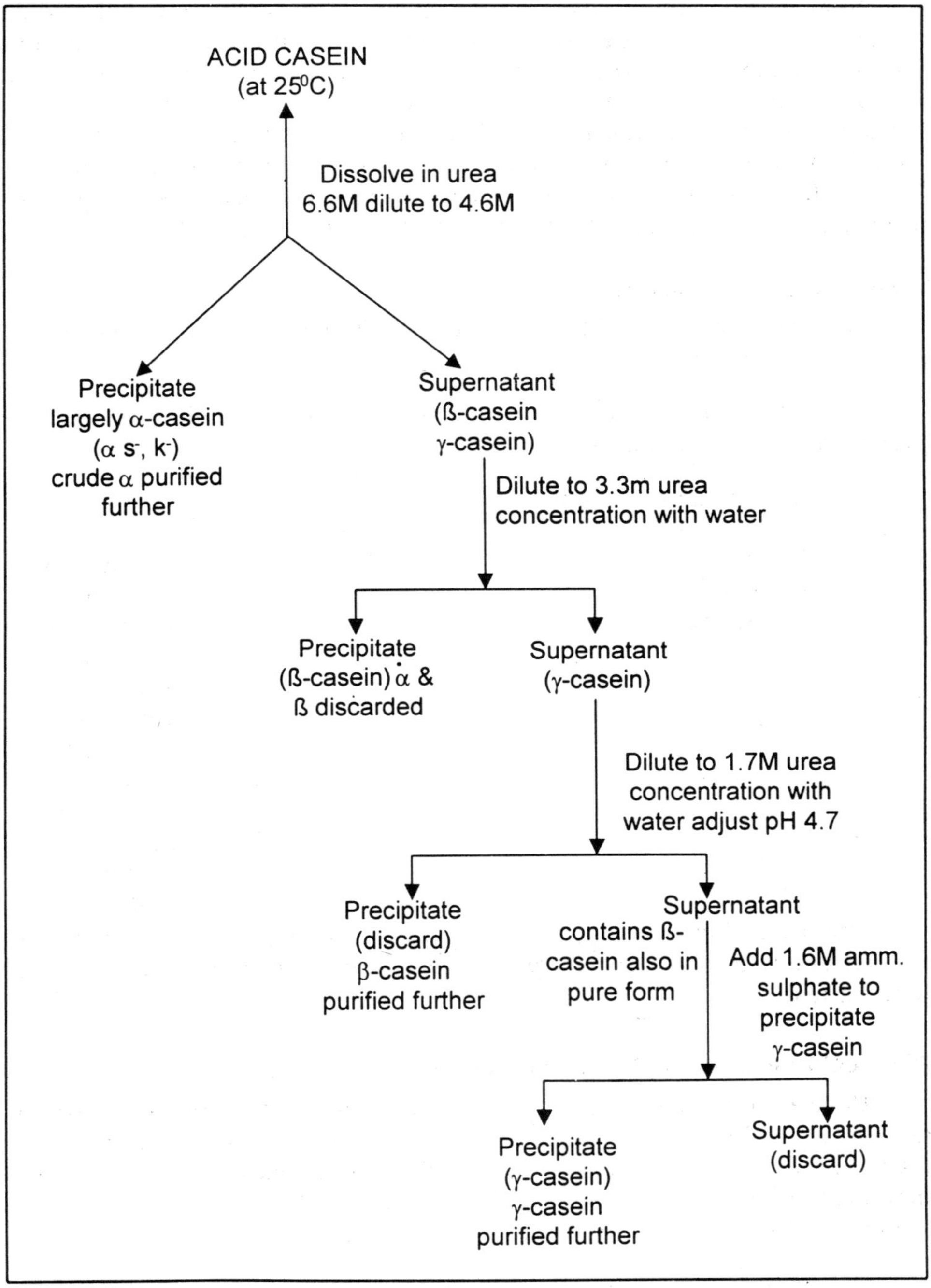

Flow Sheet-3.5: Fractionatin of casein by Hipp *et al* (1952) based on urea solubility

Skim milk is made to 2M with calcium chloride. It is centrifuged at 30000 rpm which results in three layers. Out of the three layers, two bottom layers are suspended in 0.07M $CaCl_2$ and 0.08M NaCl. Potassium oxalate or potassium citrate is used to remove calcium from the resulting solution. Ion exchange resins or dialysis may also remove calcium. The resultant solution containing casein may be recentrifuged. Ultracentrifugation of milk at 10000, 20000 or 30000 rpm can be used to prepare micellar casein or casein in native form. Casein settles at the bottom of centrifuge tube.

Fractionation of Casein

Based on urea solubility Hipp *et al.* (1952) have fractionated casein as given in Flow Sheet 3.5.

One of the principal and important fraction K-casein can also be prepared directly as follows :

	dissolved in	ppt		
Casein	6.6M Urea +	(discard)		
	7 NH_2SO_4		Add	Supernatant
	pH 1.3-1.5 and	Supernatant	$(NH_4)_2SO_4$	(discard)
	centrifuge		@ 1M i.e.	Precipitate
			13.2g/100ml	K-casein
			centrifuge	dialyse with water and further purified with ammonium acetate and alcohol

Other fractions like α^{-}_{s-1}, α^{-}_{s-2}, α^{-}_{s-0} may also be prepared by chemical and chromatographic procedure. However, these are beyond the scope of present text.

Physical and Chemical Properties of Major Milk Proteins

Casein: It can be defined as phosphoproteins obtained from raw milk at pH 4.6 at 20°C. Till recently casein was classified as a mixture of αs^{-}, b^{-}, and K^{-} and v^{-}-casein based on gel electrophoresis according to their mobility, but now it is classified on the basis of their primary structure (amino acid sequence) into α_{s-1}^{-}, α^{-}_{s-2}, and K^{-} casein.

Casein is known since 1830. The different casein fractions identified by electrophoresis and electron microscopy elucidate heterogenity. Different techniques have been used like electrophoresis, chemical fractionation etc. to separate these fractions. Exact number of fractions are not known or definite. According to Wake and Baldwin (1965) casein contains as many as twenty fractions or components. New components may be present in casein, can arise under experimental conditions by association between components, dissociation of hydrogen bonds or other weak bonds at low pH or in solution. Substances may be decomposed into units which are never found in nature.

Properties of Casein

Elementary Composition: Based on the various studies, casein contains the following elements (in per cent of casein) carbon 52.6–54.0, Hydrogen 6.75–7.10, Nitrogen 15.51–15.91, Sulphur 0.71–0.83, Phosphorus 0.71–0.85.

Phosphorus in casein is bound chiefly if not entirely in ester linkage with the –OH group of serine and thereonine.

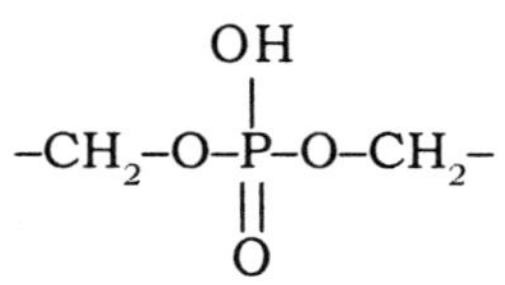

diester

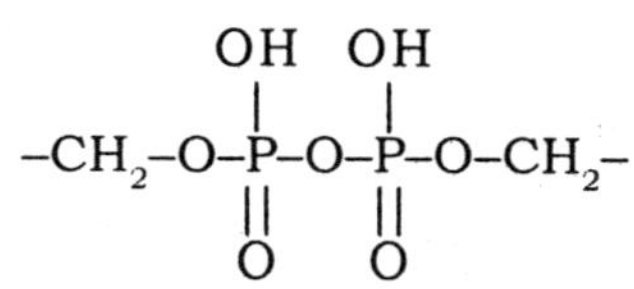

pyrophosphate

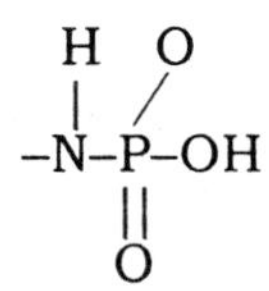

Phosphamide

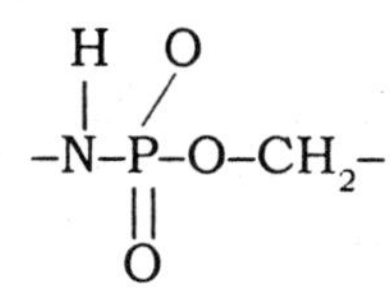

phosphamide ester

Casein as Glycoprotein: The main sugar in casein is sialic acid (N-acetyl neuraminic acid) having the formula.

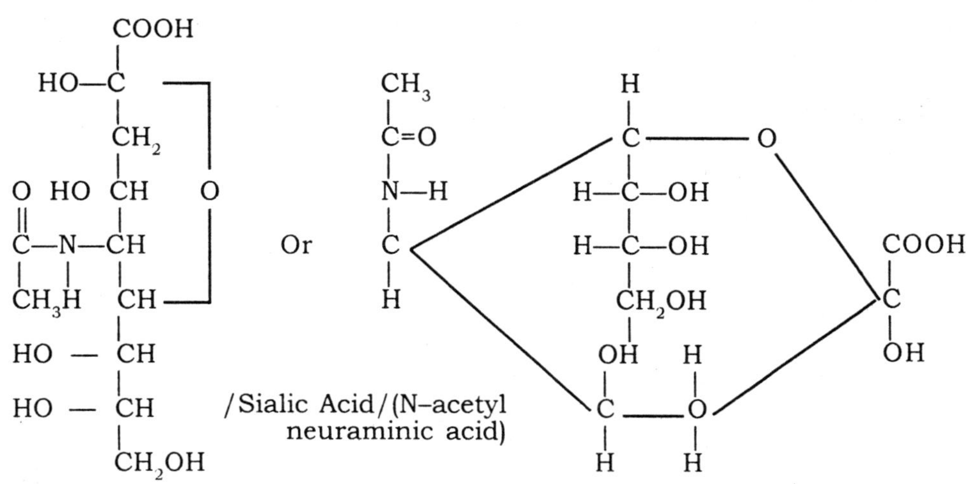

/Sialic Acid/(N–acetyl neuraminic acid)

Other sugars present include galactose, glucose, mannose, hexosamine and galactosamine (hexose).

Amino Acid Composition: Casein is a complete protein. It contains all essential amino acids. In all, 19 amino acids Glycine, Alanine, Valine, Leucine, Isoleucine, Proline, Phenylalanine, Cystine, Cysteine, Methionine, Tryptophan, Arginine, Histidine, Lysine, Aspartic acid, Glutamic acid, Serine, Threonine and Tyrosine are present.

Sulphur in Casein: Casein contains 0.78 per cent sulphur; the amino acids cystine and methionine accounts for 0.09 per cent and 0.69 per cent, respectively.

Size of Particles in Solution: It is difficult to determine the size of casein particles in solution because casein exhibits a strong tendency to aggregate which depends upon temperature, pH and salt concentration. Particles have apparent molecular weights of several millions - $\times 10^8$ daltons.

Electrical Properties: The isoelectric point of casein is pH 4.6. This value is on the basis of electrophoresis and solubility techniques. At the pH that of milk, casein carries negative charge and, therefore, migrates towards anode in an electric field.

Solubility: α- and ß-casein are insoluble at isoelectric point (4.7, 4.9) but dissolve readily on either side of it. They can be separated on the basis that α-casein is less soluble at pH 4.2 and 2°C while ß is less soluble at pH 4.2 at 2°C. Casein binds calcium and other bivalent cations very tenaciously at pH values alkaline to the isoelectric point. This causes aggregation and lowering of solubility. α-casein of course is appreciably soluble in 50 percent ethanol solution.

Optical Rotation: The caseins have high specific rotations resembling those of the denatured globular proteins. The specific rotations of casein are not affected by "denaturing treatment".

Oxidation: Amino acid of casein is oxidised to Imino acid which subsequently forms keto acid.

$$\underset{\text{Amino acid}}{R - \underset{H}{\overset{H-N-H}{C}} - COOH} + O \rightarrow H_2O + \underset{\text{imino acid}}{R - \overset{NH}{\overset{\|}{C}} - COOH}$$

$$\text{Imino acid} + H2O \text{-----------}> \underset{\text{Keto acid}}{R - \overset{O}{\overset{\|}{C}} - COOH} + NH_3$$

This oxidation may be caused by oxidising agents such as hydrogen peroxide, oxygen, bromine, potassium permanganate, potassium dichromate, perbenzoate, periodate, ferricyanide, porphyrin, O-iodosobenzoate, iodine and performic acid under different conditions. The oxidisable group are - SH (sulfhydryl), phenol and indole group of tyrosine and tryptophan.

Reduction: Sulfhydryl groups are formed when disulfide group of casein are reduced. A myriad of reagents, viz. cysteine, glutathione, thioglycolic acid, mercaptans, mercaptoethanol, hydrogensulfide, sodium sulfide and lithium aluminium hydride cause such reduction.

Disulfide group of protein (-S - S-) ⟶ Sulfhydryl (-SH)

COOH Group of protein (free) LiAlH alcohol group (-OH)

Alkylation: At various places of casein micelle alkyl group can enter. Sulfhydryl (SH), amino group (NH_2), phenolic (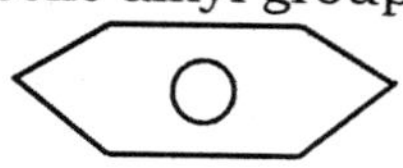 - OH)

and indolyl group) 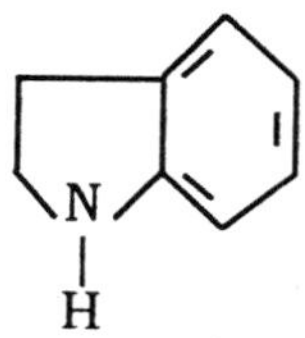

) forms alkyl derivatives when these react with Iodoacetamide.

Arylation: Arylation of casein takes place when it reacts with 1-fluoro 2:4 dinitrobenzene (FDNB). This is called Sanger's method.

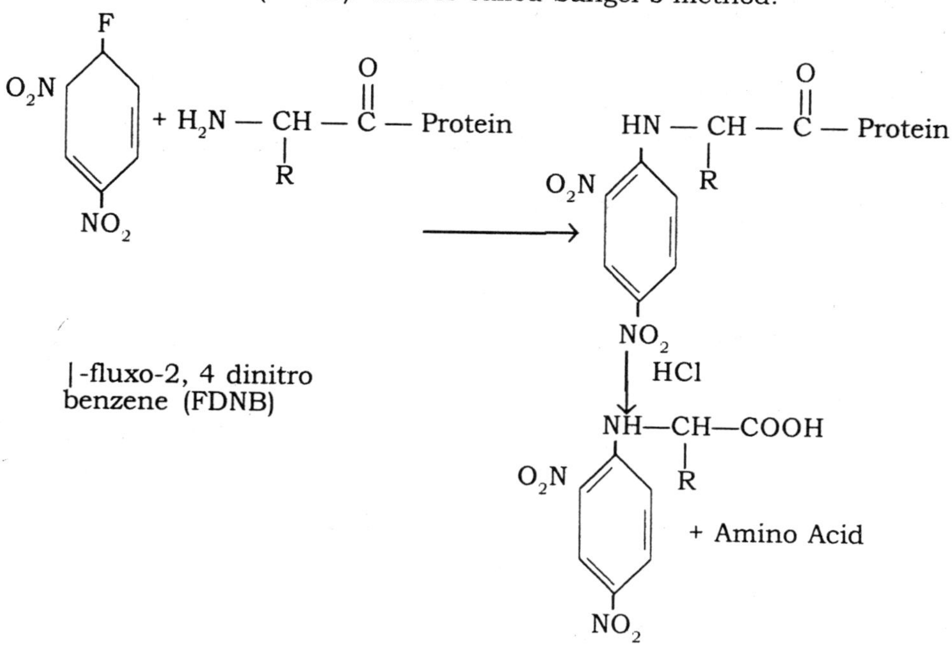

|-fluxo-2, 4 dinitro benzene (FDNB)

2,4 dinitrophenyl amino acids (Bright yellow colour)

Acylation: Acetic anhydride reacts with amino group as follows :

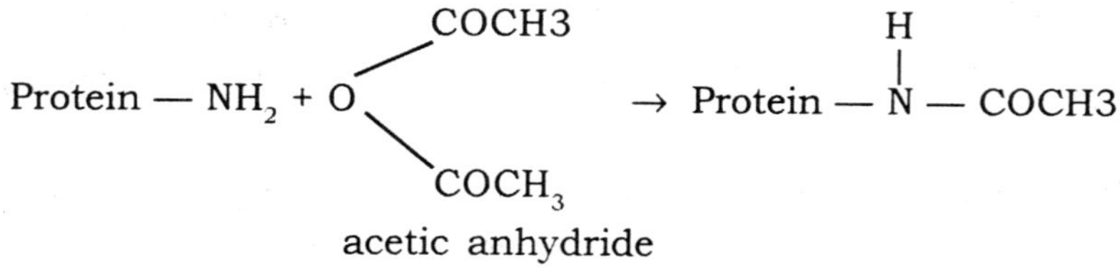

Substitution: Phenyl isocynate or benzoyl chloride can be substituted in NH_2 group of casein as follows :

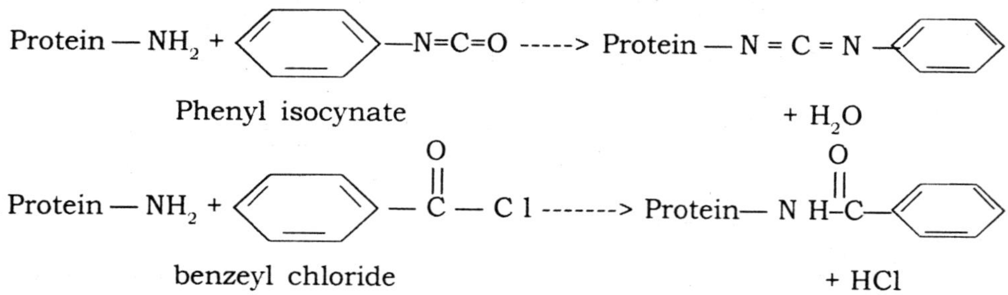

Reaction with Formaldehyde: Primary amino group, amide group and guanidyl group of casein react with formaldehyde. However, there is no reaction of formaldehyde with secondary amid groupi.e. peptide linkage. In between charged and uncharged forms, amino groups exist in equilibrium. The reaction goes to the right when formaldehyde reacts with the uncharged form. Though the proportion of N-monomethylol, N-dimethylol and N-methylene derivatives is not known but the alkalinity of -amino acid, imino acid, and epsilon amino group is definitely eliminated. The behaviour of formaldehyde treated amino acid is just like fatty acid and titration of the carboxylic group can be done with an alkali. The reaction is as follows :

$$RCH_2\ NH_3^+ \rightleftharpoons H^+ + RCH_2NH_2$$

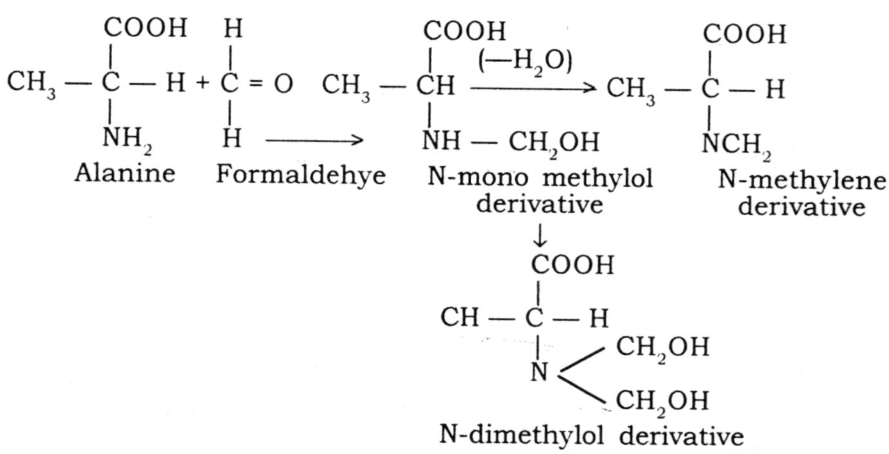

$$R-CH(NH_2)-COOH + H-C(=O)-H \longrightarrow R-CH(N=CH_2)-COOH + H2O$$

Amino acid Formaldehyde

$$R-CH(NH_2)-COOH + H-C(=O)-H \longrightarrow R-CH(NH_2CH_2O)-COO^- + H^+$$

Complex

The amino groups of milk proteins react with neutral formaldehyde and an equivalent amount of protons (H^+ ions) is released which is then titrated against an alkali.

Deamination with Nitrous Acid: The basis for determination of amino acids by the Van Slyke method is the reaction of nitrous acid with amino acid with the release of amino nitrogen. Rapid reaction takes place with Alpha-amino group of polypeptide chain whereas the reaction is slow with -amino group of lysine. The reaction with the guanidyl group of arginine is very slow.

$$R-CH(NH2)-COOH + HONO \longrightarrow R-CH(OH)-COOH + N_2 + H_2O$$

Iodination: Thyroid hormone 'thyroxine' type of compound is formed when protein is iodinised. Isolation of thyroxine can be done from iodinated casein. But it is not clear how thyroxine is produced from iodinated protein.

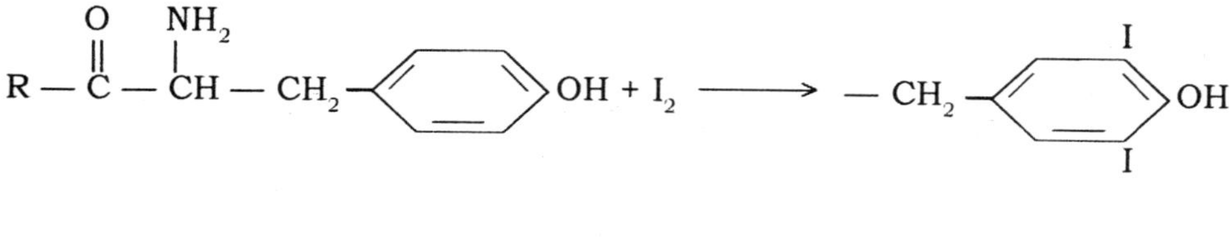

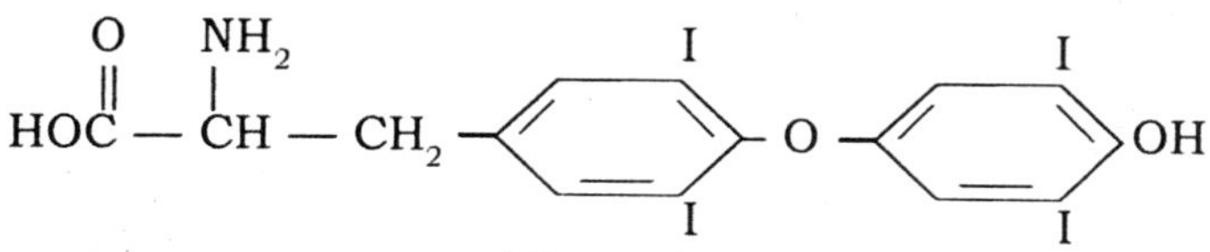

Thyroxine

Colour Test

Biuret Test: When a solution of copper sulphate is added to a strongly alkaline solution of the protein the biuret reaction occurs. A purple violet colour results. Any compound which has two peptide or amide groups separated by a carbon or a nitrogen atom gives this test. Biuret is NH_2-CO-NH-CO-NH_2. The colour complex that is formed is due to the coordination of cupric ions with unshared electron pairs of peptide-nitrogen and oxygen of water.

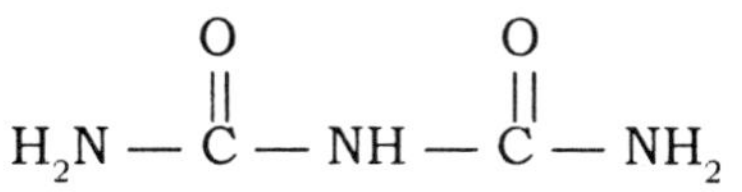

Biuret

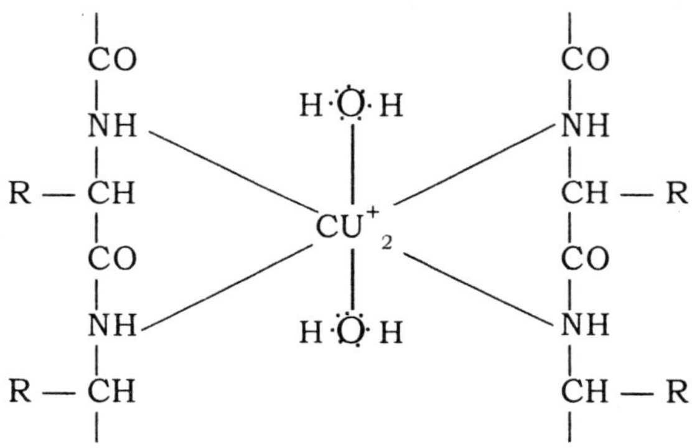

Ninhydrine Test: A general reagent for α-amino acid is Ninhydrine i.e. (Triketo-hydrindene hydrate). Purple colour is obtained when Ninhydrin in a solution of alcohol or acetone react with amino acid. The reaction is as follows :

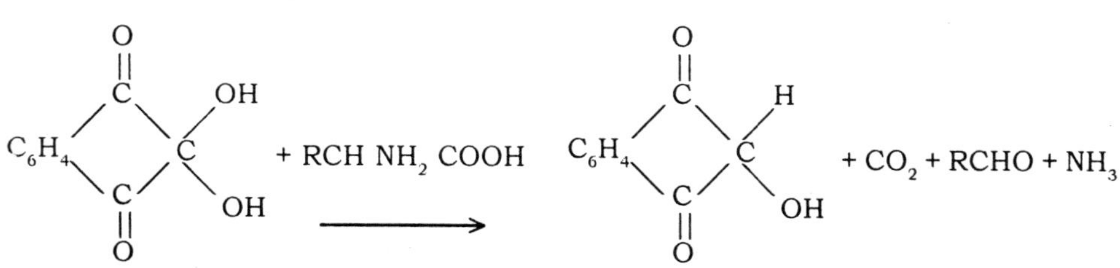

Ninhydrin
C_6H_4 $(CO)_3$ H_2O

Hydrindantin

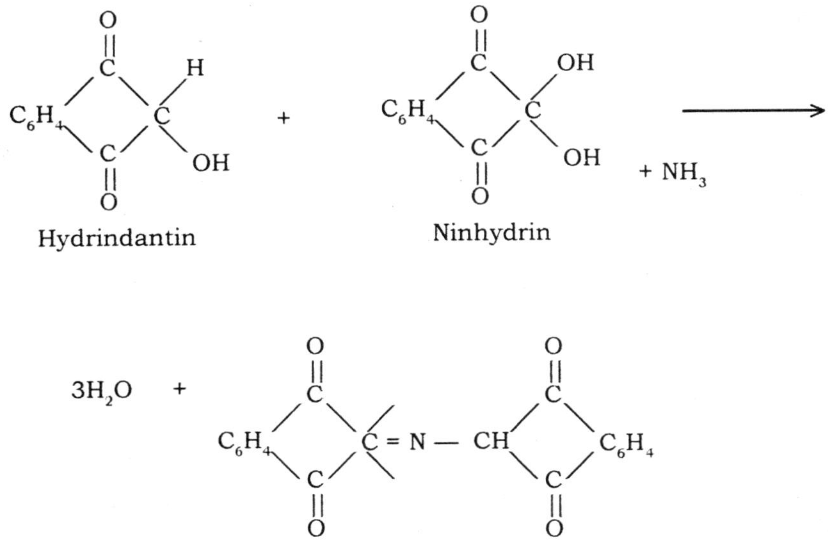

Hydrindantin Ninhydrin

Purple dye

Ruheman's purple
Triketo hydrindene hydrindamine hydrate

Hopkins Cole Test: This test is also known as Tryptophan test. When protein and aldehyde are poured into concentrated sulphuric acid, violet to blue colour is produced. The production of violet to blue colour is due to the reaction of Indolyl group of tryptophan from protein mixture with aldehyde.

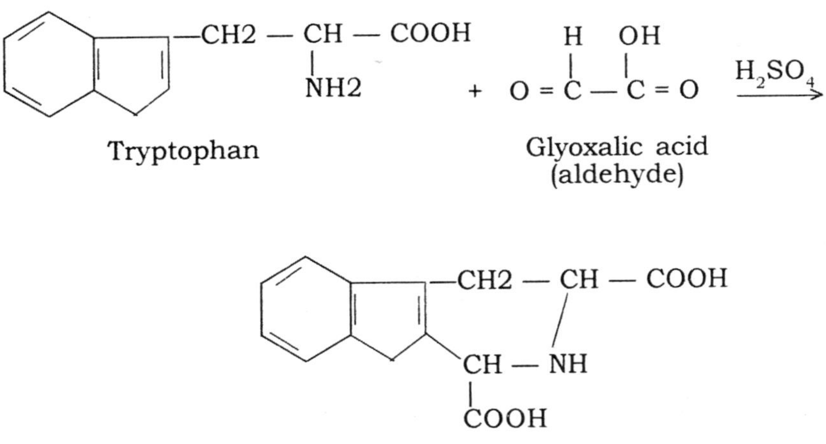

Tryptophan Glyoxalic acid (aldehyde)

Millon Test: When a solution of mercurous nitrate containing oxides of nitrogen or nitrous acid is made to react with a protein, a red colour is produced. This solution known as Millon's reagent is prepared by dissolving mercury in concentrated nitric acid. Phenol group or –OH of phenol is attacked and this reaction is specific for phenol groups or –OH of phenol.

Xanthoproteic Test: Concentrated nitric acid when made to react with a protein, yellow colour is produced which is due to the nitration of phenol and indolyl group.

Follin's Test: Follin designed a phosphomolybdotungstic acid reagent for detection and analysis of indolyl and phenol groups in protein. Distillation of sodium tungstate, sodium molybdate, orthophosphoric acid, water and concentrated hydrochloric acid produces this reagent. Lithium sulphate is then added to this distillate and finally bromine water is added. Excess bromine is removed by heating and the solution is made in 1N by the addition of distilled water. During protein hydrolysis when free tyrosine and tryptophan are generated, this method is of great use. Degradation of protein in milk and cheese can be usefully detected to some extent by this method. Alkaline protein solution containing sodium carbonate in sodium hydroxide, copper sulphate and sodium potassium tartarate solution produces blue colour in presence of Folin's reagent. Biuret reaction of protein with copper ions in alkali and reduction of phosphomolybdic phosphotungstic acid reagent of tyrosine and tryptophan present in treated protein, produces this blue colour.

Saka Guchi Test: Hypobromite and alpha-napthol react with arginine and other guanidyl derivatives of protein to produce a red coloured product which has the following chemical formulae.

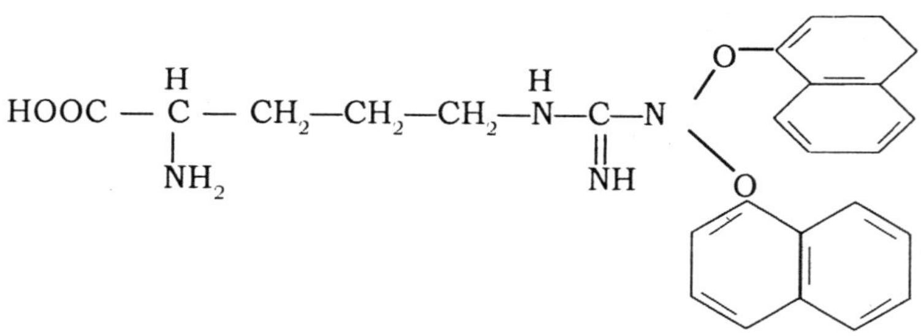

Nitroprusside Test: An intensely red but unstable colour is produced when sodium nitroprusside ($Na_2Fe(CN)_5No$) reacts with compound containing sulfhydryl groups like ß-lactoglobulin of denatured protein having free –SH groups.

Effect of Processing Variables on the Stability of Milk Proteins

Purpose of Heat Processing

The reasons for heat treatment or heat processing are the following :

(1) Pasteurisation and sterilisation fulfill the public health requirement, (2) to remove water, (3) enzymes are destroyed, (4) Ice cream mix and processed cheese are produced due to mixing and blending processes (5) to maintain the incubation temperature for cheese and cultured

dairy products (6) to produce desirable properties (7) to preserve dairy products, (8) heat treatment results in increased viscosity, reduced curd tension, resistance to oxidation, changes in flavour and colour and alteration in protein stability.

Pasteurisation: Two types of pasteurisation are recognised. (a) LTLT (low temperature, long time), (Holding method) 63°C to 66°C for 30 min., (b) HTST (high temperature and short time) 72.2°C to 72.8°C for 15 seconds. Pasteurisation causes the destruction of pathogenic bacteria, e.g. *Microbacterium tuberculosis* and most of the non-pathogenic organisms.

Sterilisation: Evaporated milk is generally sterilised by continuous agitation at a temperature of 241°F i.e. 116°C for 15 minutes. Sterilisation causes the destruction of all the micro-organisms and their spores. Amino sugar browning, i.e. Maillard reaction takes place between proteins and lactose and a brown high molecular polymer known as melanoidin results.

Forewarming: Temperature of 190° to 212°F (88–100°C) are employed for 10 minutes. Forewarming or preheating is required for the manufacture of condensed and evaporated milk. At forewarming temperature, milk enters the vaccum pan at a temperature well above boiling point to ensure removal of water.

Condensing: Condensing temperature is 110°–140°F (43–55°C). Removal of water is the main aim. In the dairy industry heating under vaccum is carried out.

Drying: There are two types of drying. The inlet temperature of air in spray drying is 160–350°F (71–177°C). In roller drying a temperature of 194°–270°F (90–132°C) is used. In spray drying and roller drying while removing water, minimum amount of irreversible change should occur mainly in the proteins, fats and carbohydrates.

Superheating: Making use of super heated steam, super heating is done at 185°F (85°C) for a period enough to give the desired increase in viscosity. It is done mainly to increase the viscosity of the fluid dairy product and to increase the properties of condensed whole or skim milk for use in ice cream mix. The milk proteins are completely destabilised when over heating is done.

Homogenisation: Lipase is inactivated when milk is heated to at least 140°F (60°C). This heating results in the breakdown of fat globules into a finer and more stable state of dispersion. The process of homogenisation leads to the liquification of fat and inactivation of lipases. Heat treatment inactivates the lipase and prevents lipolysis of fat. This prevents the production of rancid flavour when lower fatty acids are produced from the glyceride due to lipolysis.

Stability of Casein

1. The stability of casein is influenced by the time and temperature combination of heating.

2. Heat treatment causes a release in colloidal calcium to the milk serum and the resultant higher stability of milk.

3. In heat treated milk, when stored, changes occur which are not due to bacteriological or enzymatic activities but due to the behaviour of casein micelles.

4. Aggregation of heat treated milk occurs, when stored. But as a matter of fact heat treated milk due to higher surface potential shows little inclination to aggregation.

5. Heat treatment causes a release of calcium ions which results in the reduction of surface potential. This causes a reduction in the micellar charges during storage (ageing) but increases the possibility of aggregation (and gelation).

6. Serum protein of milk is denaturable. However, casein is not denaturable. The solubility, viscosity, optical rotation or casein are not changed when heated alone in solution in the range of heat treatment that denatures the serum protein. Drastic heat treatment is necessary to coagulate casein in comparison to whey proteins.

7. Heat treatment causes destabilisation of casein particles that aggregate to form a three dimensional structure entrapping some of the milk serum. Firstly, viscosity is increased and with more drastic heat treatment, coagulation of the proteins occurs with the separation of the whey.

8. Together with heat, the stability of the casein micelle is dependent on the pH, salts and serum protein.

9. Heat treatment causes breakdown of casein micelles with the concomitant release of esterified phosphate and hydrolysis of peptide linkages.

10. Caseinate particles are easily destabilised as small quanity of polyvalent cations (such as calcium, magnesium) when acting in conjunction with heat and lowered pH.

Whey Proteins

1. Heat treatment causes denaturation of whey proteins. Such coagulation is more sensitive to electrolytes with decrease in solubility.

2. Heat treatment inactivates the enzymes.

3. The agglutination power of euglobulins reduces.

4. Usually on a batch heat treatment of milk, a skin is formed due to denaturation of whey proteins which unite at the liquid/air interface and begin association with casein micelles.

5. Co-precipitates are formed on sufficiently high heat-treatment of whey proteins with casein. In this, the sulfhydryl groups of ß-lactoglobulin and K-casein join together in a sulphur-sulphur bridge.

6. Heat treatment disrupts the native configuration of whey proteins. A definite coiled configuration maintained by hydrogen bonding, hydrations and secondary valence forces exist in the native configuration of whey proteins. Heat denaturation of whey proteins causes a change in their solubility. Whey proteins are less soluble at pH 4.6 and in saturated sodium chloride. Precipitation of casein occurs in heated milk on acidification or saturation with salt.

7. Heat denaturation of whey proteins results in the evolution of H_2S, formation of cooked flavour, lowering of oxidation/reduction potential, development of antioxygenic properties and ultimately in the lowering of curd tension.

8. Normal pesteurization does not cause denaturation of whey proteins. It needs more drastic heat treatment.

9. With various heat treatments denaturation and coagulation of the whey proteins and casein occurs. Electrophoretic analysis shows that immunoglobulins are rendered acid precipitable by the least heat treatment followed by blood serum albumin, the ß-lactoglobulin and finally the lactalbumin which is very resistant to heat denaturation.

10. The sulfphydryl groups in the native protein are such as to be unreactive. Heating uncoils the native structure and results in producing more reactive sulfphydryl groups.

11. Heat treatment causes denaturation of the serum proteins and thereby reduces the curd tension with the resultant production of soft curd. Two theories have been propounded. The first is due to flocculation of serum protein producing weakness in the curd and secondly due to the interaction of ß-lactoglobulin with K-casein which results in the destruction of its capacity to form a firm curd.

12. Denatured whey protein improves the quality of breed. This may be due to the interaction of whey proteins with gluten (protein) of bread with the resultant production of soft curd.

13. Loss of cream layer formation in heated milk is due to certain globulins of milk.

14. Heat denaturation of whey proteins results in the production of amino-sugar condensation (Maillard reaction) with the ultimate formation of Melanoidin, a brown pigment.

Effect of Heat on Milk

When milk is heated near the boiling point and cooled, a film or skin is formed on the surface. This skin is due to calcium caseinate, although, other constitutents of milk are also present in this. Milk, heated to 118°C or higher cause browning. The brown colour is due to the heat effecting a change by the interaction between the casein or amino acids and the sugar lactose. Heating milk at UHT temperature (135–155°C) for a short time for seconds has the advantage of long shelf-life. Increased browning and a greater amount of whey protein denaturation were particularly noticeable. During pasteurisation (heating milk at 62°C for 30 minutes or 71°C for 15 seconds) pathogenic bacteria are destroyed and no heated taste is noticeable, however, heating for a longer peiod develops cooked flavour. This cooked flavour or cooked taste has been shown to be due to production of sulfhydryls by high temperature. These compounds came from the protein of milk, particularly ß-lactoglobulin and protein of the fat globule membrane. A sulphite salt, such as ammonium or sodium can artificially produce a cooked flavour. The same effect is obtained by adding the compound glutathione, a tripeptide consisting of cysteine, glycine and glutamic acid, which contains the sulfhydryl (–SH) linkage. Sulfhydryl compounds are readily oxidised and delays oxidation of the fat in milk or cream heated to high temperature. Oxidised flavour usually appears when the sulfhydral compounds are oxidised and cooked flavour disappears.

High temperature is required to coagulate fresh milk of low acidity. Fresh milk when heated at 137°C will curdle in 3 to 20 minutes. The development of acidity, however, lowers the temperature requirements for coagulation. Milk containing 0.25 percent acidity will not cause coagulation until heated to 82°C. When excess of citrate and phosphate are present as compared to calcium and magnesium, coagulation will occur at lower temperature. Addition of citrate and phosphate would absorb excess of calcium and magnesium and thus induce stability. Calcium and magnesium on one hand and citrate and phosphate on the other hand have opposite effects on the heat coagulation. A proper balance of these two classes of milk constitutents will prevent coagulation.

Non-Protein Nitrogenous Constitutents

Casein, α-lactalbumin, ß-lactoglobulin and proteose-peptone precipitate when 12 per cent TCA is added to milk. Non-protein nitrogen is present in the filtrate. The compounds in the filtrate are numerous, viz. creatin, creatinine, urea, ammonia, uric acid, α-amino nitrogen (amino acids), hippuric acid, orotic acid, indican, phosphoglyceroethanolamine, O-phosphoethanolamine and phenylacetylglutamine.

The compounds present in the urine of dairy animals have a remarkable similarity between those present in non-protein nitrogen fraction of milk. As a matter of fact the compounds or substances present in the urine of dairy animals are the resultant waste metabolites of dairy animal's body. It is apparent that the bulk of these waste metabolities in the urine of dairy animals originates from the blood and hence this entry or appearance and level in milk or urine is due to the protein metabolism of the animal. Intake of feed by the animals is directly proportional to the presence of these compounds or substances in milk. NPN in milk varies from season to season and has no biological value as protein. It can not be utilised by the body as a substitute of protein nor it can increase the cheese yield. Pasteurisation by itself has no effect. However, pasteurisation with homogenisation causes an increase in the non-protein and α-amino nitrogen content. Evaporated milk when stored increases NPN.

Various non-protein nitrogenous compounds in milk may be classified into 4 groups :

(a) Amines, (b) Amino acids, (c) Amino acid derivatives (d) Other compounds.

Amines: Under the amines, choline is present in largest amount. Milk contains 43.0–285.0 mg Kg^{-1} choline. Apart from choline, 1-propylamine (3.0–15.0 mg Kg^{-1}), 1-hexylamine (5.0–24.0 mg Kg^{-1}), ethanolamine (0.5–8.5 mg Kg^{-1}), putrescine, cadaverine, spermidine, and spermine are present.

Amino Acids: All the twenty amino acids are present in milk. Also, hydroxyproline, ornithine, citrulline and -amino butyric acid occur in milk.

Amino Acid Derivatives: Literature lists about ten amino acid derivatives. These are, N-methyl glycine, hippuric acid (31.0–64.0 mg kg^{-1}) salicyluric acid (0.2 mg kg^{-1}), creatine (80.0 mg kg^{-1}), creatinine (32.0 mg kg^{-1}), histamine, phenylacetyl glutamine, kynurenine, idoxyl sulfuric acid (1.2 mg kg^{-1}) and taurine 1.0–7.0 mg kg^{-1}).

Other Compounds: The other compounds present in milk are carnitine (10.0–17.0 mg Kg^{-1}), acetyl carnitine (2.0–12.0 mg Kg^{-1}), morphine and sialyl derivatives of lactose, i.e. milk oligosaccharides.

There are specific methods of determination of most of the non-protein nitrogen components, but some like phosphoglyceroethanolamine, O-phosphoethanolamine and phenylacetylglutamine have no specific method of determining. In addition to this, hippuric acid, orotic acid and indican also do not have very specific method of determination.

Immunoglobulins

An antibody (or immunoglobulin) is a protein synthesized by an animal in response to the presence of a foreign body. The antibody has a specific affinity for the foreign material that elicited its synthesis. An antigen (or immunogen) is a foreign macromolecule capable of eliciting antibody formation.

Immunoglobulin molecule consists of two polypeptide chains of ~ 25 Kdal called light (L) and two polypeptide chains of ~ 50 Kdal called heavy (H) chains. One light chain is bonded to one heavy chain by a disulphide bond. The H chains are bonded to each other by at least one disulphide bond.

Light chain (L) consists of a variable region (V) and a constant region (C). All L chains do not have the same constant region. In fact, there are two types of L-chains, called Kappa (K) and Lambda (λ), which exhibit many similarities.

Heavy chain (H), like the light (L) chain, consists of variable region (V) and a constant region (C). The variable region of the heavy chain has the same length as that of the light chain, whereas the constant region is about three times as long.

The amino-terminal parts of the light and heavy chains contain the variable regions, which are responsible for the binding of antigen. Differences in amino acid sequence yield combining sites that have different specificities. The rest of each chain contains the constant region which mediates effectors functions such as the binding of complement. The same effector functions are mediated by the antibodies of very different specificity. Thus, antibodies consist of regions of unique sequence that confer antigen binding specificity and regions of constant sequence that mediate common effector functions. A striking structural feature of immunoglobulins is that the variable and constant regions are sharply demarcated in the amino acid sequence of the light and heavy chains (Fig. 5).

Limited proteolytic action of the enzyme papain cleaves the 1g G molecule into three active 50 Kdal fragments. Two of these fragments bind antigen. They are called Fab (ab stands for antigen binding, F for fragment). Each Fab contains one combining site for hapten or antigen, which has the same binding affinity for hapten as does the whole molecule. However, Fab does not form a precipitate with antigen because it is uni-valent (i.e. contains only one combining site), in contrast with an intact immunoglobulin G antibody molecule, which contains two identical antigen-binding sites. Immunoglobulin G can form a precipitate with an antigen that contains more than one antigenic determinant. The amount of precipitate formed is largest when the antibody and antigen are present in equivalent amounts.

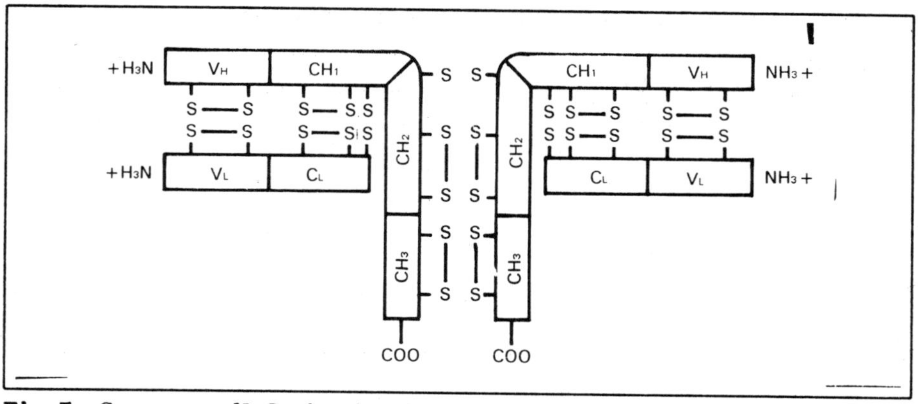

Fig. 5. Structure of IgG, showing the pattern of disulfide bonds and the regions of homologous sequence

The other fragment (also 50 Kdal), called FC, does not bind antigen, but it has other biological activities. This fragment can cross the placental membrane, whereas Fab cannot. There is also a site for the binding of complement on Fc. The binding of an antibody to an antigenic determinant on the surface of a foreign cell often leads to the lysis of that cell. This sequel of the combination of antibody and antigen is mediated by a group of proteins known collectively as complement. The binding of one of these proteins to the antigen-antibody complex triggers the reaction that lyses the foreign cell. The Fc fragment derived its name from the fact that it is readily crystallizable. This finding provided the first hint that Fc is homogenous in contrast to Fab, which usually cannot be crystallized because it is heterogenous.

Classes of Immunoglobulins

There are five classes of immunoglobulins, each consisting of heavy and light chains. The constant regions of the heavy chains differ from one class to another class, whereas the light chains are the same, either λ or K. The heavy chains in immunoglobulin G are called ν chains, whereas those in immunoglobulins A,M,D and E are called α, μ, δ and ε, respectively. These distinctive heavy chains give the five classes of immunoglobulins different biological characteristics. Immunoglobulin M (IgM) is the first class of antibodies to appear in the serum after injection of an antigen, whereas immunoglobin G (IgG) is the principal antibody in the serum. Immunoglobulin A (IgA) is the major class of antibodies in external secretions such as saliva, tears and milk. Thus, IgA serves the first line of defence against bacterial and viral antigens.

Properties of different classes of immunoglobulins alongwith their concentration in milk is given in Table 3.2.

Table 3.2. Characteristic of different immunoglobulins

Immunoglobulins	Approximate percent of skim milk protein	Sedimentation constant $(S_{20}W)^d$	Molecular weight
IgG1	1.2–3.3	6.5–7.0	161000
IgG2	0.05–0.1	6.5–7.0	146000–154000
IgA	0.05–0.15	10.8–11.0	385000–417000
IgM	0.05–0.1	18.2–19.8	1000000
Free Secretory Component (FSC)	0.02–0.1	4.0	79000

Simple Basic Techniques of Identifying Antibodies

Secondary Ring Test: In a narrow capillary tube containing serum with antibody, antigen solution is layered. The tube is left undisturbed for sometime. A ring of precipitate appears at the interface of the two fluids showing a positive reaction. Control containing saline gives a negative reaction (Fig. 6).

Oudin Procedure (Agar Diffusion): Anti serum is mixed with buffered agar and poured into a test tube and allowed to settle. Antigen solution is then layered over it. The tube is left undisturbed for sometime. The antigen diffuses into the agar and thus creates a gradient of decreasing concentration from the surface. At some point where it reaches optimum ratio with antibodies in the agar, a precipitate line is formed. This is known as single diffusion in one dimension (Fig. 6).

Mancini Technique: This is the simplest of all immuno-techniques and is used for the quantitative determination of a single antigen. Antigen sample is placed in a well punched, thin layer agarose gel containing mono specific antibody. The plate is then left for 24 hrs or more for the antigen to diffuse. As the antigen molecules diffuse outward into the gel, they bind to the antibody and soluble complexes are formed. At a certain distance from the well, the concentration of the diffusing antigen reaches equivalence with the concentration of antibody in the gel, and a precipitate is formed in a circle around the well. The area of the circle is proportional to the amount of antigen originally present. Therefore, by comparing the area of the circle produced from known concentrations of antigen with the area produced from an unknown, the concentration in the unknown can be determined (Fig. 6).

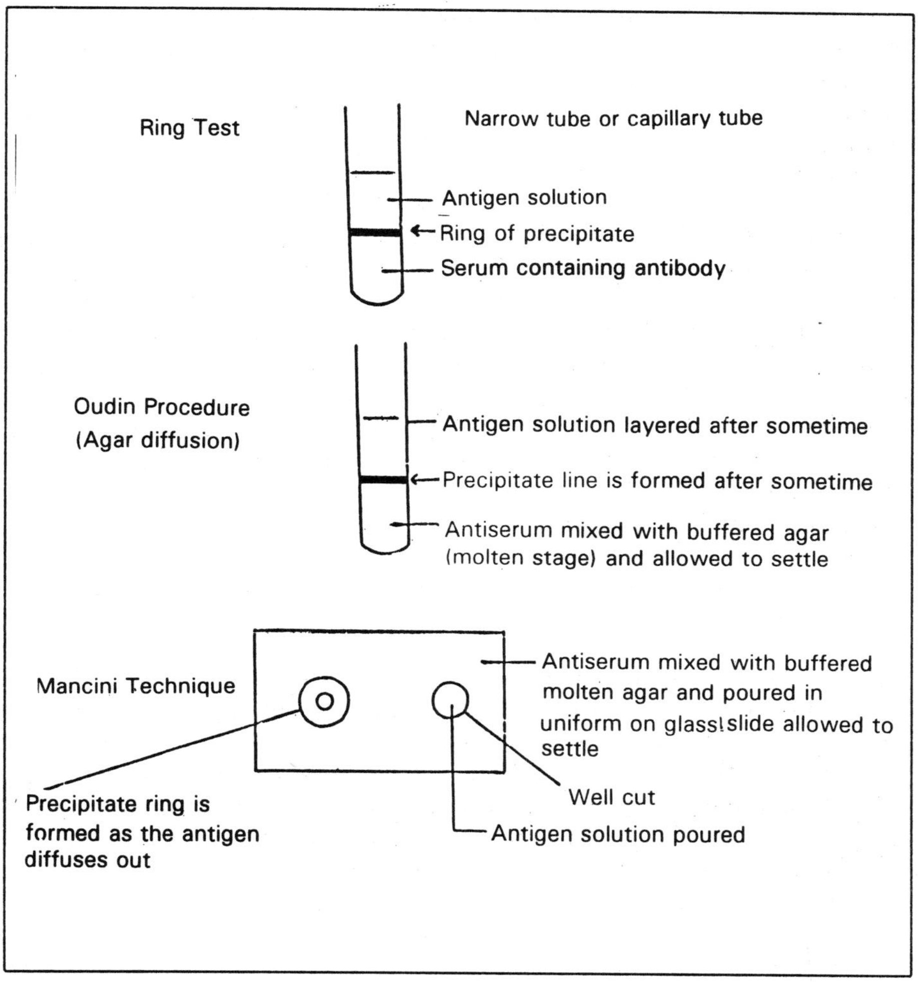

Fig. 6. Techniques of identifying antibodies

Milk Enzymes - Function, Influence of Processing Parameters and Effect on Storage

Enzymes are biological secretion of a living cell, they are protein in nature and serve as a biological catalyst. Some enzymes like pepsin, trypsin and ribonuclease are simple proteins since they consist entirely of amino acid units. Others, however, contain non-protein portions and are, therefore, conjugated proteins. Apoenzyme is the polypeptide segment of an enzyme. The non-protein organ moiety which can easily be dissociated from the proteinaceous part of the enzyme is known as the coenzyme. When the organic moiety cannot easily be removed from the apoenzyme, it is termed as the prosthetic group. The substances on which the enzyme acts are known as substrates.

Milk is a repository of myriad of enzymes which are secreted by the cells of lactating mammary tissues. About forty five milk enzymes have been purified, isolated and identified from milk. All these have been found only to be constitutents of milk. Besides these, numerous other enzymes are also present in milk. However, whether these exist solely in milk have not been established.

Enzymes in milk are not distributed homogenously. These occur in 4 distinct phases – (i) water soluble, (ii) associated with lipids, (iii) bound to casein and (iv) present in microsomal particles. These enzymes are poised in an equilibrium which can be disturbed by the chemical and mechanical or physical methods. There are three main sources of origin of these enzymes viz., (i) Blood or other organs via blood, (ii) Secretary cell cytoplasms, of which some are occasionally entrapped in fat globules by the envelope of the fat globule membrane, (iii) the fat globule membrane itself, from the membrane of golgi apparatus, the apical membrane of the secretory cell are produced and from which the fat globule membrane is produced. During late lactation and mastitis due to the greater permeability of mammary gland secretory cell membranes, the activity of milk enzymes increases.

Till today the functions of milk enzymes are exactly not known. It is not known whether they serve any purpose or are just secretory waste products of the mammary cells. Microbial enzymes in milk are not considered as original to milk, though milk contains several of these. Milk enzymes do not contribute to the nutritional well-being of young ones. Lipases and proteases act on the suitable substrates which milk has in plenty. However, others act on substrate foreign to milk. Of the total protein in milk, milk enzymes are almost negligible but they have a biological role which makes it necessary to formulate suitable methods to estimate them. The nutrition of the young and adults are not affected by the milk enzymes. Apart from the enzymes, the biosynthesis of lactose, protein and lipids, other enzymes have been extensively utilised as a yardstick for evaluating a variety of biological events operative in milk. The role of these enzymes are not known and this is a field wherein lot of studies are required.

The principal enzymes natural to milk are catalase, peroxidase, xanthine oxidase, alkaline phosphatase, acid phosphatase, amylase, protease, lipases and aldolases. The enzymes found inherently in milk may be grouped as follows :

a. Hydrolytic group of enzymes : Under this group, lipases and esterases are involved in fat hydrolysis, thereby causing rancidity in milk and milk products. Proteases proteolyze protein thereby causing gelation in milk and milk products particularly UHT milk.

b. The second group consists of catalase and lysozyme. Catalase is synonymous with udder infection and the lysozyme is related to the immune properties of milk.

c. Xanthine oxidase, cytochrome c reductase and alkaline phosphatase are associated with the microsomal particles of milk. Destruction of pathogenic bacteria in milk is indicated by the inactivation of alkaline phosphatase in milk.

d. The role of ribonuclease, salolase, rhodanase and carbonic anhydrase is not yet known.

Some of the specific enzymes and their importance are described below :

Alkaline Phosphatase: Alkaline phosphatase is inactivated by heat but reversibly reactivated when milk is chilled. *Microbacterium tuberculosis* is destroyed at a temperature wherein alkaline phosphatase is inactivated. Inactivation of alkaline phosphatase by pasteurisation (HTST or LTHT) is an index of destruction of *Microbacterium tuberculosis*. That alkaline phosphatase is inactivated by pasteurisation temperature is indicated by easy chemical methods. It can be estimated by p-nitrophenol liberated from p-nitrophenyl disodium orthophosphate. The optimum temperature for enzyme activity is 37°C and pH 9–10. It can be isolated from butter milk or skim milk. The enzyme is activated by magnesium and manganese ions while arsenite, borates, iodine, and metal complexing agents inhibit its activity. The systemamtic name of this enzyme is orthophosphoric monoesterphosphohydrolase and it is distributed in cream and skim milk.

Lipase: The cause of hydrolytic rancidity in butter fat, ghee and milk products containing fat is lipase. The systematic name of lipase is glycerol esterhydrolase. The lipases in milk can be divided into two main categories viz. plasma lipase and membrane lipase (FGM). Two optimum pH i.e. 7.0 and 8.6 are recognised. The release of free fatty acids on hydrolysis of triglycerides is an index for its estimation. Usually the substrate is tributyrin and glycerides. The enzyme is activated by Ca^{++}, Mg^{++}, homogenisation, shaking, temperature manipulation, while inhibited by trypsin, ageing, formaldehyde, heat treatment, heavy metals, hydrogen peroxide and sunlight. The sediments that form on clarification of milk serves as its source. Another source is the milk.

Catalase and Esterase: An index of udder diseases is the increase in activity of these two enzymes in milk.

The systematic name of catalase is hydrogen peroxide oxidoreductase. It occurs in milk microsomes, and is associated with casein, lactoglobulin and fat fractions. The principle for its estimation is based on the formation of O_2 from H_2O_2. The optimum pH for its activity is 7.0. It can be isolated from milk using n-butanol extraction, ammonium sulphate fractionation, DEAE sephacel chromatography and sephacryl S-300 gel filtration. Its coenzyme is Heme or Haemin. The enzyme activity is inhibited by acids, CN^-, F^-, 2:4 dichlorophenol and hydrogen peroxide above 10°C. With the increase in leukocyte count in milk, concentration of catalase increases. This enzyme is also heat stable. Mastitic milk is detected by measuring the catalase content.

Xanthine Oxidase: Often oxidative degradation of milk products is said to be caused by this enzyme. Distributed in fat globules and microsomal particles, the enzyme catalyses the addition of oxygen to a substance or removal of hydrogen from it. Quite stable to heat, its activity increases with heat and homogenisation. The coenzyme or prosthetic group is flavin adenine dinucleotide. The enzyme activity is inhibited by cyanide, arsenite, hydrogen peroxide and p-chloromercuric benzoate. The enzyme consists of protein, flavin, molybdenum and iron. The optimum pH lies between 6-9.

Lysozyme: Immunity of young and old is related to this enzyme. Also the keeping quality of milk products may be related to it. The systematic name is mucopeptide-n-acetyl neutraminyl hydrolase. It hydrolyzes ß-1, 4 linkage between N-acetyl glucosamine and n-acetyl muramic acid. The cell wall or *Micrococcus lysodekticus* cells serve as substrate for this enzyme. Its optimum pH is 9.5. Isolation of enzyme involves adsorption on amberlite, precipitation with ammonium sulphate and gel filtration on sephadex. The enzyme is heat stable at low pH. It is bacteriocidal, thereby increasing the keeping quality of milk.

Ribonuclease: Associated with the microsomal fraction, ribonuclease influences the stability of fat emulsion.

Lactose Synthetase: This enzyme was thought to be just a single enzyme catalysing the last reaction in the lactose biosynthetic pathway wherein glucose is the acceptor and UDP-galactose is the donor. However, Ebner in Oklahoma proved that this name is a misnomer. He isolated the enzyme lactose synthetase from milk and passed it through Biogel P 30. He obtained 2 peaks, the first a major peak and the second a minor peak. He termed the first peak as A protein and second as B protein. This A protein was later found to be the enzyme galactosyl transferase and the B protein to be lactalbumin, the milk whey protein. He found out :

$$\text{UDP - gal + Glucose} \longrightarrow \text{N.R. (No reaction)}$$

Here the reaction proceeds only to the right when glucose is in very high concentration. Km value is very high.

$$\text{UDP - gel + glucose} \xrightarrow[B]{A} \text{Lactose + UDP}$$

Here the Km value is very low.

The function of α-lactalbumin in this reaction is that of a modifier protein. Its effect is allosteric, i.e. it modifies the reaction. The enzyme galactosyl transferase has an allosteric site apart from its active sites. The modifier protein α-lactalbumin attaches or binds itself to allosteric site and thereby changes the 3-dimensional configuration of the active sites of the enzyme. The active sites of the enzyme galactosyl transferase gets so contoured that they are able to form ES complex and ultimately the enzyme is released and the product is formed.

Influence of Processing Parameters

The processing parameters usually used in the dairy industry are pasteurisation, ultra heat treatment (UHT) and sterilisation, forewarming, condensing, roller drying, spray drying, etc. Most of these enzymes mentioned earlier are heat labile. However, the heat lability differs from enzyme to enzyme.

Pasteurisation: Alkaline phosphatase is inactivated during pasteurisation. However, acid phosphatase requires more drastic heat treatment. Aldolase and α-amylase are more labile than lipase which is destroyed/ inactivated below pasteurisation temperature. The activity of the enzyme aldolase decreases at 37°C and 45°C. Pasteurisation temperature does not destroy the peroxidase. A slightly higher temperature than pasteurisation (76–78°C) is required to destroy it. Catalase activity is destroyed upto 84 percent by pasteurisation. The heat stability of xanthine oxidase is more than alkaline phosphatase but less than peroxidase. On the other hand the inhibition of α-amylase takes place at 45–52°C for 30 minutes. The enzyme alkaline phosphatase, peroxidase and catalase are reversibly reactivated.

Storage: Indigenous proteinases are present in normal milk. These proteinases modify certain characteristics. Milk can be stored at 4°C for one week without any significant change if the initial microflora is low. With the increase of storage time at 4°C, upto seven days, there is increase in growth of psychrophills or psychrotrophs. These secrete heat stable proteinases. However, heat stability of milk is not affected till the growth of microflora reaches 10^9 colony forming units per ml. It is unlikely that such population is present in normal commercial practice.

UHT Treatment: UHT treatment does not necessarily inactivate the inherent proteinases or proteinases of bacterial origin that are heat stable. Proteinases surviving heat treatment have been isolated from UHT milk. The proteinases from the microflora that survive UHT treatment and grow at sub-optimum temperatures produce heat resistant proteinases. Inherent proteinases of milk are similar to blood plasmin and are not heat stable. Cooling UHT milk at 55°C and holding for 60 minutes destroys the proteolytic activity unlike alkaline phosphatases which is reactivated on cooling. Therefore, the mechanism involved is different.

Enzyme Treatment: Rennet or neurominidase treatment prior to sterilisation prevents gelation and thereby increases the stability of milk. On the other hand, heat resistant proteinase or addition of psychotrophs reversibly hastens gelation.

Rennet Treatment: On treatment with calf rennin (Chymosin) 105-106 phenylalanine-methionine bond of K-casein is broken with the result para-k-casein is formed and caseinomacropeptide or macropeptide is released. Casein is destabilised and rennet curd is formed.

Forewarming: During forewarming, a temperature of 95°–105°C is applied with an objective that lipases are inactivated or destroyed.

Homogenisation: Homogenisation brings about inactivation of lipase.

Biosynthesis of Milk Proteins by the Mammary Gland

Biosynthesis of proteins by prokaryotes and eukaryotes has been a mystery for decades and has aroused considerable interest among scientists over a long period of time. Baffling as the mechanism of protein biosynthesis might have been for scientists over many decades, slowly and steadily the erudite scientists have unveiled the curtain on biosynthesis of proteins by the cellular system. Later, studies were extended to gain more knowledge on the mechanism of milk protein biosynthesis unique to the lactating mammary gland.

Milk proteins originate from several different sources. However, the main bulk comprising the various casein fractions, ß-lactoglobulin, and α-lactalbumin are synthesised in the mammary gland on the rough endoplasmic reticulum (RER) with bound ribosomes whereas the cellular proteins are synthesised on free ribosomes in the smooth endoplasmic reticulum (SER). The milk proteins from the (RER) are then transported to the golgi apparatus wherein the prosthetic groups like phosphate and clacium form a complex with serine hydroxyl and threonine hydroxyl groups of caseins. Although, α-lactalbumin and β-lactoglobulin contain serine hydroxyl and threonine hydroxyl groups, no attachment of phosphate and calcium takes place in the golgi apparatus in these milk proteins.

There are two sources of amino acids for the biosynthesis of such milk proteins. The essential amino acids of caseins, β-lactoglobulin and α-lactalbumin come from the free amino acid pool of blood plasma or from plasma proteins, and majority of the non-essential amino acids are synthesised within the mammay glands from a number of organic compounds of the blood and from the nitrogen of a limited number of nitrogenous compounds. Lacteal secretions also contain certain blood proteins. Serum albumin and immuno globulins of milk originate from the blood. It may not be out of place to mention that 400 volumes of blood plasma flow through the cow mammary glands for each volume of milk secreted.

Our present knowledge on the mechanism of milk protein biosynthesis is mainly based on studies with radioactive tracer experiments in whole lactating animals, mammary tissue slices, organ cultures and cell free system. Tracers incorporated into possible precursors showed that the source of most of the milk proteins is amino acids mostly transported to the mammary gland by the blood. Rapid labelling of caseins, ß-lactoglobulin and α-lactalbumin over serum albumins and immunoglobulins when C^{14}-labelled acetate was injected into lactating animals categorised milk proteins into two groups, viz. (a) those synthesised in the mammary gland from the common free amino acid pool, and (b) those synthesised elsewhere and transferred intact or nearly so from the blood to the milk.

Mechanism of Biosynthesis of Milk Proteins

In protein or milk protein synthesis transcription of the genetic message takes place from DNA to RNA and the translation of the messages takes place from RNA to protein. For each protein to be synthesised, DNA molecule contains the genetic message for its synthesis. The genetic message for synthesis of each protein is then relayed to the mesenger RNA (mRNA) which is a ribonucleic acid. There is another type of ribonucleic acid which takes an active part in protein synthesis. This is known as transfer RNA (tRNA) or soluble RNA. The name of this RNA (tRNA) itself suggests that it transfers the amino acid tagged to its tail to the growing peptide chain. Ribosomes, the site of protein or polypeptide synthesis consist of nucleic acid and protein. The nucleic acid in ribosomes is known as ribosomal RNA (v-RNA). Metals like magnesium or the ammonium radical are also involved in protein biosynthesis.

The overall reactions of the protein biosyntheis are given below :

1) Amino acid + Adenosine Triphosphate —> AA ~ AMP (Aminoacyl adenytate) + P — P (pyrophosphate)

2) A.A. ~ AMP + tRNA —> A.A. tRNA (aminoacyl tRNA) + AMP (adenosine monophosphate)

3) (A.A. ~ tRNA)n $\xrightarrow[\text{r — RNA, GTP, } Mg^{++}, K^{+}]{\text{mRNA, ribosome}}$ Protein + tRNA

The first reaction is called the activation reaction. Adenosine triphosphate (ATP) is a high energy compound which activates or energizes the amino acid. This activated amino acid then reacts with tRNA to form AA · tRNA. This amino acyl tRNA, then in presence of messenger RNA (mRNA), ribosomes, ribosomal RNA (r-RNA), guanosine triphosphate (GTP), Mg^{++}, and K^+ or NH_4^+ forms protein.

The activated amino acid as told earlier get tagged to the adenosine moiety at the tail of the tRNA. The messenger RNA, which is synthesised in the nucleus from the nucleotides present there receives the genetic message from the DNA. The base sequence of mRNA so synthesised is governed by the base sequence of DNA molecule and is complementary. Thus, the mRNA in a way acting as a genetic engineer having the blue prints, moves to the ribosomes. Several tRNA molecules with different amino acids tagged to them are present in the ribosomal complex. As the mRNA gets wrapped up with the ribosomes, it comes across the tRNA molecules each carrying one particular amino acid tagged to the tail. Like this, many tRNA molecules move opposite to the mRNA wrapped up with the ribosomes and at each step an amino acid gets attached covalently to the growing peptide chain. The sequence of amino acids in a growing peptide chain is guided by the triplet codon of the mRNA molecule or the mapping on the mRNA molecule of groups of 3 consecutive purine and or pyrimidine bases spread out through the entire length of the mRNA molecule. As there are 4 different bases which include 2 purine and 2-pyrimidine bases spread out through the entire length of a mRNA molecule in triplet codons, there can be mathematically 64 different triplet codon ($4 \times 4 \times 4 = 64$). That the codon is triplet was proved mathematically and also through various sophisticated experiments. First it was proved mathematically that a single purine or pyrimidine base of mRNA could not code for 20 different amino acids because there are only 4 different bases, which would mean that such a code will only be able to utilise 4 different amino acids. A duplicate codon on the other hand can only take care of 16 amino acids (4^2 bases = 16). It was then postulated that the codon must at least be triplet in order to code for the 20 different amino acids. This hypothesis was later on proved experimentally.

Initiation of protein biosynthesis at the ribosomes begins when an amino-acyl tRNA molecule having the anticodon for the first triplet codon of mRNA molecule, moves exactly opposite to the triplet codon of mRNA molecule and releases that particular amino acid. Another tRNA molecule with an amino acid attached to its tail depending on the next triplet codon of the mRNA molecules with the formation of covalent amide or peptide linkage between the first amino acid and the second. Another tRNA molecule having a particular amino acid specific for the next triplet codon of mRNA moves to the ribosomes exactly opposite to the triplet codon of mRNA for which that particular amino acid is coded. Like this the peptide chain grows and ultimately a polypeptide or protein is completely synthesised. When a particular protein gets synthesised completely, a termination codon plays its role in preventing further elongation of the peptide chain of the protein molecule. This is because it has been found many times that the mRNA contains more triplet codons than what are necessary for that particular polypeptide or protein synthesised.

An engineer cannot accomplish any feat alone. He has to function in synchrony with others to accomplish anything. Similarly, in case of protein biosynthesis the tRNA and mRNA cannot accomplish the feat alone. Protein biosynthesis can only take place if ribosomes, RNA, GTP, Mg^{++}, K^{+} NH_4^{+} are present (Flow diagram 3.6).

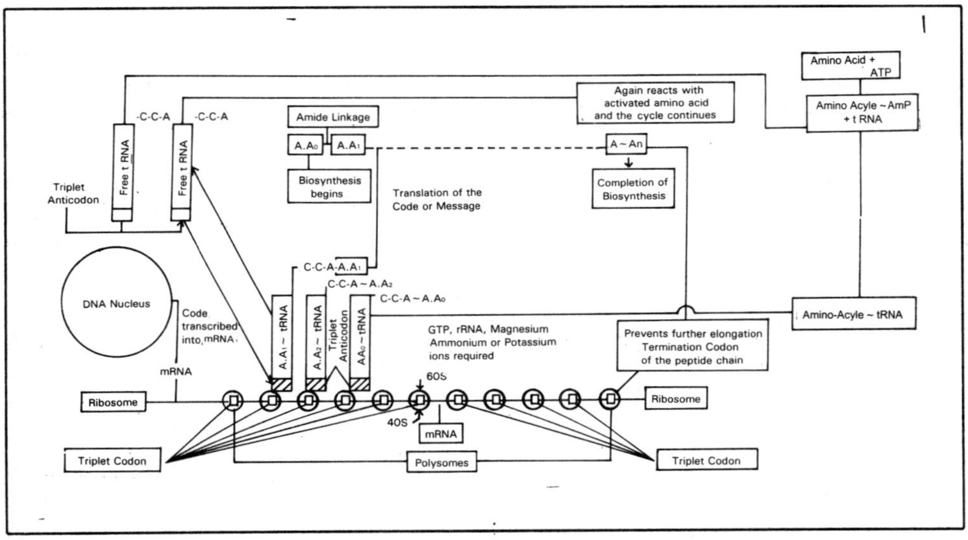

Flow diagram 3.6: Metabolic events in protein biosynthesis involving amino acid activation, transcription and translation of the code

Genetic Polymorphism of Milk Proteins

Before discussing the genetic polymorphism of milk proteins, let us review certain basic terms of genetics. Let us first discuss about chromosome. Chromosomes have been defined as a DNA-histone thread residing in the nucleus of the cell and bearing the genes in linear order. The next question that appears is what is a gene? Genes are short chromosomes made up of DNA. Each gene has a specific place or locus in the chromosome. Each gene as a functional unit contains a discrete segment of the giant DNA molecule of the chromosome. Genes normally occur in pairs in all cells except gametes as a consequence of the fact that all the chromosomes are paired except the sex chromosomes (XY) of the male. Genes are of two types genotypes and phenotype. Genotype are not visible to naked eye. The phenotype B/B (e.g. B-Bold) are visible to the naked eye.

Let us now discuss about allele. Allele may be defined as one of the pairs of series of genes which are situated in the locus or the same position in a specific chromosome. As autosomal chromosomes are paired, each autosomal locus is represented twice in normal somatic cells. If the same autosomal chromosome occupies both loci, the individual or cell is homozgous for this autosomal chromosome. If the autosomal chromosomes at the two loci are different, the individual or cell is heterozygous for both the autosomal chromosomes. Now the question comes what is phenotype? It may be defined as the character of an organism due to the interaction of the genotype and environment. We shall not try to define what is genotype. It is the genetic constitution expressed or latent for the particular set of genes present in the cells as opposed to phenotype - the properties that may be observed in an organism.

The term 'polymorphism' was first given by Prof E.B. Ford of Oxford University. Polymorphism was defined as the condition of occurrence together in the same habitat of two or more discontinuous forms of a species in such proportions that the rarest of them cannot be maintained merely by recurrent mutation. Polymorphisms have been studied using a number of different techniques like, visual observation, immunological techniques, biochemical techniques, biophysical techniques and various other miscellaneous techniques.

The studies on polymorphisms have a number of different practical applications. These include (1) parentage determination in disputed cases, (2) studies on major genes in populations, (3) analysis and definition of different breeds of domestic animals (4) recognition of different types of twins (MZ v/s DZ) which is of importance in scientific studies and practical animal husbandry (e.g. recognition of "Free martins", (5) association with pathological or production traits e.g. transferrins and performance in dairy cattle.

When mutation at molecular level occurs, genetic polymorphs occur in the biosynthesised milk proteins. Such mutation takes place one in million. The biosynthesis of mRNA takes place in the nucleus, is governed by the base sequence of DNA molecule and is complementary. Transcription of the message takes place from DNA to RNA and the translation of the message occurs at the ribosomes wherein in conjunction with tRNA, rRNA, GTP, Mg^{++} and NH_4^+, polypeptide or protein synthesis takes place. The mRNA contains the triplet code and the tRNA with a particular amino acid tagged to the adenosine moiety at the tail of tRNA which also contains the anticodon for the mRNA, biosynthesis of milk protein initiates. At the time of transcription of the message from DNA to mRNA there might be deletion or substitution or addition of the codons. For example at the time of transcription of the message from DNA to mRNA, a particular codon might be substituted by another codon which will give a different amino acid (deletion of base pairs), codons have been shown to be deleted. During meiosis, chromosomes under physical stress may break. Upon recombination, the chromatids may not create two reciprocally recombinant chromatids but rather cross over products which have a segment of DNA deleted. The deleted segment may be a single base pair or as many as 39 base pair e.g. αS_1-casein A. Consequently, when the transcription of the message takes place from such DNA, the mRNA will contain different base pairs with the resultant biosynthesis of αs_1-casein A having a different amino acid sequence for that particular segment of the αs_1-casein. This αs_1-casein will be named as αs_1-casein A and thereby different from αs_1-casein B and C. The third type of mutation is the insertion of one or more base pairs in the gene. This process is similar to deletion with an exception of breakage and union of chromatids, a portion of DNA is added to the crossover products. Such type of polymorphism does not occur in milk proteins.

Hardy Weinberg Law and Gene Frequency Distribution: The distribution of the genetic polymorphs follow the famous Hardy Weinberg Law. This law states that if there are two alleles of the type A/A and B/B, their frequency distribution in particular herd taken as p_2 and q_2, as the frequency distribution of two alleles will be p^2 AA + q^2/BB + 2 pqAB. The frequency distribution of various genetic polymorphs observed in different breeds of cattle all over the world gives a fair amount of agreement between the observed and the expected phenotypes in the cattle (Table 3.2).

Definition and Identification of Genetic Variants

The existence of genetic polymorphism in milk proteins was first reported on the occurrence of two forms of ß-lactoglobulin A and B in cow milk. The main characteristic which enables the identification of genetic species of milk proteins is the notable difference in the mobility of the same class of proteins in an electric field. Techniques like paper electrophoresis, starch gel electrophoresis and polyacrylamide gel electrophoresis have been used. In addition to mobility in an electric field the milk proteins may differ in Sedimentation constant and amino acid composition.

Genetic Variants of Casein

α s_1-Caseins

At least five genetic variants of αs_1-casein have been identified, but in Western breeds of cattle the B variant is predominant. The protein comprises a single polypeptide chain of known amino acid sequence which is found in milk as a major and a minor fraction. The major fraction, previously designated αs_1-casein, contains 8 mol P/mol (αs_1-CN-8P), whereas the minor component (previously s_0-casein) has an additional phosphorylated serine residue at position 41 and is now designated αs_1-CN-9P. According to the rules of nomenclature the variant is specified by a latin letter before the number of P atoms/mole. The major B variant is identified as αs_1-CN B-

Table 3.2. Genetic variants in milk proteins

Protein	Compared Variants	Amino acid substitution	Possible codon	Total number of Variants/specificity
α S_1-casein	B—>A B—>C B—>D B—>D	Loss of 13 residue (14-26) 192 Glu—>Gly Pro—>Ser 53 Ala—>Thr.P	Deletion of 13 codons GAA/G—>GG A/G CC U/C A/G—>UC UC/A/(GCU—>ACU	Five genetic variants (A,B,C,D,E) B (*Bos taurus*) C (*Bos indicus*) *Bos grunniens* (yaks)
α S_2-casein				Four genetic variants (ABCD), A,D (European breed) B (*Bos taurus*) *Bos indicus*, Zebu of Africa, C (yaks)
β-Casein	A_2—>A_1 A_1—>B A_1—>C	67 Pro—>Bis 122 Ser—>Arg 37Glu—>Lys 35Ser—>Ser	CCU—>CAU AGU—>AGA GAA—>AAA	Seven genetic variants $A^1 = A^2 = A^3 > B > D_1$ $E > C$
K-casein	B—>A	14S Ala—>ASP 136 Ileu—>Thr	GC U/C—>GA U/C AU U/C/A—>AC U/C/A	Two genetic variants A and B
α-lactalbumin	B—>A	Arg—>Gln	CGA/G—>CAA/G	Two genetic variants (A,B) A (African fulani, African & Indian Zebu cattle, B (western cattle only), African & Indian Zebu
β-lactoglobulin	B—>C B—>D B—>A	Gln—>His Glu—>Gln 1 Gly—>Asp 2 Ala—>Val	GA A/G—CA U/C GA A/G—>CA A/G GG U/C—>GA U/C GC U/C/A/G—>CU U/C/A/G	Seven genetic variants (A,B,C,D,E,F,G) E (yak) EFG Bali (Banteng) Cattle, BOS (*Bibos*) javanicus, BC, Bovine milk. Draught master variant has carbohydrate moiety in it similar to A variant

Source: Compiled from different sources.

8P and as_1-CN B-9P, whereas the rare D variant, which contains a phosphorylated threonine residue at position 53, is αs_1-CN D-9P and αs_1-CN D-10P.

The largely hydrophobic residues 14-26 are deleted in the A variant, resulting in a marked change in its physico-chemical properties compared to αs_1-CNB.

α s_2-Casein

Four genetic variants have been identified. The variant A is the most common in Western breed. Although, all the sites of phosphorylation in the amino acid sequence are not identified, it is thought that αs_2-casein exists, at least predominantly, in four phosphorylated forms containing 10–13 mol P/mol.

β-Casein

A total of seven genetic variants of B-casein are known, of which three (the A^1, A^2 and A^3 variants) predominate in Western breeds.

The γ-casein fractions and proteose peptone fractions 5, 8-slow, and 8-fast occur in milk due to partial gene duplication. However, identical fractions can be produced by incubation of β-casein with the natural milk proteinase plasmin and this appears to be a more reasonable explanation for their presence.

K-Casein

Two genetic variants are known, A and B, with the former prevalent in most breeds of Western Cattle.

Chymosin, the principal proteinase of calf rennet, preferentially hydrolyzes the Phe^{105}-Met^{106} bond to produce the para K-casein fragment (f 1-105), which is the same whatever the degree of phosphorylation or glycosylation, and the macropeptide fragment (f 106-169), containing all the actual or potential sites of genetic variation apart from the rate C variant. Normally only Ser-149 is phosphorylated and the seven or more bands of K-casein seen in alkaline urea gel electrophoresis in the presence of 2-mercaptoethanol originate mainly from glycosylation through O-glycosidic linkages between N-acetyl-galactosaminyl residues and either seryl or theonyl residues. A second site of phosphorylation (Ser- 127) has also been identified and there is a small degree of phosphorylation at a third, unknown, site.

Glycosylation has been observed at Thr-131, Thr-133, Thr-135 (or Thr-136) and Ser-141 (or Thr-142). The carbohydrate side chains are incompletely defined but include structures I-III.

Neu N Ac α — 2 → 3Gal β — 1 → 3 Gal N Ac β — 1→

(I)

Gal β — 1 → 3 Gal$_6$ N Ac β — 1 →
↑
Neu N Ac α — 2

(II)

Neu — N — Ac α — 2 → 3 Gal β — a → 3 Gal6 N Ac β — 1 →
↑
Neu N Ac α — 2

(III)

Because structures I and III normally predominate, it is thought that the terminal sialic acid group is added relatively rapidly. It has been shown that there is considerable heterogenity in side chain composition and point of substitution of the most sialylated forms of K-casein. More abundant, complex, and variable carbohydrate side chains have been recognised in cholesterol K-casein. The microvariants of K-CNB include unglycosylated K-CN-1P, K-CN-2P, and K-CN-3P, together with K-CN-1P species containing from one to nine sialic acid residues/moles. In general, the abundance of the fractions decreased with increasing number of side chains (zero to four) and, for a given number, the more sialylated form was present in greater amounts. It has been proposed that the distribution of glycosylated fractions follows a Poosson distribution, which could be expected for a sequence of independent glycosylation events in which side chains are attached to the protein, irrespective of whether it has previously been glycosylated. In contrast, phosphorylation of Ser-127 appears to inhibit completely subsequent glycosylation.

Whey Proteins

The most abundant whey protein is ß-lactoglobulin. It is a globular protein. Its nutritional value is high and its molecular weight is 35,000 daltons. It gets heat denatured and its isoelectric pH is 5.1. It is insoluble in water or pH 5.2 and soluble in dilute salt solution. The polypeptide chain of the simple form consists only of amino acids with the free SH groups of cysteine residue.

It has been observed by Peterson that the molecular weight differs at different pH. He has found out the molecular weight to be 40,700 dalton based on the sedimentation data. Aschaffenburg and Drewry have found out two genetic variants of ß-lactoglobulin which they named as A and B. On the other hand, McKenzie has found out that ß-lactoglobulin has three genetic variants, i.e. A,B and C. In buffalo and goat milk one genetic variant of ß-lactoglobulin has been found. However, now on the basis of decreasing mobility on starch gel electrophoresis and PAGE, 5 genetic variants, i.e. A,B,C, D and ß-lactoglobulin (Drought master) have been established. Additional genetic variants have also been observed and named as E,F,G.

The A variant contains 162 AA residue. The SH group lies between 119 and 121. The disulfide bridge occurs between 106 and 121 or 106 and 119. Its M.W. is 18362 daltons. The other disulfide bridge is between 66 and 160. The B variant differs from A at position 118 and 64 wherein valine and aspartic acid have been replaced by alanine and glycine, respectively. Whereas the C variants contain histidine for glutamine at position 59. The ßLgD differs from B by replacement of glutamine for glutamic acid at position 51. The Drought master variant contains carbohydrate moiety like N-acetyl neuraminic acid, glucosamine, galactosamine, mannose and galactose. While refering to ß-lactoglobulin, Cheeseman asserts "it is unlikely that it is solely an amino acid carrier and some secretory function may yet be ascribed to it, similar to that recently described for α-lactalbumin. β-lactoglobulin in goat consists of 162 amino acid residues as in bovine. The goat ß-lactoglobulin differs from bovine ß-lactoglobulin B at six positions.

ß-lactoglobulin bovine	Goat ß-lactoglobulin
Leu	Ile
Asp-53	Asn
ASP-30	Lys
Ser-150	Ala
Glu-158	Gly
Ile-162	Val

However, goat ß-lactoglobulin is less stable than bovine ß-lactoglobulin to denaturation. ß-lactoglobulin of buffalo is similar to B-variant of cow ß-lactoglobulin. In sheep, pig and reindeer, there are two genetic variants.

α-lactalbumin is the most important of the ruminant whey protein and is rich in human milk wherein by its presence it modifies the synthesis of large amount of lactose. Since 1966 its role in lactose biosynthesis has been determined. It acts as a regulatory protein in the biosynthesis of lactose α-lactalbumin is heat-labile and soluble in saturated magnesium sulphate. It is a simple protein having molecular weight of 16000 dalton. The molecular weight of bovine α-lactalbumin A is 14146 dalton and that of α-lactalbumin B is 14147 dalton. It contains no phosphorus or carbohydrate, rich in tryptophan about 7 per cent and sulphur, 1.9 per cent, i.e. from cystine and methionine but contains no free SH group. Its nitrogen content is 15.86 per cent of which 97 per cent may be accounted for amino acids and amide N. In the N-terminal it has glutamic acid and leucine is the C-terminal amino acid. There are two genetic variants of α-lactalbumin, i.e. α-lactalbumin A,B. The B variant is slower moving on starch gel electrophoresis.

Both A and B are available in African Fuhni, African and Indian Zebu cattle. The B variant is mostly found in western cattle milk. The B variant has 123 amino acid residue. The A variant differs from B in substitution of glutamine residue for arginine residue at position 10. The biological role of α-lactalbumin is discussed elaborately in our discussion on biosynthesis of lactose.

REVIEW QUESTIONS

1. Define proteins.
2. What is isoelectric point? What is the PI of different proteins? How casein exists in milk?
3. List the major and minor milk proteins of milk. Which are the two main whey proteins of milk?
4. How casein is precipitated from milk? Discuss in brief the various methods of preparations.
5. What are the important fractions of casein? How are these separated?
6. (a) Briefly discuss the properties of protein taking casein as an example.

   (b) Is casein a very heat stable protein? If so, why?
7. Discuss briefly the effect of processing parameters on milk proteins.
8. Write short note on "Denaturation".
9. (a) List the possible non-protein constitutents of milk.

   (b) What are conjugated proteins? Give examples.

   (c) Write notes on tertiary calcium phosphate.
10. What is antibody? Describe the typical structure of an immunoglobin.
11. Briefly mention the role of some important milk enzymes in Dairy Industry.
12. Discuss how the various processing parameters bring the inactivation of milk enzymes.
13. Briefly mention the sequence of protein synthesis.
14. What do you understand by the term "Genetic polymorphism"?

REFERENCES

Aschaffenburg, R. and Drewry, R., 1959. New procedure for the routine determination of the various non-casein proteins of milk. *Int. Dairy Congr.* 15th London **3**:1631.

Datta Roy, D. 1981. Biosynthesis of milk proteins by the mammary gland. *Dairy Guide* **3**(9):29.

Eigel, W.N., Butler, J.E., Ernstrom, C.A., Farrell, H.M. Jr., Harwalkar, V.R., Jenness, R. and Whitney, R.MCl 1984. Nomenclature of proteins of cow's milk : Fifth Revision. *J. Dairy Sci.* **67**:1599.

Fox, P.F. 1989. Development in Dairy Chemistry – 1. Proteins Applied Science Publishers, Chapter 1.

Ganguli, N.C. 1974. Milk proteins. Publication. *Indian Council of Agricultural Research, New Delhi.*

Hipp, N.J., Groves, M.L., Custer, J.H. and McMeekin, T.L. 1952. Separation of α,β-, and γ-casein. *J. Dairy Sci.* **35**:272.

Jenness, R. and Patton, S. 1959. Principles of Dairy Chemistry. John Wiley & Sons. Inc. New York. Chapter 4.

Mellander, O. 1939. Electrophoretische Untersuchungvon Casein, *Biochem. Z.*, **300**:240.

Pedersen, K.O. 1936. Ultracentrifugal and electrophoretic studies on the milk proteins I. Introduction and preliminary studies with fractions from mim milk. *Biochem. J.* **30**:948.

Rowland, S.J. 1938. The determination of the nitrogen distribution in milk. *J. Dairy Res.* **9**:42.

Wake, R.G. and Baldwin, R.L. 1961. Analysis of casein fractions by zone electrophoresis concentrated urea. *Biochim Biophys. Acta* **47**:225.

Webb, B.H., Johnson, A. and Alford, J.A. 1978. *Fundamentals of Dairy Chemistry.* The AVI Publ. Co. Inc. Westport, Connecticut, Chapter 3.

Whitney, R, MCL, Brunner, J.R., Ebner, K.E., Farel, H.M., Jr. Josephson, R.V., Morr, C.V. and Swaisgood, H.E. 1976. Nomenclature of the proteins of cow's milk : Fourth revision. *J. Dairy Sci.* **59**:785.

4

CARBOHYDRATES OF MILK

Major Milk Carbohydrates

CARBOHYDRATES are defined as polyhdroxy aldehydes or ketones and undergo all the organic chemistry reactions common to the carbonyl groups and hydroxyl groups. Decades back carbohydrates have been defined as hydrates of carbon, a rather incomplete definition.

The major milk carbohydrate is lactose, a disaccharide which on acid hydrolysis or treatment with enzyme ß-galactosidase yields, a molecule of glucose and galactose. Glucose and galactose are six carbon monosaccharides having the aldehyde group at the number one carbon atom and the methyl hydroxyl group at the sixth carbon.

Oligosaccharides are also present in milk, especially in human milk. These oligosaccharides on acid hydrolysis produce two to six or two to eight monosaccharide units. For example, lacto N-tetraose, fucopentaose I, II, III, fucohexaose I, II, III and sialic acid or N-acetyl-neuroaminic acid derviatives. Milk oligosacchardies on acid hydrolysis yield glucose, galactose, fucose, N-acetyl-glucosamine, and NANA (sialic acid). However, no polysaccharide is present in milk. Polysaccharides on acid hydrolysis yield many monosaccharide units.

Coming back to the discussion on glucose and galactose which are the hydrolytic product of lactose. Kiliani synthesis or treatment with excess of hydrogen cyanide of glyceraldehyde produce these two six carbon sugars, which are optical isomers but not mirror images.

To be an optical isomer a carbohydrate must have an asymmetric carbon atom, i.e. a carbon atom that has no symmetry while satisfying its four valencies. Let us cite an example.

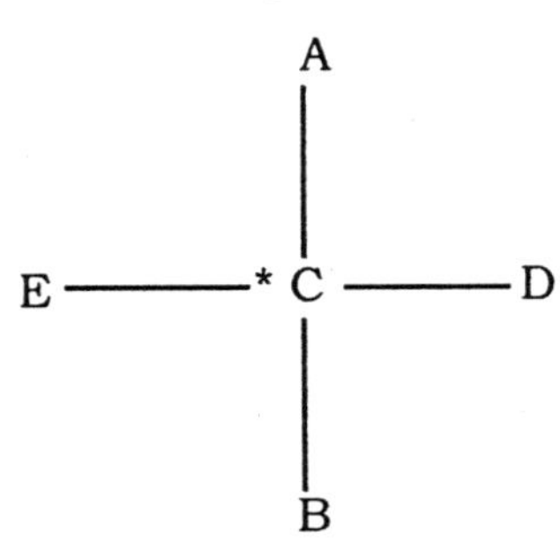

This carbon is asymmetric because it has four different atoms or groups attached to it. Now, if this compound is projected before the mirror, we have the mirror image isomer of it as illustrated below :

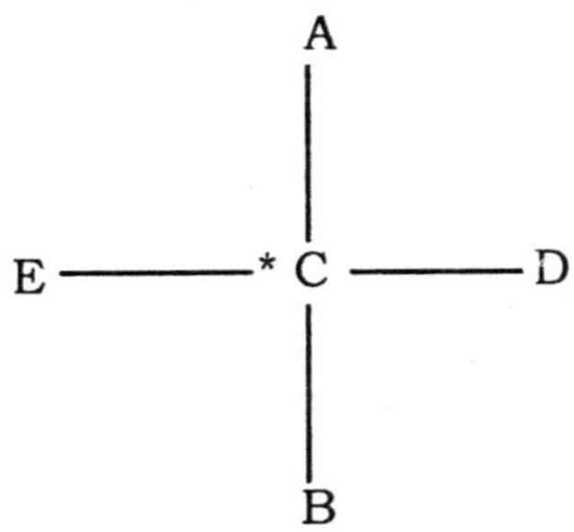

Here we see that the group or groups of atoms represented by A and B are fixed but that of E and D have interchanged positions. Thus, these two compounds are mirror image isomers or optical antipodes or a pair of enantiomers.

The molecular structure of the glyceraldehyde molecule is given below:

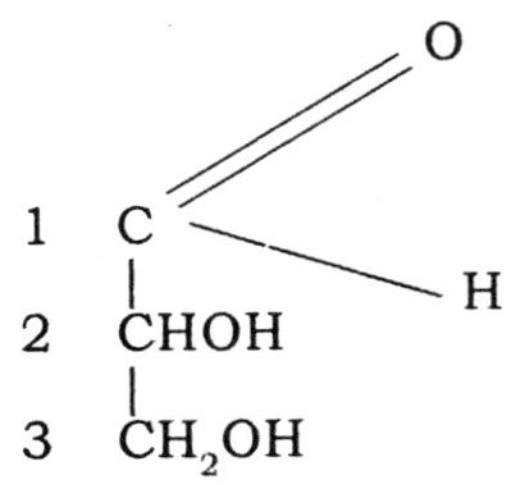

From the above structure we may conclude that the two optical isomers of this compound exist in nature as the number two carbon atom is asymmetric. Applying the Van't Hoff formula of 2n, where n stands for the number of asymmetric carbon atom present. We obtain two possible optical isomers of the glyceraldehyde molecule.

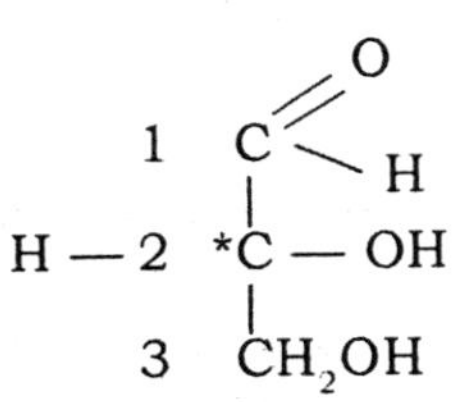

D — glyceraldehyde

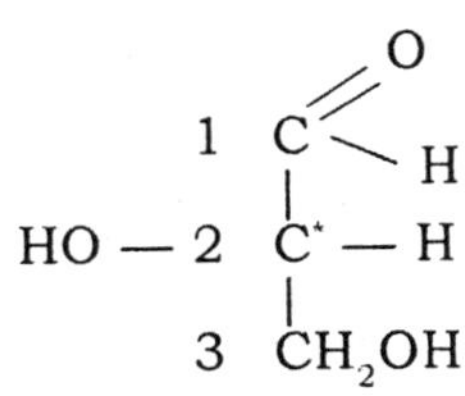

L — glyceraldehyde

Emil Fischer, a famous German scientist concluded from this that when the hydroxyl group at the penultimate carbon atom, i.e. the number two carbon atom is to the right of the plane, it is capital D-glyceraldehyde. When it is to the left of the plane, it is capital L-glyceraldehyde. Capital D and L only suggest about the spatial configuration of the hydroxyl group with respect to the carbon atom, i.e. the arrangement in space of the hydroxyl group with respect to the carbon atom. When the hydroxyl group is to the right of the plane at the penultimate carbon atom, it is D-sugar and when the hydroxyl group is to the left of the plane, it is capital L-glyceraldehyde. He also asserted that the capital D and L has nothing to do with small d and l which suggest the optical rotation of sugar molecules. Both capital D and L glyceraldehyde are optical isomers and mirror image isomers. They are also known as optical antipodes or an enantimoric pair. D and L-glyceraldehyde are 3-carbon atom sugars, and in carbohydrate chemistry they are designated as trioses. As the number one carbon atom of D and L-glyceraldehyde carry an aldehyde group, D and L-glyceraldehyde are known as aldoses (aldotriose) 1. Both D and L-glyceraldehyde are the starting point for aldoses having 4,5,6 carbon atoms.

Beginning with D and L-glyceraldehyde, these 4,5,6 carbon atom aldoses may be synthesised by the Kiliani synthesis method, i.e. treatment with excess hydrogen cyanide. The family tree of D-aldoses through sixth carbon atom is given in Flow sheet 4.1.

From the D-aldose family tree we notice that, D-glucose and D-mannose and also D-galactose and D-tagatose are epimers as depicted below :

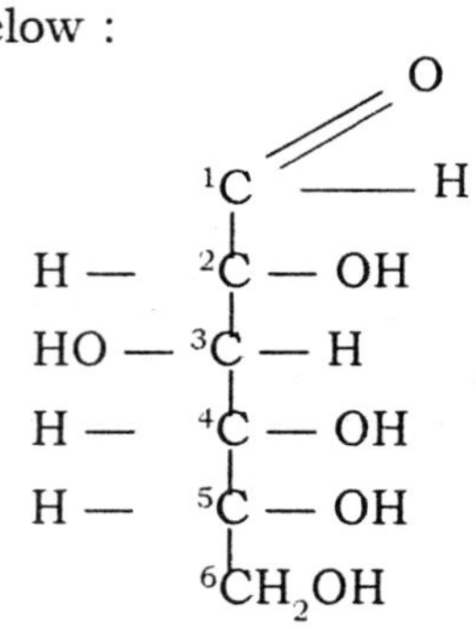

D — glucose

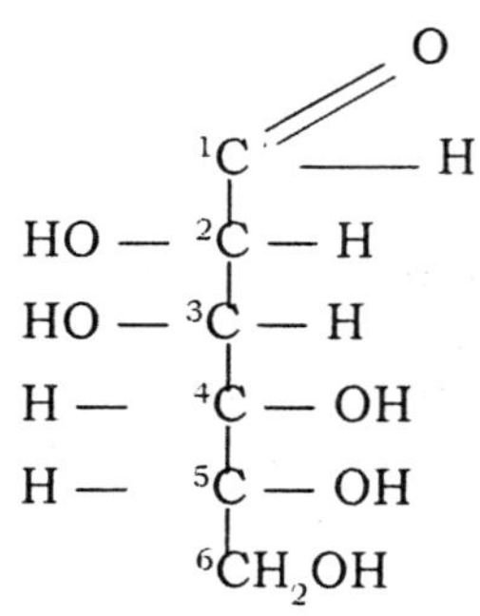

D — Mannose

On observing the open chain structures of D-glucose and D-mannose, we find that in both structures of D-glucose and D-mannose, the arrangement in space of the hydroxyl group with respect to the carbon atom number two of D-glucose, differ from that of the carbon atom number two of D-mannose. In case of D-glucose, the hydroxyl group at the number two carbon atom is to the right of the plane where as exactly opposite is the case

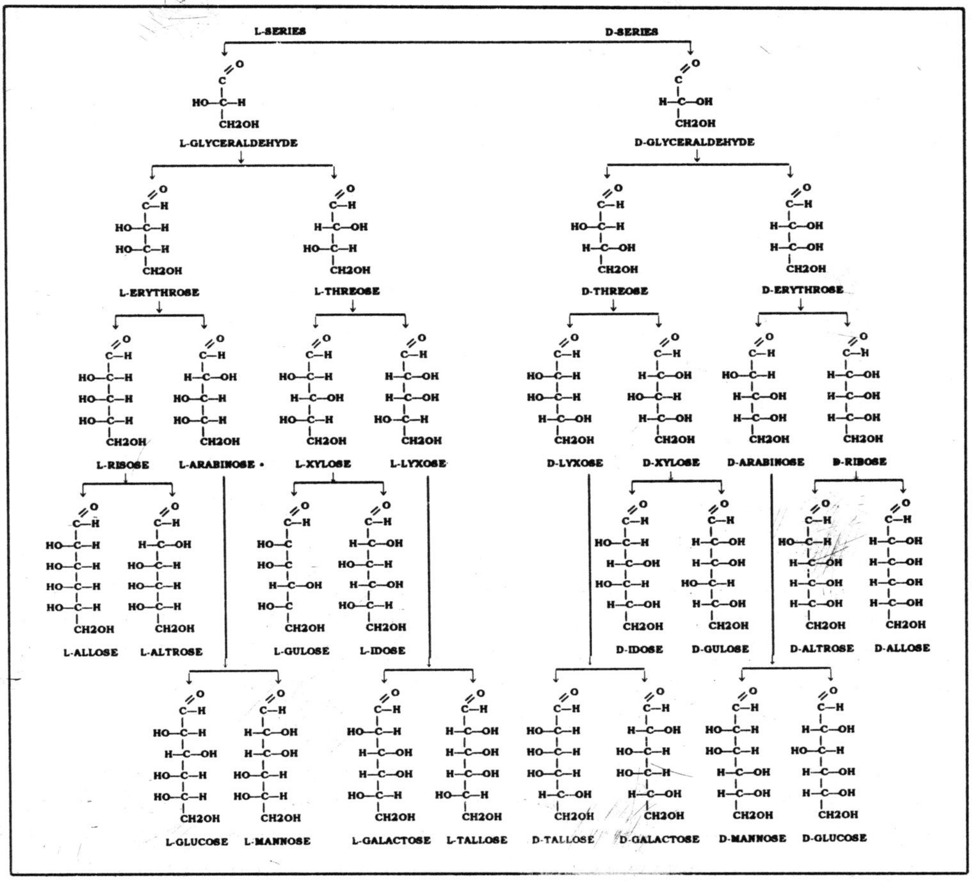

Flow Sheet 4.1 : Family Tree of Aldose

for D-mannose. However, the rest of the asymmetric carbon atoms in D-glucose and D-mannose have the same configuration.

The respective structures of D-galactose and D-tagatose are given below.

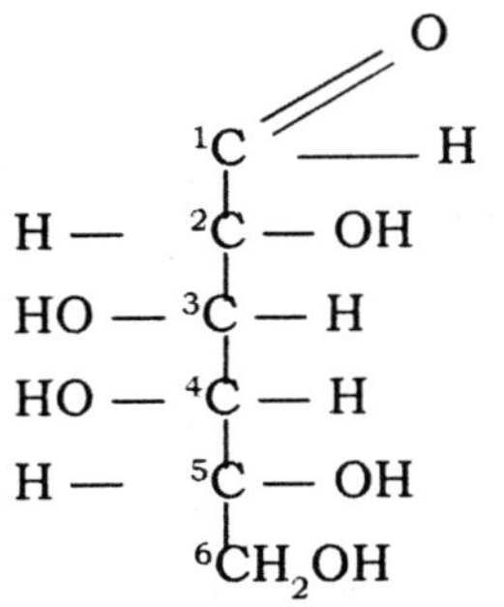

D — galactose

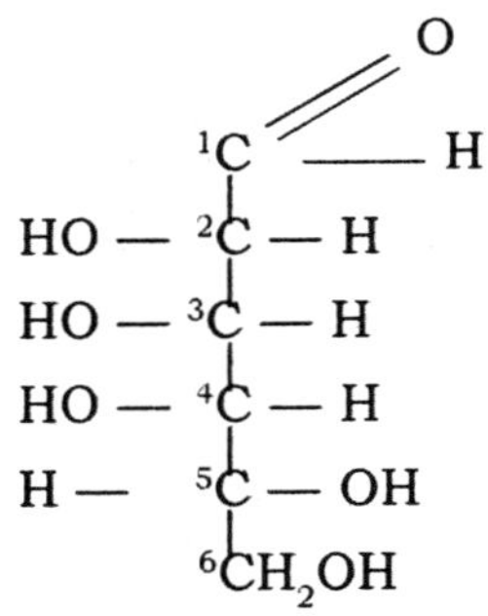

D — tagatose

On scrutinising the two structures, it was observed that the arrangement in space of the hydroxyl group with respect to the carbon atom number two is different in D-galactose from that of D-tagatose. D-galactose has the hydroxyl group at the carbon atom number two at the right of the plane while D-tagatose has it to the left of the plane.

We, therefore, conclude that D-glucose and D-mannose are related as epimers and the same is for D-galactose and D-tagatose. So, an epimer may be defined as a pair of monosaccharides which differ in their spatial configuration of atoms or groups of atoms at position number two of the pair of monosaccharides keeping in view of the fact that the rest of the carbon atoms have the same configuration in space.

We often see that D-glucose, D-galactose and D-lactose are described, respectively as a-D-glucopyranose, ß-D-glucopyranose, a-D-galactopyranose, ß-D-galactopyranose, O-ß-D-galactopyranose (1——>4)-α-D-glucopyranose and O-ß-D-galactopyranosyl (1——>4)-ß-D-glucopyranose. Now what does the nomenclature a-and ß-and the term pyranose stand for?

For this we have to review the subject of cyclic forms of sugars.

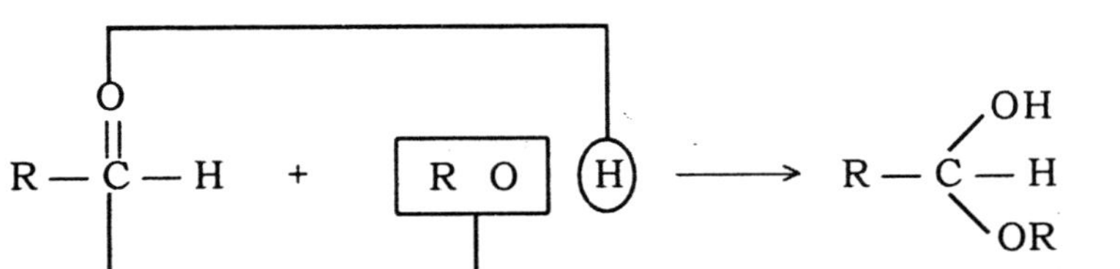

Hemiacetal

This organic chemistry reaction is intermolecular. However, in carbohydrate chemistry intramolecular reactions take place.

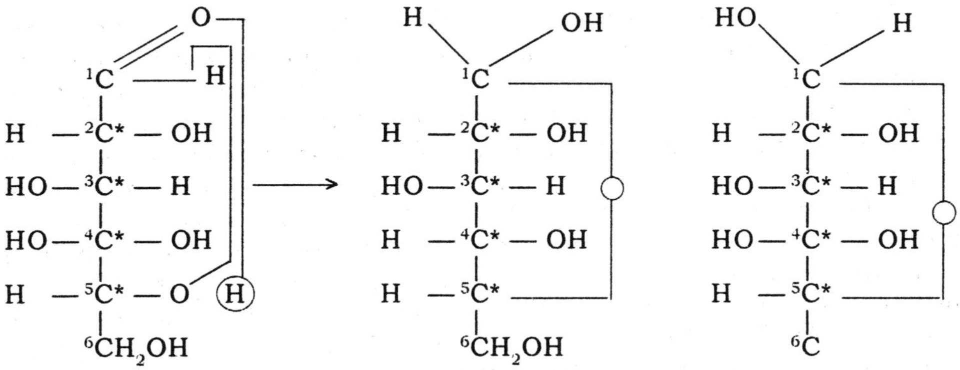

D - glucose open chain form undergoing cyclization As a result number one carbon atom also becomes asymmetric

α-D—glucose cyclic form

β-D-glucose cyclic form

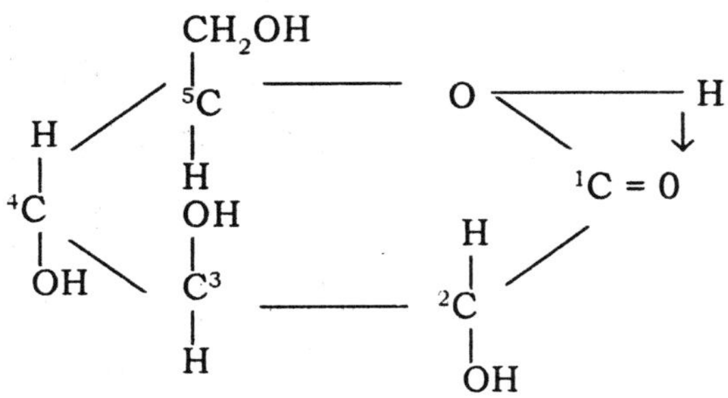

D—glucose

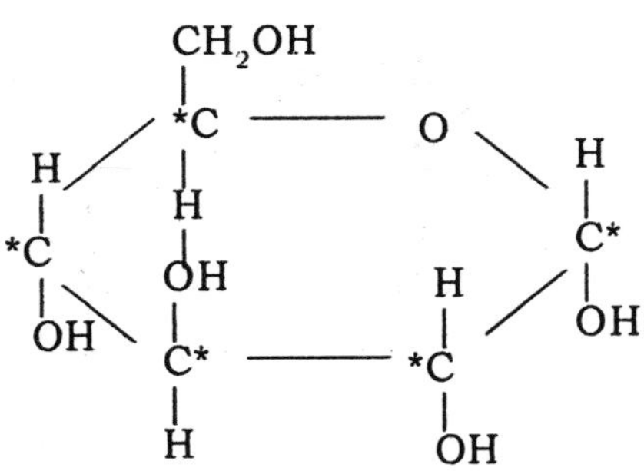

α—D—glucopyranose cyclic form

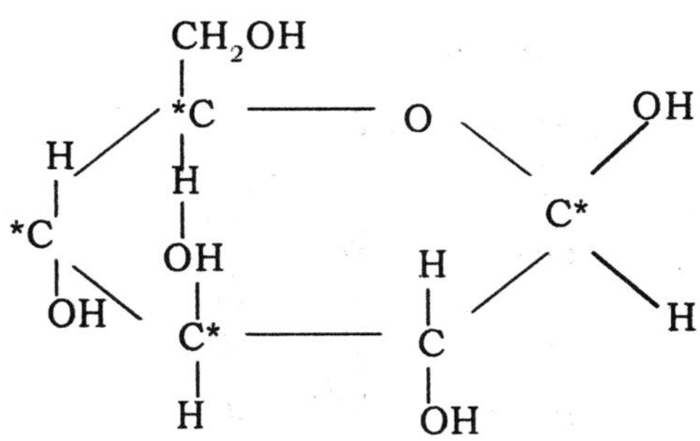

β—D—glucopyranose cyclic form

On studying the structures above, we find that as a result of cyclisation, the number one carbon atom of D-glucose has become asymmetric and also there is formation of a pyranose ring structure.

With the conversion of number one carbon atom to an asymmetric carbon, and simultaneous formation of hemiacetal hydroxyl group and pyranose ring structure, the cyclic forms of sugars become a structural isomer of the open chain form. As number one carbon atom has now turned asymmetric, the hemiacetal number hydroxyl group that forms at the number one position of carbon atom can occupy two positions, viz., it could be to the right of the plane or to the left of the plane, in case of Fischer structure and for Haworth structure, it could be below the plane of the ring or above the plane of the ring. Considering the Fischer structure, the hemiacetal hydroxyl group is α-when it is to the right of the plane and it is ß-when it is to the left of the plane. In case of Haworth structure, when the hemiacetal hydroxyl group is below the plane of the ring, it is α-and when above the plane of the ring it is ß.

The cyclic forms of D-galactose and D-lactose are as follows :

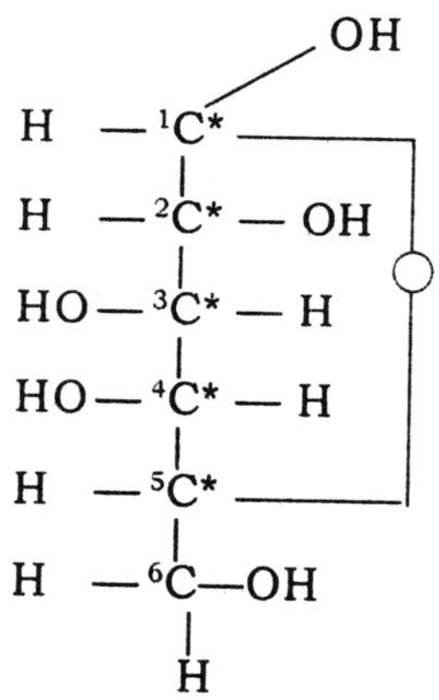

α–D–galactose

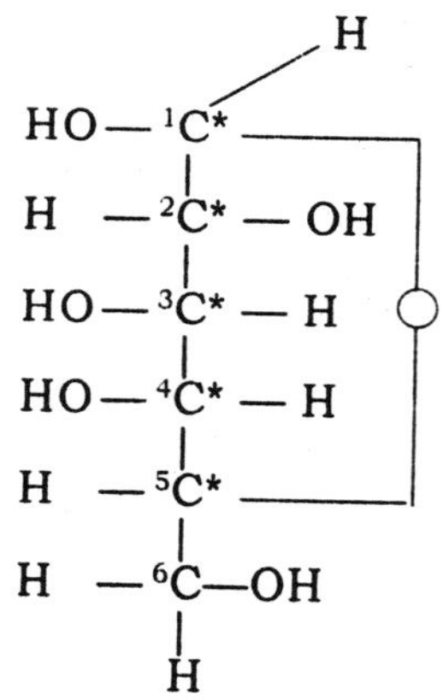

β–D–galactose
Fischer structure

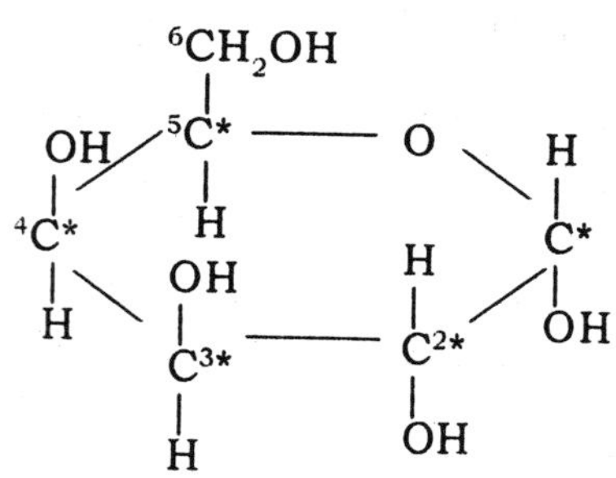

α–D–galactopyranose

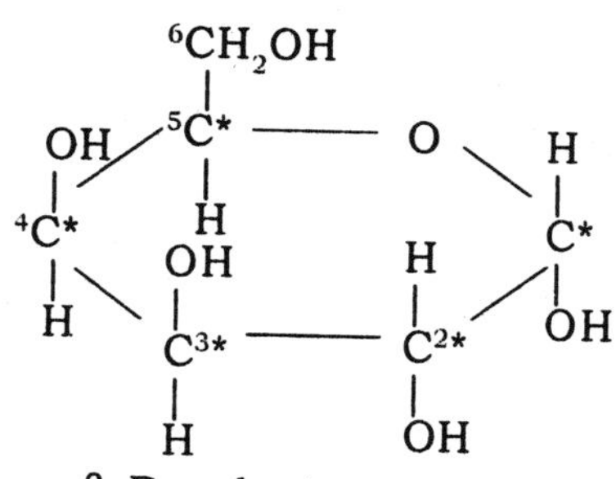

β–D–galactopyranose

HAWORTH STRUCTURE

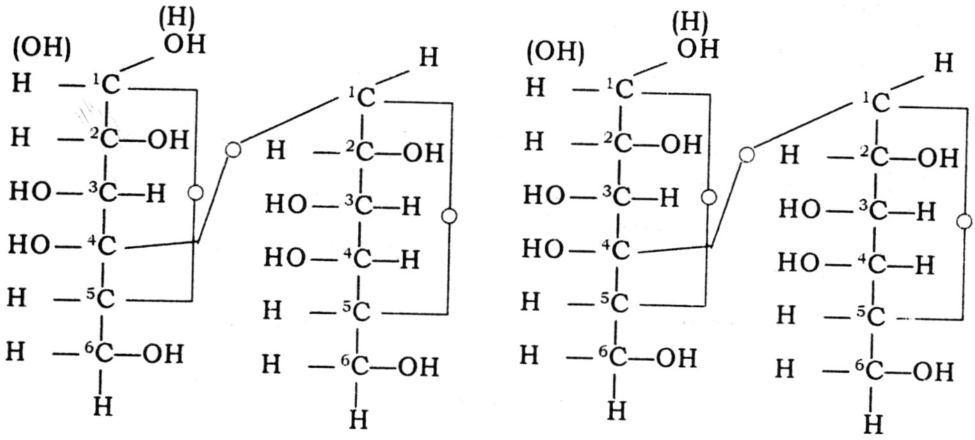

α, β–D–lactose Fischer Structure β, α–D–lactose

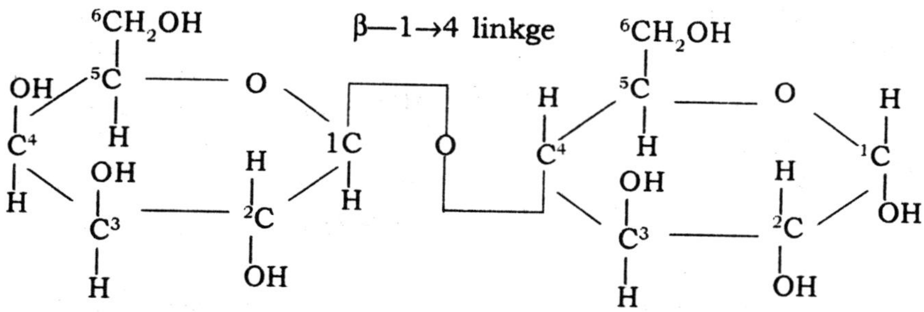

α–lactose or β–D–lactose

O–β–D–galactopyranosyl—(1→4)–α–D–glucopyranose

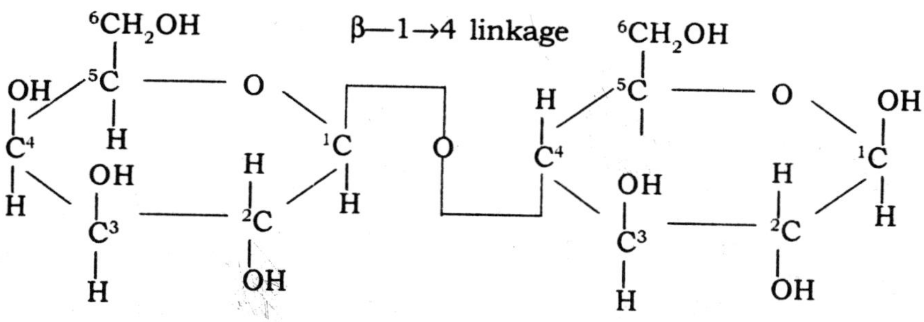

β–lactose or β–D–lactose

O–β–D–galactopyranosyl—(1→4)—β–D–glucopyranose

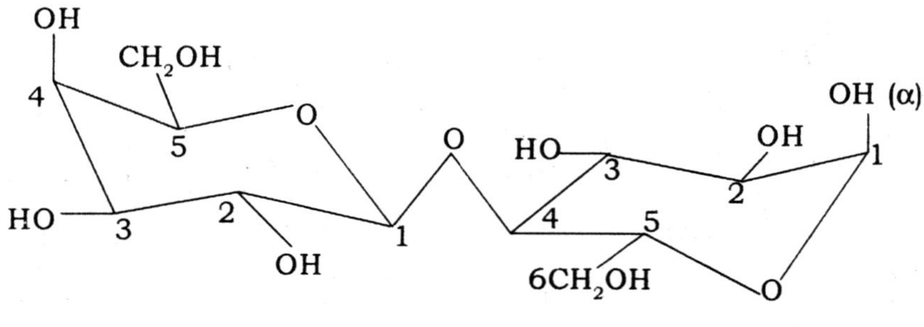

D—galactose D—glucose

α—lactose

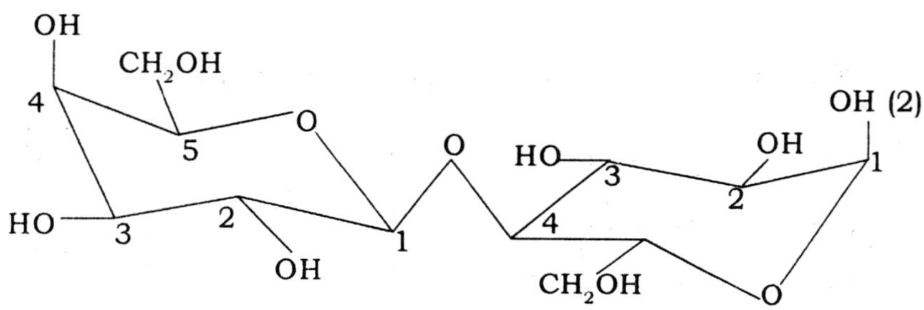

D—galactose D—glucose

β—lactose

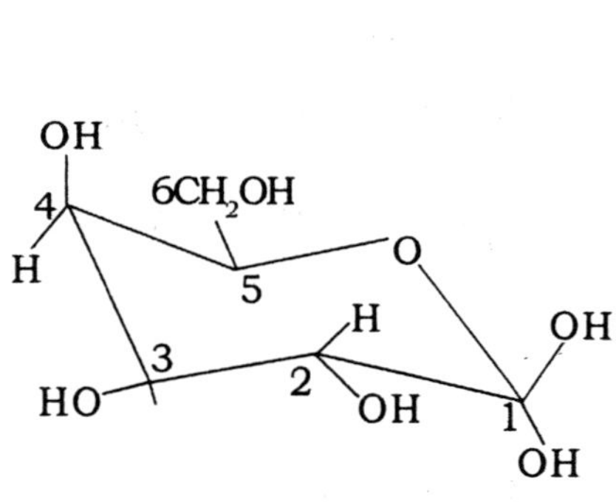

β—α—D—galactose

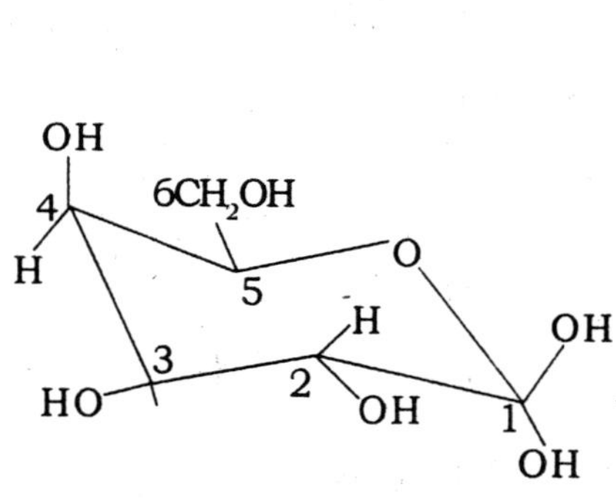

β—α—D—glucose

It has to be noted that all bulky groups like OH and CH_2OH are in equatorial position of the ring where there is least crowding and maximum stability. The hydrogen atoms are in axial position.

All monosaccharides and some disaccharides which exist in α and ß-form, predominantly exist in nature in cyclic forms. However, in solution together with the cyclic forms, there is presence of open chain form in negligible amount.

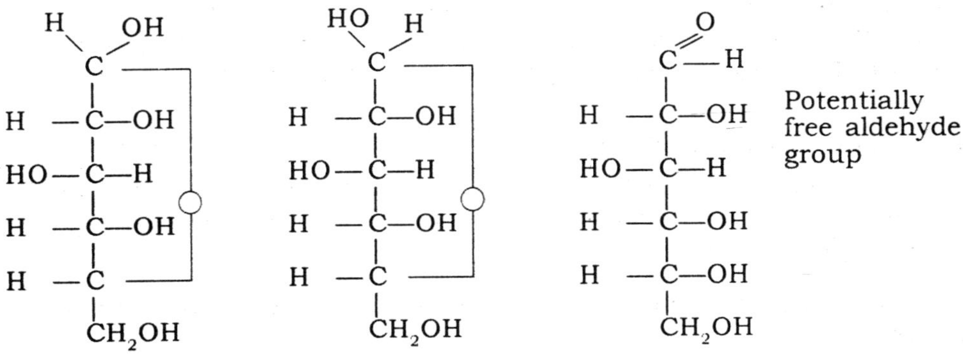

α–D–glucose β–D–glucose negligible amount of open chain form

The open chain forms have a potentially free aldehyde group wich the cyclic forms do not have as a result of cyclisation. Therefore, any reaction involving the carbonyl group will take place only with the potentially free aldehyde group; that is present in solution in negligible amount when such a reaction takes place between the potentially free aldehyde group of the carbohydrate, there is a shift in the equilibrium from the α- and ß-forms of sugar. As the reaction between the potentially free aldehyde group, proceeds, there is gradual conversion of α-and ß-forms to the open chain form and ultimately the solution is completely depleted of the α- and ß-forms of carbohydrate. This will also result in the complete exhausation of the open chain form from the solution. On this principal cupric ions are reduced to Cu_2O (cuprous oxide), a red precipitate with the ultimate formation of oxidised forms of sugars. The definition of reducing sugars is based on this principle.

Reducing sugars are able to reduce an alkaline solution of cupric ions, without itself getting hydrolysed, to cuprous oxide which precipitates as a red precipitate and the resultant formation is oxidised forms of carbohydrates.

$$Cu^{++} + \text{Reducing sugar} \xrightarrow{OH^-} Cu_2O \downarrow + \text{Oxidised form of sugar}$$

All monosaccharides and some disaccharides which exist in α- and ß-forms are reducing sugars. Therefore, glucose, galactose and lactose are reducing sugars, but sucrose is not.

It was already shown that glucose, galactose and lactose exist in α-and ß-forms. In the equilibrium solution of glucose, galactose and lactose, negligible amount of the open chain form of these sugars is present, i.e. the potentially free aldehyde groups are present. However, such is not the case with sucrose which does not exist in α- and ß-forms. Therefore, we cannot expect the presence of any potentially free aldehyde group in a solution of sucrose. This explains why sucrose is not a reducing sugar.

Proof of Lactose Structure

The structure of lactose has been already discussed earlier in the chapter. The systematic name of lactose is O-ß-D-galactopyranosyl- (1——>4)-D-glucopyranose.

The structure and the systematic name of lactose disclose four points clearly :

(1) The reducing group lies at the number one carbon atom of glucose, (2) It is a ß-galactoside, (3) Both the galactose and glucose moiety of lactose have pyranose ring structures, (4) the linkage between the galactose moiety with that of glucose is (1———>4).

However, as a chemist we would like to prove the aforesaid statements. How to proceed in such a venture?

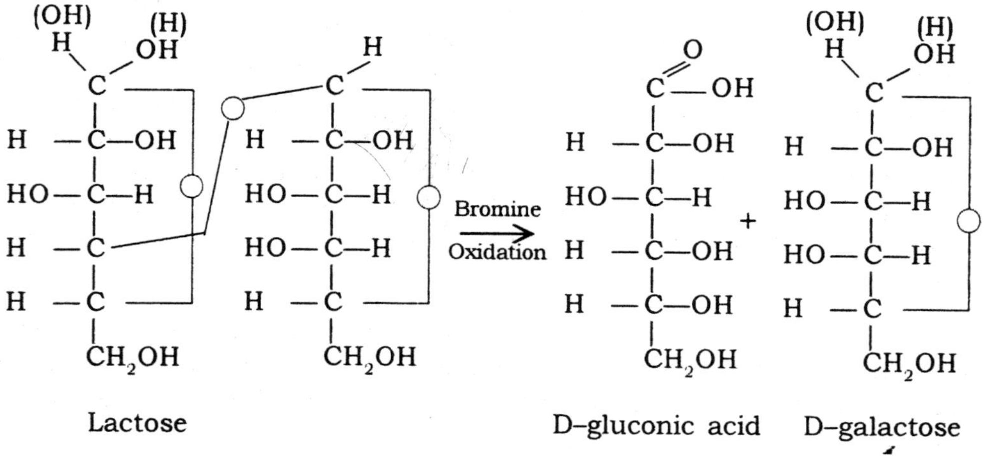

Lactose D–gluconic acid D–galactose

It is observed here that the potentially free aldehyde group of D-glucose moiety of lactose has been oxidised to a carboxylic acid group with the resultant production of D-gluconic acid. However, as the galactose moiety of lactose did not have a potentially free aldehyde group, it escaped oxidation. The production of D-gluconic acid with the simultaneous release of D-galactose as products of the reaction, suggests very clearly that the reducing group of lactose lies in the number one carbon atom of D-glucose moiety of lactose.

Our next aim is to prove that lactose is a ß-galactoside. We proceed in our attempt in the following manner :

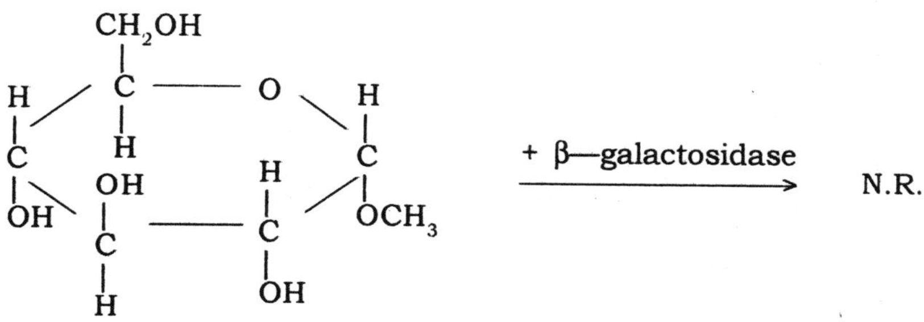

Methyl—α—D—glucoside

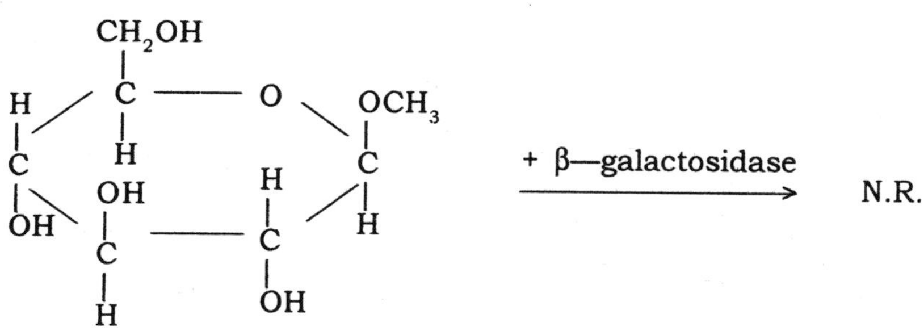

Methyl—β—D—glucoside

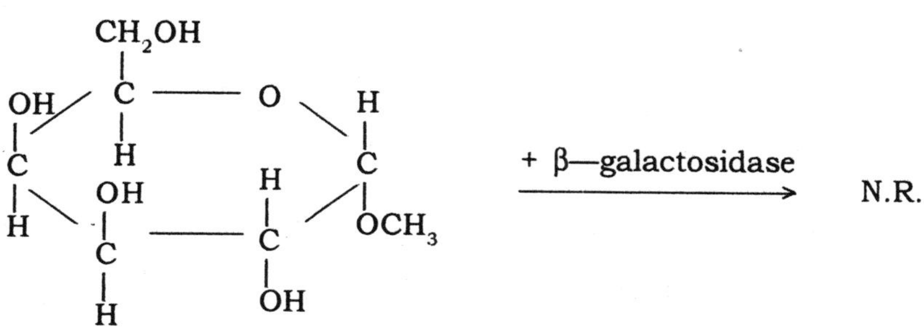

Methyl—α—D—galactoside

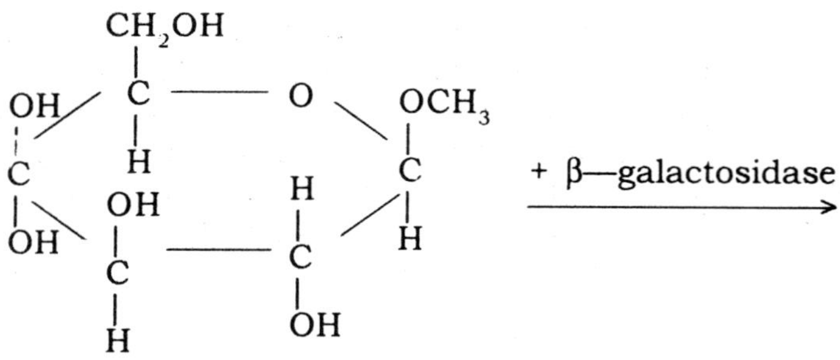

Methyl—β—D—galactoside

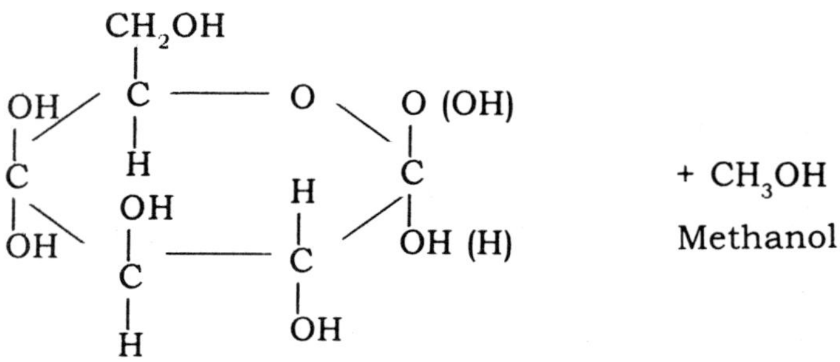

D—galactopyranose

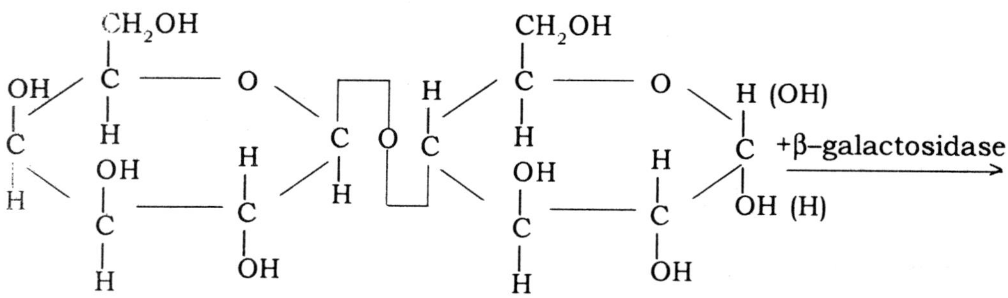

Lactose

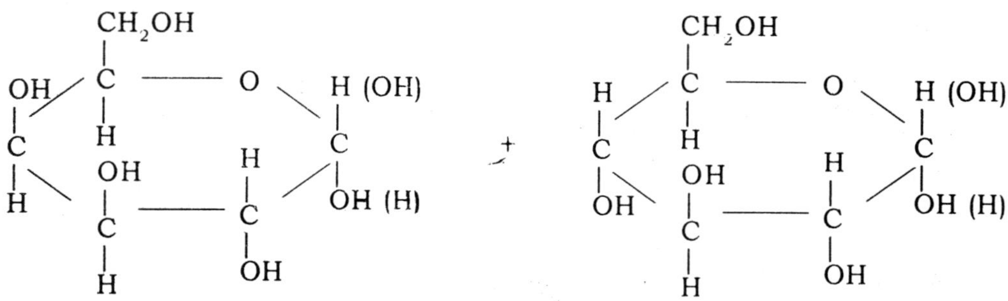

D—galactopyranose D—glucopyranose

We observe here that B-galactosidase hydrolyzes methyl-B-D-galactoside to methanol and D-galactopyranose. We also notice that B-galactosidase hydrolyzes lactose to D-glucopyranose and D-galactopyranose. From the above considerations, we conclude that lactose is a B-galactoside.

We will have to prove now, that the glucose and galactose moiety of lactose have pyranose ring structure. Let us suppose that the glucose and galactose moiety of lactose have pyranose ring structure. We may proceed that to prove in the following manner :

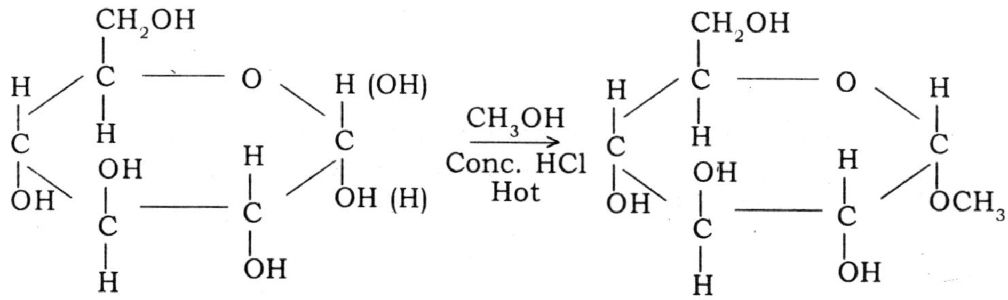

D-glucopyranose Methyl-α-D-glucoside

+

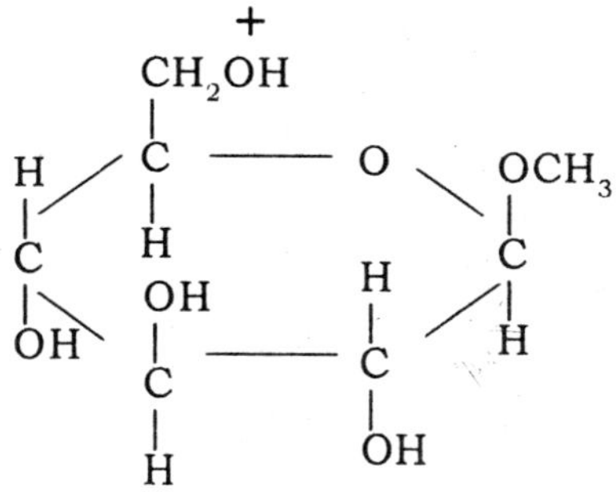

Methyl-β-D-glucoside

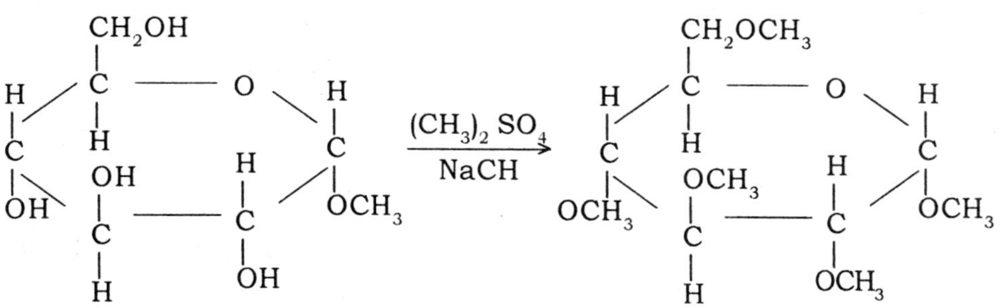

Methyl-α-D-glucoside methyl-tetra-O-Methyl-α-D-glucoside

↓ dil HCl

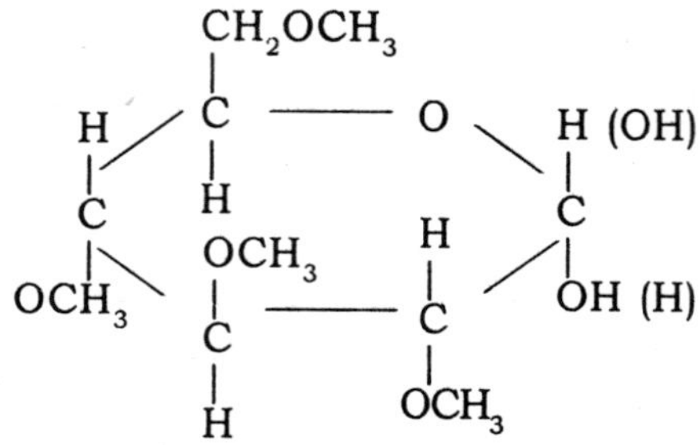

2, 3, 4, 6–tetra–O–methyl–
D–glucopyranose

As no methylation is there at the fifth carbon atom of D-glucose, we conclude that α-D-glucopyranose has a pyranose ring structure. Now let us take methyl-ß-D-glucoside.

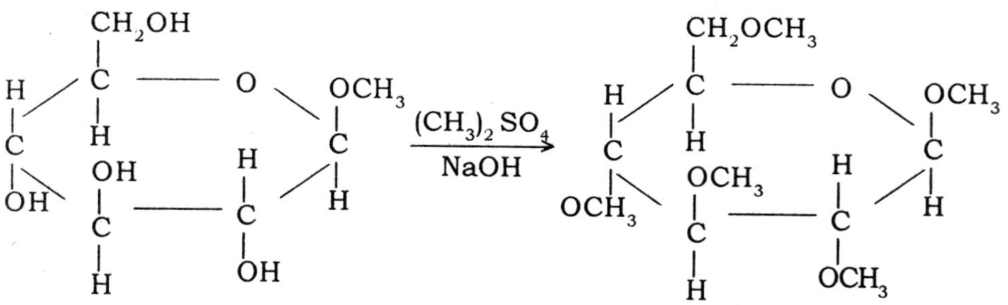

Methyl–β–D–glucoside

Methyl–tetra–O–
Methyl–β–D–glucoside

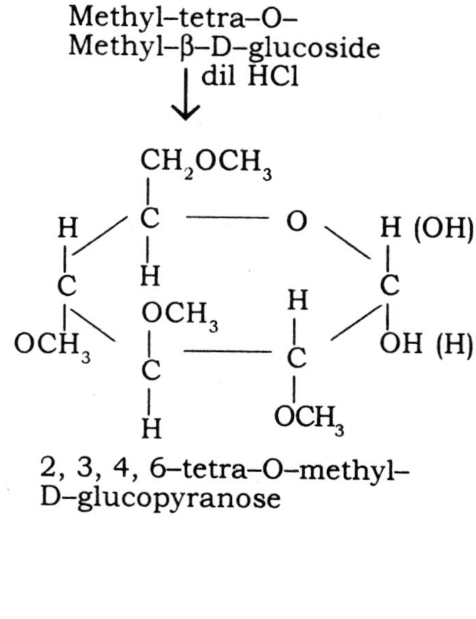

2, 3, 4, 6–tetra–O–methyl–
D–glucopyranose

We observe that no methylation has taken place at the fifth carbon atom of D-glucose. We, therefore, come to a conclusion that α-D-glucopyranose has a pyranose ring structure.

Let us now take B-D-glucopyranose.

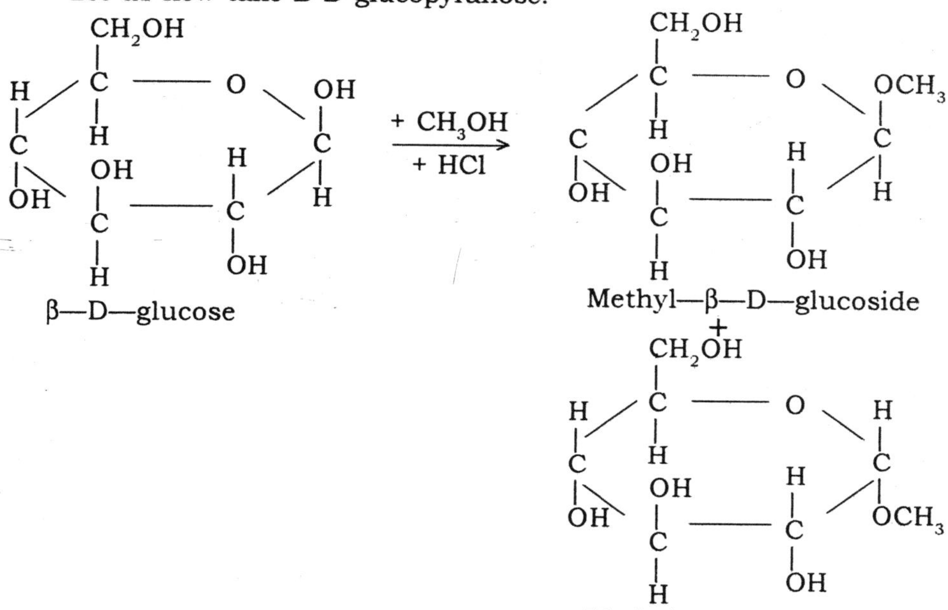

β—D—glucose

Methyl—β—D—glucoside

+

Methyl—α—D—glucoside

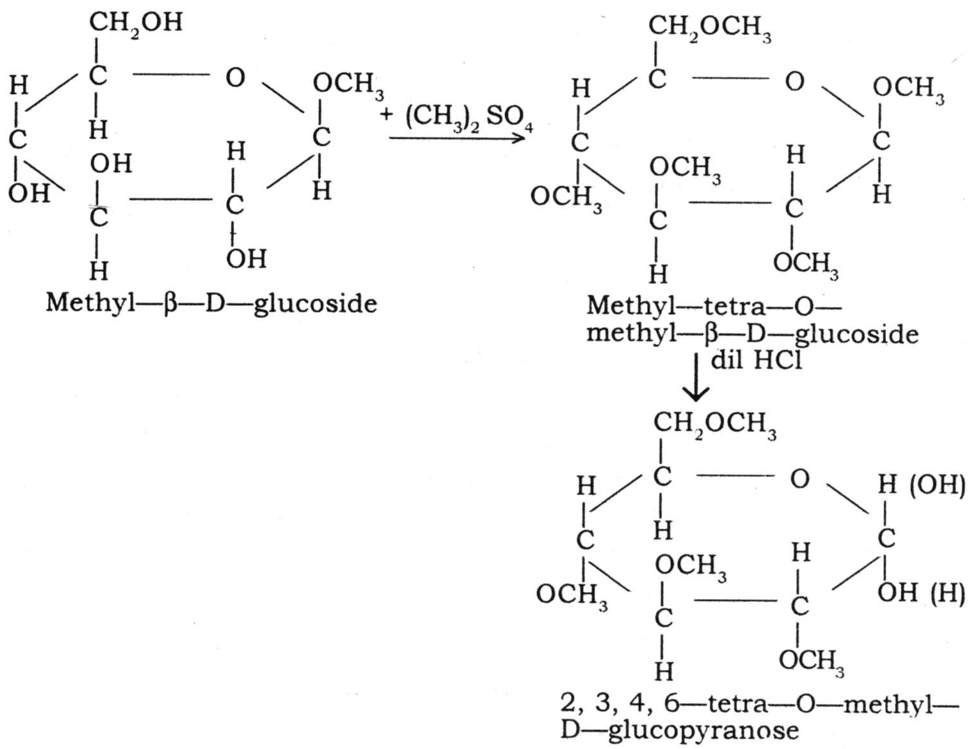

Methyl—β—D—glucoside

Methyl—tetra—O—methyl—β—D—glucoside

2, 3, 4, 6—tetra—O—methyl—D—glucopyranose

From the above structure, we derive at a conclusion that ß-D-glucopyranose has also a pyranose ring structure.

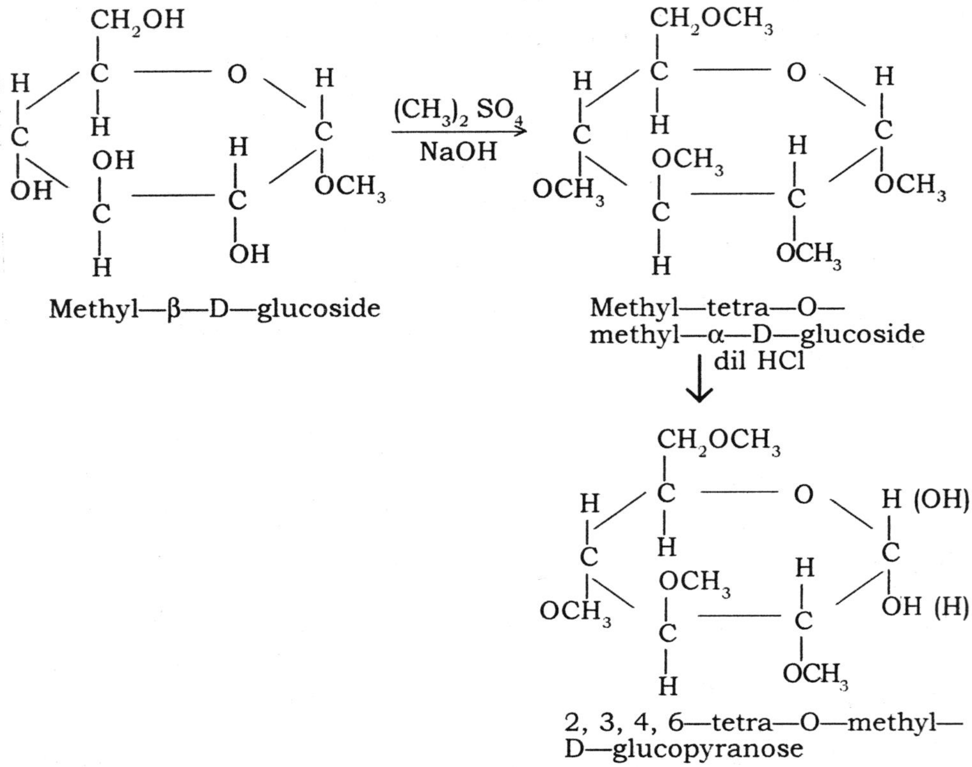

Methyl—β—D—glucoside

Methyl—tetra—O—methyl—α—D—glucoside

2, 3, 4, 6—tetra—O—methyl—D—glucopyranose

We observe that there is no methylation at the number five carbon atom of D-glucose. We, therefore, conclude that ß-D-glucose has a pyranose ring structure.

Let us suppose that D-galactose of lactose has a pyranose ring structure.

The procedure adopted to prove that D-glucose has a pyranose structure may be also followed to prove that D-galactose has a pyranose ring structure.

We have already proved that lactose has, (1) the reducing group at the number one carbon atom of the glucose moiety, (2) it is a ß-galactoside, (3) both D-glucose and D-galactose of lactose have pyranose ring structures.

Now, we have to prove that the linkage is 1———>4 between the galactose and the glucose moiety of lactose.

Let us draw the structure of lactose and carry out exhaustive methylation.

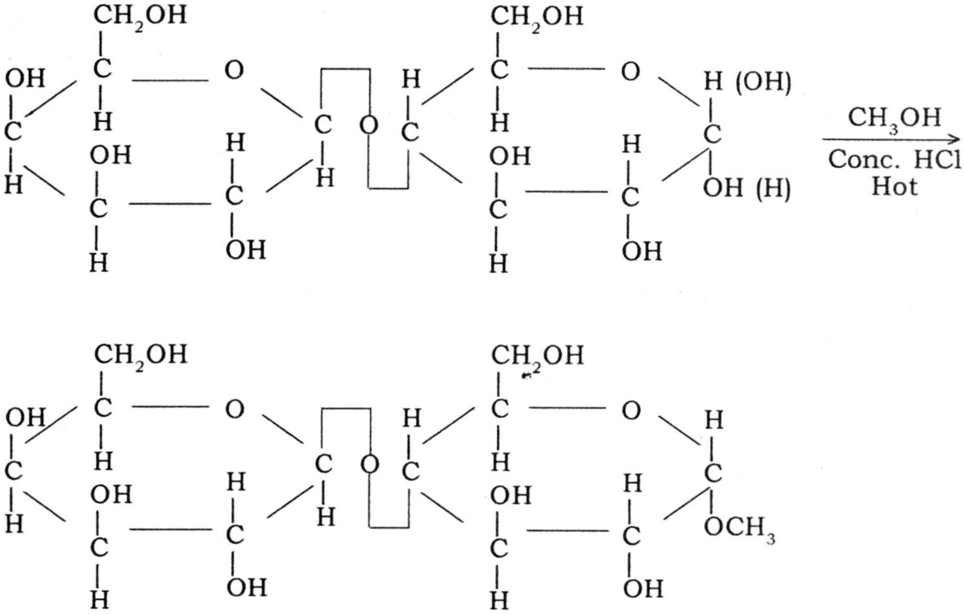

Methyl—α—lactoside

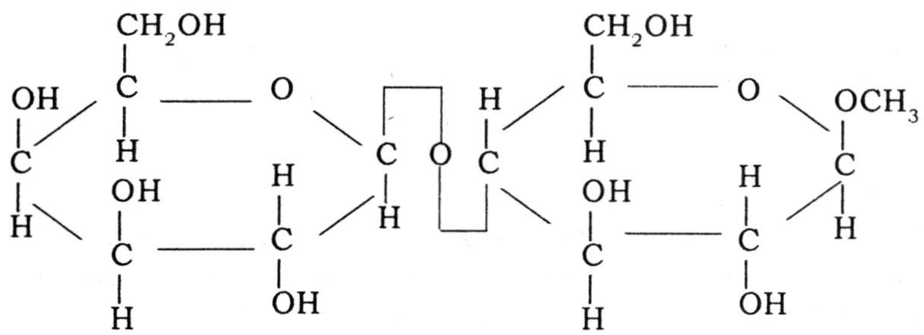

Methyl—β—D—lactoside

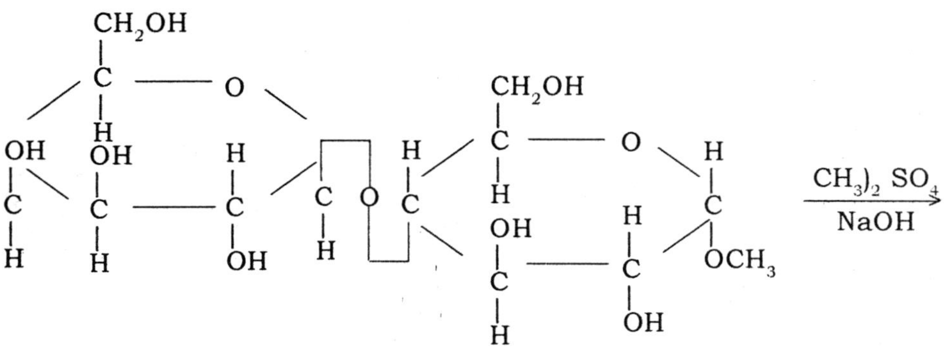

Methyl—α—D—lactoside

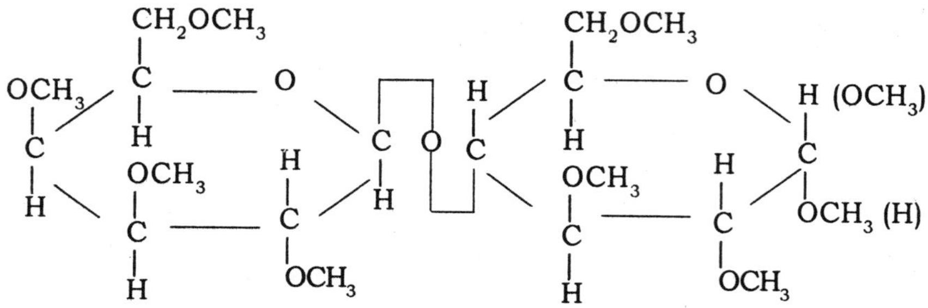

Completely methylated derivative of lactose

↓ dil HCl

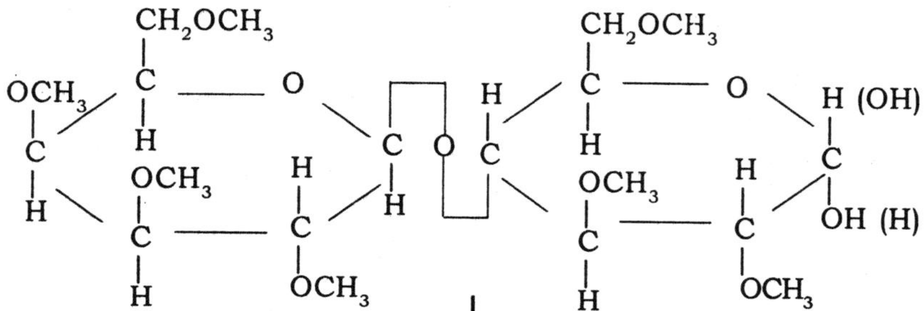

↓ treatment with mineral acid

2, 3, 4, 6—tetra—O—methyl D—galactopyranose + 2, 3, 6—tri—O—methyl D—glucopyranose

As there is no methylation at the number one and fifth carbon atom of D-galactose moiety, and in case of glucose moiety there is no methylation at number one, fourth and fifth carbon atom, we conclude that the linkage between the D-galactose and the D-glucose moiety of lactose is 1——>4.

Hence the lactose structure is proved.

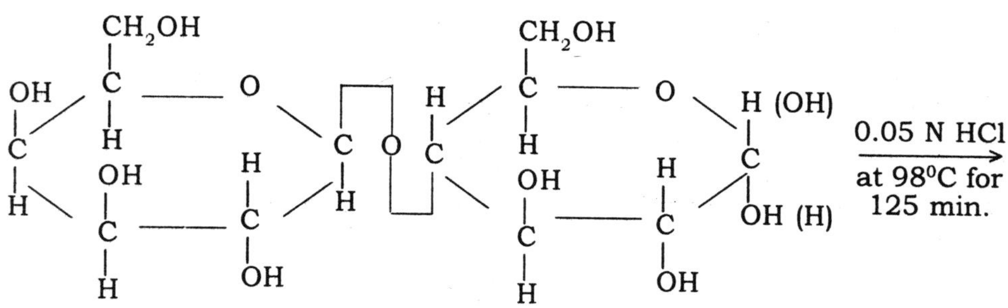

Lactose

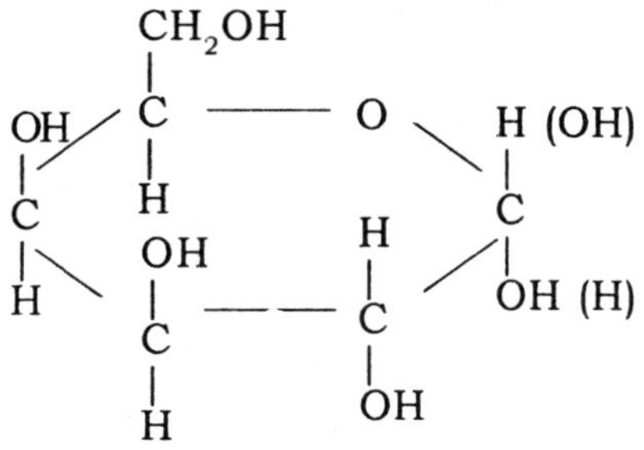

D—galactopyranose

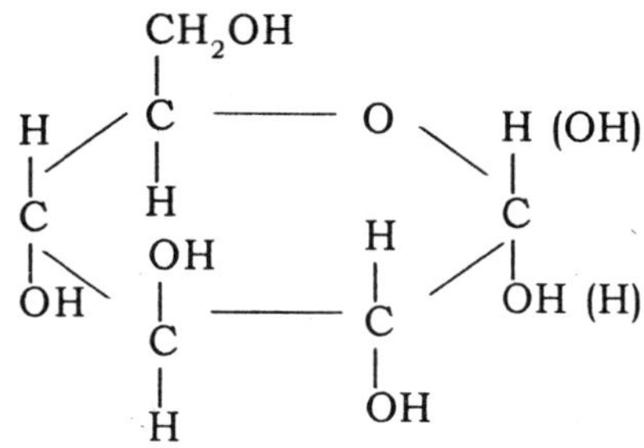

D—glucopyranose

By a process of reversion oligosaccharides are formed.

Physical Forms of Lactose

Lactose is a disaccharide and exhibits mutarotation. It occurs in two crystalline forms, and α-hydrate and ß-anhydrous forms. It also exists in an amorphous or glassy state. The most common form is the α-hydrate ($C_{12}H_{22}O_{11}.H_2O$). When a super saturated solution of lactose below 93.5°C is allowed to crystallise, the α-hydrate crystallises out. Crystallisation of a super saturated solution of lactose, above 93.5°C, results in the formation of crystals of ß-anhydrous type.

In the table below some physical properties of the two common forms of lactose are delineated.

Table 4.1. Some physical properties of the two common forms of lactose.

Property	α-Hydrate	ß-Anhydrous
Melting point[1]	202°C	252°C
Specific rotation[2] $(\alpha)20_D$	+89.4°C	+35°
Solubility (gm/100 ml)		
Water at 20°C	8	55
Water at 100°C	70	95
Specific gravity (20°C)	1.54	1.59
Specific heat	0.299	0.285
Heat of combustion (Cal/gm)	3761.6	3932.7

1. Values vary with rate of heating, α-hydrate loosing H_2O (120°C).
2. Values on anhydrous basis, both forms mutarotate to +55.4°.

Under normal conditions, the α-hydrate form is the stable one, and other solid forms will spontaneously change to that form provided sufficient water is present. At equilibrium and at room temperature the ß-form is much more soluble whereas α-form is less soluble. However, because of its lower solubility, the α-hydrate will crystallise out, and the equilibrium will shift to convert ß-into α-hydrate.

When either of the two forms are dissolved in water and the solution is permitted to reach equilibrium, the specific rotation of the equilibrium mixture is +55.4°.

This gradual change in optical rotation which continues for 24 hrs till equilibrium is estabished, is known as mutarotation and is common in all monosaccharides which exist in α- and ß-forms.

Let x be the percentage of the α-isomer expressed as a decimal.

Let y be the percentage of α-isomer expressed as a decimal.

Then we have

$$x + y = 1$$

$$x\,(89.4) + y(35.0) = 55.4$$

From this simultaneous equation, the percentage of α- and ß-forms of lactose at equilibrium can be calculated.

The solubility of lactose is less than that of most other sugars and this may present problems in a number of foods containing lactose. When milk is concentrated 3:1, the concentration of lactose approaches its final solubility. When this product is cooled or when sucrose is added, crystals of α-hydrate may develop. Such lactose crystals are very hard and sharp and when left undistrubed may develop to a large size, causing a sensation of grittiness or sandiness in the mouth. This same phenomenon limits the amount of milk solids that can be incorporated into ice-cream.

The crystal of α-hydrate lactose usually occur in a prism or tomahawk shape. The latter is the basic shape and all other shapes are dervied from it by different relative growth rates of the various faces. These faces grow at different rates. Growth rate of faces depend on the orientation of the faces towards ß-direction.

Amorphous or glassy lactose is formed when lactose containing solutions, are dried quickly. The dry lactose is non-crystalline and contains the same ratio of alpha/beta as the original product.

This holds true for the spray or roller drying of milk products and also during drying for moisture determination. The glassy lactose is extremely hygroscopic and takes up moisture from the atmosphere. When the moisture content reaches about 8 percent, the lactose molecules recrystallise and form α-hydrate crystals. As these crystals grow, powdered product may cake and be lumpy.

A commonly used procedure to prevent the slow crystallisation of lactose in some dairy products and the resulting sandiness is to use seeding. Finely ground lactose crystals are introduced into the concentrated product and these provide numerous crystal nuclei. Many small crystals are formed rapidly, therefore, there is no opportunity for cystals to slowly grow in the super saturated solution until they become noticeable in the mouth.

Oxidation

Let us recall that primary alcohols on oxidation first give aldehyde and then carboxylic acid whereas secondary alcohols give ketone.

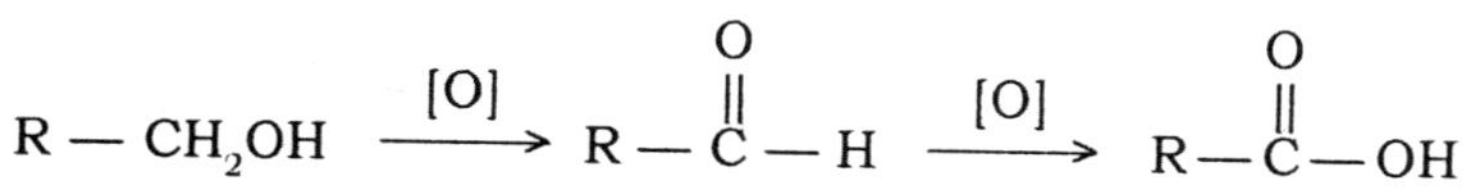

aldehyde carboxylic acid

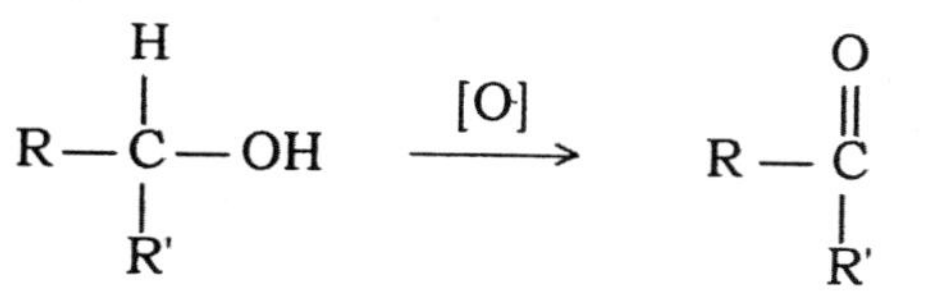

Secondary Alcohol Ketone

We, therefore, see that in case of primary alcohol, it is first oxidised to aldehyde and on further oxidation the carbonyl group is oxidised to carboxylic acid. Lactose is polyhydroxy aldehyde and the methyl hydroxyl groups and the aldehyde groups can be oxidised to a carboxylic acid groups with the simultaneous hydrolysis of the galactosidic linkage.

Lactose —On strong oxidation—> D-glucaric acid + D-galactaric acid
"Saccharic acid" "Mucic acid"

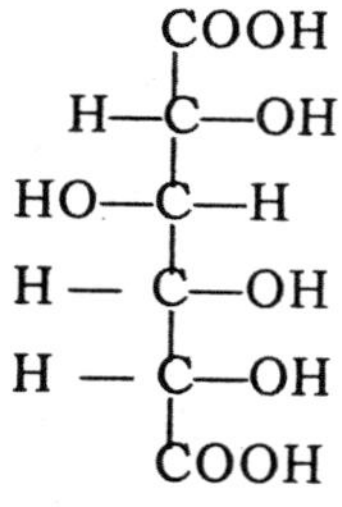

D—gluconic acid
"Saccharic acid"

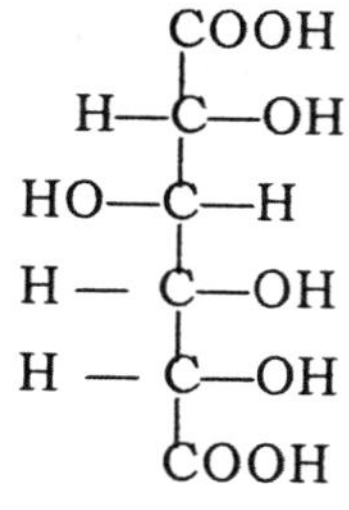

D—galactaric acid
"Mucic acid"

on further oxidation

tartaric acid

oxalic acid

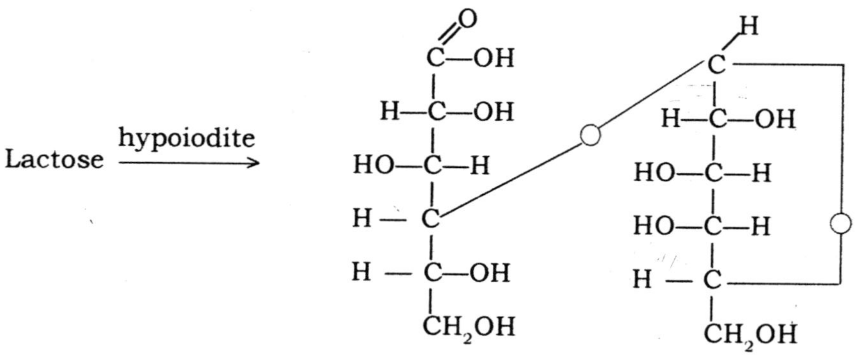

Lactobionic acid

Reduction

We know that a carboxylic acid may be reduced to its alcohol in the presence of lithium aluminium hydrate.

Similarly, an aldehyde group may be reduced to its alcohol in presence of lithium aluminium hydrate.

As lactose is a polyhydroxy aldehyde, it may also undergo similar reduction reaction :

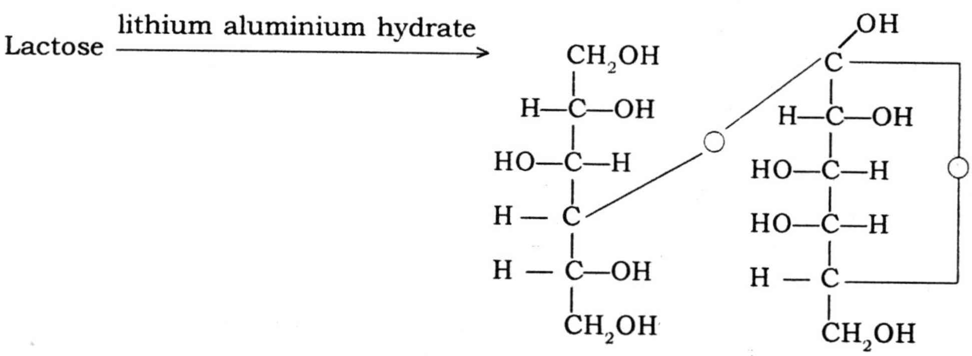

Status of Lactose in Milk

Lactose occurs in milk of all species as a true solution. The amount of percentage of lactose in milk of various species is given in Table 4.2.

Table 4.2. Percentage of lactose in various species

Species	Percentage Lactose
Buffalo	5.5
Camel	5.0
Cow	4.9
Donkey	6.1
Goat	4.1
Horse	6.2
Reindeer	2.8
Sheep	4.8
Women	7.0
Yak	4.6

While studying the percentage of lactose in the 10 species mentioned above, we find that human milk contains the highest percent of lactose. This is indeed a gift of nature to humanity. The galactose moiety of lactose is a part and parcel of cerebrosides of brain and myelin sheath of nerves. This may be a reason why mankind has progressed so much in multifarious directions.

Fermentation of Lactose and Citrate

The importance, utility and necessity of milk in human diet have been known to mankind since the birth of human civilisation. When mankind knocked at the door of civilisation, the prudents observed the phenomenon of milk souring and curdling after the milk is drawn out from the udder and particularly when left unboiled. In fact cultured dairy products came into existence in the kitchen more through accident than to any other reason. The beginning though was accidental, cultured dairy products seeped into every household of ancestral civislised man, particularly in the middle east, central Asia and the Indian subcontinent.

However, till the beginning of the 19th century, the biochemistry and bacteriology behind the production of fermented and cultured milk products were not understood. Slowly it was realized and perceived that the souring was due to lactic acid, three-carbon polyhydroxy carboxylic acid having a hand at the second carbon to hold a hydroxyl group and another hand to hold a hydrogen atom. The third carbon atom has two hands to hold one hydrogen atom each and another hand to hold a hydroxyl group. The sweet aroma and flavour produced was caused by some other compound or compounds. It was reported in the literature that two types of micro-organisms are necessary to produce lactic acid and aroma simultaneously. This was the beginning of the modern concept of preparing "starter cultures" having two types of micro-organisms, with specific assignments of one producing lactic acid and the other synthesizing the aroma producing compounds.

The modern biochemical approach to the process of fermentation in cultured and fermented milk products began with the concept that lactose is the precursor (forerunner) for production of lactic acid in milk and fermented milk. It was also realised that the metabolic pathways involved i n the fermentation of lactose and citric acid, a six carbon atom precursor for aroma production were extremely complex. It was also observed that lactic acid is not the only product of fermentation. There are other catabolic and anabolic products also. The catabolic processes led to the formation of biological energy which was required for anabolic metabolism.

Studies have revealed that there are two types of micro-organisms in milk responsible for the fermentation of lactose, the homofermentative and the hetrofermentative microorganism. The homofermentative bacteria produce mainly the lactic acid as the sole end product. These type of microorganisms can ferment lactose to lactic acid in the absence of air, i.e. anaerobically by the Embden Meyerhoff Parnas Pathway or the glycolytic pathway. The heterofermentative bacteria, and 2nd type do not have the enzyme aldolase, one of the principal enzymes of the glycolytic pathway and hence ferment lactose by a different pathway. The pathway has a broad resemblance to the pentose phosphate pathway in the first few steps of fermentation. The heterolactic bacteria ferment lactose by this pathway to produce carbondioxide, ethanol and acetic acid in addition to lactic acid. The overall reaction may be summed up as follows :

Lactose (12 carbon sugar) ———> $2CO_2$ + 2 Ethanol or

2 Acetic acid + 2 lactic acid

In the presence of air, CO_2, ethanol, and lactic acid are produced. When extraneous oxygen is present, instead of ethanol, acetic acid is produced. Acetic acid is produced from acetyl phosphate in presence of the enzyme acetokinase and adenosine diphosphate. In the process adenosine diphosphate is phosophorylated to adenosine triphosphate cleavage of whose terminal phosphate liberates energy.

Ethanol is produced via acetaldehyde from acetyl phosphate. Acetyl phosphate is reduced and cleaved to acetadehyde in presence of the enzyme acetaldehyde dehydrogenase and a reduced coenzyme Nicotinamide adenine dinucleotide phosphate (NADPH + H^+). Acetaldehyde so formed is reduced to ethanol in presence of alcohol dehydrogenase and another molecule or reduced NADP. The electron acceptor NADPH so formed is utilized in the first two reactions of the pathway wherein glucose-6-phosphate undergoes oxidation to 6-phosphate-gluconate and in the next step undergoes oxidative decarboxylation to Ribulose-5-phosphate and CO_2.

The sweet aroma of cultured dairy products is due to certain particular aroma producing micro-organisms working on citric acid, a tricarboxylic acid. Diacetyl and Acetoin (Acetyl methyl carbinol) are the main end products.

It has been reported that the sweet aroma is produced in cultured dairy products when lactose and citrate are present either in combination or citrate alone but no sweet aroma was produced when lactose was present alone. In the beginning it was thought that diacetyl is produced via acetoin. Later, it was observed that an active two carbon compound arising from an excess of pool of pyruvic acid originating mainly from citric acid is the intermediate in diacetyl production. For diacetyl production excess pool of pyruvic acid is necessary. The pyruvate produced from lactose is totally utilised for lactic acid production. Absence of an excess of pyruvate in the media, makes it impossible for lactose alone to act as a precursor of diacetyl production. But, fermentation of citrate produces an excess of pyruvate in the media which is then utilised for diacetyl production.

The active two carbon compound also plays a role in the production of acetoin. However, acetoin is not the precursor for diacetyl production but instead can be produced from diacetyl also.

To conclude, it may be summed up that two types of microorganisms are necessary for producing lactic acid and diacetyl in citrate and lactose fermentation and two types of micro-organisms ferment lactose to lactic acid, acetic acid or ethanol and carbondioxide. The homofermentative bacteria ferments lactose via the classical E.M.P. pathway while heterolactic bacteria do not possess the enzyme aldolase and hence cannot ferment lactose by the classical E.M.P. pathway. These organisms ferment lactose by a pathway which resembles in the first three steps the pentose phosphate pathway or the hexose monophosphate shunt pathway. By this pathway the heterofermentative bacteria prefer to produce ethanol instead of acetic acid because by following such a pathway there is regeneration of $NADP^+$ which is utilised in the first two steps of the pathway. This is a considerable waste of energy because when acetic acid is produced simultaneously ATP is also produced.

In the production of diacetyl and acetoin, the active two carbon compound, i.e. active acetaldehyde plays a pivotal role. This compound arises from an excess of pyruvate pool originating from the breakdown of citric acid. It may be mentioned that lactose also produced pyruvate. A question arises why then lactose alone cannot be a precursor for diacetyl production. This is because the pyruvate produced from lactose fermentation is totally utilised for lactic acid production in order to regenerate the NAD^+ used in earlier reaction thereby making it possible for the cycle to run. Only from citric acid an excess of pyruvate pool is produced which later is utilized in the production of the active two carbon compound, a precursor for diacetyl production.

Methods of Estimation of Lactose

Methods of estimation of lactose are elucidated in Chapter 10.

Milk Oligosaccharides

Milk, a delicately balanced biological fluid, contains almost all the essential nutrients necessary for human growth and maintenance. New born babies thrive almost entirely on milk. Milk contains water, fat, proteins, carbohydrate (namely lactose and lactose containing oligosaccharides about 0.3 percent to 0.6 percent) and minerals.

Bovine milk contains 1.2 g kg^{-1} of oligosaccharides; colostrum has considerably more. These are predominantly bound to proteins. The five oligosaccharides that have been isolated from colostrum contain nitrogen as N-acetyl or N-glycolyl neuraminic acid, and one of them has N-acetyl glucosamine unit as well.

Human milk is rich in milk oligosaccharides. About thirty of them have been isolated from human milk, ranging from three to fourteen saccharide units per molecule. The total concentration of these oligosaccharides is in the order of 10 g kg^{-1} in midlactation milk and 25 g kg^{-1} in colostrum.

In the scientific world sometimes spectacular research achievements are made due to sheer accident. This is true in case of the isolation and identification of lactose containing milk oligosaccharides. Western workers while trying to isolate the "bifidus factor" a factor responsible for the growth of intestinal microflora of breast-fed babies, were the first to isolate and identify lactose containing milk oligosaccharides.

How did the story begin? It all started when Tisier discovered that the intestinal microflora of infants fed on human milk was rich in *Lactobacillus bifidus* while the intestinal microflora of infants fed on cow's milk did not contain it. Schonfeld found that mother's milk contains a growth factor for *Lactobacillus bifidus.* The growth factor resembled a carbohydrate and was associated with lactose fraction in whey and in human milk. It was neither a mineral, protein or lipid in nature. Kuhne classified the lactose containing oligosaccharides into three types, namely – (a) those containing no nitrogen; (b) those containing 2-acetamido-2-deoxy-D-glucose and (c) the sialic acid derivatives.

The oligosaccharides of milk listed in the literature are :

e) Lacto–N–fucopentaose III

$$\text{gal} \xrightarrow{\beta-1,\,4} \text{glc NA}_C \xrightarrow{\beta-1,\,3} \text{gal} \xrightarrow{\beta-1,\,4} \text{glc}$$

$$\uparrow \alpha-1,\,3$$

$$\text{fuc}$$

f) Lacto–N–fucopentaose IV

$$\text{fuc} \xrightarrow{\alpha-1,\,2} \text{gal} \xrightarrow{\beta-1,\,4} \text{glc NA}_C \xrightarrow{\beta-1,\,3} \text{gal} \xrightarrow{\beta-1,\,4} \text{glc}$$

g) Lacto–N–fucopentaose V

$$\text{gal} \xrightarrow{\beta-1,\,3} \text{glc NA}_C \xrightarrow{\beta-1,\,3} \text{gal} \xrightarrow{\beta-1,\,4} \text{glc} \xleftarrow{\alpha-1,\,3} \text{fuc}$$

h) Lacto–N–difucohexose I

$$\text{fuc} \xrightarrow{\alpha-1,\,2} \text{gal} \xrightarrow{\beta-1,\,3} \text{glc NA}_C \xrightarrow{\beta-1,\,3} \text{gal} \xrightarrow{\beta-1,\,4} \text{glc}$$

$$\uparrow \alpha-1,\,4$$

$$\text{fuc}$$

i) Lacto–N–difucohexaose II

$$\text{gal} \xrightarrow{\beta-1,\,3} \text{glc NA}_C \xrightarrow{\beta-1,\,3} \text{gal} \xrightarrow{\beta-1,\,4} \text{glc} \xleftarrow{\alpha-1,\,3} \text{fuc}$$

$$\uparrow \alpha-1,\,4$$

$$\text{fuc}$$

j) Lacto–N–triose

$$\text{glc NAC} \xrightarrow{\beta-1,\,3} \text{gal} \xrightarrow{\beta-1,\,4} \text{glc}$$

3) Sialic acid derivatives

a) 3^1–sialyl lactose

$$\text{NANA} \xrightarrow{\alpha-2,\,3} \text{gal} \xrightarrow{\beta-1,\,4} \text{glc}$$

b) 6^1–sialyl lactose

$$\text{NANA} \xrightarrow{\alpha-2,\,6} \text{gal} \xrightarrow{\beta-1,\,4} \text{glc}$$

c) Disialyl lacto–N–tetraose

$$\text{NANA} \xrightarrow{\alpha-2,\,3} \text{gal} \xrightarrow{\beta-1,\,3} \text{glc NA}_C \xrightarrow{\beta-1,\,3} \text{gal} \xrightarrow{\beta-1,\,4} \text{glc}$$

$$\uparrow \alpha-2,\,6$$

$$\text{NANA}$$

Biosynthesis of Lactose by the Mammary Gland

The biosynthesis of lactose by the lactating mammary gland has been studied extensively by scientists over many decades. However, till the fifties there had been no major breakthrough in this field. Studies on the biosynthesis of lactose by the lactating mammary gland gained momentum in the early sixties through the work of Hassid and Babad in California who isolated the enzyme lactose synthetase from bovine milk. This enzyme was found to transfer galactose from UDP-galactose to glucose.

Further work showed that there are five steps in the biosynthesis of lactose and reaction in each step is being catalysed by different enzymes. It was also reported that the ultimate reaction in this five-step biosynthetic process of lactose is unique to the lactating mammary gland. Evidence was also established that the enzyme lactose synthetase consists of two sub units.

Ebner and his co-workers in Okhlahoma worked on this particular problem and observed that the enzyme lactose synthetase consists of two proteins which they named as A and B. The isolated lactose synthetase enzyme was resolved into A and B protein by eluting the enzyme through Bio-gel-P-30 column. Two peaks were obtained. The first peak which was of higher molecular weight was named as A protein while the second peak was termed as B-protein. A protein was later characterised as galactosyl transferase found also in tissues other than the mammary gland and B-protein was identified as a-lactalbumin, a milk whey protein, occurring only in the lactating mammary gland and its secretion milk.

Role of α-lactalbumin in the Biosynthesis of Lactose

In the beginning the concept was that the B-protein is a sub-unit of the A protein, galactosyl transferase. Later it was established that the effect of α-lactalbumin, the B protein, on galactosyl transferase is allosteric. In other words, α-lactalbumin acts as a modifier of the galactosyl transferase reaction. It was postulated that α-lactalbumin attaches itself to the allosteric site of galactosyl transferase and thereby altering the three dimensional conformation of the enzyme glycoprotein, galactosyl transferase, with the resultant alteration of three-dimensional conformation of the active sites of the enzyme. This enables the substrate to bind itself to the active sites of the enzyme galactosyl transferase, form an enzyme-substrate complex and ultimately produce the product lactose which is actually biosynthesised in the Golgi apparatus of mammary secretory cells. The observation that the absence of α-lactalbumin in a galactosyl transferase reaction catalysing the transfer of galactose to glucose from UDP-galactose to form lactose only succeeds in increasing the Michaelis-Menten constant (Km) to 1.4 M. Whereas its presence in the reaction mixture considerably reduces the Km value of 5 mM definitely supports the above observations. Km is a dissociation constant. This means that higher the Km value lower the specificity of the enzyme for the substrate and lower the Km value, higher is the specificity of the enzyme for the substrate. In other words in the presence of α-flactalbumin, a very low concentration of glucose is required for the reaction to proceed whereas in the absence of α-lactabumin a very high concentration of glucose is required for the reaction to continue. Also, Mn^{++} is a prerequisite for the galactosyl transferase activity while mg++ was found to be 25 per cent effective.

The control of lactose biosynthesis is depicted in the scheme below.

UDP-galactose + glucose —— A Protein ——> Reaction proceeds to the right only when the concentration of glucose is very high Km value = 1.4 M.

UDP-galactose + GlcNAc —— A Protein ——> Lactosamine, Km value = very low

UDP-galactose + glucose —— A / B ——> Lactose reaction proceeds to the right even when the concentration of glucose is very low. Km value = 5 mM

UDP-galactose + glucose —— B Protein ——> No reaction.

The scheme for lactose biosynthesis is given in diagram (Flow diagram 4.2 and 4.3).

The glucose and galactose moieties of lactose originate from two molecules of glucose which enter the mammary gland via the blood stream. The blood glucose in turn arises from the digestion of carbohydrates present in the digestive tract and also from stored polysaccharides like glycogen found mainly in muscle tissues and liver.

In conclusion it may be said that, lactating mammary gland only has the technical know-how for the biosynthesis of lactose because, in the final step of the lactose biosynthetic pathway, α-lactalbumin a whey protein plays a steller role by acting as an allosteric activator of the enzyme galactosyl transferase. Mammary gland is unique in this sense that it synthesises α-lactalbumin, a whey protein which in its turn controls the biosynthesis of lactose, a major milk constitutent. The importance of α-lactalbumin in lactose biosynthesis cannot be overlooked. It is the Golgi apparatus where biosynthesis of lactose takes place. The biosynthesis of lactose in the Golgi apparatus is directly proportional to the concentration and rate of flow of α-lactalbumin through the Golgi apparatus. The concentration and the rate of flow of α-lactalbumin through the Golgi apparatus is controlled, in it's turn, by lactogenic hormones.

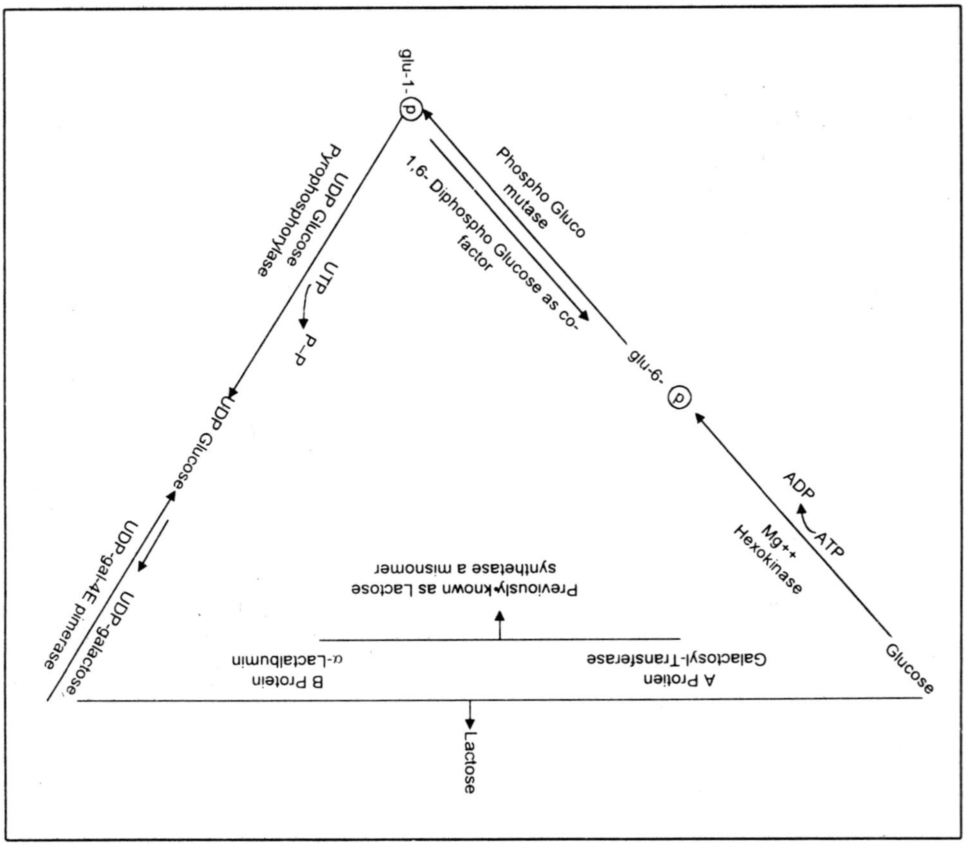

Flow Diagram 4.2 : The pathway for Lactose biosynthesis.

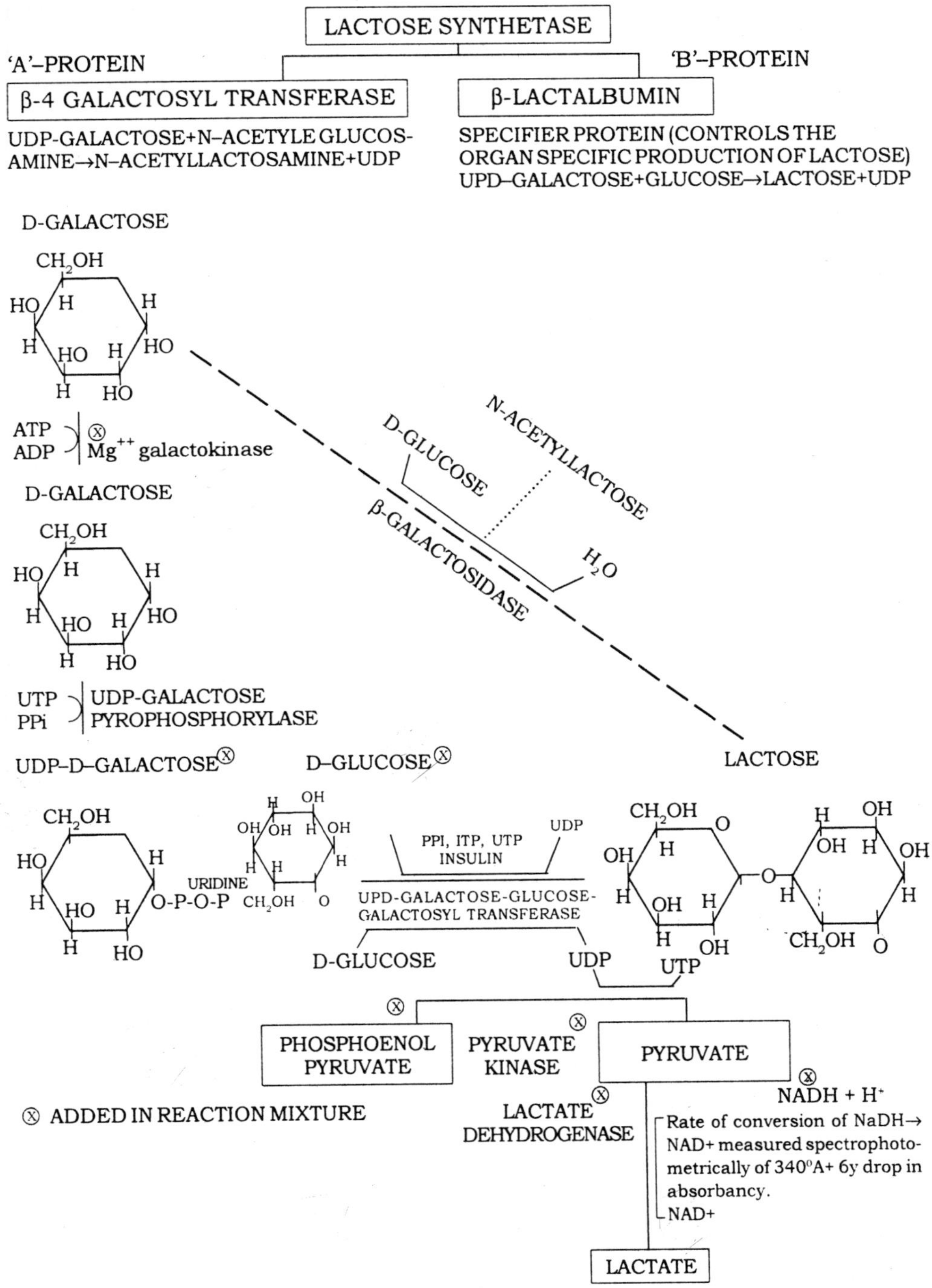

Flow Diagram 4.3 : Lactose Synthetase

Lactose Hydrolysis

The hydrolysis of lactose into glucose and galactose takes place at the oxygen bond between glucose and galactose. The oxygen bond of lactose is very stable, i.e. organic acids, which break-down, for example, saccharose, are ineffective against lactose. Only strong mineral acids break down lactose.

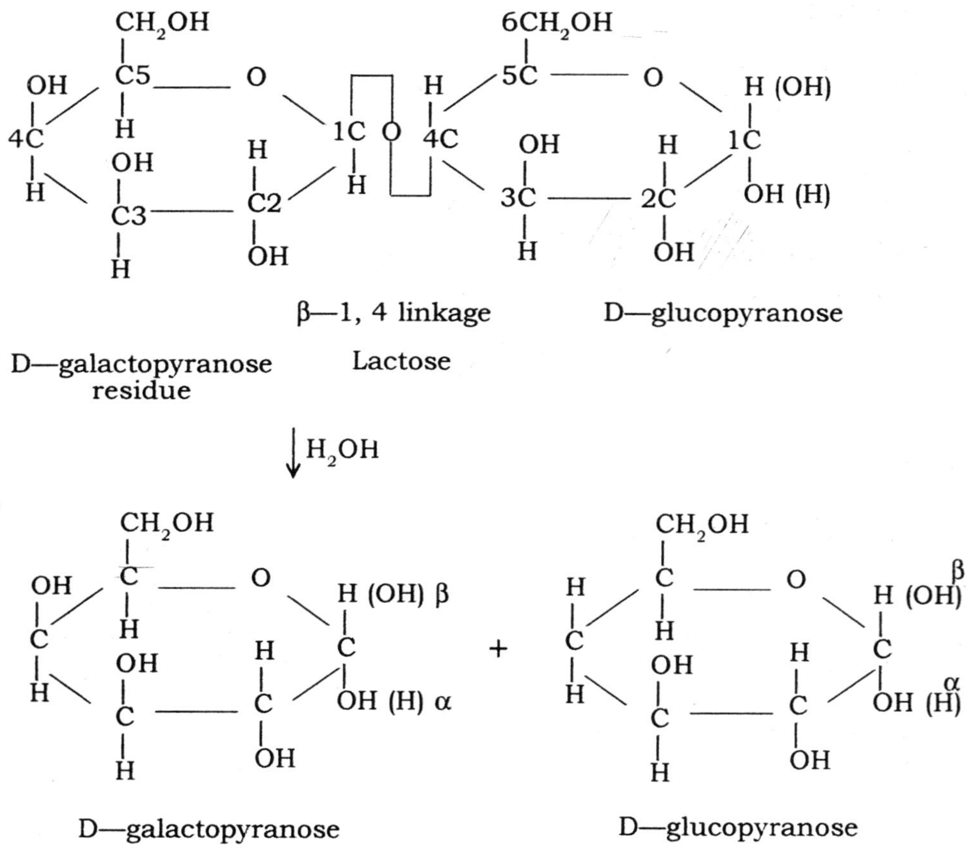

For the hydrolysis of pure lactose or lactose in milk and whey products enzymatic hydrolysis prevails in food technology. The hydrolysed products glucose and galactose are sweeter than lactose.

Physical Properties of Hydrolysed Lactose

Solubility

Contrary to lactose solubility which is barely 20 g/100 ml water, the solubility of hydrolysed lactose increases upto 60 g/100 ml water, thus resulting in special application, properties of particular value in ice-cream.

Viscosity

The viscosity is about 380 cp at 20°C in saturated solution of 60 percent d.m.

Sweetness

Hydrolysis of lactose increases its sweetening power by almost three times to about 0.70, which especially favours application in soft drinks based on milk whey.

Browning

Hydrolysed lactose is influenced far more by Maillard reactions, especially at elevated temperatures. As empirically determined, one can assure three times more browning in confectionery and caramels. For dried milk products with hydrolysed lactose this fact must be taken into account.

Stability and Commercial Forms

Hydrolysed lactose is a syrup with a dry matter content of about 70 percent, which shows good stability due to high osmotic pressure. The growth of microorganisms is very limited. However, products used in practice are not pure hydrolysed lactose, but are whey products and permeates in which the lactose is hydrolyzed to 75-90 percent.

Manufacturing and Production

Lactose hydrolysis is catalysed by a ß-galactosidase or lactase. The enzyme is found in the intestines of young mammals and in plants, fungi, yeasts and bacteria. For technological purposes only microbial enzymes are used due to availability and costs. The following table shows commercially available enzyme sources for ß-galactosidase.

Table 4.3. Commercially available enzyme sources of ß-galactosidase

Source of enzymes	pH optimum	Temperature optimum
Aspergillus niger	3–4	55°C
Aspergillus oryzae	4.8	46°C
Escherichia coli	6.9–7.5	45°C
Saccharomyces fragilis	6.5	50°C
Saccharomyces lactis	6.5–7.2	45°C

The hydrolysis processes are essential factors for profitability of hydrolysed lactose in pure form or preferably in whey, permeate and milk products. While hydrolysis of mineral acids finds no application, the following processes are competitive : ion exchange resins; mobile enzymes; and immobile enzymes. The mobile enzymes are added batchwise to a hydrolysed lactose. The immobile enzymes which are fixed on a carrier can be used repeatedly.

Use of Hydrolysed Lactose in Food Products

In application, two types of hydrolysed lactose must be differentiated:

— Hydrolysed pure lactose solutions and deproteinised/demineralised permeates, i.e. protein-free saccharide solutions.

— Hydrolysed lactose in whey, demineralised whey, skimmed milk and whole milk.

The pure hydrolysed lactose permeates are in their functionality similar to inverted sugar or glucose syrups without any outstanding special properties. The hydrolysis of lactose in milk, and preferably in whey, has resulted in the following additional applications due to the special value of the protein present, toffee, ice cream, soft fruit drinks, and dairy desserts and yoghurts.

Toffee

In toffee production, a large part of the sweetened condensed milk can be substituted by use of hydrolysed, partly or totally demineralised milk. Cost savings are the most important factor here. An increased browning reaction must be taken into consideration.

Ice Cream

In ice cream production, very good results are achieved with hydrolysed whey as a substrate for fat-free milk dry matter. With pure milk-containing recipes the fears of "sandiness", i.e. recrystallisation of lactose, is diminished or eliminated by lactose hydrolysis.

Soft Fruit Drinks

In soft drinks the cost savings of added sweeteners are most important, while in fruit drinks especially, hydrolysed wheys are suitable owing to the additional acid-stable protein that they provide. A hydrolysed milk is intended for the demands of people suffering from lactose intolerance.

Dairy Desserts and Yoghurts

In these products replacement of traditional sweeteners is in the foreground, with partial substitution of fat-free milk dry matter.

Products with hydrolysed lactose will always be used when functional advantages can be achieved in addition to partial savings of traditional sweeteners. Such applications are :

— Whey proteins in toffee for the intensification of caramelisation with partial substitution of the milk dry matter.

— Broken-down lactose in ice cream for suppressing sandiness and partial substitution of milk dry matter.

— Additional sweetening power in soft drinks on milk/whey basis.

Sweetness and Lactose

The sweetening power of different sweeteners are elucidated in Table 4.4.

Table 4.4. Sweetening power of different sweeteners

Sweetener	Sweetening Power
Sucrose	1.0
Fructose	1.2
Dextrose	0.69
Galactose	0.72
Maltose	0.35
Lactose	0.27
Hydrolysed Lactose	0.70
Lactitol	0.4
Mannitol	0.5
Sorbitol	0.5

From the above table it is evident that lactose has the least sweetening power whereas fructose has the highest followed by sucrose. As carbohydrates are polyhydroxy aldehydes or ketones, we surmise that a sugar possessing a keto group imparts more sweetness to the taste buds. It is generally agreed that there are only four basic, or true tastes, sweet, bitter, sour and salty. The sensitivity to taste is located in taste buds of the tongue. The taste buds are grouped in papillae, which appear to be sensitive to more than one taste. There is undoubtedly a regional distribution of the four kinds of receptors at the tongue. The areas of sensitivity are as follows :

- The sweet taste at the tip of the tongue
- The bitter taste at the back of the tongue
- The sour taste at the edges of the tongue
- The salty taste at both edges and tip of the tongue

All compounds which bring about a sweet taste response possess an electronegative atom A, such as oxygen or nitrogen. This atom also possesses a protein attached to it by a single covalent bond. AH can represent a hydroxyl group, an imine or amine group, or a methine group. Within a distance of about 0.3 nm from the AH proton, there must be a second electronegative atom B, which again can be oxygen or nitrogen. Investigators have recognised that sugars that occur in a favoured chair conformation yield a glycol unit conformation with the proton of one hydroxyl group at a distance of about 0.3 nm from the oxygen of the next hydroxyl group.

Extensive experiments with a large number of sugars, indicate that the fourth hydroxyl group of glucopyranosides is of unique importance in determining the sweetness, possible by donating the proton. Apparently, the primary alcohol group is of little importance for sweetness.

As the molecular weight of saccharides increases, their sweetness decreases. This is best explained by the decrease in solubility and increase in size of the molecule. Apparently, only one sugar residue in each disaccharide is involved in the interaction at the taste bud receptor site. The relative sweetness of mixtures of sugar changes with the concentration of the components.

Browning Reactions in Milk and Milk Products

Columbus on attempting to discover India, landed in America while Maillard in trying to establish the mechanism of protein biosynthesis, observed and reported the Maillard Reaction between reducing sugars and amino compounds.

As a matter of fact, years before Maillard established the browning reaction in milk and milk products, observations were made in the estimation of lactose in milk and milk products. Low lactose values are reported in samples of lactose wherein, there is an interaction between protein and lactose. Lactose when estimated as such gave the correct value while in the presence of proteins when heat induced processing was involved, the reading of the amount of lactose present were lower.

It was also established that heated milk and milk products gave lower lactose values. Also, stored milk products gave still lower lactose values. Milk and milk product processing are of 4 types (1) heat-induced processing, (2) rennet-induced processing, (3) microbe-induced processing, (4) acid-induced processing.

Browning reactions in milk and milk products are the ultimate effects of heat-induced processing coupled with storage conditions, moisture content, relative humidity and temperatures of processing and storage of milk and milk products. Browning reactions are not evidenced in unstored pasteurised milk whereas sterilised milks on storage browns readily.

Recently UHT (ultra high temperature) milk has become very popular all over the world. In UHT plants milk is heated either by a direct or an indirect process to 130–150°C for a few seconds and it is then aseptically packaged, usually in cartons. Fresh UHT milk is not brown. However, on storage it becomes brown. Considerable heat treatment is involved in the manufacture of roller and spray-dried milk powders, condensed milk, evaporated milk and processed cheese. All these products become brown perpetually on storage or even during heat induced processing.

Apart from Maillard type of browning, i.e. amino-sugar browning another type of browning occurs in some milk products. This type of browning is known as caramelisation, i.e. non-amino sugar browning. Caramelisation occurs practically on the lactose content of milk and milk products and protein due to severe heat treatment and increases as a function of pH and buffers. The indigenous milk product "Palang tor" derives its basic taste from caramalised flavour and odour. Another type of browning which occurs in nature is oxidative browning. It may be enzymic or non-enzymic.

The Flow Diagram 4.4, 4.5 and 4.6 depicts browning reactions in milk and milk products.

It is evident from the diagram that the ultimate brown pigment that forms is known as "Melanoidin". As we proceed in our discussion on the diagram, we will find that Melanoidan is an ultimate complex of a variety of organic compounds arising out of the myriad metabolic pathways of browning reactions.

The first reaction in the Maillard pathway is between lactose and amino compounds in an equimolar ratio.

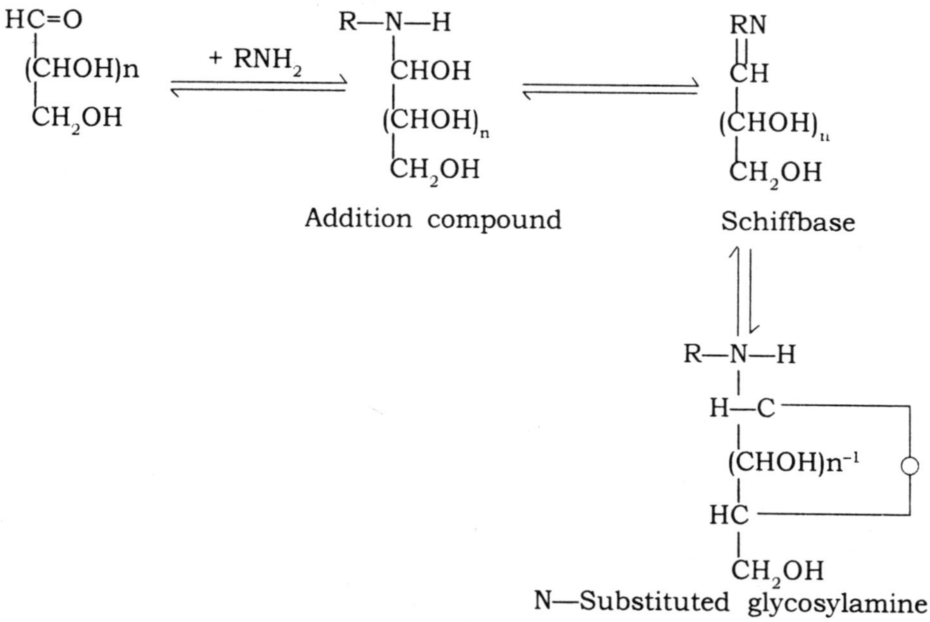

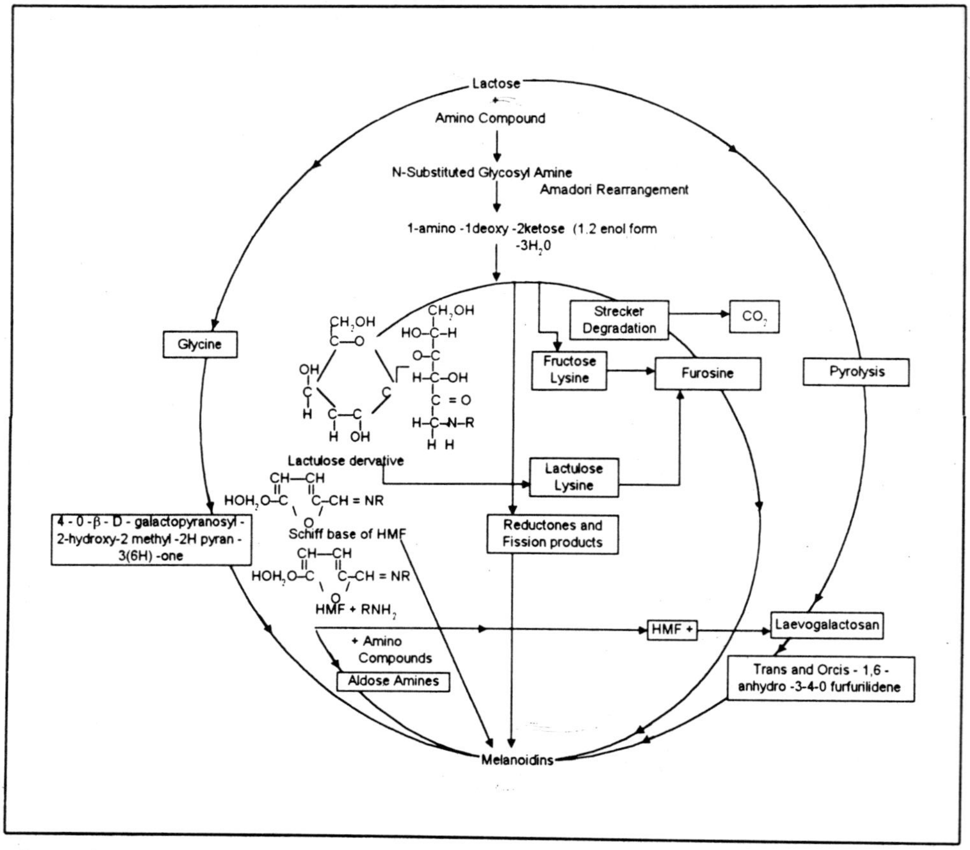

Flow Diagram 4.4 : Browning Reaction in Milk and Milk Product

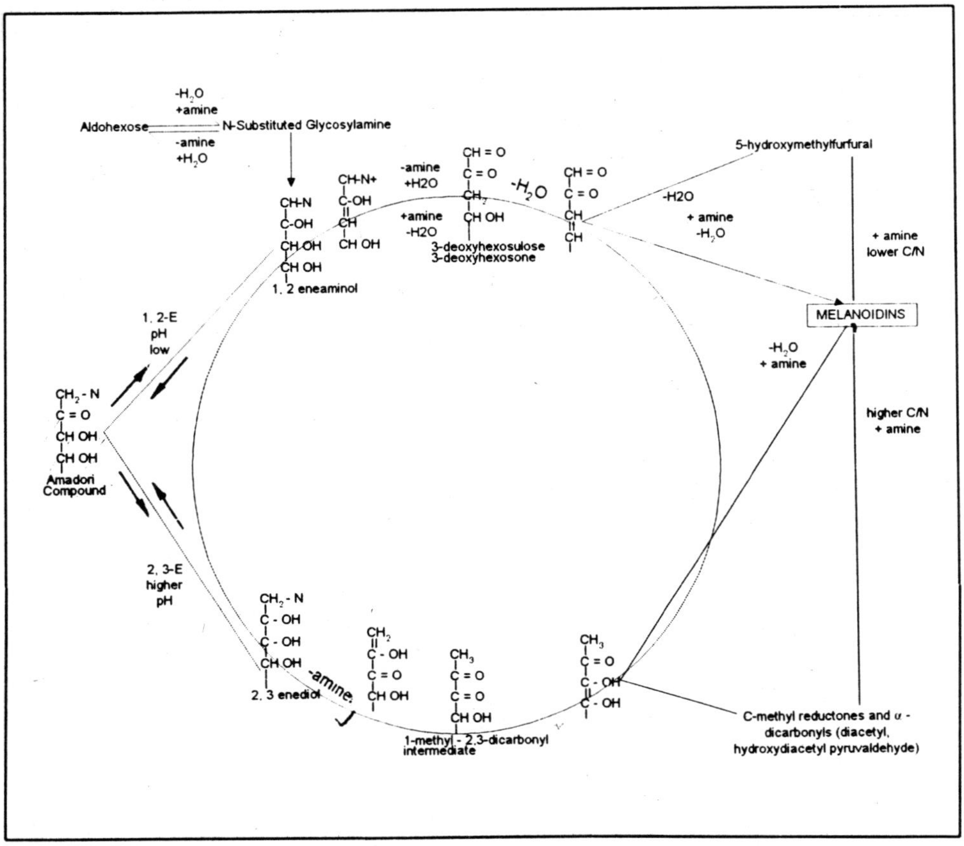

Flow Diagram 4.5: Maillard Reaction: Two Major Pathways from Amadori Compounds to Melanoidins (Based on Hodge, 1953 and Adapted from Nursten, 1986)

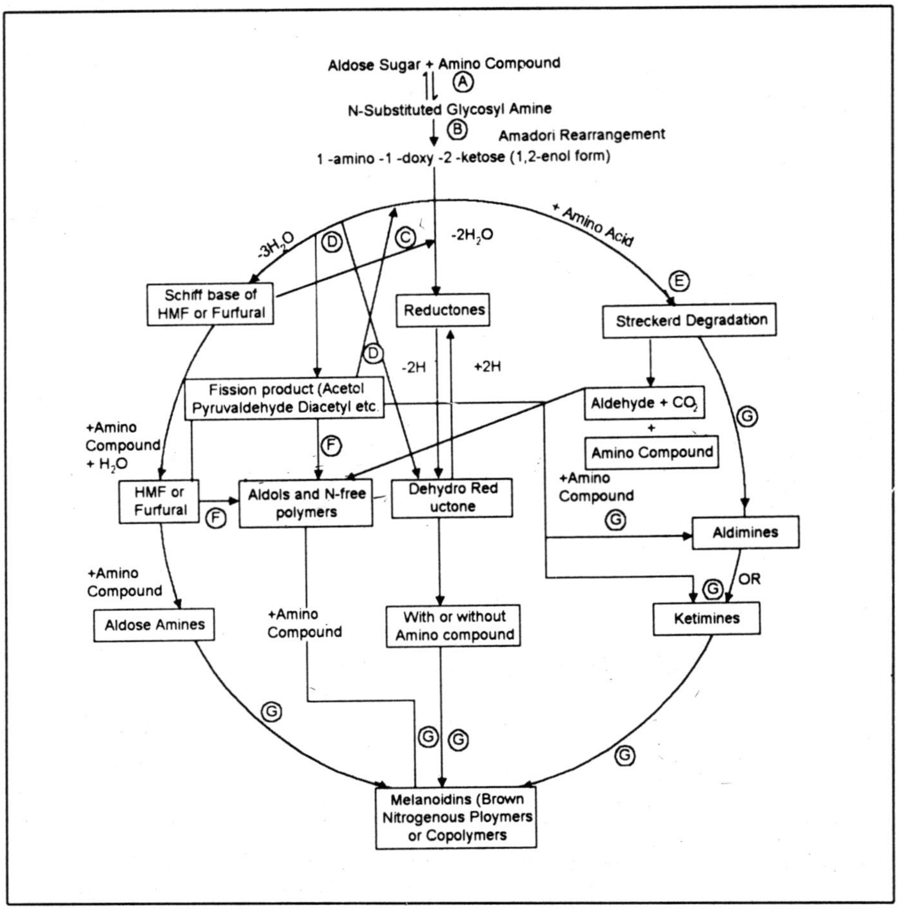

Flow Diagram 4.6: Formation of Melanoidins (Brown Nitrogenous Polymers) or Copolymers

The first few reactions are reversible. However, later, irreversible reactions completely dominate. Formation of n-substituted glycosylamine is followed by Amadori rearrangement.

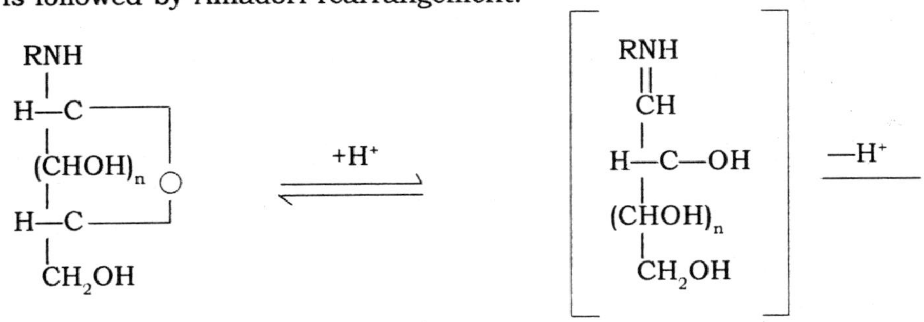

N—substituted aldosylamine Cation of Schifbase

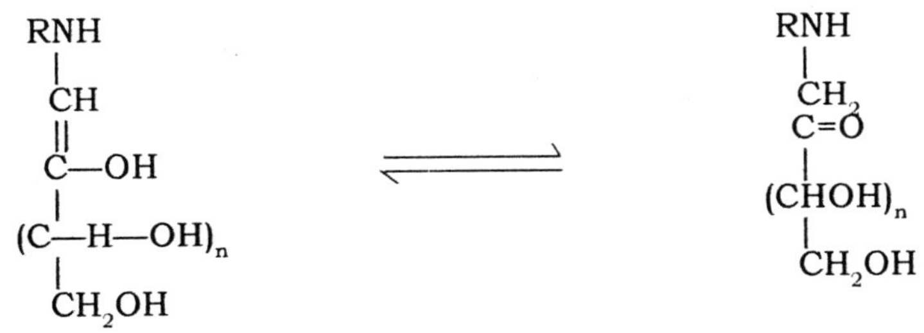

N—substituted—l-amino—l-deoxy—2
Ketose, Keto-form

Following the formation of N-substituted-1-amino-1-deoxy-2-ketose, a lactulose derivative is formed via a dehydration reaction. Formation of the lactulose derivative is followed by the production of Schiff base of HMF then HMF and the ultimate formation of the complex Melanoidin (Flow Diagram 4.4).

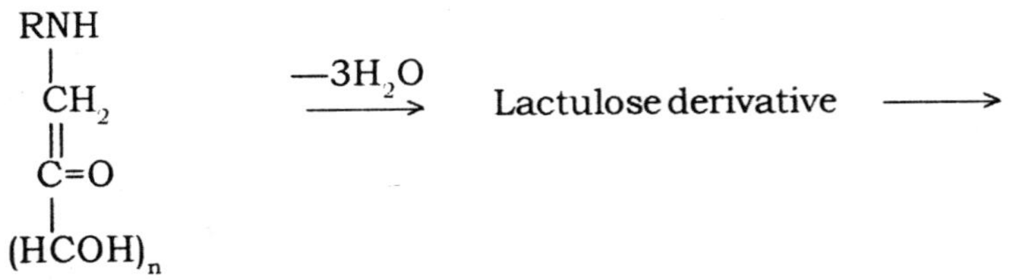

⟶ Schiffbase of HMF ⟶ HMF + RNH_2

↓

MELADOIDIN

Meladoidin is also produced via Reductones and
fission products from 1—amino—1—deoxy—2—Ketose (1,2—enol-form)
1—amino—1deoxy—2—Ketose
(1, 2 enol form)
↓ $-2H_2O$
Reductones and fission compounds ⟶ MELANOIDIN

In browning reaction of milk and milk products, Strecker degradation takes place. In such degradation an aldehyde is formed and CO_2 is evolved. The aldehyde reacts with amino compounds to form aldimines which finally enters the Melanoidin complex.

1—amino—1—deoxy—2—Ketose
↓ $-3H_2O$
STECKER DEGRADATION ⟶ CO_2 ↑
aldehydes
↓ + amino compounds
aldimines
↓
MELANOIDIN

When the reactant amino acid is lysine, the organic compound that is formed is N-substituted-glycosyl-amine which on amadori rearrangement yields fructose lysine. This compound on acid hydrolysis yields, furosine, pyridosine and lysine.

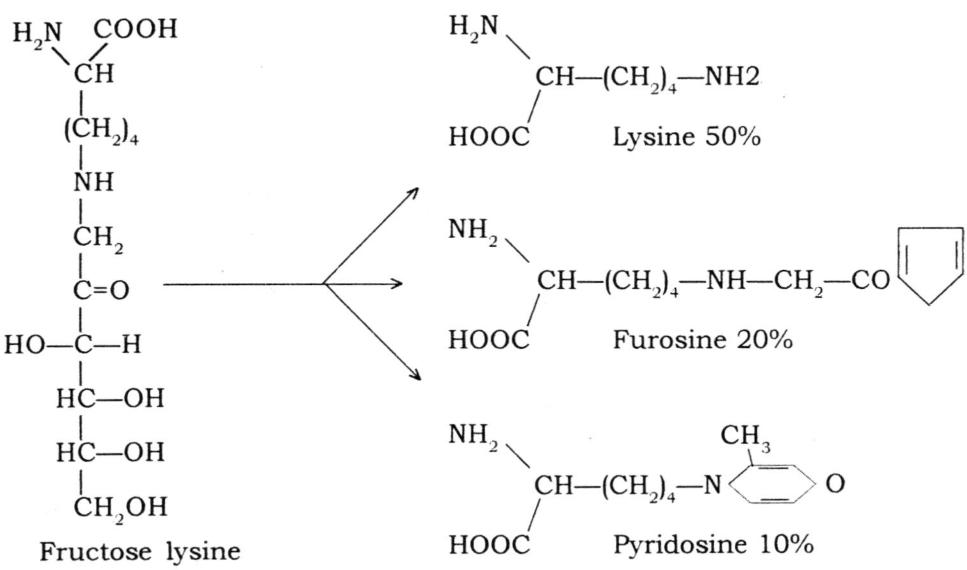

With amino acid lysine, the lactulose derivative from 1-amino-1-deoxy-2-ketose is lactulose lysine, and the structure is as follows :

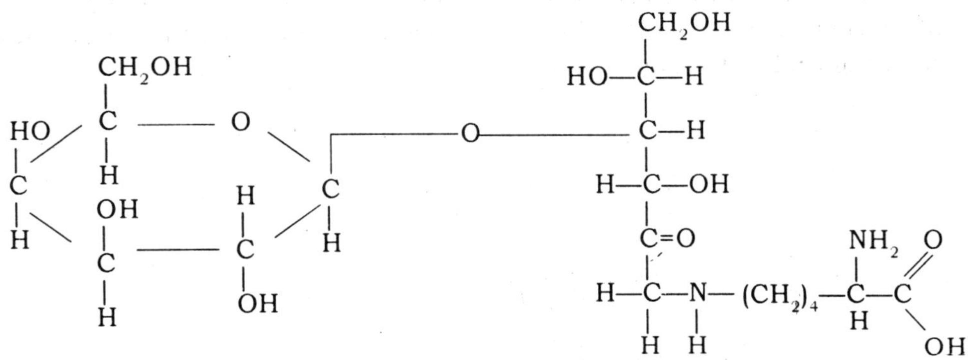

Lactulose Lysine

Lactulose lysine on acid hydrolysis yields, furosine, pyrodoxine and lysine. Furosine has been identified as a secondary product of the amadori compounds–fructose lysine or lactulose lysine in heat damaged milk samples. Furosine can be estimated by fusing amino acid analysis, high performance liquid chromatography and by gas liquid chromatography. Furosine is an ideal indicator of heat damage in the initial stages of the Maillard reaction, but is non-linear in the advanced stages. On these findings furosine cannot be considered to be a quantitative indicator in all instances of protein sugar damage, but can be recommended for evaluation of heat damage in milk and milk products. But, overall furosine is a good indicator of expressing the extent of Maillard reaction.

Proof and identification of a new Maillard compound in heated milk was demonstrated in 1986. A glycine/lactose mixture (1:10 by wt.) was heated in 0.2M phosphate buffer (pH 8.6) at 121°C and 2 atm for 60 min (1:1 v/v). When the extract was chromatographed on kiselgel, a new Maillard product was detected by polarography in one of the eluted fractions and isolated by multistage chromatographic procedure. U.V., I.R., 13 C-NMR, 1H-NMR and mass spectroscopy identified the compound as 4-O-ß-D-galactopyranosyl-2-hydroxy-2-methyl-2H-pyran-3 (6H)-one.

Recently, a new furfurylidene derivative was separated and identified. This compound tastes bitter. It was separated from lactose heated at 200°C for 2 hr. by ethyl acetate extraction and preparative TLC. The compound was identified as the trans isomer of 1,6-anhydro-3-4-O-furfurilidine-B-D-galactopyranose by U.V., I.R., mass and H-NMR spectroscopy. The compound and its CIS-isomer were identified in the reaction mixture of HMF (Hydroxy methyl furfural) and laevogalactosan heated at 175°C for 2 hr. with nitric acid. The compound was also detected after heating anhydrous galactose under the same conditions.

Laevogalactosan

+

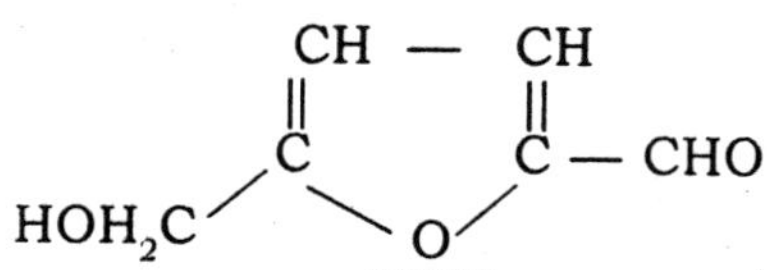

HMF

Trans and / or cis—1, 6—anhydro—3—4—O—furfurilidene.

Amino compounds itself in heated milk and milk products give rise to methyl sulphide.

When lactose is heated to 120°C for 20 minutes decomposition occurs. The sugar fragments are primarily those resulting from alkaline degradation of sugars. These sugar fragments are mainly acetic, lactic, formic, and pyruvic acids, carbondioxide, acetol methyl glyoxal, acetaldehyde, furfural, furfuryl alcohol, HMF and maltol.

In the late 50's interesting observations have been made on the formation of formic acid, furfuryl alcohol and maltol in milk heated to 121°C, for 4 hours. Prior to heating of milk, lactose-1-C^{14} was added. Radioactivity was found in formic acid and maltol but not in furfuryl alcohol. The scheme proposed is as follows :

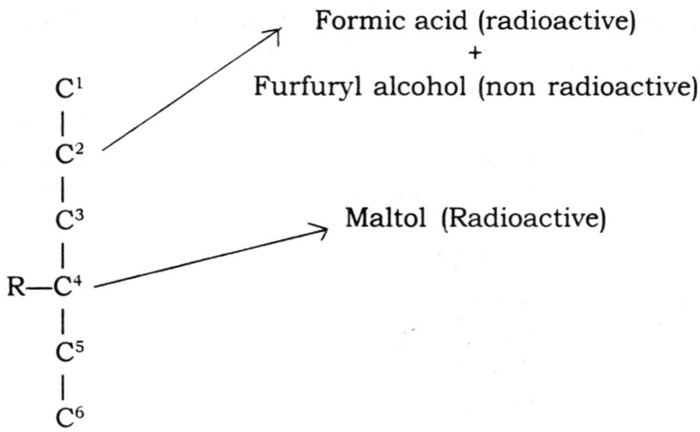

As a matter of fact decomposition of lactose largely is responsible for the gradual decrease in pH of milk during heat treatment. Observations have been made that lactose caseinate systems are of darkest colour. Also, in lactose-caseinate systems when milk powder is added browning takes place very rapidly during processing. Reconstitution of instant milk foods causes rapid browning. They contain higher level of HMF and reductones. These observations suggest that instantising process promotes the development of browning reactions. Processed cheese browns readily. This can be prevented by reasonably fast cooling. However, browning of roller-dried and spray dired powders evaporated milk and other heat treated milk and products can be reduced by lower temperature of processing, lower moisture levels, R.H. and less storage temperature and period.

Flavours Produced by Maillard Reactions

The flavours produced by the Maillard reaction also vary widely. In some cases, the flavour is remniscient of caramelisation. The Strecker degradation of α-amino acids is a reaction that also significantly contributes to the formation of flavour compounds. The dicarbonyl compounds formed react in the following manner with α-amino acids:

$$R{-}\overset{\overset{O}{\|}}{C}{-}\overset{\overset{O}{\|}}{C}{-}R + R^1\ CH\ NH_2 \longrightarrow R^1\ CHO + CO_2 + R{-}CH{-}NH_2{-}\overset{\overset{O}{\|}}{C}{-}R$$

The amino acid is converted into an aldehyde with one less carbon atom. Browning flavour has usually been described as corny, nutty, bready, and crackey aroma compounds that consist of planar unsaturated heterocyclic compounds with one or two nitrogen atoms in the ring. Other important members of this group are partially saturated N-heterocyclics with alkyl or acetyl group substituents. Compounds that contribute to pungent, burnt aromas are listed in the following Table 4.5.

Table 4.5. Aroma and structure classification of browned flavour compounds

Aromas	Burnt (pungent, empyreumatic)	Variable (aldehydic, ketonic)
Structures	Polycarbonyls (α,ß-unsaturated aldehydes–C:O–C: O–=C–CHO)	Monocarbonyls (R–CHO, R–C:O–CH_3)
Examples of Compounds	Glyoxal	Strecker
	Pyruvaldehyde	Aldehydes
	Diacetyl	Isobutyric
	Mesoxalic	Isovaleric
	Dialdehyde	Methional
	Acrolein	2–Furaldehydes
		2-Pyrrolealdehydes
	Crotonaldehyde	C_3–C_6–Methyl Ketones

These are mostly vicinal polycarbonyl compounds as α,ß-unsaturated aldehydes. They condense rapidly to form melanoidins. The Strecker degradation aldehydes contribute to the aroma of bread, peanuts, cocoa, and other roasted foods. Although acetic, phenylacetic, isobutric, and isovaleric aldehydes are prominent in the aromas of bread, malt, peanuts, and cocoa, they are not really charactersitics of these foods.

Lactitol

Lactitol is a sugar alcohol derived from milk sugar (lactose) by hydrogenation of the glucose part of the disaccharide lactose. Lactitol is crystallised under special condition as a lactitol monohydrate. Lactitol is also available as a liquid (54 per cent solution). The structure of lactitol is as follows :

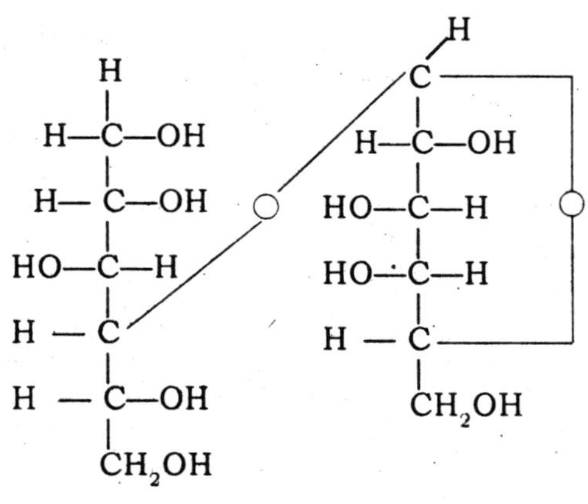

Lactitol

Lactitol delineates the specifications for lactitol monohydrate and lactitol solution.

Table 4.6. Specifications for lactitol monohydrate and lactitol solution

	Lactitol monohydrate	Lactitol 54 percent solution	
Description	White crystalline powder	Description	Clear colourless liquid
Taste	Sweet	Taste	Sweet
Odour	Odourless	Odour	Odourless
Lactitol content	93.5–96.0%	Lactitol content	50–54%
Water (Karl Fischer)	Max. 5.5%	Water (Karl Fischer)	45–47%
Other polyols	Max. 2.5%	Other polyols	Max. 2.5%
Reducing sugars	Max. 0.05%	Reducing sugar	Max. 0.1%
Specific rotation $(\alpha)_D$	+14.0° to +14.6°	Specific rotation $(\alpha)_D$	+13.5° to +15.5°
Melting range (microscope)	115–125°C		
pH of 10% solution	4.5–7.0	pH of 10% solution	4.5–7.0
Sulphate	Max. 20 ppm	Sulphate	Max. 100 ppm
Ash content	Max. 0.05%	Ash content	Max. 0.05%
Heavy metals	Max. 10 ppm	Heavy metals	Max. 5 ppm
Arsenic	Max. 1 ppm	Arsenic	Max. 1 ppm
Nickel	Max. 1 ppm	Nickel	Max. 1 ppm

Lactitol monohydrate has the physical property of being non-hygroscopic. It is less hygroscopic than sorbitol, xylitol but similar to mannitol.

Lactitol has good solubility in water. At 25°C, 149 g lactitol-monohydrate will dissolve in 100 g water. Figure 7 shows the solubility of lactitol-monohydrate at different temperature.

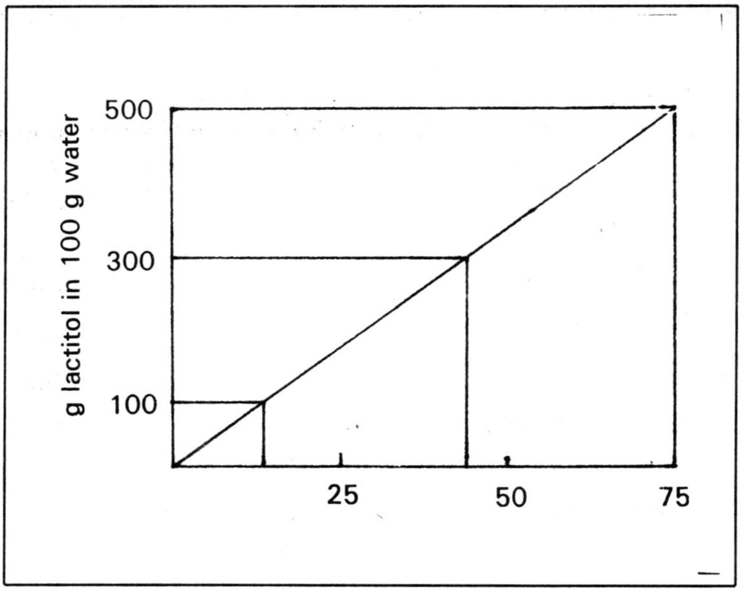

Fig. 7. Solubility of Lactitol monohydrate at different temperatures

The viscosity of aqueous lactitol solutions is similar to those of sucrose solutions at equal concentrations. The viscosity of lactitol at different concentrations is shown in Figure 8.

The sweetening power of lactitol, which depends on concentration, is 0.4 compared to sucrose = 1 in solution. Lactitol has a clean, sweet taste, closely resembling sucrose in sweetness character.

As lactitol has no reducing group, it is chemically more stable than lactose. Lactitol is highly resistant to pH fluctuation and elevated temperatures. This is important for conditions during baking processes or manufacturing of hard boiled sweets.

Production

Lactitol is produced technically by the hydrogenation of lactose in the presence of nickel catalysts. A lactose solution will be hydrogenated at ~ 100°C and ~ 40 bar. The solution will be clarified by filtration and ion exchange. Available types of lactitol are monohydrate and lactitol solution with 54 per cent dry matter.

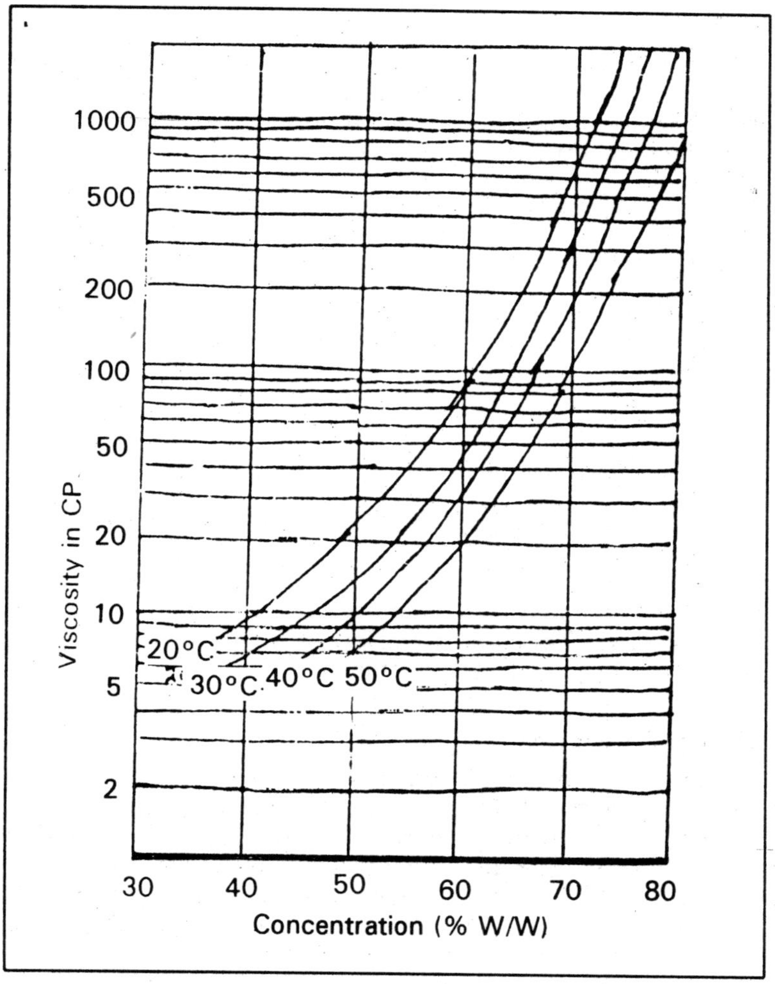

Fig. 8. Viscosities of lactitol solution

Table 4.7. Splitting of lactose and lactitol by human intestinal disaccharidases

Substrate	After 1 hour	After 2 hours	After 3 hours
Lactose	1.60	2.54	3.44
Lactitol	1.022	0.045	0.071

Source : Jagerstadt, M., University of Lund, Sweden, 1985.

Biological Characteristics

The enzyme ß-galactosidase, which splits lactose into glucose and galactose, hydrolyses lactitol very slowly. Table 4.7 shows that less than 1 percent of the lactitol will enter the bloodstream, and this is excreted unchanged in the urine. Most of the the lactitol passes the small intestine and is converted by the intestinal microflora to organic acids and CO.

Lactulose

Lactulose is a disaccharide with the structure of 4-ß-D-galactosido (1,4)-D-fructose. Lactulose has not yet been isolated in its genuine form under normal condition, although small amounts can be detected in heated human milk. The structure is as follows :

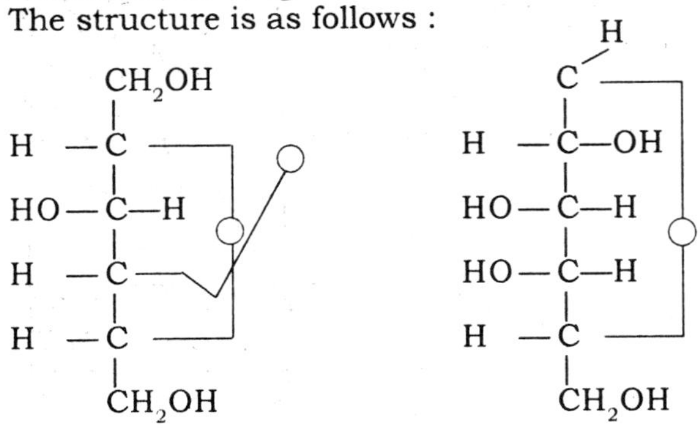

β—Galactosido—fructose (Lactulose)

Production

Lactulose is produced from α-lactose-monohydrate by isomerisation with calcium hydroxide. The commercial form of lactulose is a fluid hydroscopic concentrate with 65 per cent dry matter.

Biological Characteristics

Lactulose is not broken down in the small intestine and is hardly absorbed at all, as there are no adequate enzymes to split the di-saccharide into the two easily reabsorbable monosaccharides, galactose and fructose. However, lactulose is generally broken down by the action of *Bacterium bifidum* and acidophillus bacteria with lactic acid as the by-product. With the aid of lactulose breakdown the acidophilus bacteria create an acid colon environment in which these bacteria will thrive, thus suppressing alcalophil intestinal bacteria. The resulting lactic acid can also stimulate intestinal peristalsis. Lactulose reduces serum cholesterol and paradontosis prophylaxis.

The Chair and Boat Conformations of Milk Sugars

In the 19th century it was thought that all the atoms in the ring lie in the same plane. Therefore, the bond angles of a three-membered, four-membered, five-membered and six-membered rings would respectively be 60°, 90°, 108° and 120°C. However, saturated carbon atoms are characterised by the tetrahedral angle of 109.28^0. This is recognised as the normal angle between the four bonds to a saturated carbon atom. Let the formula of a saturated carbon atom be AX_4 and let the diagram be drawn and described it as a tetrahedral.

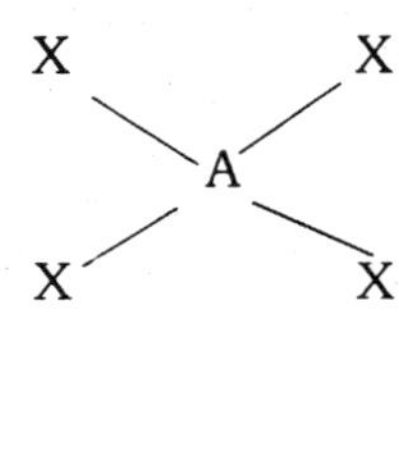

A tetrahedron is a pyramid with the four faces each an equilateral triangle. A tetrahedron is used for representation of the carbon atom and with nucleus of the atom inside the pyramid and with the four single bonds extending to the corners of the pyramid.

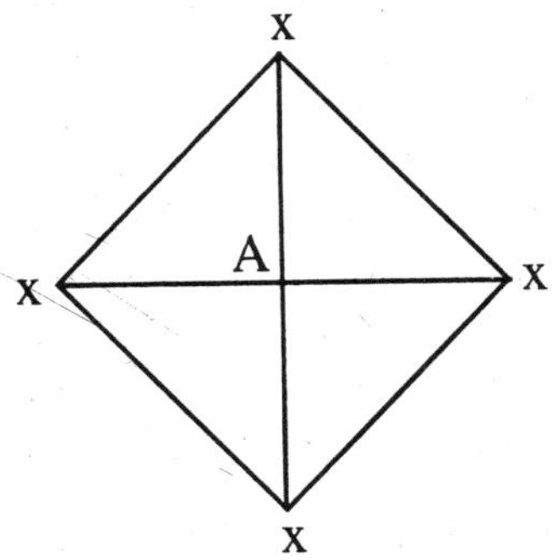

In 1885, Bayer suggested that, greater the deviation of the bond angle in the ring from the tetrahedral angle, the strain is greater and the stability of the ring is lower.

According to Bayer, the c-c-c angle in cyclopentane differs least from the tetrahedral angle. However, this is not the case with the three-membered rings and large rings and if the rings are planar, the strain is greatest. Cyclopentane, therefore, must have the lowest energy and cyclopropane and large rings should have the highest energy. Indeed, the cyclopropane ring was found to be most unstable whereas cyclobutane was considerably stable. As if to support Bayer's strain theory cyclopentane was found to be extremely stable. But, Bayer's strain theory remained explained as regard to cyclohexane which according to the theory, should be most unstable but is actually very stable.

In the later part of 19th century, scientists, particularly Sachse, started having a sneaking hunch that cyclohexane could have a non-planar arrangement of ring atoms. Sachse's postulation was substantiated with evidence by Mohr in 1918 and suggested the spatial arrangement of atoms in cyclohexane and proposed two possible models (two so-called puckered or folded forms) for its ring structure, now known as boat and chair forms and are free from the Bayer angle strain theory. This proved why cyclohexane is not inferior in strength and case of formation to cyclopentane. Also, the heat of combustion values per methylene group and energy values of cyclopentane ring and large rings were very close to one another. Whereas according to Bayer's strain theory, the reactivity or energy content must gradually increase with increasing ring size. Also, stability of larger rings did not differ from the stability of cyclohexane. Later, the structure of cyclohexane was established through spectroscopic data. The ring structure of cyclohexane is twisted in such a manner as to make the angles almost equal or close to the tetrahedral angle.

In cyclic compounds hydrogen atoms are partly or completely in eclipsed position. As a matter of fact hydrogen atoms are almost in contact with each other. Like charges repel and unlike charges attract and in case of hydrogen atoms lying so close together causes a mutual repulsion. Therefore, to the angle strain energy, the energy of mutual repulsion is added in cyclopropane. In cyclobutane, besides angle strain theory, and the energy due to mutual repulsion, there is an additional strain due to the interaction between the first and fourth carbon atom. This strain is known as bond opposition strain. It is universally accepted now that one of the ring atoms in cyclobutane is slightly out of the plane of the other three atoms. The total strain energy of cyclobutane is decreased due to this out of plane position. The strain of the molecule, which twists one of the atoms out of the plane of the ring, is known as torsional strain.

By assuming boat and chair conformations the cyclohexane ring avoids angular strain. In the chair form there are two parallel planes each containing three carbon atoms whereas in the boat conformation four carbon atoms lie in the same place and two atoms are in the out-of-plane position. In nature, cyclohexane is found in one of the two identical "chair" conformations in which all groups are staggered whereas in the boat conformation the hydrogen atoms are so close that there is mutal repulsion, i.e. eclipsed interaction occurs thereby making the boat forms very unstable.

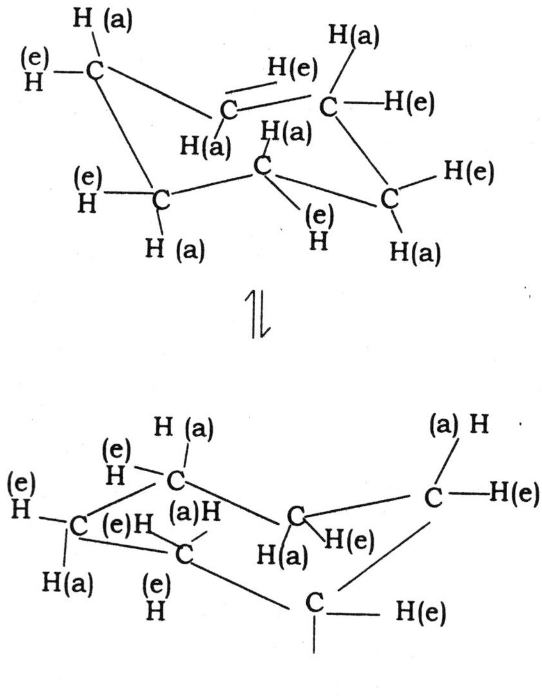

Chair conformation

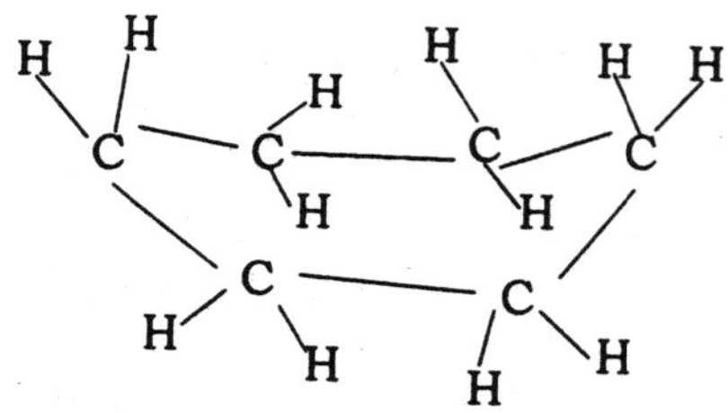

Boat conformation

Chair Skeletons Boat

The configuration of hydrogen atoms in a given chair conformation falls under two distinct categories viz. axial and equatorial. Half of the hydrogen atoms have their C-H bonds parallel to an imaginary axis of the chair form and lie above or below the plain of the ring and the other half in a belt around the periphery of the ring and their C-H bonds are parallel to an imaginary "equatorial" line drawn across the ring. These two types of hydrogen can be very easily differentiated through nuclear magnetic resonance of cyclohexane at very low temperatures. But at ordinary temperature the configuration of the hydrogen atoms of two different types of chair conformation undergo interconversion in such a manner as to have half of the hydrogen atoms as axial and the other half as equatorial.

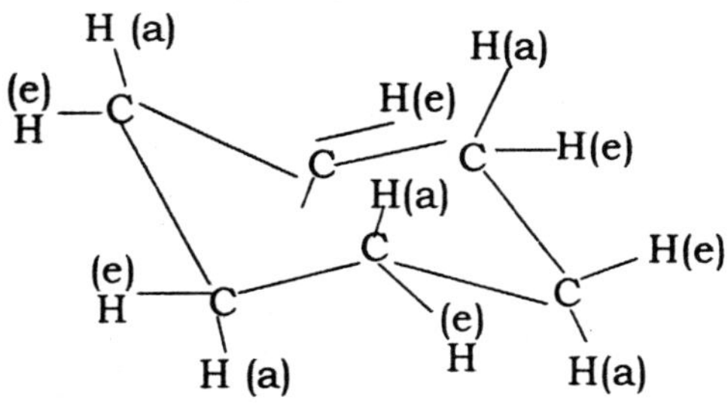

one chair

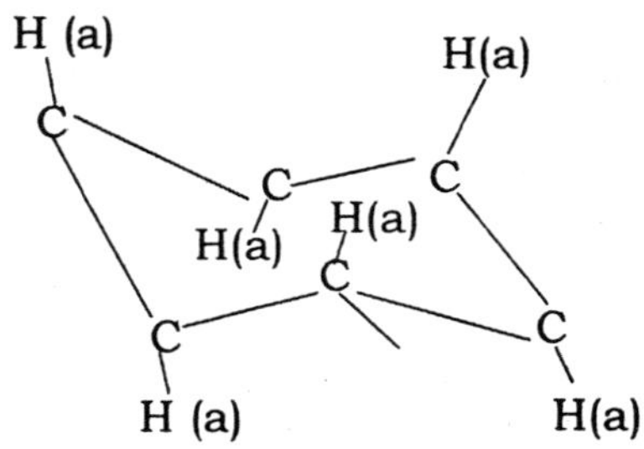

axial H's

⇌

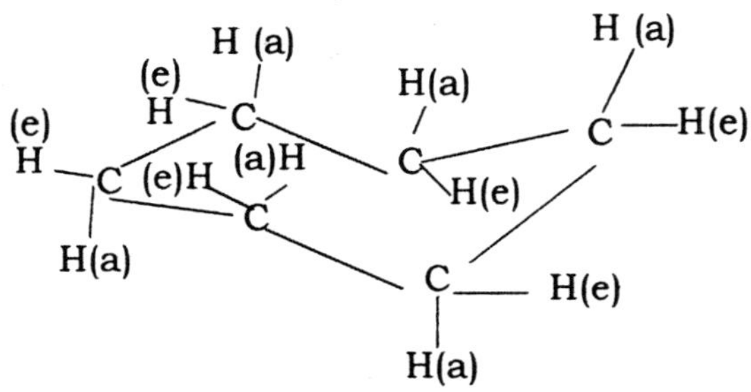

other chair

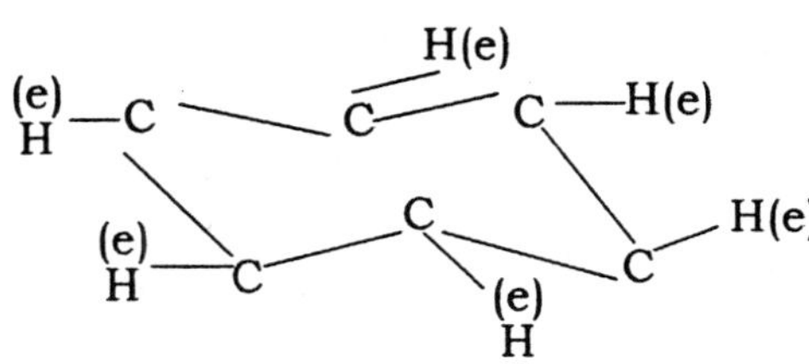

equatorial H's

Both the chair and boat forms are free of angle strain but differ in their energy content. Combination scattering spectra, infrared spectra and electron diffraction studies have shown that the cyclohexane molecules are mainly in the chair conformations. The chair conformation is, therefore, energetically more favourable, i.e. more stable than the boat form which is energetically less favourable, i.e. less stable and more flexible. One important consequence of the difference between axial and equatorial spatial arrangement is that bulky substitutents prefer to be in the equatorial conformation because such conformation is energetically more favourable thereby bringing about stability in the molecule. Also in equatorial position, the bulky substitutents are less crowded and would not bump against each other.

D-glucose and D-galactose (pyranose ring) just as the cyclohexane ring is not flat but exists in a chair or a boat conformation. Energetically more favourable is the chair form than the boat form and hence the chair form is more stable.

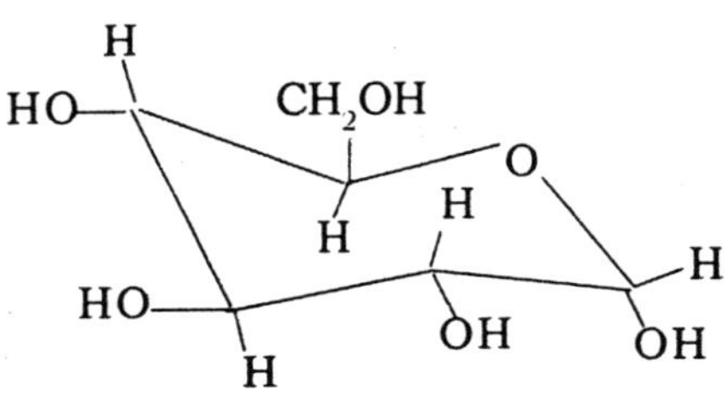

α—D—glucose

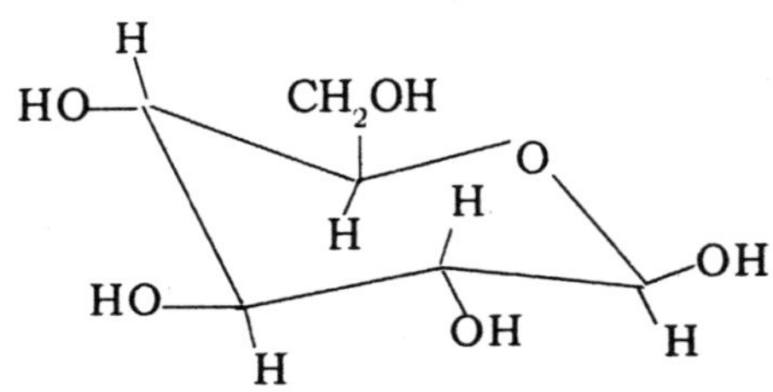

β—D—glucose

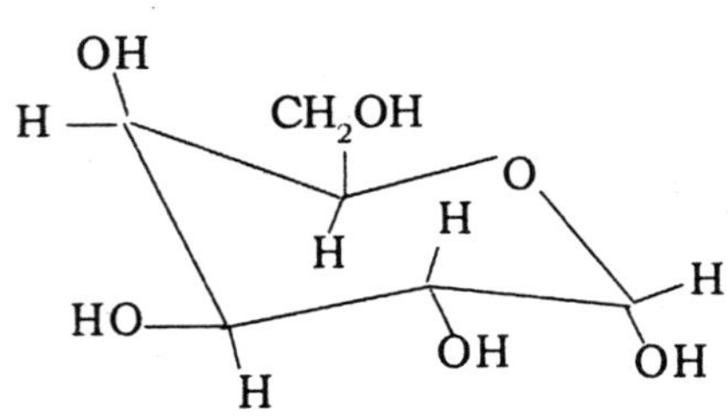

α—D—galactose

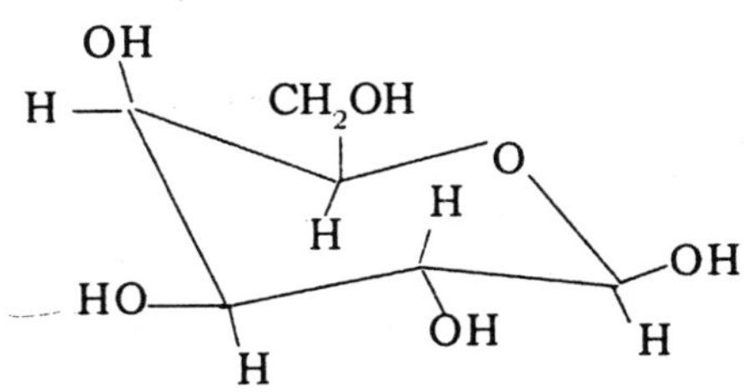

β—D—galactose

In additon to the chair conformation of six-membered rings less stable "boat" conformations are also possible. The six boat forms are all freely interconvertible through intermediate "skew" forms.

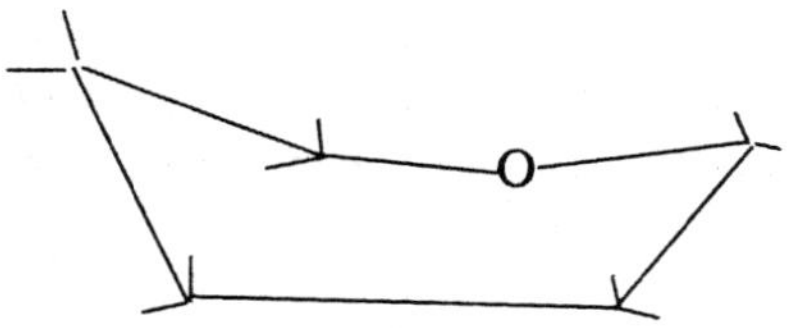

Boat

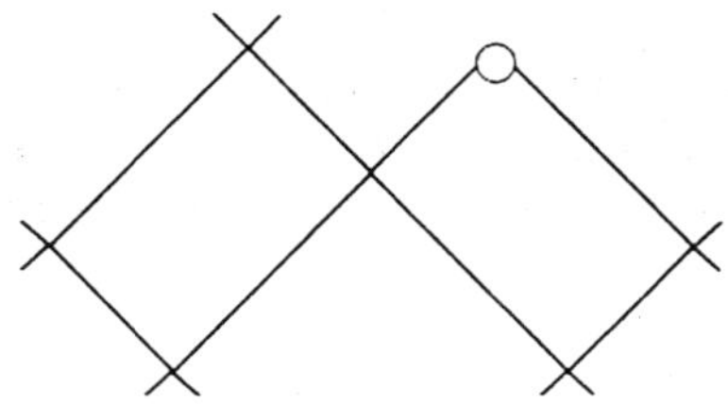

Skew

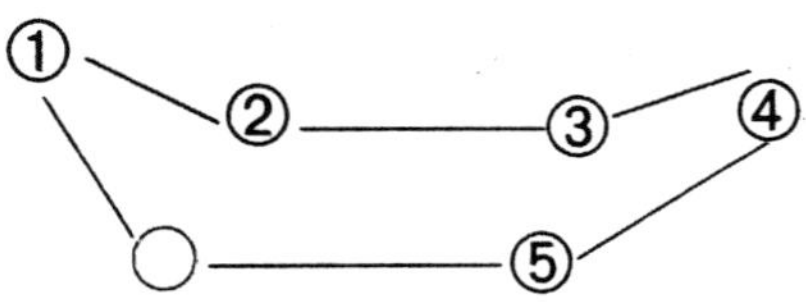

a

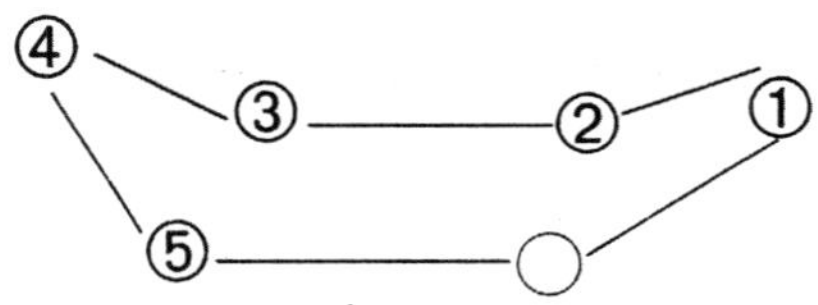

a′

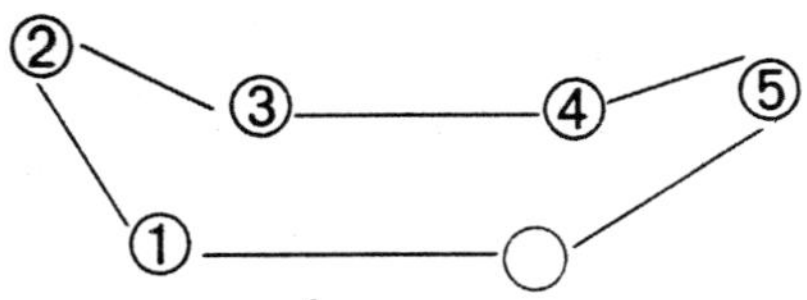

b

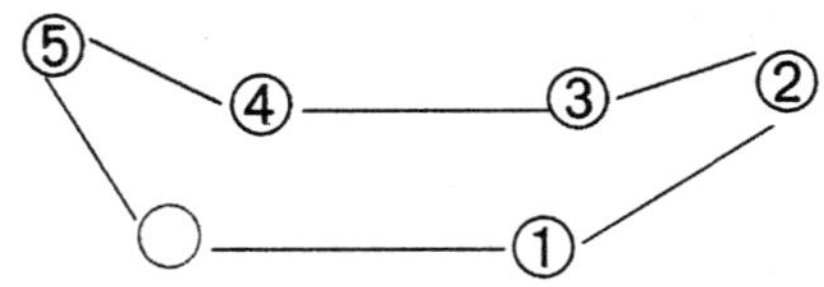

b′

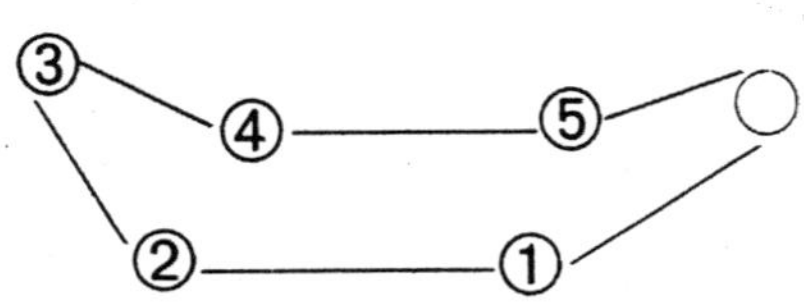

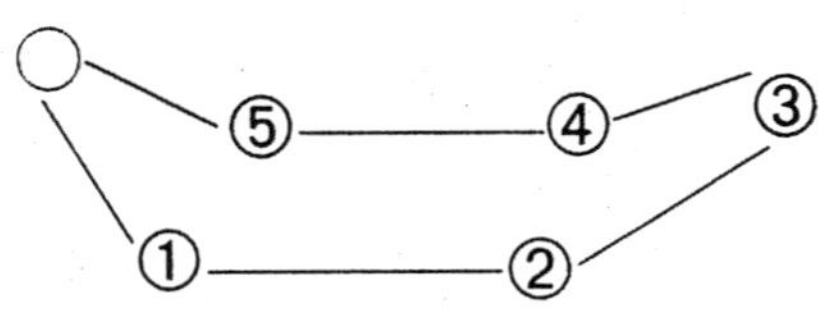

In case of the chair conformations of α-D-glucose, ß-D-glucose, α-D-galactose and ß-D-galactose it is to be noted that all bulky substituents OH and CH_2OH are in equatorial positions. They project out from the ring where there is least bumping of the bulky groups and they interfere least. All the hydrogens attached to the ring carbons project downward or upward in axial positions. Owing to the presence in the pyranose ring of an oxygen atom, which changes the ring symmetry, it may have larger number of conformations than those that exist for cyclohexane. For example, two different chair conformations are possible for glucose.

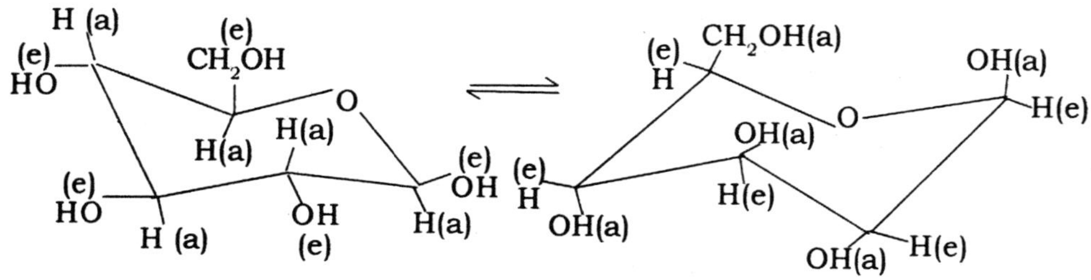

D–glucose D–glucose

The preferred conformation is determined by the ratio of the sizes and the number of the substitutents which are equatorial and axial positions. The stability of the glucose molecules depends on the fact that the bulky substitutents would take equatorial positions and hydrogen should have axial spatial arrangement. Therefore, D-glucose only exists in the conformation drawn on the left. As the carbon number one of D-glucose is anomeric, i.e. the spatial arrangement of the hydroxyl group at number one carbon atom may be α or ß, it is easy to draw the chair structure of D-glucose as follows.

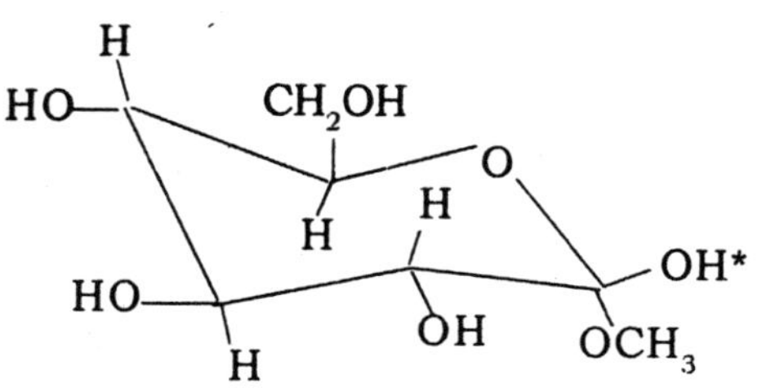

*asterisk means that the configuration in this carbon is not specified

The structure of methyl α–D–glucoside is given

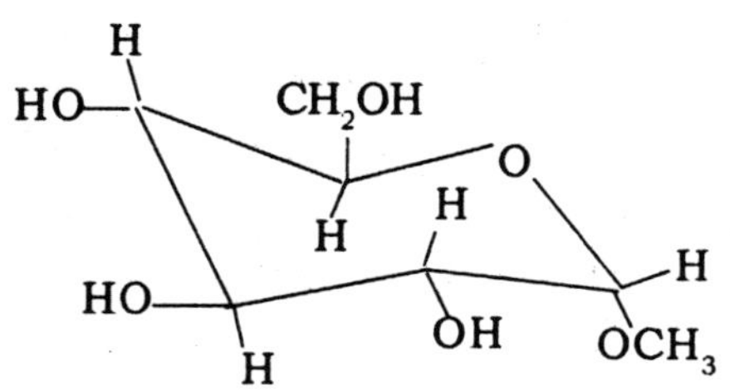

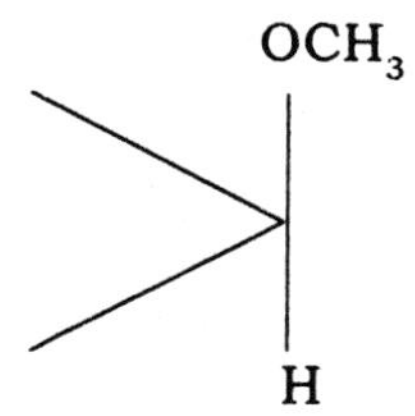

Or

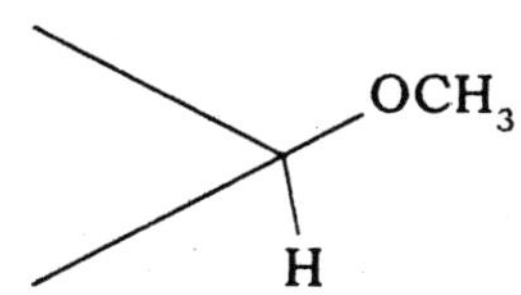

β–configuration

The systematic name of lactose, the principal milk sugar is 4-O-ß-D-galactopyranosyl-D-glucopyranose. Hence, the chair structure of lactose would look like this :

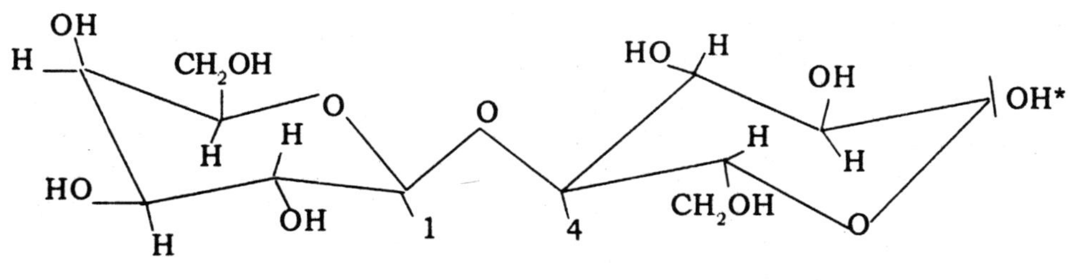

D—glucose Lactose D—glucose

*asterisk means that the configuration in this carbon is not specified

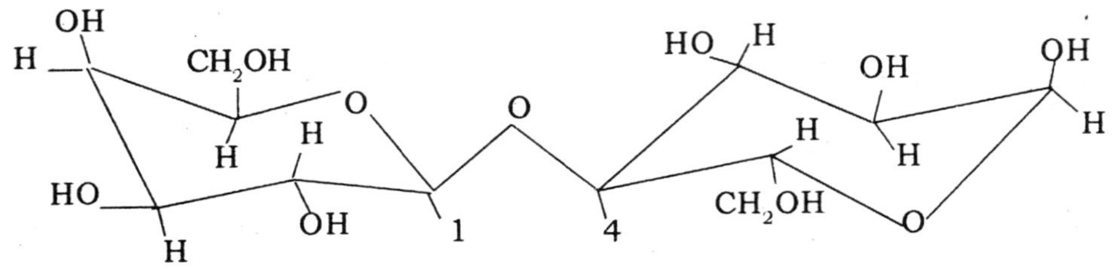

D—glucose α—Lactose D—glucose

Note—α—configuration at this "reducing end" of the molecule of α—lactose

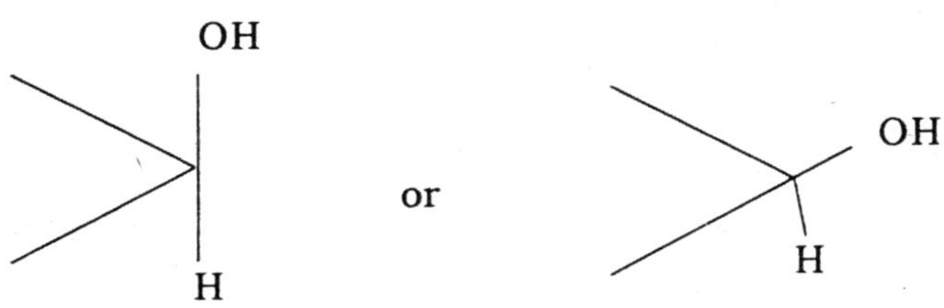

β—configuration

The α- and ß-forms of free sugars can readily undergo interconversion. However, in case of lactose, the hemiacetal hydroxyl group of D-galactose is fixed or frozen due to glycosidic linkage. In such a case the D-galactose moiety of lactose is only found in the chair form. However, the glucose ring of lactose is free and opens up to aldehyde form. In solution, therefore, the glucose ring can equilibrate to various other ring forms.

Production and Processing of Lactose

The raw material source of lactose is whey. Whey is a by-product of the production of cheese, quarg and casein in Northern Hemisphere, Latin American nations, Australia and New Zealand. In the Indian subcontinent it is a by-product of paneer and channa based sweets. The whey obtained from cheese is sweet whey. Paneer, Channa and casein are obtained by coagulation of milk at its isoelectric pH 4.6. The whey produced on such coagulation is known as acid whey. Among the many possibilities for further processing of whey, the manufacture of lactose is a process that has been made possible only by modern industrial engineering techniques. The steps in the production of lactose from whey are given in the figure.

The liquid whey, with a dry matter content of about 6.5 per cent, is defatted, clarified and concentrated in falling film evaporators in the

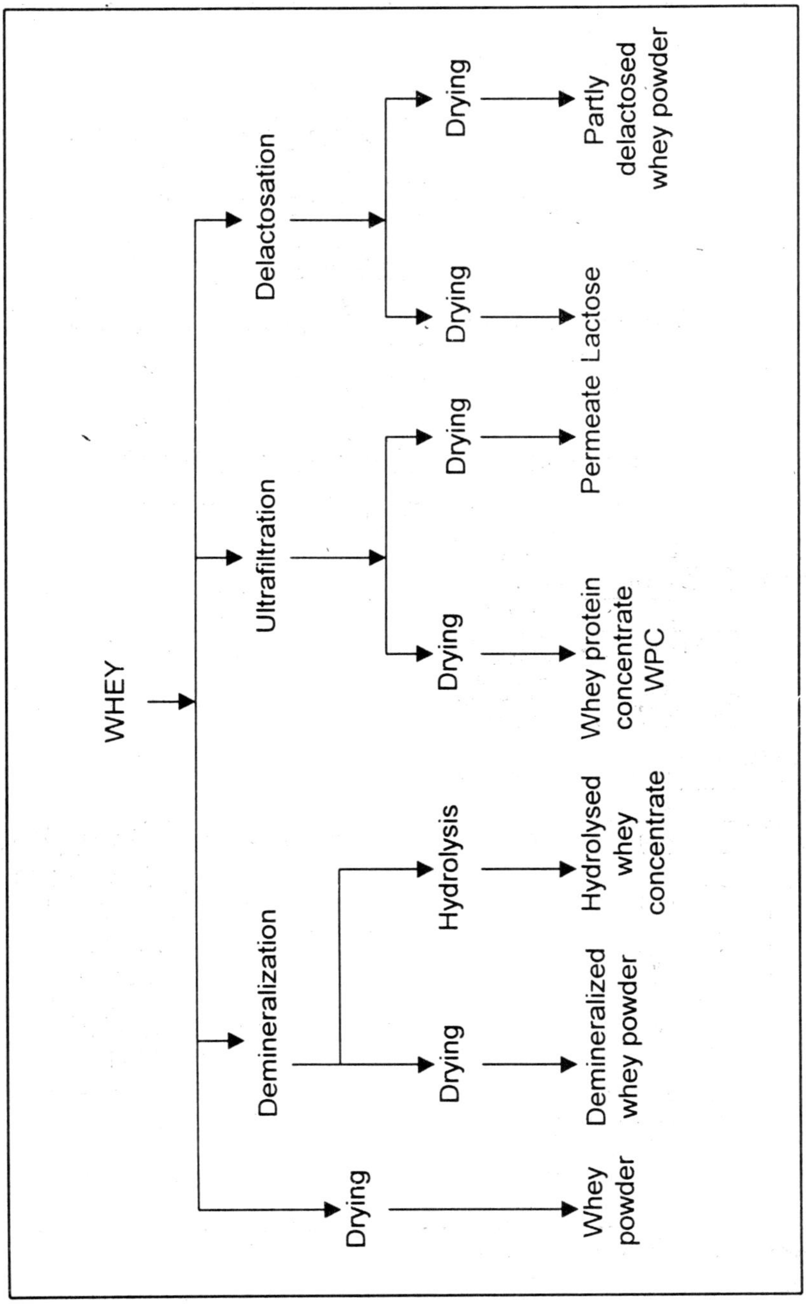

Flow Sheet-4.7 Processing of Whey

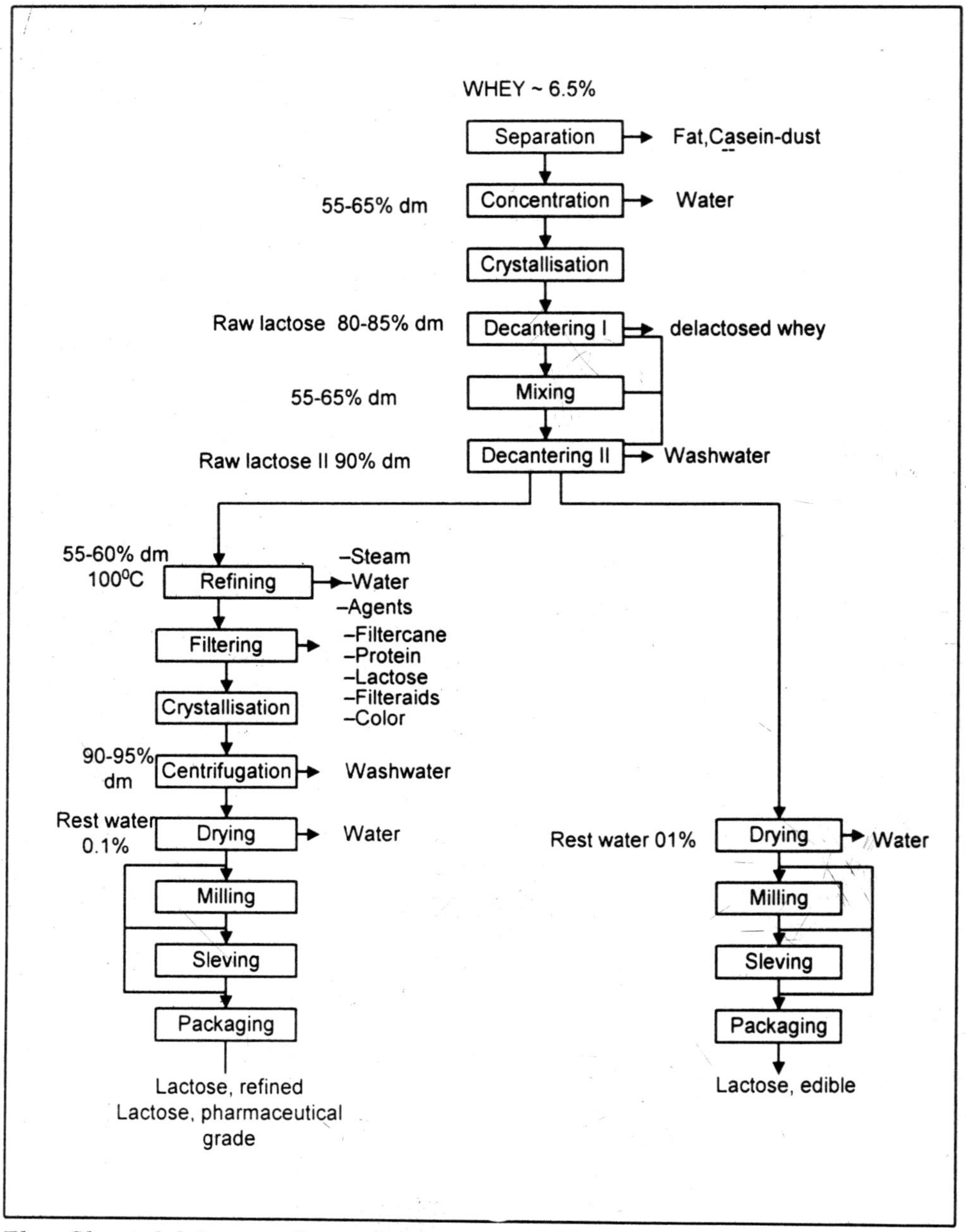

Flow Sheet-4.8 Lactose Processing

temperature range 55–65°C. During cooling, the greater part of the lactose crystallises. The crystalline product is separated in the first decanter into raw lactose and mother liquor. Through further washing processes yellow, edible quality lactose and, by recrystallisation, refined or pharmaceutically pure quality (Flow Sheet 4.7, 4.8) is obtained.

Physiological Properties of Lactose

Calcium absorption can be remarkably improved by the presence of lactose in food. An explanation for this improvement is that the formation of an acid pH in the digestive tract increases the solubility and availability of calcium for absorption. The consumption of milk with its high lactose content provides the pre-conditions for good calcium absorption.

Lactose is broken down to glucose and galactose by enzyme action only after reaching the intestines. The lactic acid resulting from the subsequent microbiological decomposition creates the required acid environment in the intestine by which putrefactive germs are suppressed.

The cause of lactose intolerance is normally a greatly reduced lactase activity in the mucosa of the small intestine. In Europe and North America, less than 8 percent of the population are affected, while in Asia, Africa and South America large parts of the population have difficulty in absorbing lactose.

The lactose fraction of milk and milk products has no, or hardly any carcinogenic effect. Lactose forms no dental plaque. With diabetics, blood sugar levels rise by only 74 mg/100 ml after consumption of lactose (compared with 146 mg/100 ml after intake of glucose). Because of this delayed resorption there are no objections to diabetics taking daily lactose in quantities of about 35–50 g spread throughout the day.

Use of Lactose

Besides it's application in modified baby foods and in tablet making, lactose is today an economically profitable sugar for the food industry in general and the confectionery industry in particular, and is a basic raw material for technical applications. Thus, the hydration of lactose to produce lactitol gains more and more importance, and processes of lactose hydrolysis using free or immobilized enzymes are currently being developed.

Pharmaceutical Industry

In its most highly purified form (see European Pharmacopoeia), lactose is one of the most suitable carriers for tablet making. Well-known producers of lactose have developed a whole range of lactose varieties which differ in granular distribution, free flowability and bulk density, to meet the increasing demands of the pharmaceutical industry. Lactose—

monohydrate with its storing and reaction stability is well established. Some producers directly offer compressible lactose qualities, such as Tablettose + Fastflow, which save the tablet producers the costly process of wet granulation.

The use of lactose in culture solutions for fermentation (antibiotic products, etc.) offer advantages due to the delayed decomposition of lactose.

Food Industry

Lactose is added to soups, sauces, sauce binders, instant drinks, spice mixes, meat products, etc. where the prime benefits of using lactose are reduced sweetness, aroma enhancement, long storage life and the acceptable price.

Reduced sweetness, fortification of aroma and, depending on application, improved shelf life are among the most important reasons for using lactose along with sucrose, dextrose and fructose in confectionery products. In chocolate and sugar confectionery in particular, lactose is an established ingredient. The delayed crystallisation brought about by adding lactose to high-sucrose-concentrate solutions offers considerable advantages in the coating of sweets and dragees.

REVIEW QUESTIONS

1. What is an assymetric carbon atom?
2. What for the symbols, D,L,d, l stand for ?
3. What is an epimer?
4. Define mutarotation, epimeric diastereomers, optical rotation, "specific rotation"
5. Why lactose acts as reducing agent?
6. Briefly describe the fermentation of lactose
7. Outline various methods of lactose estimation in milk
8. What are milk oligosaccharides?
9. Write a note on how lactose is synthesised
10. What do you understand by Browning reactions in milk?
11. What do you understand by cyclisation and boat and chair forms of sugars

REFERENCES

Fox, P.F. 1989. *Developments in Dairy Chemistry*-3. Lactose and Minor constitutents. Elsevier Applied Science. London & New York.

Jenness, R. and Patton, S. 1959. *Principles of Dairy Chemistry*. John Wiley & Sons. Inc. New York. Chapter–3, Lactose.

5

LIPIDS IN MILK

LIPIDS are one of the large groups of organic compounds which are of great importance in the food we eat. Definition of the term lipids is difficult due to the heterogenous nature of the substances it embraces. Bloor (1943) has characterized lipids as (1) insoluble in water but soluble in fat solvents such as ether, chloroform, benzene (2) actual or potential esters of fatty acids and (3) utilizable by living organisms. These three criteria are generally acceptable, although there are some exceptions to each. For example, sphingomyelins and cerebrosides are insoluble in ethyl ether and phosphatides are insoluble in acetone, fatty acid moiety of the cerebrosides is present as an amide and not as esters of fatty acids. In addition, certain lipids may contain ether linkage rather than ester linkage. Mineral oils (not considered as lipid) are utilized by strains of bacteria found around oil wells.

Classification

Lipids are classified as simple lipids (which yield fatty acids and alcohol on hydrolysis) and complex lipids (which may contain phosphate, various carbohydrates, and nitrogenous components such as sphingosine, ethanolamine, serine hexosamines).

A. Simple Lipids

1. Triglycerides (fat and oils).
2. Waxes
3. Sterol esters (sterol).

B. Complex Lipids

1. Glycerophosphatides (phospholipids) derived from α-glycerophosphoric acid, lecithin (phosphatidyl choline), cephalin (phosphatidyl ethanolamine), phosphatidyl serine, phosphatidyl inositol, plasmalogen.

2. Sphingolipids (derived from sphingosine and phytosphingosine), ceremides, sphingomyelins, cerebrosides, gangliosides, and other complex oligosaccharide derivatives of ceramides (for example, phytoglycolipid).

Lipids in milk provide a major source of energy and essential structural component for the cell membranes of the new born in all mammalian species. They also confer distinctive properties on dairy foods that affect processing. Table 5.1 shows an average lipid (fat) content in various species.

Table 5.1.-The fat content of milk from various species

Species	Fat (g/litre)
Cow	33-47
Buffalo	70
Sheep	40-99
Goat	41-45
Human	38
Pig	68
Rat	103
Rabbit	183

Chemical Nature of Milk Lipids

Milk lipids are found in three different phases of milk, namely, the fat globule, the membrane surrounding these globules, and the milk serum. The fat globules which average approximately 2 to 5 μ in size consist almost entirely of triglycerides and is coated with an adsorbed layer of material known as fat globule membrane which contains complex lipids like phospholipids and proteins. Table 5.2 presents various lipid materials in milk.

Table 5.2. Lipids in milk

Constitutents	Location in milk
Triglycerides	Fat globules
Phospholipids (lecithin, cephalin, sphingomyelin)	Fat globule-membrane and serum
Sterols (cholesterol, lanosterol)	Fat globulem, globule membrane and milk serum
Free fatty acids (various)	Fat globules and milk serum
Waxes	Fat globule
Squalene	Fat globule
Fat-soluble vitamins (Vitamin A, Carotenoids, Vitamin E Tocopherols) Vitamin D, Vitamin K.	Fat globule

Source : Fundamentals of Dairy Chemistry Webb, Johnson & Alford. Principles of Dairy Chemistry, Jennes & Patton.

The triglycerides are the major lipid class, accounting for 97-88 percent of the total lipids in many species. The triglycerides are almost invariably accompanied by small amounts of di and mono-acylglycerols, free cholesterol and cholesterol esters (10:1), unesterified (free) fatty acids and phospholipids. In addition, a number of minor simple lipids and glycolipids have been found.

Triglycerides can be defined as esters of glycerol and three moles of fatty acids. Triglycerides can be hydrolysed under suitable conditions to yield component fatty acids and glycerol.

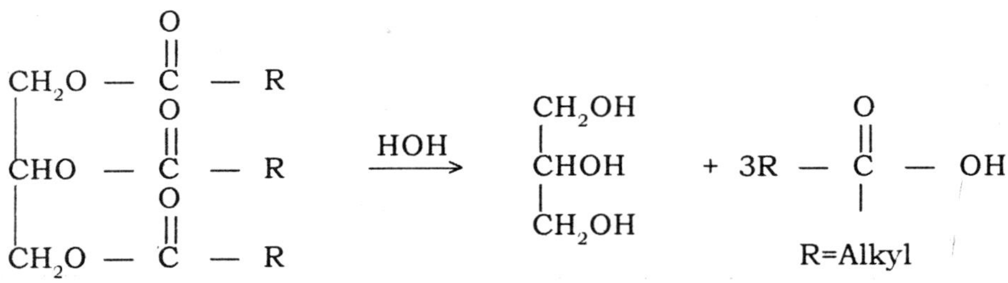

A triglyceride molecule containing only one fatty acid (R = R[1] = R simple triglyceride) would have to be a pure compound. Such fats do not naturally exist except by synthesis. Natural fats contain triglycerides composed of different fatty acids. Assuming that there are 18 different fatty acids constituting the triglycerides of milk fat, it has been shown by Sommer that 5,832 different triglycerides are possible. The random distribution of fatty acids in triglycerides indicate the complex nature of milk fat. These acids differ not only between glyceride molecule but within the same molecule as well.

Like other natural fats, milk fat also contains oleic and palmitic acids in greatest amounts representing one third and one fourth of the total respectively. According to Hilditch (1956) oleic acid found in greatest amount is mainly distributed between four types of glycerides as follows :

CH_2 oleic	CH_2 oleic	CH_2 oleic	CH_2 oleic
|	|	|	|
CH palmitic	CH palmitic	CH oleic	CH stearic
|	|	|	|
CH_2 (R)	CH_2 stearic	CH_2 palmitic	CH_2 (R)

where, R represents one of the lower fatty acid, first group is the most common out of the above four.

Fatty Acid Profiles and Glycerides

Fatty Acid Profiles of Milk Fat

Milk fat consists almost entirely of triacylglycerols or triglycerides, which constitute nearly 97-98.5 percent of the total lipid. Small quantities of di- and mono-acylglycerols are also present in traces. These glycerides are formed by the esterification of the hydroxyl groups of glycerol with the fatty acids. The nature of these fatty acids varies vastly and has a significant influence on the chemical, physical and organoleptic properties of the fat.

A list of as many as 437 different fatty acids from bovine milk has been compiled. Though such variety of fatty acids have been identified, only a handful of these are present in significant quantities and are of nutritional, chemical and physical importance. Narrowing the field further, only five fatty acids between them make up nearly 80 percent of the total fatty acids. These are oleic, palmitic, butryic, stearic and myristic acids. The remaining fatty acids are present in small quantities. Nearly 50 of the 80 commonly encountered fatty acids together make up hardly 1% of the total. Some of them have been isolated and studied while some others have been identified only on the basis of analytical data.

There are different ways of classifying the myriad types of fatty acids encountered in milk fat. The generally used classes are based on saturation (saturated, mono-unsaturated, polyunsaturated fatty acids); geometric isomerism (straight chain, branched chain); chain length (short, medium and long), etc. Polyunsaturated fatty acids are further classified as either cis or trans (geometric isomerism), or conjugated–non conjugated (positional isomerism). Not very often, fatty acids are also termed as either even or odd numbered fatty acids depending on the number of carbon atom. The odd numbered fatty acids are rare and not commonly found in milk fat. A convenient method of notation of fatty acids involves two numbers designating the carbon chain length and the unsaturation (number of double bonds). Thus, $C_{4:0}$ denotes butryric acid which has 4 carbon atom and is saturated without any double bond. Similarly $C_{18:1}$ or $C_{18:2}$ denotes oleic acids or linoleic acid having a carbon chain length of 18 with 1 and 2 double bonds, respectively. Milk fat contains a wide range of fatty acids varying in chain length from C_4 to C_{26}.

Considering the fact that fatty acids of such a wide range of molecular weights are present, their representation as weight percentage would present a distorted and erroneous picture. This is because fatty acids with a low molecular weight like butyric acid even if present in greater numbers would form a lower weight percent than high molecular weight fatty acids present in smaller numbers. To overcome this discrepancy the percentages are normally expressed as molar percentage. In this way we can know the number of molecules of particular fatty acid out of 100 molecules of the total fatty acids presents (Table 5.3).

Table 5.3. Fatty acid profile of milk fats of different species (Mole %).

Component mole %	Fatty acid	Cow	Buffalo	Goat	Sheep	Camel
Butyric	4:0	8.8	11.4	7.6	8.4	5.9
Caproic	6:0	3.5	3.1	4.5	5.4	1.9
Caprylic	8:0	2.0	1.0	6.2	5.8	1.1
Capric	10:0	3.0	1.6	11.1	10.1	2.1
Lauric	12:0	3.8	2.6	5.1	6.0	5.7
Myristic	14:0	9.9	10.6	11.2	11.8	7.9
Palmitic	16:0	26.1	30.3	21.5	20.4	28.3
Stearic	18:0	9.1	10.5	7.3	5.4	9.7
Higher Saturation	20-26.0	1.0	0.7	0.1	1.3	–
Unsaturated (Mono)	10:1–14:1	1.8	1.0	–	–	–
Lower	16:1	3.6	3.6	–	–	–
Unsaturated (poly)	18:1	26.2	21.6	24.2	22.2	34.1
		3.5	2.0	1.2	3.2	3.3

Lower Saturated Fatty Acids ($C_{4:0}$ to $C_{12:0}$)

The lowest member of this category viz $C_{4:0}$, butyric acid is the most prominent and constitute nearly half the strength in this group. Butyric acid is unique to milk fat. Other fats, particularly coconut fat, contains the other members of this group viz. $C_{6:0}$ to $C_{12:0}$ but not $C_{4:0}$.

Butyric acid attains importance due to its influence on the flavours and off flavours of milk fat. Goat and sheep milk contain nearly equal quantities of the other lower saturated fatty acids with particular accent on capric and caprylic which may be responsible for the gouty flavour. Camel milk fat contains nearly equal quantities of $C_{4:0}$ and $C_{12:0}$ (Table 5.3).

The lower saturated fatty acids are steam volatile and are responsible for the relatively high Reichert-Meissl and Polenske values of milk fat. They are also responsible for the soft texture and low melting of milk fat though it has a relatively low iodine value.

The lower saturated fatty acids are present more in ruminant milk than in non-ruminant milk fat because the former can synthesise them in the rumen from the precursor present in the diet. Of the higher saturated fatty acids only $C_{16:0}$ and $C_{18:0}$ are of any significance. The concentration of these acids is fairly constant and palmitic acid ranges from 25–30 per cent while stearic acid constitutes around 10 per cent.

Unsaturated Fatty Acids

The unsaturated fatty acids are of considerable importance in the butter fat. They contribute a lot to the flavour, spoilage/off flavour, chemical and physical properties (Table 5.4).

Table 5.4. Unsaturated fatty acids

Mono-unsaturated	Mole %
$C_{10:1}$	0.27
$C_{12:1}$	0.14 9 cis, 11 cis or 11 trans
$C_{14:1}$	0.76 seven different acids
$C_{15:1}$	Trace
$C_{16:1}$	1.77 19 different acids
$C_{17:1}$	0.28
$C_{18:1}$	29.8 16 acids; cis 9 oleic 25.5 per cent Trans 9 elaidic 0.5 per cent cis 11 vaccenic 0.67 per cent
$C_{20:1}$-$C_{26:1}$	Traces
Polyunsaturated	30 acids
Dienoic	$C_{14:2}$ $C_{16:2}$, $C_{18:2}$ (linoleic 2.11 mole per cent)
Trienoic	$C_{18:3}$ (linoleinic acid 0.4 mole per cent)
Tetraenoic	$C_{20:4}$ (Arachidonic acid 0.14 mole per cent)

The following fatty acids are present in cow's milk fat–even and odd saturated acids from 2:0 to 28:0, even and odd monoenoic acids from 10:1 to 26:1, with the exception of 11.1 and including positional and geometric isomers; even unsaturated fatty acids from 14:2 to 26:2 with some conjugated geometric isomers; polyenoic even carbon atom acids from 18:3 to 22:6 including some conjugated trans isomers; monobranched fatty acids 9:0 and 11:0 to 25:0. Some iso and ante-iso (iso acids have a methyl branch on the penultimate carbon, ante-iso, on the next to penultimate carbon atom, e.g.,

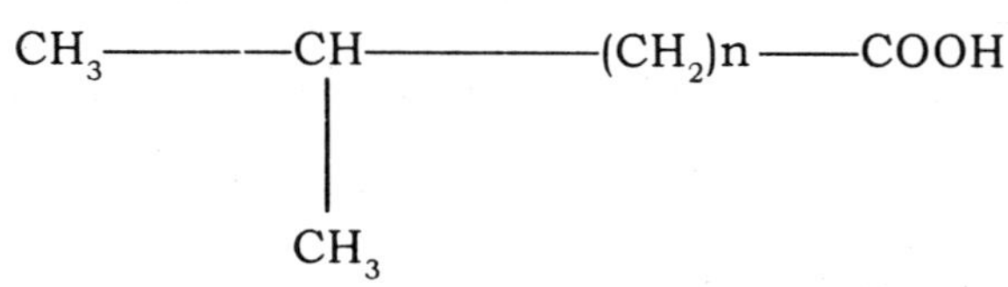

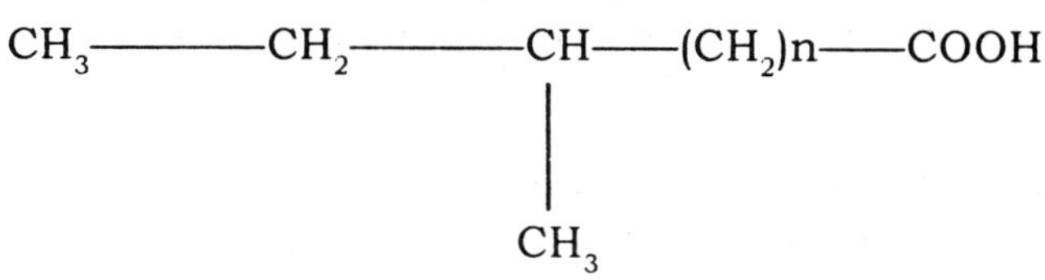

There are multi branched acids from 16:0 to 28:0 both odd and even with three to five methyl branches; and to close the list, a number of keto, hydroxy, and cyclic acids.

It is impossible to determine all of the constitutents of milk fatty acids by a normal chromatographic technique, because many of the minor component fatty acids are either not resolved or are covered by peaks of other major fatty acids.

Table 5.5. Fatty acid composition of milk fat as reported by different investigators

Richardson (1968) Wt (%)		de Man (1968) Wt (%)		Shehata *et al.* (1970) Wt (%)	
4:0	3.98	4.0	3.6	4.0	3.61
6:0	2.36	6.0	1.8	6.0	1.95
8:0	1.36	8.0	1.0	8.0	1.18
10:0	2.76	10.0	1.9	10.0	2.21
10:1	0.32	10.1	0.4	10.1	0.22
12:0	3.14	12.0	2.5	12.0	2.28
13:0(12:1)	0.14	14.0	10.1	14.0	8.79
14 br	0.12	A[1]	4.5	16.0	23.89
14:0	8.39	16.0	31.4	16.1	2.28
14:1(15 br)	1.84	B[1]	4.6	18.0	12.95
15:0	1.34	18.0	8.5	18.1	28.16
16 Br	0.35	18.1	26.5	18.2	3.78
16:0	30.05	18.2	1.2	18.3	1.91
16:1 (17 br)	2.80	18.3	1.0	others	6.75
17:0	1.00				
17:1 (18 br)	0.37				
18:0	11.74				
18:1	24.93				
18:2	1.78				
18:3 (20:0)	1.23				

A[1] refers to fatty acids with retention times between 14:0 and 16:0 B[1] to those between 16:0 and 18:0.

The Table 5.5 shows the fatty acid composition as reported by different investigators and indicates several ways of reporting the minor components.

In most natural fats the double bonds of unsaturated fatty acids occur in the cis configuration. In milk fat a considerable proportion is in the trans configuration. These trans bonds result from microbial action in the rumen where polyunsaturated fatty acids of the feed are partially hydrogenated. Catalytic hydrogenation of oils in the fat industry also results in trans isomer formation. The level of trans isomers in milk fat has been reported as 2 to 4 percent. Since the total content of unsaturated fatty acids in milk fat is about 34 percent, trans isomers may constitute about 10 percent of total unsaturation. The complexity of the mixture of different isomers is demonstrated by the distribution of positional and geometric isomers in the monoenoic fatty acids of milk fat.

Table 5.6 shows positional and geometric isomers of bovine milk lipid monoenoic fatty acid.

Table 5.6. Positional and geometric isomers of bovine milk lipid monoenoic fatty acids (wt percent)

Position of	Cis Isomers				Trans Isomers	
double bond	14:1	16:1	17:1	18:1	16:1	18:1
5	1.0	Tr			2.2	
6	0.8	1.3	3.4		7.8	1.0
7	0.9	5.6	2.1		6.7	0.8
8	0.6	Tr	20.1	1.7	5.0	3.2
9	96.6	88.7	71.3	95.8	32.8	10.2
10	–	Tr	Tr	Tr	1.7	10.5
11	–	2.6	2.9	2.5	10.6	35.7
12	–	Tr	Tr	–	12.9	4.1
13	–	–	–	–	10.6	10.5
14	–	–	–	–	–	9.0
15	–	–	–	–	–	6.8
16	–	–	–	–	–	7.5

The Table 5.7 shows the location of double bonds in unconjugated 18:2 isomers of milk lipids.

Table 5.7. Location of double bond in unconjugated isomer of milk lipid.

Cis, Cis	Cis trans or trans,	Cis Trans trans
11, 15	11, 16 and or 11, 15	12,16
10, 15	10, 16 and or 10, 15	11, 16 and or 11, 15
9, 15	9, 15 and or 9, 16	10, 16 and or 10, 15
8, 15 and or 8,12	8,16 and or 8,15	9, 16 and or 9, 15
7, 15 and or 7.12	and or 8, 12	and or 9, 13
6, 15 and or 6, 12		

The iodine value of milk fat is in the range of 30 to 35 much lower than that of lard, shortening, or margarine which have similar consistencies.

Some fatty acids with other functional groups have also been identified. These are the keto acids with a keto group prominent, among these is the keto stearic acid (C_{10}) with a keto group at 8 or 9.

Hydroxy Acids

These have a hydroxy group in addition to the terminal carboxyl group. The hydroxy groups are attached to the 4th, 5th or 6th carbon atom and are capable of forming lactones, which are important flavour constituents of butter and ghee.

Isolation of Fatty Acids

The fatty acids in milk fat are esterified with the hydroxyl groups of glycerol and exist as tri, di or mono-glycerides. Therefore, the first step in the isolation of fatty acids would be to separate them from glycerol and release them as freey fatty acids. Normally they are separated as their esters and this process is called the saponification.

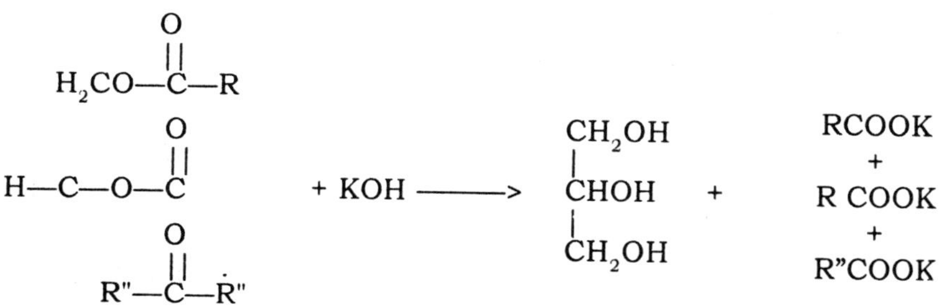

Saponification yields one mole of glycerol and one mole of each of the fatty acids as its potassium salts. This process is commercially used in the manufacture of soaps.

Alternatively, potassium hydroxide can be replaced by an alcohol like methanol. In the presence of a suitable catalyst like sodium methoxide, this reaction will yield the methyl esters of the individual fatty acids.

Separation of Fatty Acids

The fatty acids from the saponified fat can be separated by a variety of methods which include distillation, crystallisation, chromatography, etc.

Distillation: Distillation is used as a means of separating fatty acids with a wide difference in boiling points. Ordinary distillation can separate saturated or long chain fatty acids with a high b.p. Fatty acids with closer or less widely separating boiling point can be separated by fractional distillation. The lower, short chain fatty acids like butryic, caprylic, caproic, etc. are steam volatile and hence can be separated by steam distillation. In this group, the lowest members viz. butryric and caproic acids are water soluble, steam volatile and are represented by the Reichert-Meissl value while the Polenske value represents caprylic, capric, and lauric acids which are steam volatile but not water soluble. However, this is not a strict separation since some caprylic acid may also come into the former and a few higher acids may also be distilled alongwith the Polenske value estimation. Moreover, both these separations are subject to stringent specification and conditions laid out for this method.

Generally the methyl esters are preferred to the free fatty acids, since they offer several advantages. The boiling points of the esters are much lower than that of the free acid as can be seen from the Table 5.8.

Table 5.8. Boiling point of free fatty acids and their esters

	Free Acid °C	Methyl ester °C
$C_{4:0}$	163.5	102.0
$C_{6:0}$	205.0	150.0
$C_{8:0}$	238.0	193.0
$C_{10:0}$	270.0	224.0

The longer chain fatty acids have boiling points greater than 300°C which is practically difficult to achieve with normal equipment. However, the methyl esters have lower boiling point. Another advantage is that the free-COOH group is protected to a certain extent by esterification. The methyl esters are more stable towards heat than the free acids, especially in the case of unsaturated fatty acids.

Crystallisation: Advantage can be taken of the wide difference in the melting points to effect a separation of fatty acid by crystallisation. This method involves the formation of a super saturated solution of the fatty acid in a suitable solvent by dissolving the mixture in the solvent and heating and removing excess of solvent. By gradually cooling the solution the fatty acids can be crystallised out and separated from the solution, in the decreasing order of their melting points. Acetone, or petroleum ether is a suitable solvent and by stepwise lowering of the temperature upto –70°C, separation can be achieved.

Unsaturated fatty acids have much lower melting points than saturated acids and the latter can be crystallised readily from a solvent containing a mixture of the two.

Chromatography: Chromatographic separation of fatty acids is based on their relative distribution between two phases. Normally a stationary phase is solid or liquid while the mobile phase is liquid or gas. A number of techniques are available for the separation, identification and quantification of fatty acids or their methyl esters by chromatography.

Distribution of Fatty Acids in the Glyceride Structure

Glycerol forms the backbone of the glyceride structure and the different fatty acids are esterified. The distribution of the fatty acids can be studied from two angles.

i. Intermolecular distribution of fatty acids between different glycerides.

ii. Intramolecular distribution of fatty acids within the glyceride structure.

(i) Intermolecular Distribution: From the point of distribution of saturated and unsaturated fatty acids the following four possibilities can occur

	Saturated		Saturated
Saturated		Unsaturated	
	Saturated		Saturated
Tri-saturated (G S_3)		Disaturated monounsaturated (G S_2U)	

Saturated	Unsaturated
Unsaturated	Unsaturated
Unsaturated	Unsaturated
Monosaturated	Triunsatured
Diunsaturated (G SU_2)	(GU_3)

Out of these four possible structures a lot of different permutations are possible for the second and third triglyceride structure.

An approximate distribution of these different glycerides has been reported

Glyceride	Cow (mole %)	Buffalo (mole %)
GS_3	35.1	39.7
GS_2U	36.2	37.5
G SU_2	28.7	22.8

(ii) Intramolecular Distribution of Fatty Acid: Triacylglycerols are synthesised in the mammary gland by enzymic mechanisms that exert some selectivity in the esterification of different fatty acids at each position of the L-(or Sn⁻) glycerol moiety. Within the glyceride structure fatty acids may favour a certain postion. Butyric acid prefers to occupy an outer or Sn_1 position in the triglyceride in which it is present. Caproic, stearic and oleic acids also like to occupy a terminal position Sn_3 on the triglyceride. Myristic and palmitic acid, however, prefer the inner or Sn_2 position in glyceride structure. Table 5.9 shows the distribution of fatty acids esterified to each position of the triacyl-Sn- glycerols in the cow milk.

Table 5.9. Fatty acid composition (Mole % of the total)

Fatty Acid	Cow		
	Sn_1	Sn_2	Sn_3
4:0	–	–	35.4
6:0	–	0.9	12.9
8:0	1.4	0.7	3.6
10:0	1.9	3.0	6.2
12:0	4.9	6.2	0.6
14:0	9.7	17.5	6.4
16:0	34.0	32.3	5.4
16:1	2.8	3.6	1.4
18:0	10.3	9.5	1.2
18:1	30.0	18.9	23.1
18:2	1.7	3.5	2.3

Interspecies Variation of Component Fatty Acids

The milk fat of different species differs considerably with regard to the component fatty acids. These variations are broadly, typical of the type of species, like herbivorous (ruminant or non ruminant) omnivorous or carnivorous. This is because the fatty acids in milk area derived from two sources – either from the plasma lipids or from synthesis in the mammary gland. The fatty acid pool in the plasma is replenished or enriched by the diet and also occasionally from the lipolysis of the adipose tissue. Therefore, the nature of the diet has an important bearing on the nature of fatty acids in the milk fat. This is highlighted in the table 5.10.

Table 5.10. Variation of component fatty acid among different species

Species	Component fatty acid (weight percent)					
	Saturated		Mono unsatu-rated	Poly unsatu-rated	Conju-gated	Trans
	C_4-C_{10}	C_{12}-C_{18}				
Cow (herbivore ruminant)	9.2	49.1	33.1	3.8	1.1	5.5
Buffalo (herbivore ruminant)	6.3	61.0	30.5	0.9	–	–
Horse (herbivore non-ruminant)	9.4	31.6	29.9	28.8	–	–
Human (omnivore)	1.93	43.6	39.5	10.1	–	1.0–3.0

Normally a significant increase or decrease in the concentration of a particular fatty acid in the diet is reflected by a corresponding change in its concentration in the milk fat. However, ruminants have an abundant pool of micro-organisms in the rumen, which are capable of modifying or altering the fatty acids chemically. The common modifications are biohydrogenation (i.e. saturation or hydrogenation of unsaturated bonds), desaturation, conjugation etc. When the food ingested by the animal passes through the rumen, these changes take place. This factor explains the low concentration of poly unsaturated fatty acid (PUFA) in the milk fat of cow and buffalo. To overcome this problem sometimes recourse is taken to protect the fat with a protective protein coating while it passes through the rumen. These are called protected fats and by feeding them an increase in the PUFA content of ruminants can be achieved. This is important with respect of linoleic ($C_{18:2}$) acid which is an essential fatty acid. This acid cannot be synthesised in the body but is essential and has to be assimilated from the diet. That is why sometimes infant milk formulae are supplemented with vegetable fats.

Ruminants are thus capable of transforming the dietary fatty acids while non-ruminants lack that capability and can only assimilate the fatty acids from the diet. The concentration of short chain fatty acids is more in te milk fat of ruminants. This is because of the ready availability of large quantities of the precursor, ß-hydroxy butryic acid in the rumen. The ruminal microflora take up and metabolise this from the diet. The high concentration of the short-chain saturated fatty acids is a unique feature of the milk fat of ruminants. This category of fatty acids is not present in the other tissues but is seen exclusively in the mammary gland. They probably help in maintaining the liquidity of milk fat.

The fatty acid profile of milk fat undergoes remarkable changes during extended periods of fasting. Due to non-availability of dietary sources the animal will take recourse to the stored body fat or adipose tissue which gets depleted. The short-chain fatty acids decrease drastically while the longer chain acids and unsaturated acid increased sharply.

Influence of Diet

Fats and Oils: The high calorific content of fats and oils compared with alternative foodstuffs makes them an obvious choice for inclusion in diets if increased energy input is required. In practice, the supplementation of ruminant diets with fats to achieve an increase in energy intake is limited because fats at concentrations above 5 to 7 per cent of the basal ration impair digestions and metabolism in the rumen and reduce feed intake. Despite this limitation, many investigations have been undertaken to determine the effect of fat supplementation on the composition and yield of bovine milk fat.

Fat and oils in the diet of ruminants influence the lipogenic pathways either directly or indirectly leading to milk fat secretion in the following ways:

- the synthesis *de novo* of fatty acids, such as 16:0 and 18:0, by micro-organisms in the rumen may be reduced, leading to a reduction in blood lipids and mammary gland uptake and to reduction in the yield of long chain fatty acids in milk.
- The amount of long-chain fatty acids released in the rumen by lipolysis may be increased, leading to an increase in the blood lipids, mammary gland uptake and the yield of long-chain fatty acids in milk.
- The intermammary synthesis of fatty acids may be reduced thus decreasing the yield of short and medium chain fatty acids in milk. Two mechanisms have been suggested for this reduction – firstly, a reduction in the amounts of acetic acid and butyric acids produced in the rumen, leading to decreased supplies of acetate and ß-hydroxy-butyrate to the mammary gland; and secondly, an inhibition of mammary gland acetyl - CoA carboxylase.

These effects appear to be influenced by factors such as :

- Stage of lactation.
- Roughage - to - concentrate on the ratio and fat content of the basal diet.
- Amount composition and physical form of fat added to the basal diet.

Possibly the clearest insight into the effect of dietary fat on milk fat synthesis has been gained by examining the inclusion of specific fatty acids in the diet. In Table 5.11, the influence of palmitic, stearic and oleic acids on the yields of fatty acids in bovine milk fat is delineated.

Table 5.11. Influence of dietary supplements of palmitic, stearic, and oleic acids on the yields of fatty acids in bovine milk fat.

Fatty acid	Yield g/day				
	Basal diet	Basal diet + palmitic acid (4% of diet	Basal diet + stearic acid (4% of diet	Basal diet	Basal diet + oleic acid (4% of diet
4:0 to 8:0	26.6	25.9	29.8	32.4	15.6
10:0	1.9	0.5	2.2	1.4	0.6
12:0	8.0	4.2	5.2	6.2	2.4
14:0	43.3	29.1	39.7	45.8	20.6
14:1	1.9	1.9	0.9	–	–
16:0	147.0	280.7	119.5	174.1	90.0
16:1	4.9	16.2	5.2	–	–
18:0	38.4	19.9	80.7	40.4	33.9
18:1	80.9	65.6	129.9	98.8	133.1
18:2+18:3	11.0	7.9	6.0	14.6	9.3
Total	364	452	419	414	306

Protected Fats and Oils

Scientists in Australia have developed a technique to enable large amounts of fats and oils to be fed to ruminants without causing a reduction in dietary intake. This has been accomplished by treating fats or oils so that they are not released into the rumen and thus do not affect the metabolic processes in the rumen nor are they subject to ruminal lipolysis and biohydrogenation. Protection is achieved by coating minute droplets of oil with a layer of formaldehyde treated protein which resists the action of rumen micro-organisms. The oil is released in the acidic conditions prevailing in the abomassum thereby permitting its subsequent digestion and absorption. Maximum encapsulation is achieved by emulsifying the fat in a protein phase prior to treatment with formaldehyde. Direct treatment of oilseeds with formaldehyde does not give adequate protection of the lipid. Provided the release of lipid into the rumen is prevented, upto 1500 g/day of lipid can be effectively digested by the cow. Studies on serum lipids demonstrate the ability of ruminants to absorb, transport and metabolise large quantities of dietary fat without any metabolic stress.

The utilization of long-chain fatty acids, without intervention of rumen reactions, provides an energy source which can be used more efficiently than the major source of digestible energy normally available namely cellulose and other carbohydrates via the volatile fatty acids. Despite this the feeding of protected fat supplements does not always result in increased energy intake. Lactational effects and the use of inadequately protected supplements may offer explanations for this response. Diets supplemented with protected lipids produce a considerable increase in the daily yield of milk fat and amounts 15 to 30 percent greater than that from control diets have been recorded. This is a consequence of the direct absorption of large quantities of long-chain fatty acids, coupled with the lack of depression of intramammary synthesis normally associated with the inclusion of non-protected fat supplements in the diet. One of the most interesting facets of feeding protected lipids is that the fatty acid composition of milk fat is a direct reflection of the component fatty acids of the supplement in a manner similar to that found in non-ruminants. Therefore, the technique enables the fatty acid composition, and hence the physical properties of milk fat to be dramatically altered by inclusion of the appropriate fat or oil. Within 24 to 48 hrs of commencement of feeding a protected supplement containing an 18:2–rich oil, such as sunflower-seed-oil, the 18:2 content of milk fat begins to rise rapidly and may reach levels in excess of 15 percent, with the transfer of 18:2 to milk in the range of 20 to 40 percent.

The feeding of protected lipids was introduced as a means of producing bovine milk fat with elevated 18:2 levels, which may be regarded as desirable from a nutritional view point and which provides the opportunity to produce a butter which is spreadable at low temperature. Feeding of protected non-polyunsaturated fat may be used at selected times during the dairying season to maintain milk fat production and reduce weight loss.

Influence of Lactation and Season

It is agreed that when the plane of nutrition is maintained at a high level, the amount of milk secreted daily increases slightly during the first four to six weeks of lactation and thereafter declines, whereas the proportion of fat in milk initially declines and then increases later in lactation. As a consequence of these trends, the yield of milk fat gradually declines as lactation progresses.

The yield of fatty acids vary with the stage of lactation in the following manner :

- 4:0 gradually declines throughout lactation.
- The acids 6:0 to 4:0 rise from low levels at the beginning of lactation to a maximum six to ten weeks after calving and thereafter decline.
- 16:0 shows little variation as lactation progresses.

– 18:0 and 18:1 decline rapidly for the first six to eight weeks and thereafter at a slower rate.

The seasonal variation in the fatty acid composition of milk fat is affected by a number of factors such as composition of diet-plane of nutrition and stage of lactation. Studies in different contries have shown a seasonal pattern of fatty acid and variation which consistently re-occurs. In countries, where the general pattern of dairy husbandry is such that cows are grazed on pasture throughout the year lactation commences in early spring the following trends in the proportions of fatty acids of milk are observed :

– A general decline in 4:0 throughout the season.

– 6:0 to 14:0 increase from low values at the beginning of the season in early spring to a maximum value during early summer and thereafter decline through the rest of the season.

– Values for 16:0 rise during early summer and then decrease in autumn.

– Both 18:0 and 18:1 show a seasonal trend which is the reverse of that exhibited by 16:0.

The seasonal change in the fatty acids of chain length C_4 to C_{14} appears to be caused by lactational effects since the variation is the same as that mentioned above. In contrast, the seasonal trend in 16:0, 18:0, and 18:1 does not fit the lactational changes shown by these acids. Rather these appear to be diet-related. It is established that milk fat from cows grazed on mature summer rye grass, which has less lipid than new spring and autumn growth, contains lower proportions of 18:0 and 18:1 and a greater proportion of 16:0 than milk fat from cows fed immature rye grass. The finding that the seasonal trends in 16:0, 18:0 and 18:1 are a consequence of dietary effects is not surprising when account is taken of the sources of these fatty acids. Although the short and medium chain fatty acids are synthesised in the mammary glands, variations in fermentation reactions in the rumen brought about by seasonal dietary effects may also be important in leading to variations of these fatty acids in milk which may be superimposed upon lactational effects.

These findings are in contrast to results obtained in North America and Europe which generally show the following trends in the proportions of fatty acids :

– 4:0 to 14:0 show no appreciable variation throughout the season.

– 16:0 is lower in summer than in winter.

– 18:0 and 18:1 are higher in summer than in winter.

The lack of seasonal pattern in the fatty acids 4:0 to 14:0 may be attributed to the practice of not confining calving to the early spring while the variations in 16:0, 18:0 and 18:1 may result from a change in feeding conditions, since it has been shown that the change from a winter feed of hay and concentrate to a diet of fresh grass causes an increase in the proportions of fatty acids containing 18 carbon atoms and a decrease in the proportions of 14:0 and 16:0.

Effect of Cottonseed Oil on Milk Fat

Cottonseed oil, being rich in oleic and linoleic acid, occupies a middle position between saturated and polyunsaturated oils. The practice of feeding cottonseeds/cottonseed cake is quite common in certain regions of India and the milk of animals fed cottonseeds is claimed to have certain merits especially in regard to the quantitative secretion of milk and milk fat. The oil part of this feed appears to be a significant contributor to this impact on the quality and quantity of milk and milk fat. Further, the cotton tract ghee is known to differ markedly in its physico-chemical properties from ghee prepared from milk of animals not fed with cottonseeds. In fact, the cotton-tract ghee resembles, in many respects, ghee adulterated with animal body fats such as lard and tallow. This has further aggravated the already acute problems of detecting adulterants like animal body fats in ghee.

Effect of feeding cotton seed oil in conjugation with normal ration have been studied :

- There was no significant change in the level of fat percentage and milk fat yield.
- However, milk yield per cow per day increased by 0.47 kg (significant at 10 percent level) during the third week, a week after the oil feeding was stopped.
- R.M. value decreased (4 units and 6 units respectively during the first week and second week of oil feeding) significantly.
- Polenske value decreased significantly in the first week but not in the second week of oil feeding.
- A significant increase (5.7 units and 7.2 units respectively during the first and second weeks) observed in I.V. and B.R. readings also showed significant rise during the feeding period.
- However, no significant change was observed in the saponification value, acid value, unsaponifiable matter and total cholesterol contents.

- All the fat constants, except polenske value, returned almost to the normal values a week after the oil feeding was stopped.

- Fatty acids such as 10:0, 12:0, 14:0, decreased while fatty acids such as 16:0 and 18:0 + 18:1 increased. However, $C_{18:2}$ fatty acids showed no significant change.

Phospholipids

The phospholipids or phosphatides are lipids which contain phosphorus present as esterified phosphoric acid. Phospholipids occur naturally in all living cells, particularly in nerve and brain tissues and egg yolk. Cell protoplasm is a complex compound formed by the union of sterols, proteins and phospholipids. The phospholipids have a strong affinity for water. However, because of fatty nature they are soluble in fat and fat solvents. The phospholipids are classified as follows:

Phosphatides Yielding Glycerol Upon Hydrolysis :

Glycerophosphatides or phosphoglycerides – (a) phosphatidyl cholines (lecithins), (b) phosphatidyl ethanolamines (cephalins), (c) phosphatidyl serine (in cephalins fraction), (d) acetal phospholipids or phosphoglyceracetals, (e) cardiolipins.

Phosphatides Yielding Inositol Upon Hydrolysis-Phosphoinositides or Inositol Lipids :

(a) Monophosphoinositides, (b) Diphosphoinositide (from brain).

Lipids Yielding Sphingosine upon Hydrolysis (Sphingolipids) :

(a) Phosphosphingosides (sphingomyelins), (b) Glycosphingosides (cerebrosides), (c) sulfolipids, (d) gangliosides, (e) hematosides and globosides, (f) strandin.

The phospholipids of milk are of three types – lecithin, cephalin and sphingomyelin in the approximate ratio of 9:5:1. Both lecithin and sphingomyelin contains the choline base but sphingomyelin also contains another nitrogenous base sphingosine, in addition to choline. Cephalin differs from lecithin in that instead of choline as the nitrogenous base, amino ethyl-alcohol $CH_2OH.CH_2.NH_2$ is attached to phosphoric acid.

Structure and Properties

Lecithin: It is one of the most common phospholipid of milk (60 per cent of milk phospholipid). Lecithin group of phospholipid consists of glycerol, fatty acid, phosphoric acid and choline. The structure of lecithin is represented as follows :

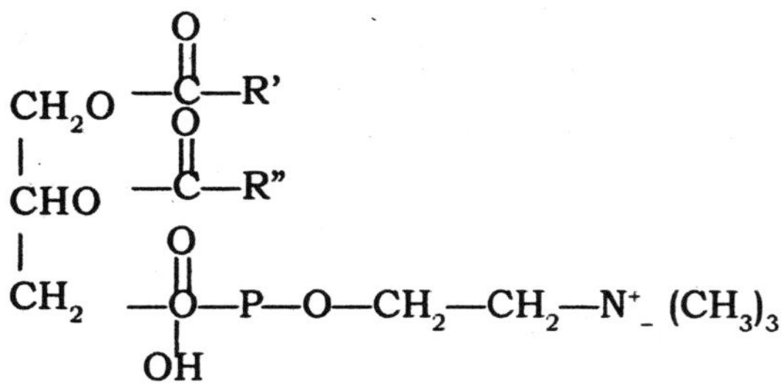

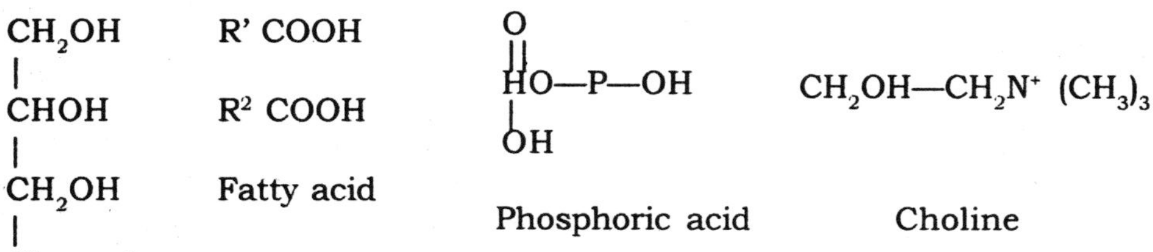

Glycerol Fatty acid Phosphoric acid Choline

Lecithins are soluble in most organic solvents. Being fat soluble and hydrophilic it acts as a good emulsifying agent and stabilises the milk fat emulsion. It forms lecithino-protein complex and brings metals in contact with fat and initiates fat oxidation. The water binding capacity assumes considerable importance in moisture relations of dried milk. Lecithin is evenly distributed in the fat and aqueous phase. It is a heat labile phospholipid. It is partly responsible for the rich flavour of butter milk and other dairy products. It contains myristic, palmitic, stearic, arachidic acid as saturated and oleic acid as unsaturated acid.

Lecithin contains a strong acidic phosphate group and a quaternary nitrogen atom of basic strength, the zwitterion formula for its structure. The isoelectric point is pH 7.5. Purified lecithin preparations moderately effective in lowering the surface tension of aqueous solution, but when lecithins are either combined chemically with, or absorbed to substances such as proteins and carbohydrates, they are remarkably active and they constitute valuable agents for the emulsification of fats and oils.

Cephalins: The cephalins are structurally identical with lecithin with the exception that the base ethanolamine is present instead of choline. The structure of phosphatidyl ethanolamine may be represented as follows :

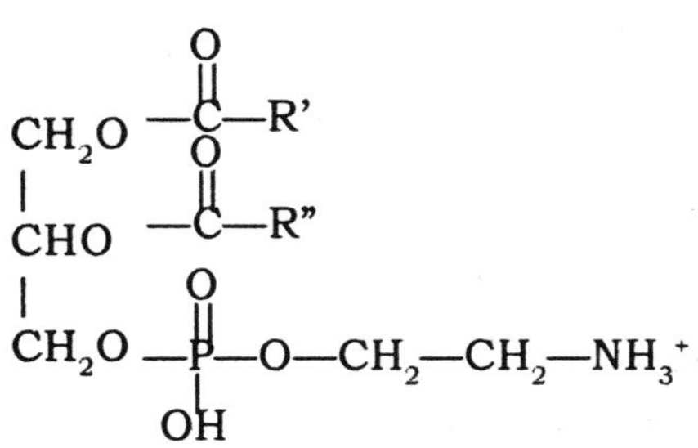

The ethanolamine group is a more weak basic than the choline of lecithin. At physiological pH, phosphatidyl ethanolamine probably exists in combination with various metal ions, since its isoelectric point should be relatively low. It is somewhat more soluble than lecithin towards the polar end of the solvent spectrum (i.e. in the methanol, ethanol, glacial acetic acid and moist ether). Like lecithin, it auto-oxidizes in air with darkening. It appears to be somewhat more stable towards hydrolysis at the ethanolamine-phosphate linkage than in lecithin at choline-phosphate linkage.

The cephalins, like the lecithin forms internal salts or zwitterions. The cephalins are hydrolysed by boiling with alkalies and dilute mineral acids.

Sphingomyelins: The sphingomyelins upon hydrolysis yield the unsaturated nitrogen alcohol, sphingosine, choline, a fatty acid and phosphoric acid. It can be represented by the formula

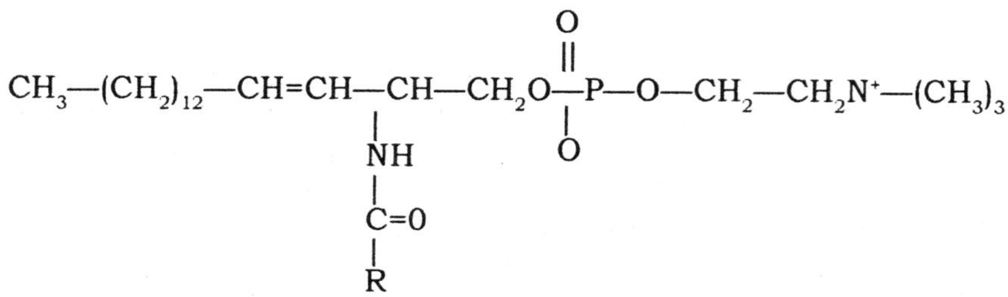

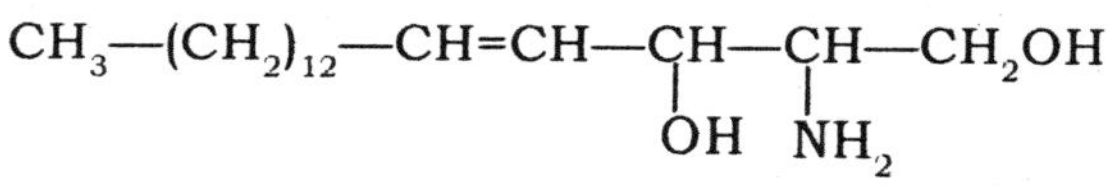

Sphengosine

Regarding basic and acidic groups, sphingomyelin parallels lecithin. The sphingomyelins are neutral compounds with isoelectric point of pH 7.0. The fatty acid range between the C_{20} to C_{24}. It is much less susceptible to auto-oxidation and is relatively stable in air. Since it is more stable, it would be expected to contain less unsaturation and lower iodine number than lecithin and cephalin. Sphingomyelin have approximately 30 iodine value. It is soluble in benzene, glacial acetic acid and warm ethanol. It has high melting point (209°C). They are present in large amount in brain and nerve tissues. They are said to form the essential parts of protoplasmic structure.

Importance of Phospholipid

According to Sommer, phospholipids are the primary stabilising force responsible for the control of fat globule size. Triglyceride molecule on synthesis is secreted by mammary gland and is soluble in aqueous medium. So, they combine with similar molecules and build up a fat globule. This is due to the Vanderwall forces. Triglycerides have no polarity and they form the building blocks for fat globule. Phospholipids are also synthesized at the same time and are present in fluid. Triglyceride has two components : main (non polar) and tail (polar). Phospholipids deposit on triglyceride molecule with an orientation, polar end towards triglyceride and non-polar towards aqueous phase.

Deposition of phospholipid is quite random and continued till the whole fat globule surface is covered by it. Further, the fat globule is sealed off. So, phospholipids are the main components responsible for preventing fat globule growth. Phospholipids can have an antioxidant effect in foods with some species being more effective than others. Although the mechanism of this anti-oxidant action is not well understood. Phospholipids may bind metals, regenerate other anti-oxidant and provide synergism with phenolic anti-oxidant via the nitrogen function. On the other hand, unsaturated fatty acids usually associated with phospholipids in just apposition with pro-oxidants in membranes are quite susceptible to auto-oxidation.

Active Oxygen Species

Singlet Oxygen

The properties of triplet-state oxygen ($3O_2$) result from the distribution of its 12 valence electrons upon application of molecular orbital theory. The atomic orbitals of oxygen (O) are combined to form the molecular orbital of dioxygen containing the appropriate bonding and antibonding orbital. The triplet nature (unpaired electrons with the same or parallel spin) of ground-state dioxygen flows necessarily from the distribution of the 12 valence electrons required by application of the Aufban and Paul exclusion principles and of Hund's rules. The two unpaired electrons degenerate orbital (equal energy) having parallel spins yield a paramagnetic, triplet-state oxygen (i.e. in a magnetic field three possible electronic configurations are observed as triplet absorption bands). Mechanistically, it is useful to think of oxygen as a diradical (O=O) with a border of two. The unique behaviour of $3O_2$ stems from the distribution of electrons whereby the absorption of energy during photo-oxidations results in an inversion of electron spin to yield singlet oxygen (O_2) in which spins are antiparallel to yield two types of diamagnetic singlet oxygen (only a singlet absorption band is observed in a magnetic field). The first type is a very energetic $\Sigma\, O_2$ (37.5 K Cal mol^{-1}) with antiparallel electrons unpaired in different degenerate orbitals, and the second type is of lower energy. The Σ state is too unstable to be of consequence for initiating oxidative reactions. An energetic Δ state is sufficiently stable to survive relxation to the bond state long enough to react with singlet organic molecules. In addition to reacting chemically, singlet oxygen can be physically quenched on interaction with solvent or bulk-phase molecules as well as by other solutes such as the tocopherols or carotenoid pigments containing at least one conjugated double bond. ß-carotein is one of the most potent quenching agents known for O2, with one molecule capable of quenching, on an average, 250 molecules of O_2. By comparison, it has been calculated that a molecule of α-tocopherol can quench approximately 40 molecules of O_2 before being destroyed. The nature of the bulk phase markedly affects the physical stability of O2. The lifetime O_2 in water (~ 2 μs) is markedly less than in non-polar (~ 45–700 μs) or lipid phases. In aqueous lipid biphasic systems, the longer half-life of O_2 in the lipid phase favours oxidation of compounds that partition into the lipids. In contrast to triplet oxygen, singlet oxygen is very electrophilic and reacts readily with double bonds in unsaturated lipids as shown below:

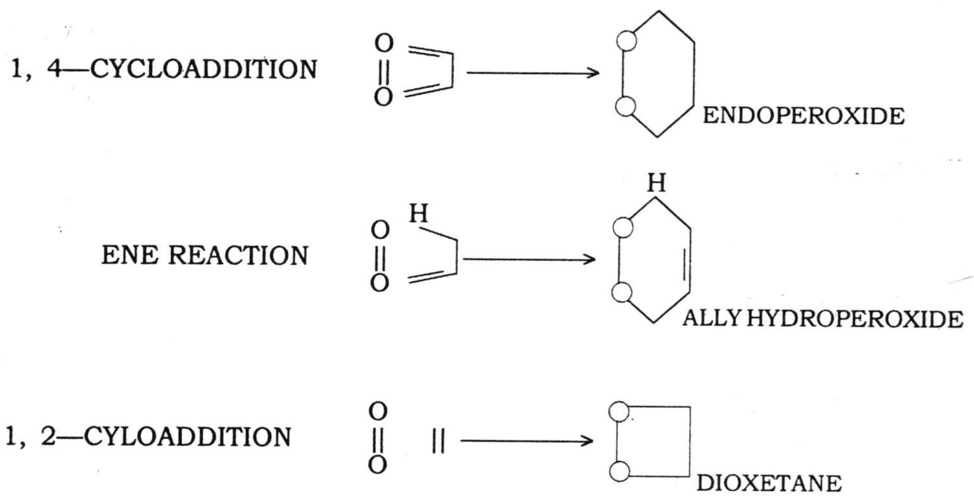

Consideration of the geometry of possible six membered transition states in the reaction between O_2 and linoleate leads to the predicted formation of conjugated or non-conjugated *cis-trans* products because of favoured geometry in the *cis-trans* reactive complex. This exothermic reaction provides a convenient mechanism whereby the oxidation of lipids may be initiated through the formation of hydroperoxides which can decompose homolytically to initiate new free-radical chain reactions. It is important to remember that an organic free radical is not further constrained by spin conservation and can now react very rapidly with abundant O2 to form peroxy radicals. Thus, initiation of oxidation by very low levels of O2 is sufficient to paridly propagate large number of reaction chains involving ground state oxygen.

In addition the electrophicity of O2 is exemplified by its rapid reaction with dienes as in substituted furans. This dienophilic character of O2 is exploited in studies where inhibition of oxidative reactions by substituted furans is considered as evidence for participation of O_2 in promoting the oxidation. Electrophillic O_2 readily reacts with the nucleophiles, such as amino groups and sulphides, common in food constitutents. In addition, the inhibition of oxidative reactions by physical quenchers that "de-energise" O_2 (e.g. ß-carotene) is often presented as additional evidence for participation of O_2 in promoting the oxidations. However, inhibition of oxidative reactions in the presence of chemical traps and physical quenchers for O_2 may not be taken as the absolute proof of its intermediate participation.

Based primarily on the effects of chemical trapping agents and physical quenchers, a singlet oxygen has been implicated in the oxidation of raw milk lipids.

Five possible ways of generating O2 in dairy products as a result of processing and storage come to mind :

- Chemically by reaction of residual hypochlorite with hydrogen peroxide, which may both appear in milk in small quantities :

  $H_2O_2 + {-OCl} \longrightarrow O_2 + H_2O + Cl^-$

- Chemically or enzymically by reactions involving metalloprotein

  A. $P \longrightarrow Mn^{+1} + O2 \longrightarrow O_2$ - like complex.

  B. Peroxidase - $OCl^- + H_2O_2 \longrightarrow O2$

- Photochemically via riboflavin or another sensitizer :

  $1\ sens \longrightarrow 1\ sens^* \longrightarrow 3\ sens^*$

  $3\ sens^* + 3O_2 \longrightarrow 1\ sens + O2$

- Self-reaction of secondary peroxy radicals

  $2R{-}HCOO \longrightarrow 1O_2 + R{-}HC{=}0 + R{-}HCOH$

- Oxidation of superoxide by a restricted number of oxidising agents can lead to O_2

Saponifiable and Unsaponifiable Matter of Lipid

Saponifiable matter of lipid are those portions of lipid which can be hydrolysed with base. These include triglycerides, phospholipids and other esters such as those of cholesterol. Free fatty acids are also included in this group sometimes because they bind the base and are disposed with the saponifiable matter. Saponifiable matter is in larger proportion than unsaponifiable matter. The unsaponifiable matter of milk is rather complex. The known principal components are cholesterol and related sterols, the fat soluble vitamins A, D, E and K, traces of squalene, unidentified waxes, and homologous series of n-alkyl methyl ketones containing odd numbers of carbons, C_3 to C_{15}.

The unsaponifiable matter is usually determined by first hydrolysing a weighed amount of fat in an excess of approximately 0.5N alcoholic base by boiling for 1 hour. The hydrolysed matter or unsaponifiable matter is then extracted by dilution with water and several time extraction with ethyl ether which is dried. The dried residue or unsaponifiable matter represent 0.80 to 0.45 percent by weight of milk fat.

Milk Fat Constants

For the purpose of characterisation, certain well known physical and chemical constants have been derived for the more common fats. These constants serve as an indication of the type of component fatty acids present in fats. They also enable the detection of fat adulteration qualitatively and in some instances quantitatively.

The characteristic milk fat values are presented in Table 5.11.

Table 5.11. Range of characteristic milk fat values

Refractive Index	1.4527 – 1.4566
Iodine number	26–35
Saponification number	210–233
Reichert-Meissl number	17–35
Polenske number	12–24

Refractive Index

The basic principle in determining refractive index concerning fat is the degree of binding of light waves passing through a liquid or transparent solid will be a characteristic for the particular liquid or solid. It is the ratio of the velocity of light in vaccum to the velocity of light in the sample medium. More generally, it is expressed as the ratio between the sine of the angle of incidence to the sine of the angle of refraction when a ray of light of a definite wave length (usually 589.3 mμ the mean of the D-lines of sodium) passes from air into fat.

In case of milk fat the constant may be determined readily with an Abbe refractometer. The reading is normally made at 40°C. The refractive index of milk fat generally ranges between 1.4527 and 1.4566. This value is low in comparison to that for other fats and oils mainly because of the greater number of saturated glycerides and short chain acids in milk fat. The refractive index of a fat is influenced by both the molecular weight and the degree of saturation of the component fatty acids. Its determination may be useful as an indication of adulteration but would seldom, in itself, be conclusive.

Saponification Number

The saponification number is equivalent to the number of milligrams of potassium hyroxide required to saponify 1 gm fat. This value may range from 210 to 233 for milk fat and more often falls within the range of 225 to 230. In essence, this constant is an indication of the average molecular weight of the fatty acid present. With the exception of coconut and palm kernel oil, this constant for milk fat is well above those for other fats and oils. Saponification value is more useful in detecting the presence of mineral oils such as liquid paraffins in ghee as they are not acted upon by alkali and such a sample does not form a homogenous solution on saponification. The saponification value is related to the molecular weight of ghee and from it can be calculated the saponification equivalent, which is the amount of ghee saponified by one gram equivalent of potassium hydroxide and is equal to 56100 divided by the saponification value.

Iodine Number

The iodine number is the number of grams of iodine absorbed by 100 grams of fat under specified conditions. This constant is a measure of the unsaturated linkages present in a fat. The iodine number for milk fat falls within the range of 26 to 35 which is low in comparison to most other fats.

The absorption of IBr (Hanus method) and ICl (Wij's method) gives a measure of the unsaturated acids present in a fat. One molecule of halogen compound is absorbed by each unsaturated linkage, and the absorption is expressed as the equivalent number of grams of iodine absorbed by 100 grams fat.

Reichert-Meissl Number

This value is the number of millilitres of decinormal sodium hydroxide or aqueous alkali solution required to neutralise the steam volatile water soluble fatty acids distilled from 5 grams of fat under the precise conditions specified in the method. This constant for milk fat is quite significant since it is primarily a measure of butyric acid and caproic acid. The value for milk fat ranges between 17 to 35 which is well above that for all other fats and oils. Milk fat contains more of these acids than does any of the fats from which it might be desirable to distinguish it.

Polenske Number

The polenske value is the number of ml of 0.1N aqueous alkali solution or deci normal sodium hydroxide required to neutralise the steam volatile water insoluble fatty acids distilled from 5 gram of fat under the precise conditions specified in the method. Caprylic acid, capric acids which are somewhat steam volatile but largely insoluble in water, are indicated mainly by the polenske number. The polenske number for milk fat ranges from 12 to 24.

As most of the volatile acids obtained from milk fat are soluble, this test helps to distinguishh between this fat and one of high caprylic acid content (e.g. coconut oil).

Kirschner Value

The Kirschner value is the number of ml of 0.1N aqueous alkali solution or deci normal alkali required to neutralise the steam volatile water-soluble fatty acids (R.M. value Distillate) which form water-soluble silver salts distilled from 5g fat under the precise conditions specified in the method.

Melting Point

Melting range from 30° to 41°C are reported in the literature for milk fat.

Oxidation and Its Control

Rancidity of Milk Fat: Milk Fat is commercially the most valuable component of milk. At the same time it is also the component which is most susceptible to oxidative spoilage or rancidity which in turn rapidly decreases the commercial value of the products. With the onset of rancidity, the normal pleasant appetising flavour and aroma of the product is replaced by undesirable and unpleasant flavours. This is due to the formation of volatile compounds which can impart noticeable off flavours even when present in very small quantities.

There are three types of rancidity development in milk fat viz. the hydrolytic rancidity, the ketonic rancidity and the oxidative rancidity.

Hydrolytic Rancidity: In this type of rancidity the fatty acids which are esterified with glycerol to form the triglycerides are broken off from the glycerol and released as free fatty acids. This leaves behind a diglyceride, or monoglyceride depending on the extent of hydrolysis. Fresh butter or ghee contains very small amount of free fatty acids (FFA) but as a result of hydrolysis, the free fatty acids level increases. This type of hydrolysis requires the presence of moisture and lipolytic enzymes, commonly known as lipase. Both these agents are commonly found in butter, but to a much lesser extent in ghee.

The nature of the free fatty acids released on lipolysis influence the rancidity. It was generally thought that butryic acid is responsible for off flavours but it is now accepted that all the lower free fatty acids from C_4 to

C_{12} i.e. butyric, caproic, capric, caprylic acids have powerful mal-odours which are responsible for rancidity. Milk fat, characteristically contains an abundance of these short-chain saturated fatty acids and is, therefore, readily susceptible to this type of rancidity. The long chain fatty acids, whether saturated or unsaturated, do not possess any distinct odour or aroma of their own, and by themselves do not significantly contribute to rancidity. They are however subject to transformation which in turn leads to off flavoured products.

Ketonic Rancidity: Free fatty acids are present in milk fat either inherently or due to hydrolysis of glycerides. These free fatty acids are more prone to degradation by moles etc. through the ß-oxidation. This deterioration is brought about by the enzymes elaborated molds like pencillium, aspergillus etc. or by bacteria like micrococci, bacillus etc. In ß-oxidation the carbon atom at a ß position with respect to the carboxyl group is activated and undergoes transformation. The methylene group loses hydrogen to form a double bond (unsaturation) which then adds on a molecule of water before losing a molecule of hydrogen to give keto acids. The keto acid further loses a molecule of CO_2 to yield a ketone which is responsible for imparting off flavours. For example butyric acid is converted to acetone as shown below :

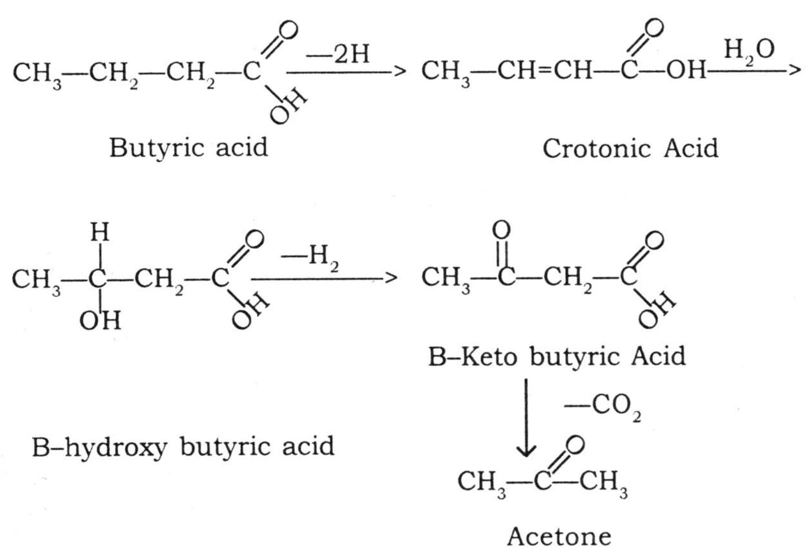

The type of off flavour imparted depends on the aldehyde or ketone produced. Glycerides on deterioration yield either saturated or mono-unsaturated aldehyde or shorter chain length which impart tallowy or oily flavours. Phospholipids on the other hand produce longer aldehyde with one or two double bonds and are responsible for metallic or fishy flavours.

Phospholipids are much more vulnerable to ß-oxidation than the glycerides.

Oxidative Rancidity: The most common type of rancidity encountered in the deterioration of fats is oxidative rancidity. Oxygen is implicated as the main causative factor in this process. The oxygen could be present either in the air in contact with the fat or in the dissolved air within the fat. The unsaturated fatty acids in the lipid are the main targets of oxidation leading to the formation of products which impart unpleasant off flavours and also cause a loss of nutritive and commercial value of the product.

Oxidation of fat is often called autoxidation due to the fact that once the process is initiated its reaction rate increases rapidly. This phenomenon proceeds through a free radical chain mechanism and can be classified into three distinct stages.

i. Initiation or formation of free radicals.

ii. Propagation of free radical chain reaction.

iii. Termination through formation of non-radical compounds.

Initiation

In the initiation stage an unsaturated hydrocarbon loses a hydrogen atom to form a radical

$$RH \longrightarrow R^{\circ} + H^{\circ}$$

Oxygen adds on to the adjacent double bond, a di-radical.

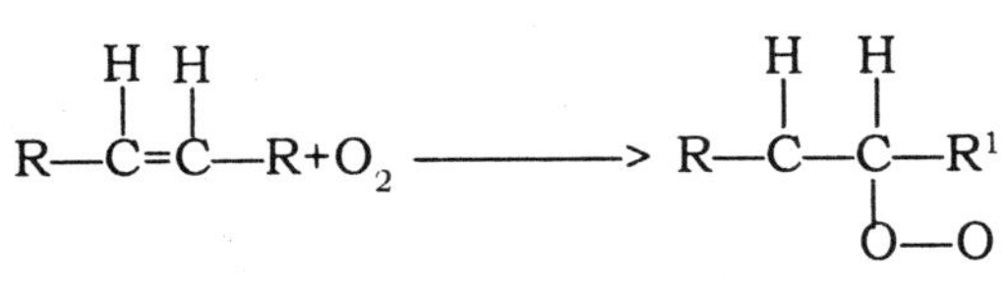

It has also been postulated that oxygen in the singlet state can directly attach between a labile hydrogen atom to form a hydroperoxide directly. This phenomenon is not very common and may not necessarily be a free radical chain mechanism.

Propagation

The hydroperoxide free radical propagates by attracting a hydrogen atom from an adjacent molecule, and thus creates a new free radical.

$$ROO^{\circ} + RH \longrightarrow ROOH + R^{\circ}$$

The new free radical formed in turn is very active and results in the formation of more peroxy radicals, hydroperoxides and new hydrocarbon radicals. The new radical formed reacts very readily with oxygen.

Termination

Termination could occur when two free radicals react.

$$R^{o} + R^{o} \longrightarrow RR$$

$$ROO^{o} + ROO^{o} \longrightarrow ROOR + O_2$$

$$RO^{o} + R^{o} \longrightarrow ROR$$

$$ROO^{o} + R^{o} \longrightarrow ROOR$$

$$2RO^{o} + 2\,ROO^{o} \longrightarrow 2\,ROOR + O_2$$

When no more free radicals are available for further reaction with oxygen, it is necessary for a new initiation reaction to occur before oxidation can continue.

Secondary Reaction

The oxidation process does not end with the formation of hydroperoxides. In fact, upto the formation of the hydroperoxides there is no loss of flavour or organoleptically perceptible deterioration of the fat. This is because the hydroperoxides are odourless substances and also quite stable. But, the hydroperoxides are readily decomposed or altered and the resulting products contribute to the development of off-flavours.

Decomposition of Hydroperoxides

Hydroperoxides are readily decomposed by high energy radiations, enzyme activity, thermal energy, or metal catalysis, depending on various factors and the system in which they exist.

When the concentration of hydroperoxide is low, decomposition could be a monomolecular reaction.

$$ROOH \longrightarrow RO^{o} + OH^{o}$$

This leads to the formation of more free radicals. Decomposition of higher concentration of hydroperoxide could be a bimolecular reaction.

$$2\,ROOH \longrightarrow RO^{o} + ROO^{o} + H_2O$$

The free radicals formed can again participate in the chain propagation. Here the more energetic ROO radicals tend to predominate. This reaction is catalysed by the presence of metallic ions, capable of existing in two valency states. The catalytic action is as follows :

$$M^{+} + ROOH \longrightarrow RO^{o} + OH^{-} + M^{2+}$$

$$M^{2+} + ROOH \longrightarrow ROO^{o} + H^{+} + M^{+}$$

$$2\ ROOH \longrightarrow RO^{o} + ROO^{o} + H_2O$$

In a system containing multivalent metal ions like Cu^{+}, Cu^{2+} or Fe^{2+}, Fe^{3+}, etc. the hydroperoxides decompose readily. Hence, extreme care is normally taken to exclude the contamination of these metal ions in stored milk fat.

The breakdown of a hydroperoxide to form RO radicals leads further to the formation of other compounds, which have been isolated from the system. The various possibilities are :

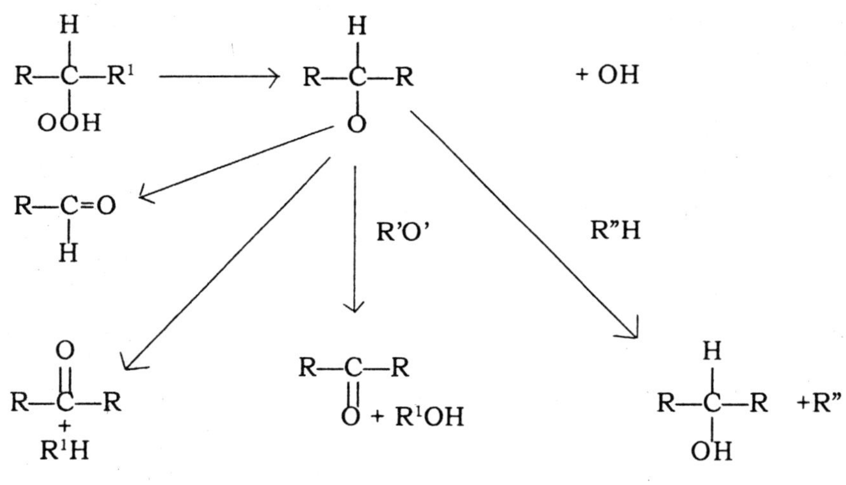

Therefore, the end products of oxidation could be aldehydes, saturated or unsaturated ketones, alcohols, hydrocarbons or semi aldehydes. These carbonyls and other products formed during autoxidation are responsible for the off flavours in oxidising the lipid system.

Measurement of Rancidity in Ghee

The deterioration of ghee follows a pattern which can be monitored by chemical tests. Before the start of spoilage, it is the induction period where there is a build up of hydroperoxides. During this phase the peroxide value increases while the iodine value decreases rapidly. This indicates that the unsaturation is being removed by oxidation leading to the formation of hydroperoxides. At the maximum autoxidation period there is a steady fall of iodine value matched by an increase in the peroxide value which then starts declining. The off flavours become noticeable with the formation of carbonyl from the hydroperoxides. These include saturated or unsaturated aldehyde, ketones, hemiacetals, etc. Out of these compounds the ketone have relatively mild flavour while aldehydes have been implicated in the deterioration of organoleptic quality.

The primary and most important method of assessing rancidity or quality is by organoleptic evaluation of taste, odour and flavour. The products which impart off flavour often have very low threshold values and can be detected by the palate if present even in very small quantities. The ghee should be acceptable to a sensitive palate. Chemical tests to measure the degree of rancidity include the peroxide value, kries test, thiobarbituric acid (TBA) test, carbonyl compounds measurement, aldehyde value and free fatty acids.

Peroxide value measures the hydroxy peroxides as the millimoles of peroxide per kilogram of ghee. In an acidic medium peroxides liberate iodine from potassium iodide, which is measured by titration against sodium thiosulphate.

Kries test measures the malonyl aldehyde $CHO–CH_2–CHO$, a secondary aldehyde produced from hydroperoxide breakdown, by the pink colour it forms with phloroglucinol in an acidic medium. 2-Thiobarbituric acid (TBA) forms a red colour with malonyl aldehyde, which can be extracted and measured at 450 or 490 nm. The free fatty acids can be measured by titration against a standard alkali, after extracting them in 95 percent ethanol.

Control of Autoxidation

After having studied the phenomenon of autoxidation the next endeavour should be to control it. There are several substances which are capable of preventing or atleast controlling the rate of auto-oxidation. These substances are called antioxidants and are effective in very small quantities.

Some of these antioxidants occur naturally and can be incorporated into the product without adverse effects. α–Tocopherol and ß-carotene are natural antioxidants which occur in many fats including ghee. It is a common traditional practice to add betel leaf or drumstick leaves to ghee during clarification/heating. By this process hydroxy charcoal (betel leaf) or gallic acid (other leaves, fruits, etc.) is imparted into the ghee during manufacture, which helps in extending it's shelf-life.

A few synthetic antioxidants are also used and have gained legal sanction. These are butylated hydroxy anisole (BHA); butylated hyroxy toluene (BHT), nor dihydroguaretic acid (NDGA), hydroquinone, catechol, pyrogallol, etc. The common feature of these compounds is that they are ortho or para substituted phenols. The presence of a tertiary butyl group in a position ortho to a hydroxy group remarkably improves the antioxidant efficiency as in the case of BHA and BHT.

Of the different antioxidants BHA is legally permitted to a level of 0.01 percent. Recently, tertiary butyl hydroquinone (TBHQ) has been reported to be very effective in preventing auto-oxidation.

The antioxidants function by the very simple and effective method of donating a hydrogen atom to free radicals and thus terminating their life. As has been seen earlier, autoxidation of fat depends on the initiation and propagation of free radicals. By providing a free hydrogen atom these free radicals can be terminated and further process of deteriorative oxidation prevented. The antioxidants provide the free hydrogen atom.

Normally a very small optimum level of antioxidant is sufficient to prevent auto-oxidation. Higher levels of antioxidant do not increase the efficiency of preservation or extend the shelf life beyond certain length. High levels of antioxidants can be counter productive by themselves becoming a source of free radicals.

Synergists

There are some compounds which by themselves possess very little antioxidant activity but are capable of significantly increasing the activity of primary antioxidants. As described under the mechanism of autoxidation, metal ion plays a major role in promoting oxidation. The synergists act by capturing or binding up the metal ions and thus prevent them from taking part in any deteriorative activities. The synergists could also function by regenerating the spent antioxidant or by preventing the antioxidant from stabilising the hydroperoxide free radical.

Common, naturally occurring synergists are phospholipids, absorbic acid, citric acid and phosphoric acid. Ethylene diamine tetra acetic acid (EDTA) is a very strong sequenstering agent and is capable of binding up metal ions. Most synergists are poly-basic acids capable of capturing or tying up metal ions. The synergists are capable of retarding autoxidation at low levels but if present in higher levels may even promote or accelerate auto-oxidation.

Triglyceride Biosynthesis by the Mammary Gland of Lactating Mammals

A quality attribute of the female mammalian species is the possession of mammary gland, a tissue unique for its multifarious metabolic activities. Cell proliferation of this gland begins at puberty due to hormonal secretion and continues till adulthood. With the ushering in of pregnancy, development of the mammary cells are dramatic and prolific mainly due to the secretion of some specific hormones. The advent of parturition is concomitant and simultaneous with the biosynthesis and secretion of milk, a phenomenon triggered by lactogenic hormones. It is, therefore, apparent that different hormones play a significant role in the proliferation of mammary cells from puberty to adulthood, in the mammary cell growth during pregnancy, and biosynthesis and secretion of milk by lactating animals.

During the growth of mammary cells from puberty till parturition, lipids are biosynthesised mainly for mammary cell proliferation. In such synthesis arterial blood brings in acetate, ß-hydroxy butyrate, propionate, glucose, chylomicra or low density ß-lipoproteins which subsequently form or help in the formation of the lipid core of the developing mammary cells of lactating mammals.

After parturition arterial blood continues bringing in acetate, ß-hydroxy butyrate and propionate, all arising from rumen fermentation and also glucose, the source of which is ingested carbohydrate and depot carbohydrate. Ingested lipids and depot fats do not fall far behind. These ingested fats and depot fats are also taken in by blood and fed into the mammary gland as chylomicra, low density ß-lipoprotein and triglycerides for subsequent metabolic conversion into milk lipids.

Basic Differences in the Lipid Biosynthesis of Ruminants and Non-Ruminants

Fatty acids which form ester linkage with glycerol, a trihydroxy alcohol, are one of the important constitutent of triglycerides. Acetyl-CoA is the key 2-carbon compound for fatty acid synthesis by the mammary gland of ruminants and non-ruminants and it now appears that the presence or absence of rumen controls its source in ruminants or non-ruminants. In ruminants, acetate, ß-hydroxy butyrate and propionate are the precursors for fatty acid biosynthesis by the mammary gland. Conversion of acetate which arises in the arterial blood from rumen fermentation into the key 2-carbon compound acetyl-CoA by the enzyme acetyl-CoA synthetase is one of the main highlights of short chain and medium chain fatty acid synthesis by the ruminant mammary gland and stands out as one of the principal points of difference between the fatty acid biosynthesis of ruminants and non-ruminants. ß-hyroxy butyrate and propionate arising from rumen fermentation are transported into the mammary gland via the arterial blood. These two precursors are converted into fatty acids comprising even and odd numbered carbon atoms respectively by the ruminant mammary gland. However, non-ruminants have very little acetate in arterial blood, and synthesise, acetyl CoA by a different pathway. In case of non-ruminants, glucose provides the carbon skeleton for synthesis of acetyl CoA. Arterial blood glucose originating from depot carbohydrates and ingested carbohydrate reaches the mammary gland where it is converted into pyruvate by the glucolytic pathway. Subsequently pyruvatepermeates through the mitochondrial membrane into the mitochondria containing all the enzymes of oxidative pathways and the electron-transport-chain and is converted to acetyl-CoA. Subsequent to the formation of acetyl CoA from pyruvate, the key 2-carbon compound condensed with oxalo acetic acid to form citrate. According to Smith and Abraham (1975), the mitochondrial membrane is permeable to citrate which then moves out into the cytosol C-soluble cytoplasm, and ultimately cleaved to acetyl CoA and oxalo acetic acid by the enzyme ATP-citrate lyase, a specific enzyme present only in the mammary gland of non-ruminants. However, Jones (1969) holds the view that citrate is not permeable to the mitochondrial membrane and it is converted to α-ketoglutarate for its subsequent transportation from mitochondria to the soluble cytoplasm where it is further metabolised to the soluble cytoplasm where it is further metabolised to citrate. It then becomes the substrate for ATP-citrate lyase action. Due to the absence of this enzyme ruminants can not produce acetyl-CoA by this pathway though glucose is present in the arterial blood of ruminants. The enzyme acetyl-CoA synthetase present in ruminants is also present in the non-ruminants but appears to be an inducible enzyme because of the absence of its substrate, i.e. acetate in non-ruminant blood is responsible for its lack of activity noticed in non-ruminant mammary gland for milk lipid biosynthesis.

About 80-100 per cent of reducing equivalents, i.e. reduced NADP requirement for fatty acid synthesis are provided by the pentose phosphate pathway in non-ruminants mammary gland. Also, reducing equivalents are provided in non-ruminants by reduction of malate to pyruvate by malic enzyme. Cleavage of citrate by ATP-citrate lyase to acetyl CoA and oxalo acetate results in the utilization of oxalo acetate through its conversion of malate which partly enters the mitochondria and is partly converted to pyruvate by malic enzymes with the concomitant production of reduced NADP, the reducing equivalent is a necessary coenzyme for fatty acid biosynthesis. The malate which crosses across the mitochondrial membrane into the mitochondria is subsequently responsible for the regeneration of oxaloacetic acid in the mitochondria thereby enabling the entire cycle to run.

As ruminants do not possess malic enzyme the reducing equivalents in case of ruminants apart from being provided by the pentose phosphate pathway are also provided by the conversion of iso-citric acid to α-ketoglutarate in presence of the enzyme isocitric dehydrogenase. In case of ruminants almost 5 percent of reduced NADP is provided by the afore mentioned pathway.

Chain Lengths of Fatty Acid Synthesised by the Mammary Gland of Ruminants

All the short fatty acids of ruminant mammary gland arise from acetate, ß-hydroxy butryate and propionate. However, the medium chain fatty acids originate from two sources i.e. these are partly synthesised by the mammary gland from acetate, ß-hydroxy butryate and propionate and also arises from the hydrolysis of blood chylomicra or low density ß-lipoproteins or triglycerides. On the other hand, the long chain fatty acids are entirely produced by hydrolytic turn over the blood chylomicra or low density ß-lipoproteins or triglycerides by the hydrolytic enzymes of the mammary gland.

Source of Glycerol for Triglyceride Biosynthesis

It has been estimated that 20 percent of glycerol arise from blood glucose which is converted to 3-phosphoglyceraldehyde and dihydroxy acetone phosphate by the glycolytic enzymes. As 3-PGA and DHAP are in equilibrium with each other in the presence of the enzyme triose isomerase and subsequent reduction of DHAP into Sn-glycerol-3-phosphate shifts the equilibrium between 3 PGA and DHAP production of DHAP. However, chylomicra or low density lipoproteins or triglycerides arriving in arterial blood from depot fats and ingested fats are the main sources of glycerol for triglyceride biosynthesis. Hydrolysis of chylomicra, low density B-lipoproteins and triglycerides produce glycerol in the ruminal surface of the mammary gland endothelial cells by the action of lipoprotein lipase. Phosphorylation of glycerol to Sn-glycerol-3-phosphate is then carried out by glycerol kinase. Sn-glycerol-3-phosphate so produced is finally utilised in the biosynthesis of triglycerides by the lactating mammary gland.

Scientific Basis of this Information

In their bid to unravel the mechanism of lipid biosynthesis by the mammary gland, scientists approached the problem in two ways. The first involved the measurement of individual enzyme activities concerned with the glycolytic and the pentose phosphate pathway. The basis for other studies was the isolation of isotopically labelled products or intermediates from specifically labelled carbon and tritium substrate.

Biosynthesis of Fatty Acids in Mammary Gland

Animals cannot convert fatty acids into glucose. Specifically acetyl CoA cannot be converted into pyruvate or oxalo acetate in animals. The two carbon atoms of the acetyl group of acetyl CoA enter the citric acid cycle, but two carbon atoms leave the cycle in the decarboxylations catalysed by isocitrate dehydrogenase and α-ketoglutarate dehydrogenase. Consequently, oxaloacetate is regenerated but it is not formed *de novo* when the acetyl unit of acetyl CoA is oxidised by the citric acid cycle. In contrast, plants have two additional enzymes enabling them to convert the carbon atoms of acetyl CoA into oxaloacetate.

Fatty acid synthesis is not merely a reversal of the degradative pathway. Rather, it consists of a new set of reaction which once again exemplifies the principle that synthetic and degradative pathway in biological systems are usually distinct. Some important features of the pathway for the biosynthesis of fatty acids are :

1. Synthesis takes place in the cytosol, in contrast with degradation, which occurs in the mitochondrial matrix.

2. Intermediates in fatty acid synthesis are covalently linked to the sulphydryl groups of an acyl carrier protein (ACP) whereas intermediates in fatty acid breakdown are bonded to coenzyme A.

3. The enzymes of fatty acid synthesis in higher organisms are joined in a single polypeptide chain called fatty acid synthetase. In contrast, the degradative enzymes do not seem to be associated.

4. The growing fatty acid chain is elongated by the sequential addition of two carboon units derived from acetyl CoA. The activated donor of two carbon units in the elongation step is malonyl-ACP. The elongation reaction is driven by the release of CO_2.

5. The reductant in fatty acid synthesis is NADPH.

6. Elongation by the fatty acid synthetase complex stops upon formation of palmitate (C_{16}) or stearate (C_{18}). Further elongation and the insertion of double bonds are carried out by other enzyme systems.

The formation of Malonyl coenzyme is the committed step in fatty acid synthesis.

Establishing the fact that bicarbonate is required for fatty acid biosynthesis has been an important clue in the elucidation of this process. In fact, fatty acid synthesis starts with the carboxylation of acetyl CoA to malonyl CoA. This irreversible reaction is the committed step in fatty acid synthesis.

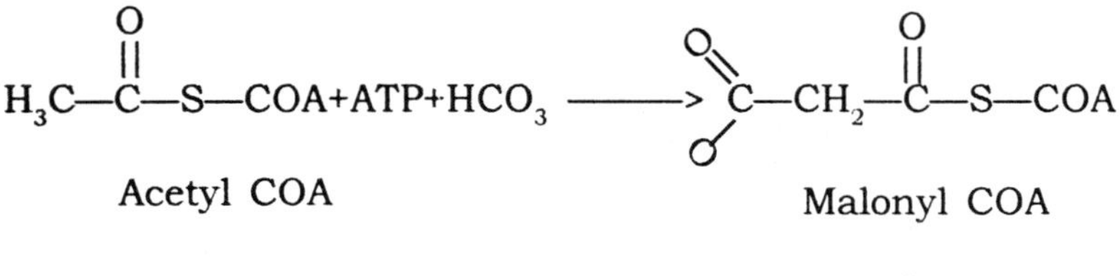

Acetyl COA Malonyl COA

+

ADP + Pi + H^+

The synthesis of malonyl CoA is catalysed by acetyl CoA carboxylase.

Carboxylase

The prosthetic group of this enzyme is biotin. The carboxyl group of biotin is covalently linked to the Σ-amino group of lysine residue of acetyl CoA carboxylase. Acetyl CoA is carboxylated in two stages. First, a carboxybiotin intermediate is formed at the expense of an ATP. The activated CO_2 group in this intermediate is then transferred to acetyl CoA to form malonyl CoA.

Biotin - enzyme + ATP + HCO_3 $\rightleftharpoons$

CO_2 ~ biotin - enzyme + ADP + Pi

CO_2 ~ biotin - enzyme + acetyl CoA ——>

malonyl CoA + biotin - enzyme

The intermediates in fatty acid synthesis are attached to an Acyl Carrier Protein (ACP). Specifically, they are covalently attached to the sulphydryl terminus of a phosphopantheine group. In the degradation of fatty acids, this unit is part of CoA, whereas, in synthesis, it is attached to a serine residue of the ACP. This single polypeptide chain of 77 residues can be regarded as a giant prosthetic group, a "macro CoA".

The enzyme system that catalyses the synthesis of saturated long-chain fatty acids from acetyl CoA, malonyl CoA, and NADPH is called the fatty acid synthetase.

The elongation phase of fatty acid synthesis starts with the formation of acetyl-ACP and malonyl-ACP.

Acetyl transacylase and malonyl transacylase catalyse these reactions

Acetyl CoA + ACP $\rightleftharpoons$ Acetyl - ACP + CoA

Malonyl CoA + ACP $\rightleftharpoons$ Malonyl - ACP + CoA

Fatty acid with an odd number of carbon atoms are synthesised starting with propionyl-ACP, which is formed from propionyl CoA by acetyl trans acylase.

Acetyl-ACP and malony-ACP react to form acetoacetyl-ACP. This condensation reaction is catalysed by the acyl-malony-ACP condensing enzyme.

Acetyl - ACP + malonyl - ACP ——> Acetoacetyl - ACP + ACP + CO_2

This condensation reaction produces a 4-carbon unit.

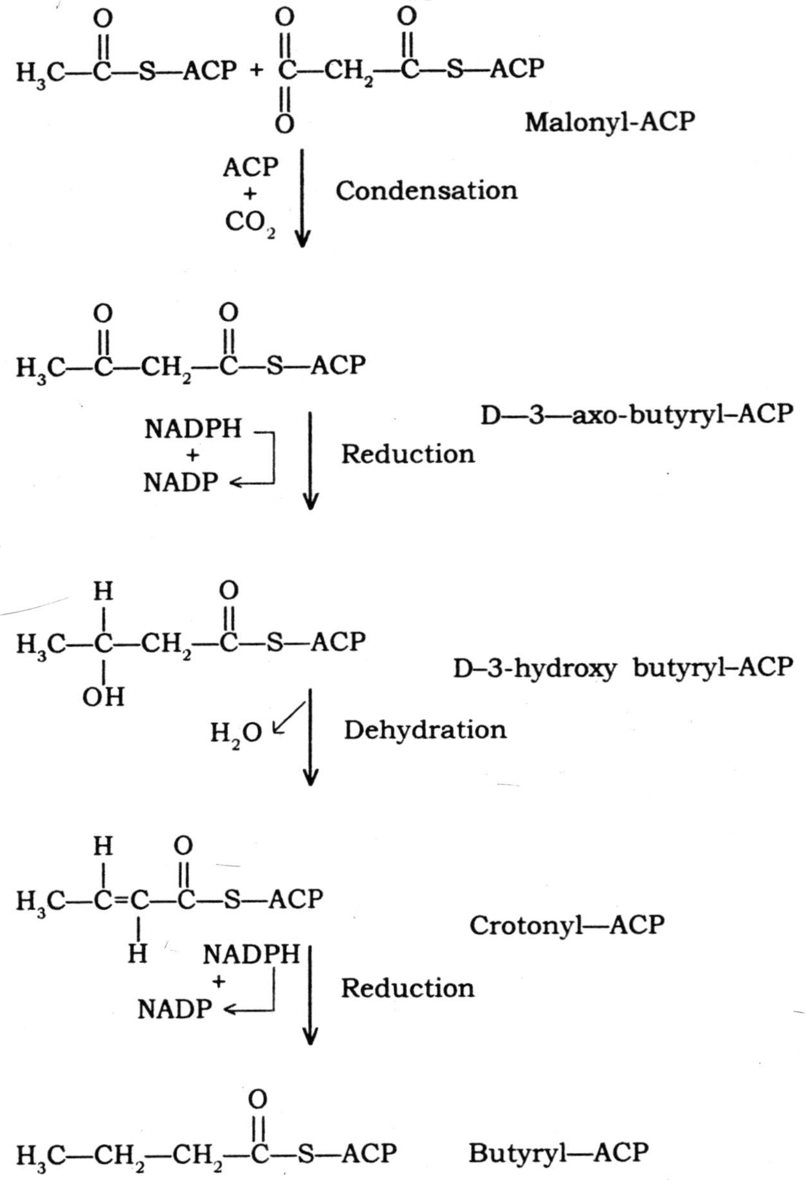

Stoichiometry of Fatty Acid Synthesis

The stoichiometry of the synthesis of palmitate is

Acetyl CoA + 7 malonyl CoA + 14 NADPH + 7H^+ ——> palmitate + 7CO_2 + 14 $NADP^+$ + 8 CoA + 6H_2O

The equation for the synthesis of the malonyl CoA used in the above reaction is

7 acetyl CoA + 7CO_2 + 7ATP——>7malonyl CoA + 7 ADP + 7Pi + 7H+

Hence, the overall stoichiometry for the synthesis of palmitate is

8 acetyl CoA + 7ATP + 14 NADPH——>Palmitate + 14$NADP^+$ + 8CoA + 6H_2O + 7ADP +7Pi

Triglyceride Biosynthesis

The principal pathway for triglyceride biosynthesis is the glycerol phosphate pathway. Another probable route for triglyceride biosynthesis is the monoglyceride pathway. However, existence of this pathway is very much doubted. Several reprots support the fact that glycerol phosphate pathway is operative in lactating mammary gland.

Smith and Abraham (1975) could not find any distinction by the enzyme acylating position 1 and 2 of Sn-glycerol-3-phosphate. They observed accumulation of diglycerides while monoglycerides were conspicuous by their absence. Studies conducted by these workers on the acylation of 1,2-diglyceride suggested no marked specificity for any particular chain length. On the specificity of acyl transferases, they observed that medium chain length acylate the Sn-glycerol-3-phosphate at position 3 while they are likely to be absent from position 1 and 2. However, their finding do not support the observed stero specific distribution of fatty acids in milk fat triglycerides.

Diagram showing the metabolic processes involved in the biosynthesis of milk triglycerides by the mammary gland of lactating mammals is shown in Flow Diagram 5.1.

Though the diagram briefly indicates almost all the metabolic processes taking place in the mammary glands of mammals, certain basic points require clarification. Arterial blood carries the precursors for lipid biosynthesis in the mammary gland of lactating mammals. Arterial blood after supplying the necessary ingredients for lipid biosynthesis transport itself out of the mammary gland as venous blood. About 400 volumes of arterial blood passes through a cow mammary gland for every volume of milk secreted. We firmly believe that probably due to a process of evolution which speaks volumes of the ability of nature to adopt to different situations, the presence or absence of rumen is a controlling factoor in the biosynthesis of acetyl-CoA and reduced NADP, i.e. the reducing equivalents in ruminants and non-ruminants. Acetyl CoA, the key-2 carbon intermediate in the fatty acid synthesis of mammals, is synthesised via an acetyl CoA, synthetase mediated reaction in ruminants. However, in non-

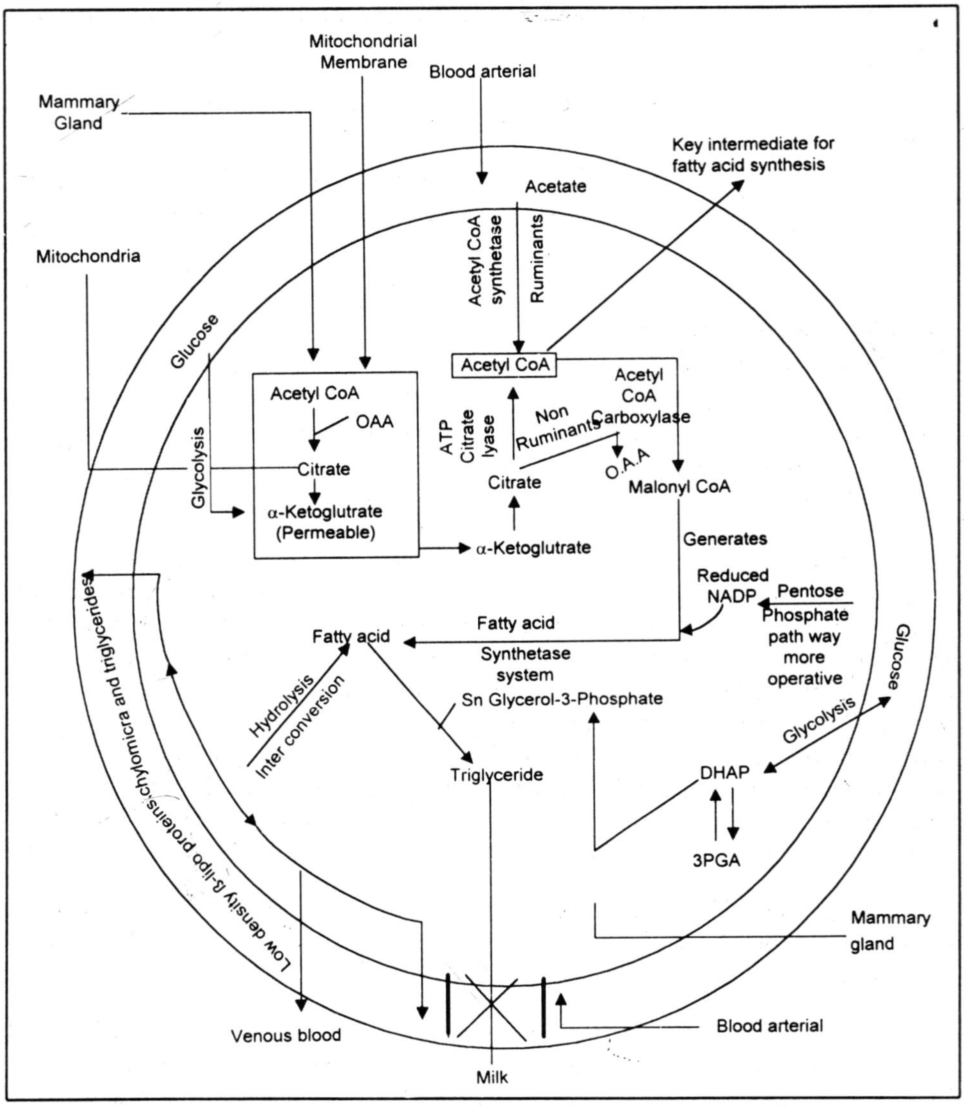

Flow Diagram 5.1 Triglyceride biosynthesis in the lactating mammary gland of mammals

ruminants, glucose is the precursor for the cytoplasmic biosynthesis of acetyl-CoA, which appears due to the split of citrate by ATP-citrate lyase, the presence of which is noticed only in the mammary gland of non-ruminants. The citrate in its turn is synthesised in the mitochondria due to the condensation of acetyl CoA and oxalo acetate. The acetyl CoA appearing in the mitochondria due to catabolic breakdown of glucose and the citrate of α-keto glutarate so formed in the mitochondira eventually crosses the mitochondira membrane and reaches the soluble cytoplasm for its ultimate metabolism into acetyl CoA. The reducing equivalents so necessary for fatty acid biosynthesis are provided by the pentose phosphate pathway and also by the conversion of isocitrate to oxalo-succinate in the presence of the enzyme iso-citric dehydrogeanse in ruminants. In non-ruminants, however, oxalo acetate, arising in the cytosol due to the split of citrate to acetyl CoA is converted to malate which finally gets reduced to pyruvate in the presence of malic enzyme with concommitant production of reduced NADP. A point worth noting is that the malic enzyme is absent in ruminants. Though not mentioned in the text, there exists a slight difference between ruminants and non-ruminants fatty acid by biosynthesis by the six enzymes of the fatty acid synthetase system.

Milk Fat Globule Membrane

Milk consists of a continuous aqueous phase with the fat distributed in it as an oil-in-water type emulsion. The fat in milk exists as globules of varying sizes, but they do not coalesce together to form larger aggregates even on standing for long periods. This is because the emulsion is stabilised and the fat globules are enveloped by a layer called the fat globule membrane. The membrane does not consist of a continuous film or sac enclosing the fat, as the term membrane suggests. Instead it consists of a well organized sequence of different components arranged according to their polarity and hydrophobicity. This layer is predominantly proteinaceous in nature, though it contains a lot of other components. These materials are absorbed and specifically oriented on the fat surface (Fig. 9).

The composition of the MFGM often varies with the method of isolation or preparation. The fat globules are separated from milk by centrifugal force. The separated globules are then washed by re-suspension in a suitable solution. The washing and harvesting is repeated a number of times to remove serum contamination. The MFGM can be released from the globule by extraction or by physical methods like freezing, thawing, agitation or sonification. Membrane fragments can than be collected by ultracentrifugation of the aqueous phase.

The composition of MFGM is not consistent and varies considerably, depending on different factors. One fact, however, remains undisputed and that is the membrane is predominantly proteinaceous in nature. Protein constitutes nearly 25-60 percent of the total dry weight. The nature of this

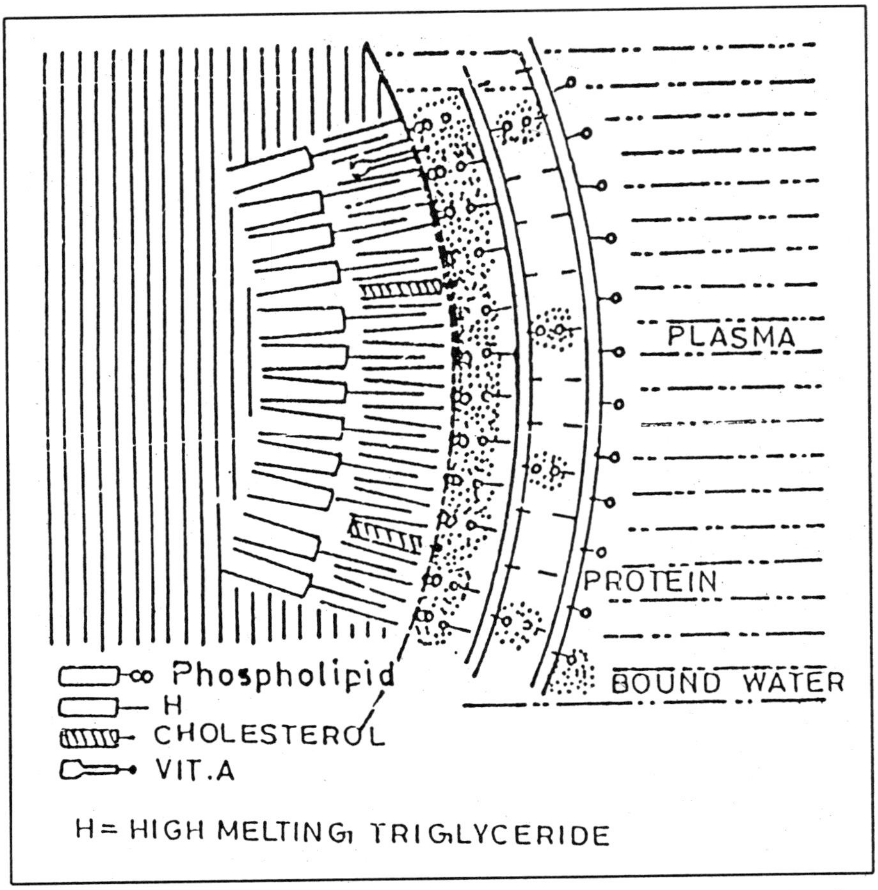

Fig. 9. Physical structure of the fat globule membrane as proposed by King

protein also varies widely and a number of enzymes have been reported at present in the membrane. Atleast 25 different enzymes have been reported and the isolated MFGM is a rich source of enzymes, including 5' nucleotidase, alkaline phosphatase; acid phosphatase, aldolase and xanthine oxidase. MFGM is the ideal starting material for isolating xanthine oxidase which has been extensively studied and characterised from the MFGM. Apart from the enzymes, the protein of MFGM could be in the form of polypeptides of varying molecular weight. The major classes identified are in the range 67,000 and 155,000. Different polypeptides varying in range of molecular weights from 11,000 to 250,000 have been reported.

Lipids are the next major constituent of MFGM and their content could vary from 0.5 to 1.2 mg per gm of protein. The protein:lipid ratio also varies considerably and different values have been reported. The lipid portion comprises of neutral lipids (56-80 percent of total lipid) or phospholipid. All the commonly occurring phospholipids have been reported. The phospholipids have an important role to play in the MFGM due to their amphiphilic nature. Glycerides constitute a major portion of the neutral lipids of the membrane. Other constitutents like cerebrosides, gangliosides, sialic acid, cytochrome and hexoses etc. have been reported.

Organization of MFGM

All the above listed materials are not distributed haphazardly but are specifically oriented on the fat surface (Fig. 9). The inner most layer consists of the high melting glycerides in contact with the fat layer. The next layer consists of phospholipids with their hydrophobic portion oriented inwards and their hydrophilic portion oriented outwards. The phospholipids are interspersed with cholesterol, vitamin A, etc. The outermost layer is made up of protein, enzymes etc. absorbed onto the surface. These proteins are in contact with the plasma and are even capable of binding water to a certain extent. The membrane constituents are very thinly spread over the surface of the fat globule and stabilises the fat in water emulsion. The breadking or disruption of this membrane also makes the lipid more susceptible to enzymatic action due to greater accessibility.

REVIEW QUESTIONS

1. Define lipids.
2. Give approximate fat content of milk of various species used for human consumption.
3. Mention the gross constituents of milk lipid.
4. Mention few examples of saturated and unsaturated fatty acids.
5. What is Isomerism?
6. What are phospholipids? Mention different phospholipids alongwith their constituents.
7. Define refractive index.
8. What is Iodine number?
9. What is rancidity? Briefly mention types of rancidity.
10. Give an account of milk fat globule membrane.
11. Describe fully the triglyceride biosynthesis in the mammary gland of mammals.

REFERENCES

Bactor, B.S. and Narayanan, K.M. 1974. *Indian J. Dairy Sci.* **27**:90.

Bloor, W.R. 1943. Biochemistry of the fatty acids. Reinhold, New York.

Chandravadana, M.V. and Dastur, N.N. 1976. Milchwissenschaft **31**:26.

Hilditch, T.P. 1956. The chemical constitution of natural fats. 3rd ed, pp. 135, 574 John Wiley & Sons, New York.

Jain, M.K. and Bindal, M.P. 1968. *Indian Dairy Sci.* **21**:10.

Smith, S. and Abraham, S. 1975. *Adv. in Lipids Res.* **13**:195.

Webb, B.H., Johnson, A. and Alford, J.A. 1978. *Fundamentals of Dairy Chemistry.* The AVI Publ. Co. Inc. Westport, Connecticut, Chapter.

Fox, P.F. 1989. *Developments in Dairy Chemistry* -2, Lipids. Elsevier Applied Science, London and New York.

Jenness, R. and Patton, S. 1959. *Principles of Dairy Chemistry.* John Wiley & Sons, Inc. New York, Chapter-2, Lipids.

6

MINERAL, VITAMINS AND MINOR CONSTITUTENTS OF MILK

Minerals

Distribution of Major Minerals in Milk

SALT constituents of milk are mineral materials including metals and inorganic radicals. In addition to this,some organic acid radicals are also present. A distinction is usually made between the major salt constituents of milk and the trace elements. The major salts include potassium, sodium, magnesium, calcium, phosphate, citrate, chloride, sulfate and bicarbonates.The minor salts include all other minerals and salts which are in trace amounts. Milk salts particularly calcium and phospahte, are of tremendous importance in nutrition. The physical state and physical stability of milk proteins, particularly the caseinate is dependant on the composition of the salt system. Heat coagulation of evaporated milk and rennet coagulation of milk depends on salt composition. Metallic elements like copper and iron catalyse oxidation of milk lipids, which leads to undesirable flavours. During ashing or incineration of milk, organic compounds are destroyed leaving the mineral constituents behind. Ash is not the true representative of salt composition but gives a rough idea of the total mineral contents. Organic radicals like citrate are destroyed during ashing. Phosphorus, and sulfur of milk proteins and lipids appear in ash. Carbonates appearing in milk ash arise in part from carbon dioxide produced by the decomposition of the organic materials. During ashing oxides of metals are formed as a result of oxidation. Some salts may be lost by volatilisation during ashing.

Table 6.1. Salt composition of milk

Constituent	Average content mg/100 g	
	Buffalo milk	Cow milk
Sodium	44.63	58.0
Potassium	103.0	140.0
Calcium	175.5	123.3
Magnesium	19.25	11.1
Phosphorus (Total)	97.6	95.1
Phosphorus (Inorganic)	–	–
Chloride	62.84	104.5
Sulfate	–	10
Carbonate (as CO_2)	–	20
Citrate (as citric acid)	163.6	176.0

Salt composition of milk is influenced by various factors such as breed, individuality of animal, stage of lactation, feed, infection of the udder, season of the year and bacterial action.

Trace Elements in Milk

Normal milk contains several elements in trace amount. These trace elements include metals and metalloids. Major trace elements are boron, copper, iron, zinc and sulphur (Table 6.2). In addition to these, detection of several other elements have not been ruled out.

Table 6.2. Trace elements in milk of three species

Elements	Cow (μg/litre)	Buffalo	Goat
Boron	100-400	0.82 (ppm)	–
Zinc	1000-6000	650 (μg/litre)	3.87 (ppm)
Iron	About 300	2404 (μg/litre)	–
Copper	0.2-1.4	212 (μg/litre)	–
Sulphur	–	239.4 (ppm)	–

These include arsenic, silicon, fluorine, bromine, iodine, aluminium, arsenic, barium, chromium, cobalt, lead, lithium, manganese, molybdenum nickel, rubidium silicon, silver, strontium, tin, titanium and vandium (Table 6.3).

Table 6.3. Average value of some trace elements in milk

Element	Average value
Chlorine	119 mg/100 ml
Sulphur	30 mg/100 ml
Iron	3 ppm
Zinc	3 ppm
Silicon	2 ppm
Copper	0.3 ppm
Fluorine	0.15 ppm
Aluminium	Traces
Manganese	Traces
Iodine	Traces
Boron	Traces

Data is lacking on the specific form and concentration of these elements. One of the main difficulty is the presence of trace amount of these elements in milk and the suitable method of detection and estimation in ppm concentration. Spectrophotometric, atomic absorption spectrophotometry and colorimetry are usually employed for their detection. These trace elements gain their entrance in milk through animal (by feed) or by subsequent contamination. Metal utensil and equipment are especially important sources of some elements such as copper, iron, nickel and zinc in milk and milk products. Most of the information on trace elements is available from cow milk of western origin. The status of trace elements in milk of Indian breeds is not well established. However, available data are presented in the Table Nos. 6.2 and 6.3.

Copper and iron are related to oxidative defects of milk lipids. They also form complexes with protein. Iron is also a constituent of enzyme xanthine oxidase, peroxidase and catalase which are found in milk. They are also bound to some extent by the fat globule membrane complex.

Salt Balance - Significance and Factors Affecting Salt Balance

Salts in milk exist in colloidal and soluble form. A large amount of salt is present in colloidal particles containing casein, calcium, magnesium, phosphate and citrate. Colloidal salts remain in equilibrium with the dissolved salt (diffusate form). Various treatments of milk may cause transfer of salts from one phase to the other. Various methods have been used to study the partition of salts between dissolved and colloidal phases. Method itself must not cause any transfer of dissolved constituents to the colloidal state or vice versa nor it should alter the equilibrium between the dissolved and colloidal state. The principal methods used for this study are pressure ultrafiltration equilibrium dialysis, high speed centrifugation, and rennin or papain coagulation. Pressure ultra filtration involves forcing dissolved phase of salts of a sample of milk through some fine pore filter having pore size to prevent smaller colloidal particle to come out. Equilibrium dialysis involves dialysing a small amount of water against a large amount of milk until equilibrium is established. This can be conveniently done with cellophane membranes. High speed centrifugation can be used to sediment the colloidal particles and leave the dissolved phase as a supernatant. Milk can be coagulated with rennin or papain to separate the colloidal caseinate complex and sera contains the whey proteins or serum proteins.All the four methods of partition of the system gives reasonably concordant result on the distribution of salts between the two phases.

Recent views on partition of milk salts indicate that about two-thirds of the calcium, and one-third of the magnesium and one-half of the Pi are colloidal in a typical milk. It is presumed that all colloidal Pi is present in the micellar calcium phosphate and it appears that all the colloidal citrate is similarly located whereas colloidal Ca and Mg are partly incorporated in milk colloidal phosphate and partly bound in a more direct manner to casein. A small proportion of colloidal Ca is bound to α-lactalbumin. β-lacto globulin can also bind Ca and Mg and there are other minor calcium binding protein. Nearly all the Na, K and Cl is diffusible (Table 6.4).

Table 6.4. The partition of milk salts

	Concentration (mM)		
	Skim milk	Diffusible	Colloidal
Ca	30.3	10.0	20.3
Mg	5.2	3.3	1.9
Cit	9.5	9.1	1.0
Pi	21.4	11.9	9.5
Na	22.8	21.8	1.0
K	39.7	37.5	2.2
Cl	31.5	31.5	0.0

The principle dissolved salt constituents of milk consist of phosphate, citrate, chloride, sulphate, bicarbonate, sodium, potassium, magnesium and calcium. At the pH 6.6 of milk, sodium and potassium are not found in any combination withother constituents and are present as the cations Na^+ and K^+ while the chloride and sulfate being anion of strong acids are present as free ion Cl^- and SO_4. However, the salts of weak acids phosphate, citrates and carbonates are distributed among various ionic forms. Calcium and magnesium form soluble complex ions with citrates, phosphates and bicarbonates. Various factors affect salt equilibria. These include effect of (a) temperature, (b) variation in acidity, (c) variation in CO_2 content, (d) concentration, (e) sequestering agent and ion exchanger.

Shifts in salt balance among various forms occur as milk is subjected to various cooling and heating treatments after it is drawn from udder at 37°C. Milk is essentially saturated with calcium phosphate and since the solubility of calcium phosphate decreases as the temperature is raised, it is not surprising that these materials are transferred from the dissolved to the colloidal phase with a rise in temperature. Addition of acid to milk either directly or indirectly by bacterial action or by heat treament causes pronounced shifts in the salt equilibria. As the pH decreases by adding acid, calcium and phosphate are withdrawn from the colloidal particles until at about pH 5.2, all the calcium and phosphate is in the dissolved state. Milk when secreted contains about 20 mg of CO_2 per 100 ml. The carbon dixoide is rapidly lost from milk due to low content of this gas in the atmosphere. It is also lost when milk is heated and agitated. When milk is concentrated, there is a tendency for calcium phosphate and calcium citrate to accumulate in the colloidal particles because the solubility is exceeded. As these materials are insolublised, hydrogen ions are liberated, lowering the pH. The net result is an increase in concentration of citrate and phosphate in both the dissolved and colloidal state. When milk is required for some particular purpose, use of reagents such as EDTA, various silicates, and artificial resins are made to alter its ionic balance for a better stability.

Effect of Heat on the Milk Salts

In order to understand the effect of heat on the milk salts, it is necessary to consider the temperature to which milk is treated. Heat induced changes in the milk salt system can be of three types (a) readily reversible shift in the ionic balance e.g. ionic conditions in hot milk at various temperatures (b) sluggishly reversible shifts in ionic balance e.g. changes in cooled milk after heating, (c) irreversible shifts in ionic balance. Fresh milk contains carbondioxide which is lost in atmosphere owing ot its low content, and also by heating and agitation. This loss is irreversible. Due to this there is also an increase in the pH and reduction in the titratable acidity. Concentration of dissolved calcium and phosphate in milk is decreased on heat treatments. The dissolved calcium and phosphate are transferred to colloidal state by heat treatment and it becomes absorbed to the colloidal micelles of caseinate-phosphate. The increase in the pH and decrease in titratable acidity due to loss of carbondioxide is counteracted upon heat treatment by hydrogen ions liberated by insolubilisation of calcium and phosphate. High temperature heat treatment results in production of acids (lactic and formic among others) by degradation of lactose. This increase in acidity greatly alters the ionic equilibria in the milk. Citrate is not destroyed by heating milk upto boiling temperature or by autoclaving. Heat has little effect on dissolved citrate in milk, however, heat treatment causes a precipitation of calcium citrate as calcium citrate is less soluble at high temperature than at low temperature.

Colloidal Phosphate System

Colloidal phosphate is an apatite-like complex of tertiary calcium phosphate - $Ca_9(PO_4)_6$ and some calcium citrate.

Not all micellar calcium is bound to phosphate but a part is directly bound to casein. Amount of calcium bound to different caseins is approximately equivalent to the number of ester phosphate groups.

The ratio of calcium and magnesium to inorganic phosphorus in the colloidal phosphate is 1.58. Since a small amount of the calcium must be bound to the micellar citrate, it is concluded that colloidal phosphate was tertiary calcium phosphate.

In protein free solution tertiary calcium phosphate cannot exist in acidic pH.

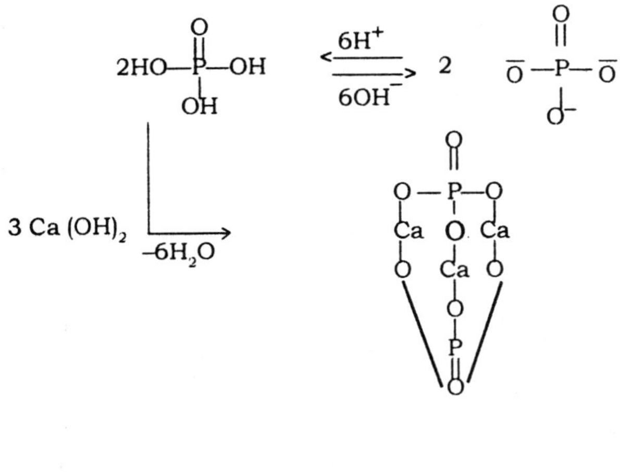

Tertiary calcium phosphate can only exist in protein free solution at a pH above 8,. Therefore, at the pH of milk, 6.7, tertiary calcium phosphate must be prevented from being transformed into a more stable form for instance hydroxyapatite. It has been well established that when solutions containing calcium and phosphate are mixed, the first reaction product is an amorphous tertiary calcium phosphate. The latter is transformed into hydroxyapatite via a solution mediated first order reaction which depends on the dissolution of the morphous tertiary calcium phosphate and the nucleotion of hydroxyapatite. This conversion appears to be retarded by small ions such as magnesium fluoride carbonate and pyrophosphate and macromolecules such as lysozyme, casein, phosvitin and polylysine. The influence of magnesium has been studied extensively and it might be concluded that at the average molar ratio of magnesium to calcium in skim milk, 0.18, conversion of amorphous tertiary calcium phosphate into crystalline hydroxyapatite is completely inhibited. Not only does casein contribute to this inhibition but it further stabilises the colloidal phosphate against ready sedimentation, a property which casein shares with many other macromolecules such as lysozyme, phosvitin, sodium polyacrylate, and poly-l-glutamic acid.

There are eight phosphoserine residues in αs_1-casein. The consequences of dephosphorylation are : (a) Ca^{2+} induced precipitation occurs via non-phosphate Ca^{2+} binding sites. Seven of these phosphoserine residues occur in a cluster between residues 43 and 80 (αs_1). This highly acidic segment also contains 12 carboxyl residues and accounts for almost all of the molecules net negative charge, (b) The second most abundant milk protein, ß-casein, is also a single chain polypeptide with five phosphoserine residues and a molecular weight of 24,500 Daltons, (c) Ability to be stabilized by K-casein in the presence of Ca^{2+} is impaired but not eliminated. Incorporation into micelles is decreased, (d) Micelle formation and stability reduced, (e) α^{-}_{s-1}, K-casein micelles are disrupted, indicating a minimum number of phosphate groups is necessary for micelles to remain dispersed in their medium, (f) α^{-}_{s-1},-casein is shifted from the micelle phase to an insoluble phase, i.e., it precipitates.

Types of linkages possible in colloid calcium phosphate are the following :

Casein—XX—Ca—PO_4 ⟨Ca, Ca⟩ (PO4)n—Ca—XX—Casein

—H
—CaOH
—Ca—PO_4=Ca

Casein—XX—Ca—PO_4—Ca—XX—Casein
|
Ca—citrate = Ca
Casein—XX—Ca—PO_4—Ca—XX—Casein
|
CaOH

(Abbreviation used XX=carboxylate or phosphoserine group)

Vitamins in Milk

Vitamins in Milk, their Nutritional Importance and Structure Vitamins may be defined as carbon compounds of diverse structures, which are not used for energy, or fixed into the tissue frame work but which are required in minute amounts for the normal functioning of the living organisms. Vitamins are also recognised as nutrients for animals obtained from different foods etc. These can not be synthesised in the body. However, a plant can synthesise these nutrients from CO_2, water, oxygen and minerals. Animals and many micro-organisms depend upon the plants for these nutrients (Table 6.5).

Table 6.5. Vitamins in Milk

Vitamins	Concentration (per litre)
Vitamin A	1590 I.U.
Vitamin D	22.1 I.U.
Vitamin E	1.0 mg
Vitamin K	0.04 mg
Vitamin B_1	0.4 mg
Vitamin B_{12}	1.5 mg
Panthothenic acid	3.0 mg
Biotin	50 μg
Niacin	0.2–1.2 mg
Pyridoxomine B_6	0.7 μg
Folic acid	1.0 μg
B_{12}	7.0 μg
Vitamin C	20 mg
Inositol	180 mg
Choline	150 mg

On the basis of their solubility vitamins are distinguished into two :

i. Fat soluble vitamins A,D, E and K, NP

ii. Water soluble vitamins, the B-complex - consisting of thiamin, riboflavin, niacin, panthothenic acid, pyridoxine biotin, choline, inositol, folic acid, P-amino benzoic acid. B_{12} and vitamin C (ascorbic acid). All the above recognised vitamins are present in milk, however, vitamin A and B_2 group vitamins are present in appreciable amounts as shown in Table 6.5.

Vitamin A (Carotenes) $C_{20}H_{30}O$

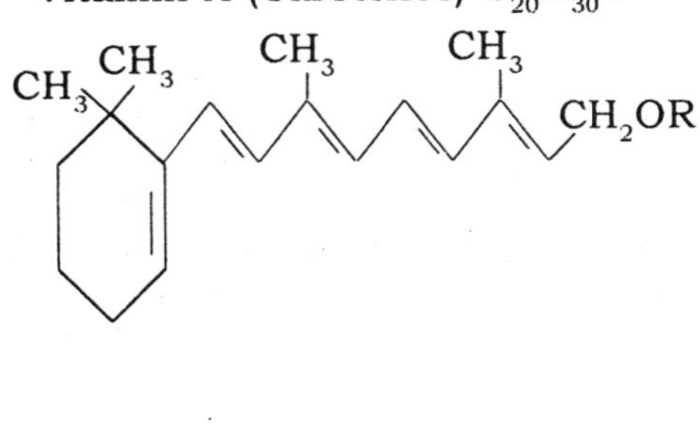

The structure shows the unsaturated character of vitamin A. The all-trans isomerase is the most active biological form. The 13-*cis* isomer is known as neo-vitamin A; its biological activity is only about 75 per cent of that of the all-trans form. The amount of neo-vitamin A in natural vitamin A preparations is about one third of the total. This amount is usually much less in synthetic vitamin A. The synthetic vitamin A is made as acetate or palmitate and marketed commercially in the form of oil solutions, stabilized powders, or aqueous emulsions. The compounds are insoluble in water but soluble in fats, oils and fat solvents.

There are several provitamin A, these belong to the carotenoid pigments. The most important one is B-carotein and some of the pigments that can be derived from it are of practical importance. These are B-apo-8' carotenal and B-apo-8'-carotenoic acid ethyl ester. For example,

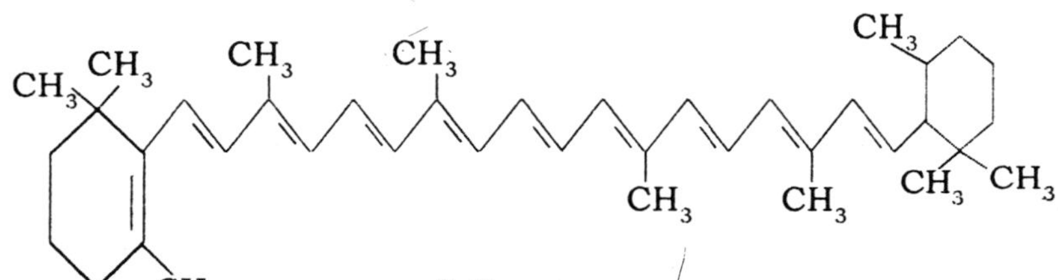

β-Carotene

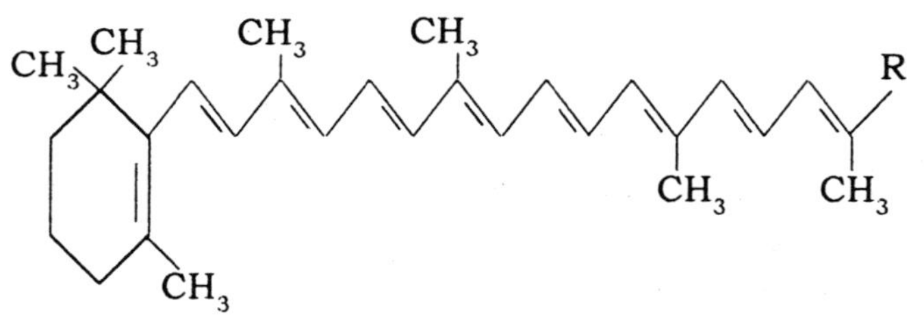

Apo carotenal (R = CHO) and apo-carotenoic acid ester (R=$COOC_2H_5$). Other provitamins are α- and γ-carotene and cryproxanthin.

Vitamin A levels are frequently expressed in International Units (I.U.), although this unit is officially no longer accepted. One I.U. equals 0.344 μg of crystalline vitamin A acetate, or 0.300 μg vitamin A alcohol; or 0.600 μg ß-carotein. The recommended daily allowance is 5000 I.U. for an adult person. In literature the human requirement of vitamin A is also quoted as about 1 μg/day. Condition of rapid growth, pregnancy or lactation, increase the need for vitamin A. Vitamin A is relatively stable to heat in the absnece of oxygen. Because of the highly unsaturated character of the molecule, it is quite susceptible to oxidation, especially in the presence of light, be that sunlight or artificial light.

Vitamin A is unstable in the presence of mineral acids but stable in alkali. Vitamin A and the carotenoids have good stability during various food processing operations. Losses may occur at high temperatures in the presence of oxygen. These compounds are also susceptible to oxidation by lipid peroxides, and conditions favouring lipid oxidation will also result in vitamin A breakdown. Pasteurisation of milk does not result in vitamin A loss but exposure to light will result in loss. It is essential, therefore, that sterilised milk be packed in light-impervious containers. Possible losses during storage of foods are more affected by duration of storage than by storage temperature.

It has been found that vitamin A added to milk, is more easily destroyed by light than the native vitamin A. This is not because natural and synthetic vitamin A are different, but because these two types of vitamin A are dispersed differently in the milk. The form in which vitamin A is added to food products may influence its stability. Vitamin A in beadlet form is more stable than that added as a solution in oil. The beadlets are stabilized by a protective coating. If this coating is damaged by wetting with water, the stability of the vitamin is greatly reduced.

Vitamin A is necessary for growth, maintenance of epithelial tissues, and proper vision. Lack of vitamin A causes night blindness, poor dark adaptation, and a disease of the eye called xeropthalmia.

Vitamin A and Vision

The mammalian organism splits carotene at the plane of symmetry between C-15 and C-15^1. The initial cleavage product is assumed to be retinat (Vitamin A_1 aldehyde) which is then reduced to the alcohol. The alcohol, simply called Vitamin A_1 (retinol), can be esterfied for storage. Another active substance, vitamin A_2, differs from A_1 by an additional double bond in the ring, in conjugation with the entire system.

Vitamin A is a growth factor. In addition to the alcohols A_1 and A_2 and the aldehyde, vitamin A acid is also active in biological assays. Perhaps the latter is the active form of the vitamin. But in the visual process the acid does not fill the role of vitamin.

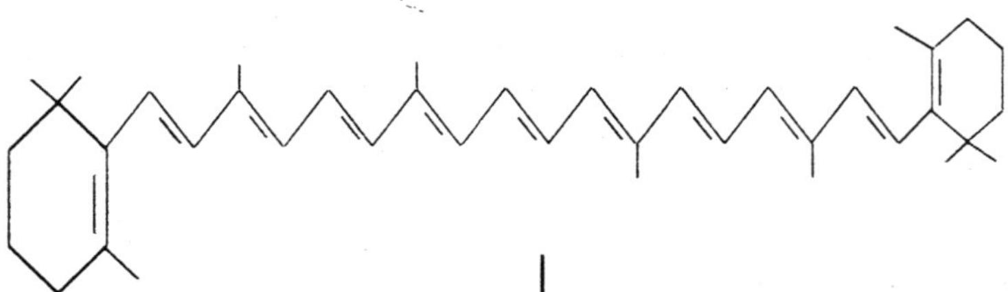

B-Carotene ↓ oxidation

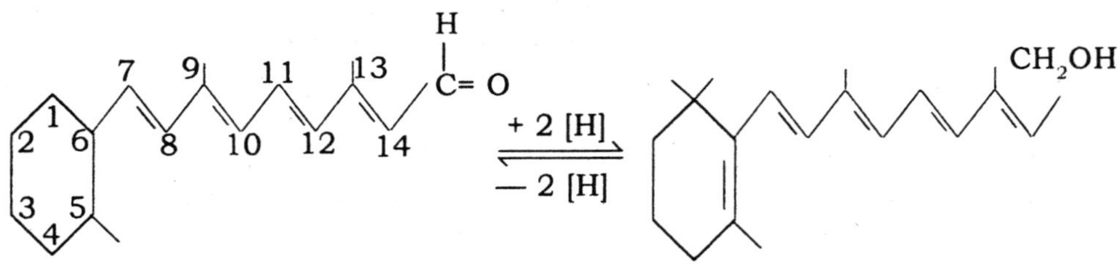

Retinol (Vitamin A_1)

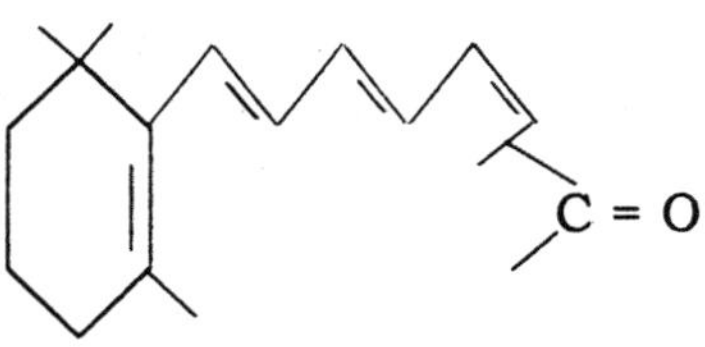

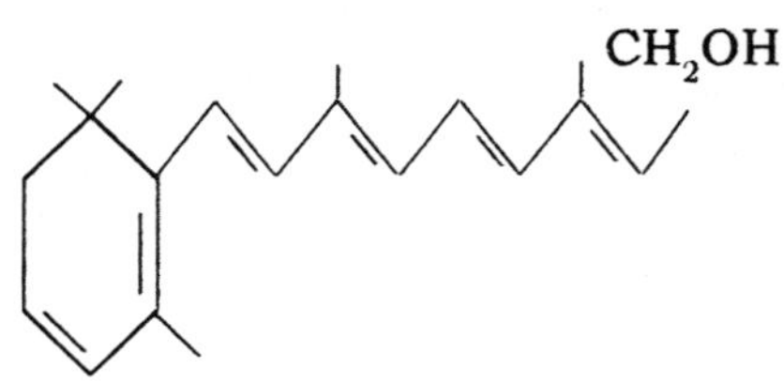

|| Cis Retinol (Neoretinal b)

3, Dehydroretinol (Vitamin A_2)

Biochemistry of the Visual Process

The photosensitive pigment of the eye is visual purple or rhodopsin. It consists of the protein opsin and neoretinal b, a stereo isomer of the all-trans-retinal (vitamin A_1 aldehyde). The protein opsin attaches only to the cis-isomer to form the chromoprotein.

During the visual process the conversion of light to nerve impulses-neoretinal b is first arranged to the more stable all-trans-configuration. Light only initiates the process. The altered molecular shape interferes with the bond to the protein. Probably, the spatical structure of the protein is affected and the all-trans-retinal dissociates. In what way the photochemical changes are translated to neutral excitation will be discussed.

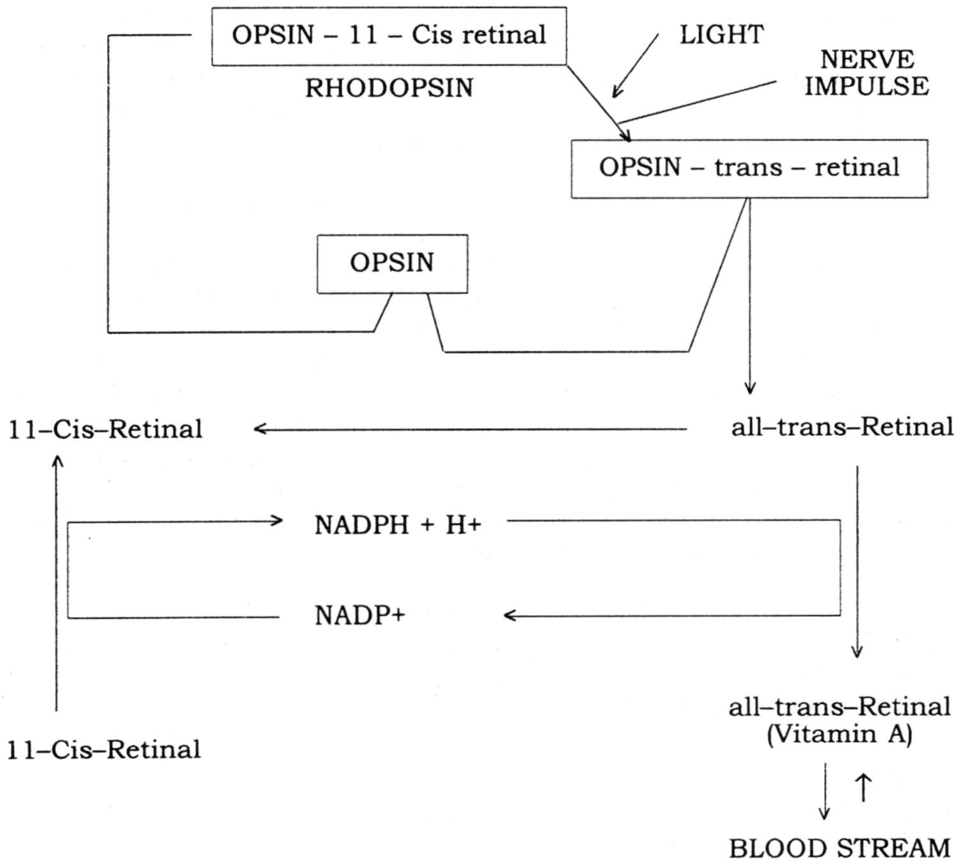

Besides opsin, there is another protein, scotopsin. Either protein can combine with 11-cis-retinal or 11-cis-3-dehydro retinal. Consequently, there are four visual pigments, which cover a broad area of the spectrum.

Rhodopsin is not the visual pigment. It is most important in vision in dim light. Lack of rhodopsin due to dietary deficiency conditions is known as "night blindness".

Vertebrates have excitable receptors that are activated by light. There are two kinds of photoreceptor cells. Because of their distinctive shape, they are called rods and cones. Cones function in bright light and are responsible for colour vision whereas rods function in dim light but do not perceive colour. A human retina contains about three million cones and a hundred million rods. These photo receptor cells convert light into atomic motion and then into a nerve impulse. The electrical signals generated by the photo receptors are processed by an intricate array of nerve cells within the retina and then transmitted to brain by the fibres of the optic nerves. Thus, the retina has a dual function to transform light into nerve impulses and to integrate visual information.

When light is absorbed, a photo receptor cell gets stimulated. In addition, the light absorbing group (called a chromophore) must undergo a conformational change after it has absorbed a photon. The photosensitive molecule in the discs of rod cells is rhodopsin, which consists of opsin, a protein, and 11-cis-retinal, a prosthetic group.

Opsin without its prosthetic group, does not itself absorb visible light. The colour of rhodopsin and its responsiveness to light depends on the presence of 11-cis-retinal, which is an effective chromophore. 11-cis-retinal gives rhodopsin a broad absorption band in a visible region of the specturm with a peak at 500 nm which nicely matches the solar output. The intensity of the visible absorption band of rhodopsin and the coefficient are also not worthy. The absorption coefficient of rhodopsin at 500 nm is 40,000 $cm^{-1}M^{-1}$, a high value. The intergrated absorption strength of visible absorption band of rhodopsin approaches the maximum value attainable by organic compounds. 11-cis-retinal has these favourable chromopheric properties because it is a polyene. Its six alternating single and double bonds constitute a long, unsaturated electron network.

11-cis-retinal is attached to rhodopsin by a Schiff base linkage. The aldehyde group of 11-cis-retinal is linked to the α-amino group of a specific lysine residue of opsin. The spectral properties of rhodopsin indicate that the Schiff base is protonated.

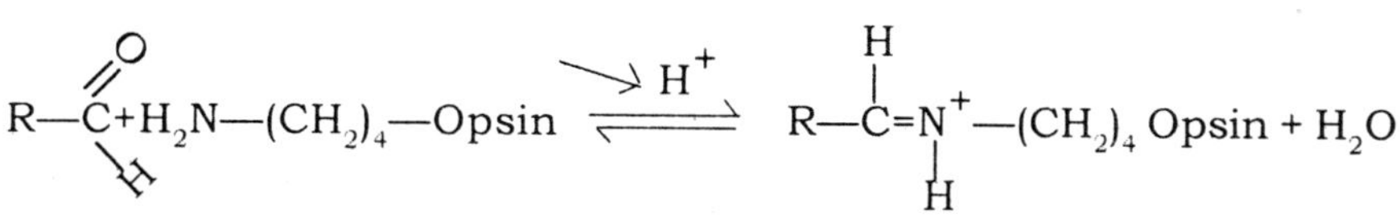

The precursor of 11-cis-retinal is all-trans-retinal (Vitamin A), which cannot be synthesised *de novo* by mammals. All-trans-retinal is converted into 11-cis-retinal in two steps. The alcohol group is oxidised to an aldehyde by retinol dehyrogenase, which uses $NADP^+$ as an electron acceptor. Then the double bond between C-11 and C-12 is isomerised from a trans to a cis configuration by retinal isomerase.

Dietary Requirement

10 I.U. per day per kilogram body weight.

Food Sources

Fish liver oils, liver, milk fat, egg yolk, yellow and green leafy vegetables, carrot and yeast. In milk, vitamin A occurs in the esterform, however, about 2 to 6 per cent is present as alcohol. Carotenoid makes up 1 to 50 percent of the total vitamin A activity. It is expressed as retinal equivalents (RE).

Milk and milk products contribute 11.82 per cent vitamin A of the total percentage of the vitamin by food groups, whereas vegetables and (meat, poultry and fish) contribute respectively 36.4 and 22.9 per cent. Oils, eggs, fruits and potatoes respectively provide 8.6, 6.8, 7.3 and 5.7 percentage of the total food groups.

The vitamin A and carotene content of some foods are delineated in the Table 6.6.

Table 6.6. Vitamin A and carotene content of some foods

Product	Vitamin A (I.U./100 g)	Carotene (mg/100g)
Beef (grilled sirloin)	37	0.04
Butter	2363-3452	0.43-0.77
Cheddar cheese	553-1078	0.07-0.71
Eggs (boiled)	165-488	0.01-0.15
Herring (canned)	178	0.07
Milk	110-307	0.01-0.06
Tomato (canned)	0	0.5
Peach	0	0.34
Cabbage	0	0.3
Broccoli (boiled)	0	2.5
Spinach (boiled)	0	6.0

Deficiency Diseases

Xerophthalmia, myctolopia, hermeralopia, keratomalacia, hyperkeratosis.

Properties

Appears as yellow oil, molecular weight 286.4, melting point 62-64°C, crystalline forms as prism, insoluble in water, soluble in acetic acid, alcohol, chloroform, ethanol, behaves as unsaturated alcohol. Important group for activity are B-ionone ring, transmethyl, alcoholic hydroxyl, stable to heat, acid and alkalies, large losses may occur through oxidation and exposure to light.

When fresh milk is exposed to day light for 8 hrs, there is a loss of 2 per cent in vitamin A content, however for milk stored in dark at 5°C for 4 to 7 months and then exposed to light for 8 to 10 hrs, there is a loss of 60 to 70 per cent in vitamin content. Vitamin A is also destroyed to the extent of 10 per cent in amber-glass bottle. Fluorescent light also causes destruction of vitamin A while ultraviolet radiation and an X-ray do not cause any destruction. Irradiation of gamma rays from Co^{60} causes loss of 85 per cent. Vitamin A and 45 per cent carotenoid in milk while losses are also observed with milk product. In general, processing parameter such as pasteurisation, sterilisation, spray and roller drying causes little or no loss of vitamin A and carotene.

Isolation

Vitamin A is isolated from fish liver oil by saponification with alcoholic potassium hydroxide, extraction with ether and crystallisation.

Nutritional Importance

Vitamin A is important in the diet as it is necessary for growth, health and reproduction. It keeps the epithelial tissues healthy and thus aids in preventing infections. Lack of this vitamin will cause night blindness. Vitamin A aids in maintaining normal glandular functions such as maintenance of adrenal cortex and steroid hormones synthesis. It acts to initiate vision from light energy. Vitamin A is present as carotene in plant and must be converted to vitamin A for its function.

Use in Dairy Products

Synthetic ß-carotene is finding increasing use in the colouring of dairy products. Micro-pulverised preparations containing upto 30 per cent of ß-carotene are now available for cheese making and continuous butter making and there is also a thioxotrophic, gel containing upto 10 per cent ß-carotene for use in batch butter making. Water dispersible forms are also available in cheese and ice-cream manufacture. While the quantities added are very low, the synthetic ß-carotene does contribute to the nutritive value of the product.

Vitamin D (D_2, D_3) calciferol (D_2) $C_{28}H_{44}O$)

Structural Formula

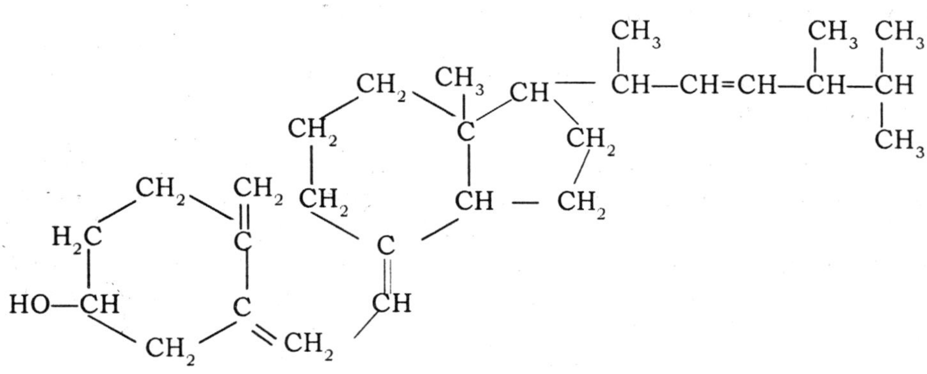

Dietary Requirement

100 I.U. per day for average growth.

Food Sources

Fish, liver oils, vitamin D concentrate. In milk both the forms of vitamin i.e. ergocalciferol (D_2) produced by irradiation of ergosterol and choplecalciferol (D_3) produced by irradiation of 7-dehydrocholesterol occurs. Conversion of ergosterol to ergocalciferol occurs when eradicated plants are fed to animals while cholecalciferol is formed by solar radiation when animals are exposed to sunlight. Average value of vitamin in bovine milk are 13.0 and 31.2 I.U. for winter and summer milk.

Deficiency Diseases

Rickets, Osteomalacia, Hypoparathyroidism.

Properties

White powder, molecular weight 384.65. Melting point 84.85°C, fine needles, insoluble in water, soluble in acetic acid, alcohol, benzene, chloroform, ethanol. Important group for activity for C_{10}-C_{18}, methylene, alcoholic OH behaves as sterol. Vitamin D is stable to heat, alkalines and oxidation. The vitamin content of milk in animals can be increased by feeding them substances rich in vitamin D, by irradiating the milk and by adding vitamin D concentrates. Sunlight affects its concentration, more in summer than winter. It is stable in milk, not affected by pasteurisation, boiling or sterilisation. Storage of fluid milk for 30 months at −17.8°C, irradiated evaporated milk for 2 to 3 years or of frozen butter for over 2 years resulted in little or not loss of the vitamin, however losses are observed in vitamin D on storage of enriched dried milk for 1 year at 24°C.

Isolation

Liver oil is irradiated by yeast and saponified. Vitamin A and sterols are removed by partitioning solvent and by adsorption chromatography. The inactive sterols are removed with digitonin and then it is crystallised as 3,5 dinitrobenzoate ester.

Nutritional Importance

Vitamin D is necessary to aid in calcium and phosphorus retention so as to make strong bones and prevent rickets. It increases tubular preabsorption, and citrate blood levels. It maintains and activates alkaline phosphatase in bones and serum cacium and phosphorus level. Vitamin D is sometimes called the sunshine vitamin because the ultraviolet radiation of the sun can convert an activer precursor of the skin (7-dehydrocholesterol) to vitamin D. Children need vitamin D for their growth and to a lesser extent mothers require this vitamin during pregnancy and lactation.

Vitamin E (tocopherols) α-tocopharol $C_{29}H_{50}O_2$

Structural formula

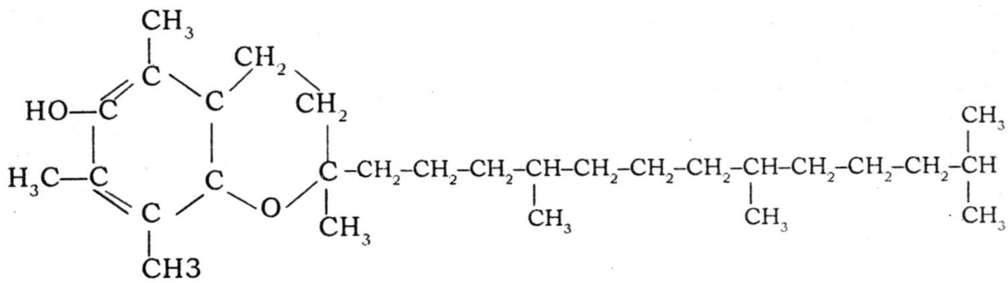

Dietary Requirement

14 mg α-tocopherol per day.

Food Sources

Vegetable oils, green leafy vegetables, milk fat and most common food stuff. In cow milk, tocopherol content ranges from 87 to 103 μg/ 100 ml milk. It is present in milk mostly as α-tocopherol, however, γ-tocopherol, -z-tocopherol have also been reported.

Deficiency Diseases (Animal)

Degeneration of reproductive tissues, muscular dystrophy, encephalomalacia, livernecrosis (Man), skin collagenosis, red cell hemolysis, xanthomatosis, cirrhosis of gall bladder, steatorrhea and creatinuria.

Properties

Appearance as yellow oil, molecular weight 430.7, melting point 2.5-3.5°C. Insoluble in water, soluble in acetic acid, alcohol, benzene, chloroform, ethanol. Chemical nature - aromaticquinoid, important groups for activity hydroxyl alcoholic. Stable to heat, light alkalines and acids, destroyed when fat turns rancid. Heat and other processing treatments involved in the production of pasteruised milk, butter, condensed milk, or evaporated milk do not appear to bring about any significant loss of vitamin E.

Isolation

Sources like wheat germoil, soyabean oil or rice oil are saponified with methanolic KOH. The non-saponifiable fraction which has vitamin E dissolved in ether, sterols are removed with digitonin precipitation and xanthophyll are removed by methanol extraction. The tocopherols are converted to allophanate esters with HCN and crystallise.

Nutritional Importance

It is a strong antioxidant, prevents oxidation of unsaturated fatty acids and vitamin A in intestinal tract and body tissues. It acts in oxidation-reduction reactions, plays role in human nutrition and is associated with reproductive factor, necessary for nutrition of muscles and better utilisation of vitamin A. It maintains integrity of vascular system and central nervous system. Vitamin E is a detoxifying agent and maintains kidney, tubules, lungs, genital structure, liver and red blood cell membranes.

Vitamin K (K_1, K_2, K_3) K_3 - Menaphthane (menadione) (2 methyl-1,4-napthoquione) $C_{31}H_{46}O_2$

Structural formula

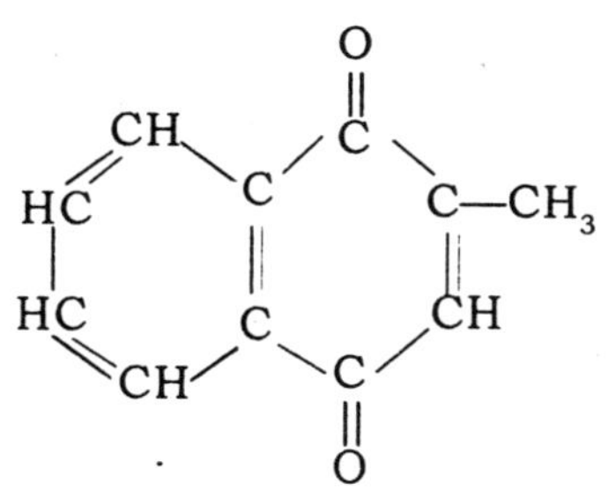

Dietary Requirement

1 mg per day for infant, 2-5 mg for mother in labour.

Food Sources

Green leafy vegetables, synthesised by micro-organism in intestinal tract. In milk, the vitamin occurs in the form available in the food of animal, mostly as phythylmenaquionone. Average concentration in bovine milk is 50 μg/litre. Traces are found in pasterurised milk, butter milk and butter oil.

Deficiency Diseases

Hypoprothrombinemia.

Properties

Yellow oil, melting point 20°C, molecular weight 450.7,, insoluble in water, soluble in acetone, alcohol, benzene, chloroform, ether, chemical nature - Quinone. Important groups for activity - menadione nucleus, phytylside chain, trans-methyl group. Stable to heat but destroyed by light and oxidising agent.

Isolation

It is isoated from fish meal and alfa alfa. From alfa alfa chlorophyll is removed by petroleum ether followed by extraction on column chromatography using zinc carbonate. The resulting product is reduced to hydroquione using sodium hydrosulfite and then extracted with petroleum ether, alkali, ether and finally oxidises hydroquinone to quinone.

Nutritional Importance

Aids in production of prothrombin, a compound required for normal clotting of blood, prothrombin synthesis in liver, blood clotting mechanisms; electron transport mechanisms and growth photosynthetic mechanism.

Vitamin B_1 (Thiamine hydrochloride) $C_{12}H_{17}N_4OSCl.HCl$.

Structural formula

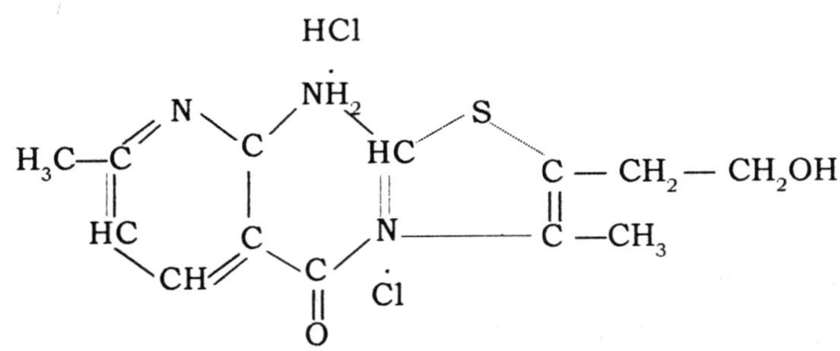

Dietary Requirement

0.23 to 0.3 mg per 1000 food calories. 1.5 mg/daily for adult, child 1 to 1.5 mg/daily.

Food Sources

Meats, legumes, whole grain and enriched cereals and bread, milk and dairy foods. Vitamin B_1 occurs in milk in three forms - free thiamine 50 to

70 per cent, phosphorylated thiamine 18 to 45 per cent and a proportion of both forms bound to protein as 5-17 per cent. The average thiamine content of whole milk is 0.43 mg per litre.

Deficiency Diseases

Beriberi, opisthotonous (in birds), polyneuritis, hyperesthesia, bradycardia edema.

Properties

White crystals, molecular weight 337.3, melting point 244°C, soluble in water, acetic acid, benzene, chloroform, insoluble in ethanol, chemical nature as base, –OH of $-CH_2-CH_2OH$ act as important group, other important group C-2 of pyrimidine and thiazole.

Thiamin in milk is destroyed by heat depending on frequency and severity of heat treatment. Pasteurisation causes a destruction between 5.5 and 25 per cent while sterilisation causes 20 to 45 per cent or more. In preparation of evaporated milk destruction between 20 and 60 per cent has been observed while in dried milk powder preparation, there is a loss of 10 per cent. Gamma radiation has no effect on vitamin B_1. Storing milk in dark bottle for 72 hours at different temperatures like 4°C, 12°C and 20°C show a loss of 24, 14 and 16 per cent. Commercial heat treatment of milk like HTST and UHT results in a ~ 10 per cent loss of vitamin B_1 activity.

Isolation

Vitamin B_1 is isolated from rice bran or wheat germ or yeast. The aqueous extract is absorbed on Fuller's earth. It is eluted with quinine sulfate, precipitated as phosphotungstate. The precipitate is decomposed and reprecipitated with $AuCl_3$, extracted with water and precipitated with EtOH as hydrochloride.

Nutritional Importance

Vitamin B_1 is necessary for growth, appetite, muscular coordination, intestinal motility, and prevention of nervous diseases as beri-beri in man and polyneuritis in animals. It acts as a coenzyme in pyruvate metabolism and carbohydrate metabolism. It is necessary for energy production. It acts as part of cocarboxylase. Vitamin B_1 aids in the removal of CO_2 from alpha-keto acids during the oxidation of carbohydrates and may function similarly as part of lipothiamide.

Vitamin B_2 (Riboflavin) 6,7 dimethyl-9-D-ribityl, isoalloxazine

$C_{17}H_{20}O_6N_4$

Structural Formula

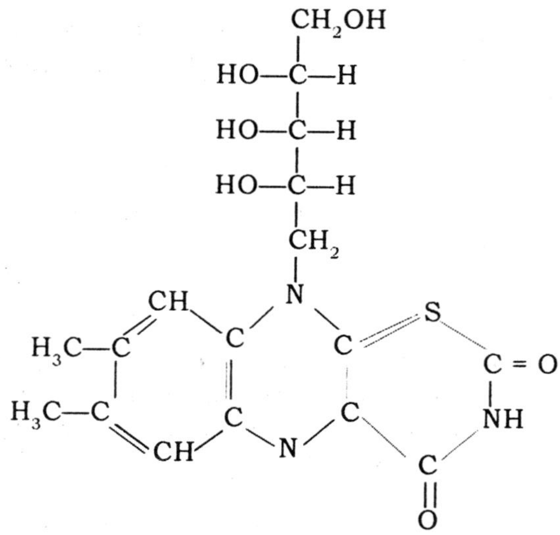

Dietary Requirement

0.5 to 1.0 mg per day.

Food Sources

Milk and dairy foods, organ meats, some vegetable enriched cereals, and breads. In bovine milk, 54 to 95 per cent of riboflavin is in free form. The remaining occurs as bound form to protein as FAD or FMN. This is 20 per cent of total vitamin present in milk. The average value of vitamin B_2 in bovine milk is 1750 µg/litre.

Deficiency Diseases

Glossitis, cheilosis, seborrhei, dermatitis, corneal vascularisation, an emia.

Properties

Orange yellow powder, molecular weight 376.4 melting point 282°C, needle crystals, soluble in water, insoluble in acetic acid, alcohol, benzene, chloroform and ethanol. Chemical nature - as reducing agent; nucleotide substituted purine. Important groups are 9-N', 5'-OH 6-7 methyl 3-N. Stable to heat, fairly stable to oxidation, acid and alkalies, but sensitive to light. It takes up sunlight readily and uses it to oxidise ascorbic acid, thus destroying this vitamin. In the absence of light the riboflavin in milk is quite stable to heat. Pasteurisation causes a negligible amount of destruction of vitamin. Heat treatment such as sterilisation or preparation of evaporated milk, dried milk or instant non-fat dried milk is without much effect. When milk is exposed to light, the riboflavin gradually disappears. Exposure to fluorescent light is less harmful than exposure to sunlight. There is no effect on vitamin when milk is stored in refrigerator.

Isolation

The usual sources are, free urine, whey, retina, tissues and egg white. The aqueous extract of tissue is treated with ether followed by fractional precipitation with picric acid. The fractions are precipitated with ammonium sulphate and absorbed on fuller's earth. These are further eluted with 0.1N sodium hydroxide and crystallised from aqueous petroleum ether-acetone mixture.

Nutritional Importance

This vitamin is necessary for growth, proper nutrition of the eyes, tissues and linings of the mouth and lips and prevention of premature ageing of the skin. It has been shown to be essential in the formation of red blood cells and haemoglobin. It shows up in whey as a greenish yellow pigment. As part of flavoprotein, aids in removal of hydrogen from breakdown products of proteins, fats and carbohydrates. It is a coenzyme in respiratory enzyme systems and is essential for the development of the foetus and part of redox system.

Vitamin - Panthothenic Acid (α, γ-dihydroxy-ß, ß-dimethyl-butyryl, - ß-alanide) $C_9H_{17}O_5N$

Structural formula

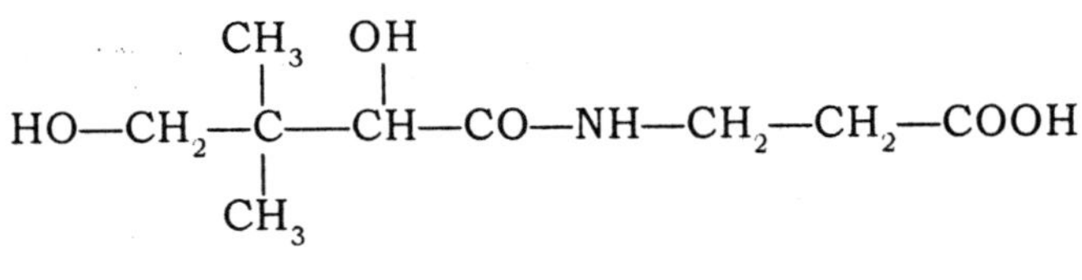

Dietary Requirement

A 2500 caloric diet of foods of animal and plant origin supplies about 10 mg pantothenic acid.

Food Sources

Present in all plant and animal foods including skim milk and whey. Milk contains 3.39 mg/l. About one-fourth pathothenic acid may be present in bound form. Cheese also contains 40-60 per cent panthothenic acid in bound form.

Properties

Little or no pantothenic acid is lost during pasturisation of milk. There is no loss of vitamin during preparation of evaporated milk, condensed milk, sterilised milk, dried whole milk or dried skim milk. The vitamin is also stable on storage at ambient temperature. Sunlight might cause some destruction. UHT treated bovine milk result in loss of 30-35 per cent when stored at room temperature for 6 weeks.

Nutritional Importance

As part of coenzyme A, functions in the synthesis and breakdown of many vital body compounds, thorough activating two carbon fragments provided by protein, fat and carbohydrates. Essential for development of central nervous system, for growth, for proper utilization of riboflavin and certain other vitamins.

Vitamin Biotin (Vit H) $C_{10}H_{16}O_3N_2S$

Structural formula

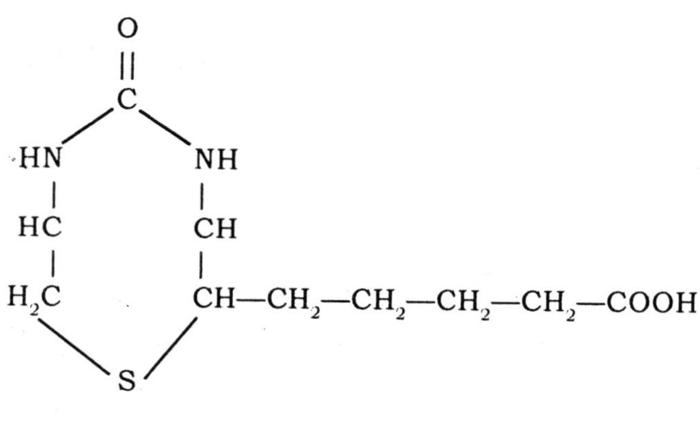

Dietary Requirement

Average diet should have 150-300 μg per day.

Food Sources

Organ meats, milk, eggs, most foods of plant and animal origin. Biotin is present in milk in the free state. The average content in bovine milk is 30 μg/litre.

Properties

Pasteurisation and sterilisation of milk results in loss of this vitamin to the extent of less than 10 per cent. In preparation of evaporated, condensed or dried whole milk, the loss is not more than 15 per cent. There is no loss of vitamin during ultra high temperature sterilisation. Storage in frozen state or sunlight does not result in any loss of this vitamin.

Nutritional Importance

Appears to be involved in the synthesis and breakdown of fatty acids and amino acids through aiding the addition and removal of CO_2 to or from active compounds, and the removal of NH_3 from amino acids. It increases resistance to diseases. It cures paralysis in small animals.

Vitamin Niacin

Structural formula

Nicotinic acid

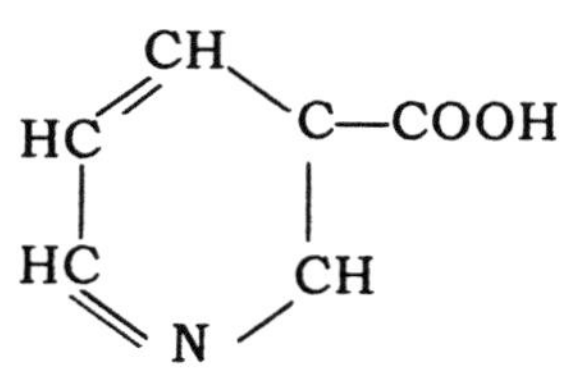

Nicotinamide

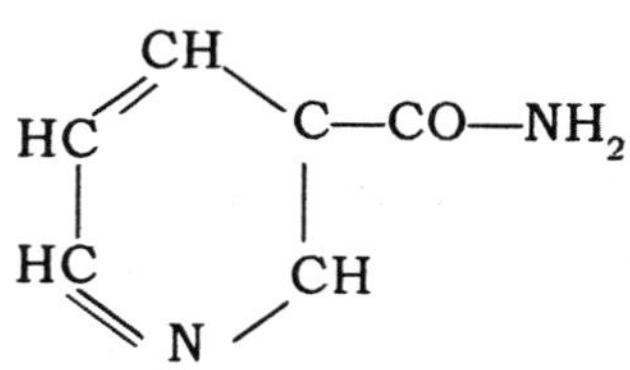

Dietary Requirement

10 to 15 mg per day

Food Resources

Many grains, meats, eggs and milk. Niacin is not present in milk in bound form. Average niacin content of bovine milk is 900 μg/litre.

Properties

Various processing treatments such as pasteurisation, sterilisation or those used in preparation of milk products has no effect upon the niacin content of milk. There is no loss when milk is exposed to sunlight or stored as frozen milk.

Nutritional Importance

As part of coenzyme I, II and III, aids in the transfer of hydrogen in the oxidation of breakdown products of protein, fat, carbohydrates, steroid hormones, cholesterol, and vitamin A. It acts as pellagra preventive factor. Prevents nervous and digestive disorders, sore mouth and tongue.

Vitamin Pyridoxamine Pyridoxal (Vit. B_6) (2 methyl-3-hydroxyl-4,5-di (hydroxymethyl pyridine $C_8H_{11}O_3N$

Structural formula

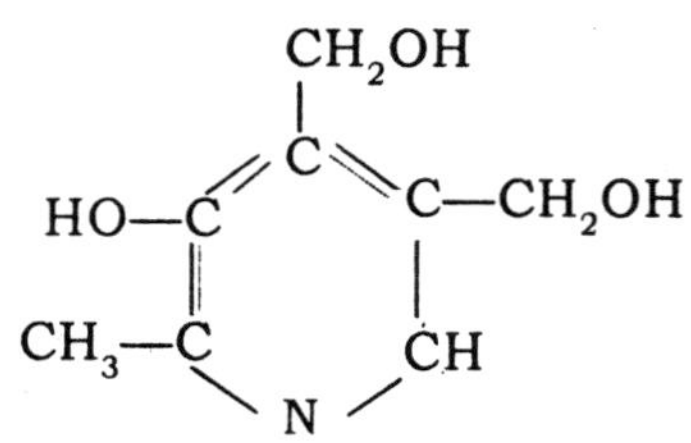

Dietary Requirement

0.15 to 3 mg per day, (1 to 2 mg readily provided by diet).

Food Sources

Cereal bran and germ, liver, milk, egg yolk, most common food stuffs. In milk 14 per cent is in bound form and 86 per cent is in free form. Average value in bovine milk is 0.60 mg/litre.

Properties

Vitamin B_6 in milk is stable to processing parameters like pasteurisation, Ultra Heat Treatment (UHT) and normal storage conditions. Light causes a destruction of 21 per cent after 8 hours.

Nutritional Importance

As a coenzyme, aids in the synthesis and breakdown of amino acids and in the synthesis of unsaturated fatty acids from essential fatty acids. Acts on removal of CO_2, transfer of NH_2, removal of H_2S, and other amino acid reactions. Appears to bear some relationship in the metabolism of both fat and protein.

Vitamin (folic acid, folinic acid (pteroyl-glutamic acid) $C_{19}H_{19}O_6N_7$

Structural formula

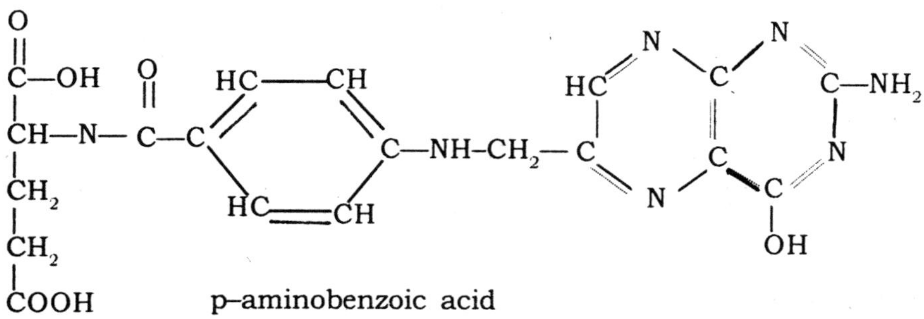

Glutamic acid Pteroic acid 2 amino-6-hydroxy pteridine

Dietary Requirement

0.5 mg folic acid per day.

Food Sources

Green leafy vegetables, organ meat, lean beef, wheat. Milk contains on an average 59 µg/litre.

Properties

The folate in milk is strongly and specifically bound to minor whey protein. The molecular weight is 38000. Sterilisation of milk destroys 40 to 50 percent of the vitamin.

Nutritional Importance

Essential for the biosynthesis of nucleic acid and also for normal fat metabolism. Involved in the metabolism of single - carbon fragments. "Active single - carbon fragments" come from limited sources, primarily choline, methionine, glycine, and serine. Folic acid along with another compound Inositol ($C_6H_{12}O_6$) appears to work in prevention of tumour growths.

Vitamin (B_{12}) Cynocobalamine, cobalamine $C_{63}H_{90}O_{14}N_{10}PCO$.

It has complicated structure involving many groups and compounds.

Dietary Requirement

About 1 µg/day

Food Sources

Milk and dairy foods, meats, eggs. The vitamin B_{12} activity in milk is due to cobalamin. It is bound to various protein fractions of milk particularly higher concentration with whey protein. Average value in milk is 4.2 µg/litre.

Properties

Pasteurisation causes only slight destruction of vitamin B_{12} in milk. Normal processing parameter do not cause any change in vitamin concentration.

Deficiency Diseases

Retarded growth, pernicious anemia, megaloblastic anemia, macrocytic, hyperchromic anemia, glossitis, spinal chord degeneration, sprue.

Nutritional Importance

As a coenzyme in nucleic acid, protein and lipid synthesis, it maintains growth, nucleic acid synthesis, protein synthesis, lipid synthesis and methylations. Maintains epithelial cells and nervous system (myelin sheath) erythropoiesis (with folic acid) and leukopoiesis). Presence of cobalt in this vitamin differentiates it from other vitamin B complex. It is a red crystalline compound, essential for normal development of red blood cells and aids in the growth of children and maintenance of proper body functions. Involved in the metabolism of single carbon fragments, essential for biosynthesis of nucleic acid and probably for normal fat metabolism.

Vitamin C

(L-ascrobic acid) $C_6H_8O_6$

Structural formula

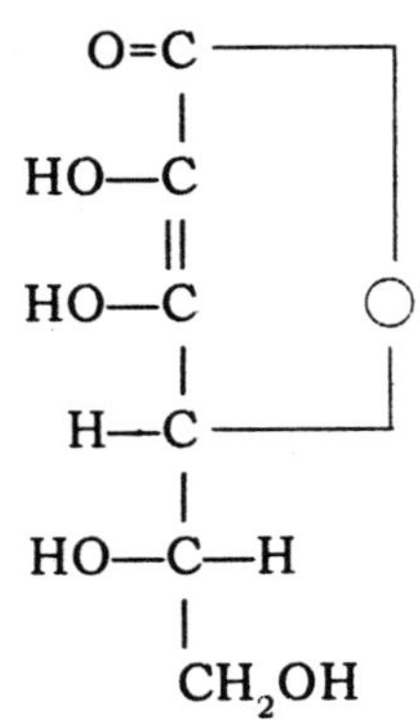

Dietary Requirements

Average 10 to 25 mg per day, adult 75 mg daily, child 50 mg daily.

Food Sources

Citrus fruits, tomatoes, melons, peppers, green and most fresh plant foods. In fresh milk it is present as L-Ascrobic acid which on storage is converted to dehydro ascorbic acid. Vitamin C is present in milk in concentration of 20 mg/litre.

Properties

During pasteurisation of milk, vitamin C is destroyed. When milk is exposed to sunlight, oxidation of vitamin C takes place. Exposure to fluorescent light is less harmful as compared to sunlight. Gamma rays also cause destruction to the extent of 77 per cent. When fresh raw milk was stored for 2 weeks at –40°C it shows ~ 30 per cent loss of vitamin C content. There is a loss of 20 per cent in vitamin C content when milk is subjected to UHT treatment.

Deficiency Diseases

Scurvy, megaloblastic anemia of infancy.

Properties

White powder, molecular weight 176.12, melting point 190-192°C (decompose), crystal form - plate needles, soluble in water, slightly soluble in acetic acid, alcohol, insoluble in benzene, chloroform, ether, chemical nature as hexose acid, important group for activity - lactose ring, lost due to oxidation, sensitive to alkalies, stable to weak acids, strong reducing agent. Ascorbic acid is a derivative of carbohydrates and functions in oxidation reduction reaction. Many reactions are aided by ascorbic acid.

Isolation

From citrus fruits it is precipitated as Pb-complex and crystallise from alcohol-petroleum ether.

Nutritional Importance

This vitamin is necessary in the diet to prevent scurvy, a disease in which the gums become swollen, bleed easily and the teeth loosen. It is also important as a stimulant to the appetite and for growth and health. During pasteurisation of milk vitamin C is destroyed. It functions as coenzyme in the oxidation of tyrosine, functionally related to folic acid and vitamin B12, essential for production of collagen, the basic substance of connective tissue, related to biosynthesis of steroid hormones. Absorption of iron, cold tolerance, maintenance of adrenal cortex, antioxidant, metabolism of tryptophan, phenylalanine, tyrosine, growth wound healing, synthesis of polysaccharides and collagens, formation of cartilage, dentine, bone teeth, maintenance of capillaries.

Flavour and Off-flavour in Milk and Milk Products

Flavour in milk is an important parameter for its acceptance as in any other food item. Food may be nutritious, wholesome and attractive to look or reasonably priced but will likely to be rejected if its flavour is defective. Milk is no exception to these attributes. Flavour in milk is dependant on various factors. Flavour in short, is a sensory perception perceived through suitable receptor like nose, mouth and related structures. Olfaction deals with the odour of the food item and once the food enters the mouth it becomes apparent as vapour pass up to the back of the nose, the olfactory area. Variation in odour is almost limitless, however, taste (gustatory) is concerned with a relatively few sensation like sweetness, sourness, saltiness and bitterness that are detected in the mouth and more particularly on the tongue by the taste buds when food is present in solution. The other aspect of flavour is tactual, the way the substance feels in the mouth. These include sensations of smoothness, graininess, tenderness and chalkiness etc. Individual vary in their sensitivity to different kinds of flavour stimuli.

Flavour is usually measured by psychometric evaluation and chemical analysis. Chemical measurement of flavour is based on the assumption that a flavourful compound is present in milk in sufficient quantity to be measured by suitable method of analysis. Psychometric evaluation are more important and variety of methods are available. Whatever may be the method of flavour determination, it has following objectives :

(a) Quantitative measurement of flavour intensity, (b) qualitative description of flavour component, (c) comparative or determining the presence and extent of difference in flavour between samples, (d) acceptance or like-dislike testing, (e) ranking or scoring products for flavours and off-flavours, (f) threshold determinations. In milk and milk products flavour compounds are present at levels of parts per million or even per billion. These compounds, are detected by methods based on charomatographic techniques, spectrophotometry, infrared spectral data and gas liquid chromatography. The discussion on the use of these techniques for evaluation of flavour component is beyond the scope of the present text.

Flavour of milk can be divided into two groups, that is the normal or natural flavour that is desirable, and the off-flavours or those that are not typical of the milk and that usually but not always are undesirable flavour.

Normal Flavour

Milk is described as having a faint characteristic flavour. It cannot be told with certainity as to what compounds are exactly responsible for the natural flavour of normal milk. However, it is known that compounds present in traces like acetone, acetaldehyde, butyric acid and certain other fatty acids contribute to flavour of milk. In addition to these, milk also contains trace amount of methylsulfide $(CH_3)_2S$ having boiling point of 38°C. The flavour threshold for this compound is 12 pp.b in distilled water. At slightly above this level it has a malty or cowy flavour. This compound has been detected in the cow's breath and its sulfone derivative in the blood stream of the animal. It is quite probable that this compound is also responsible for the flavour of the milk.

Flavour of Milk Product

The odour and flavour of certain fermented dairy products are due to carbonyl compounds and volatile acids which are the by-products of the lactic acid fermentation. Important among these include diacetyl, acetoin (acetyl methyl carbinol), acetone, acetaldehyde, butryic acid, propionic acid, acetic acid and formic acid. Of the above compounds diacetyl is most important. Optimum production of the flavour and aroma due to diacetyl is produced by a use of mix culture of *Streptococcus lactic, S. citrovorus*, and *S. paracitrovorus* The initiation of production of this compound is by citrates. In addition to these, compounds like 2,3-butylene glycol, acetoin which are intermediate during such reactions also play their part in flavour production of dairy products.

Cheese Flavour

There are many varieties of cheeses. Each cheese has its own characteristic flavour, however, the principal flavour compound in most varieties have not been elaborated. Cheddar cheese has a characteristic flavour. This flavour may be due to acetaldehyde, ethyl alcohol, butyraldehyde, ethylacetate, and ethyl butyrate or even lower free fatty acid. The flavour of mold-ripened cheese such as roquefort, stilton, blue and gorgonzola, is derived in part from free fatty acids and methyl ketones particularly 2-heptanone. The flavour in camembert and limburger cheese is due to a compound indole which has a definite feacal odour. Swiss cheese has flavour characteristic of propionic acid. The compound diacetyl is said to contribute, to some extent to the flavour of most varieties of cheese.

Off-flavours

Milk is susceptible to off-flavour problem. This off-flavour in milk can be due to the animal, feed, weed and the barn air to which the animal is exposed. The off-flavour in milk are grouped into three types (a) chemical, (b) microbiological, (c) extraneous or absorbed.

Chemical Off-flavour

This is mainly due to deterioration of chemical constituents present in milk. Compounds like phospholipids on oxidation produce oxidised off flavour. This flavour is also sometimes accompanied by other off-flavours such as cardboard or cappy flavour. Various antioxygenic procedures have been developed to prevent oxidation such as deaeration of milk. Compounds like iron, copper, ascorbic acid and light acclerate the oxidation of phospholipids and related compounds thereby producing the oxidised flavour. Hence, copper and iron containing vessel and cans and equipments are avoided from use, during the process of handling of milk.

Milk also contains rancid flavour which is due to hydrolysis of compounds like glycerides to liberate fatty acid by an enzyme lipase. Excess production of butyric acid is the principal agent responsible for rancid flavour. Rancidity is also due to processes like homogenisation, vigorous agitation and warming followed by cooling. Rancidity is controlled by pasteurisation which normally destroys lipases present in milk.

Light also exerts detrimental effect on milk. Due to light, methionine produces sunlight off-flavour which is also like hay like off-flavours. Fragmentation of milk protein in presence of riboflavin (Vit B_2) is necessary to release this amino acid. Sunlight off-flavour is mainly due to intensity of light, nature of the container and duration of light exposure.

Microbiological Flavour

Milk is supposed to be a good medium for the growth of most of the microbiological agents like bacteria, mold and yeasts. The flavour may be fruity, rancid, stale or putrid depending on the nature and chemical changes brought about by the bacteria. Some of these are shown in Table 6.7.

Table 6.7. Flavour changes in milk due to microbial agents

Flavour Type	Organism Responsible
Acidic flavour (clean)	*Streptococcus lactis*
Acidic flavour (aromatic)	*S. lactics* and *Leuconostoc* species
Acidic flavour (sharp)	*S. lactic* and *Leuconostoc* species *Clostridium* species, *Coliforms*.
Bitterflavour	*Micrococcus* species, *Alcaligenes* species, *Pseudomonas* species, *Proteus* species, Achromobacter species, *Flavobacterium* species, *Serratia* species, *Bacillus* species, *Clostridium* species.
Bitter/musty flavour	Actinomycosis species, certain type of cocci.
Burnt or carmel flavour	*Streptococcus lactis* var *maltigenes*

Barny flavour	*Aerobacter oxytocus*
Soapy flavour	*Pseudomonas sapolactica*
Turnip-like flavour	*Eschericia coli, Pseudomonas fluorescens.*
Malty flavour	Yellow micrococci
Potato flavour	*P-grayeolens* and *P-mucidolens*
Fishy flavour	*P-icthyosmia*
Fruity flavour	Yeasts
Musty flavour	*Actinomycosis* species

Sometimes the colour of milk may change due to excess of organism. These are shown in Table 6.8.

Table 6.8. Colour change in milk due to bacterial proliferation

Type of change	Organism responsible
Bluish-grey or brownish milk	*Pseudomonas syncyanea*
Brown milk	*Pseudomonas fluorescens*
Deep blue milk	*Pseudomonas syncyanes* and *Streptococcus lactis*
Red milk	*Serratia marcescens, Brevibacterium erythrogenes, Micrococcus reseus, Torula glutinis*
Yellow milk	*Pseudomonas synxantha flavobacterium*

Extraneous Absorbed Flavour

Flavour and odour compounds can be absorbed into the milk either prior to or following milking. There are two ways by which these substances enter into milk. One is by mouth-nose lungs-blood stream-udder cells-milk and other from digestive tract-blood stream-udder cell-milk. In both cases the animal acts as the odour trap. When the animal eats or smells onion or garlic, the odour is noted from the blood stream within few minutes and the same is reflected in milk. Flavour substances can also be absorbed from the feed sources and it appears in milk through digestive tract - blood stream-udder cell route. Cowy and barny flavour is a common problem in winter months. The sources for this particular flavour are tainted air of the cow shed, fermented silage and ketosis (a disease of cattle involving endogenous energy metabolism).

Pigments

Milk of different species shows physical and chemical differences. One of the difference is their pigment content. In buffalo milk, a green pigment is secreted and is bound to protein. This green pigment on reduction gives a yellow pigment and is detached from protein. This phenomenon occurs during souring of milk. Ghee prepared from fresh buffalo milk has waxy white appearance as compared to that prepared from sour milk which shows greenish yellow pigment colouration. The green pigment has been identified as biliverdin whereas yellow pigment is reported to be bilirubin.

Biliverdin in milk withstands pasteurisation. However, during high heat treatment its content gets reduced. Its presence in milk does not affect methylene blue or resazurin reduction test. Bilirubin, on the other hand has no effect on keeping quality of ghee. Though green and yellow pigments have been isolated from fresh and sour cows and goat milk, however, these differ from the bile pigments isolated and identified from buffalo milk.

The conversion of biliverdin to bilirubin in milk appears to be a non-specific reduction and a release mechanism is required to set free the bilirubin formed from the protein matrix which held the pigment. When micro-organisms bring about the conversion of biliverdin to bilirubin in milk, they seem to make dual change of reduction and release, since bilirubin formed is easily extractable in fat.

REVIEW QUESTIONS

1. List important salts and trace elements of milk.
2. What are the factors that affect "Salt balance" in milk ?
3. How does casein exist in milk ?
4. What is the effect of heat on milk ? What changes do you expect when milk is heated ?
5. List fat soluble vitamins of milk.
6. What is "carotein" ?
7. Describe normal or natural flavour and off-flavour of milk with examples.

REFERENCES

Jennes, R. and Patton, S. 1959. *Principles of Dairy Chemistry*. John Wiley & Sons, Inc. New York, Chapter-5 and 12.

Webb, B.H., Johnson, A. and Alford, J.A. 1978. *Fundamentals of Dairy Chemistry*. The AVI Publ. Co. Inc. West Port Connecticut.

Fox, P.F. 1989. *Development in Dairy Chemistry* - 3 Lactose and Minor constitutents. Elsevier Applied Science. London and New York.

7

COMPARATIVE ASPECTS OF MILK OF DIFFERENT SPECIES

MILK is a lacteal secretion of the lactating mammary gland controlled specifically and completely by lactogenic hormones. In chemical terms milk can be described as a delicately balanced biochemical fluid containing mainly protein, fat, lactose and lactose-containing milk oligosaccharides and ash. The aforesaid constituent of milk differs from species to species. The milk protein of each species bears an inverse relationship with the growth rate of each species. This relationship is depicted in Figure 10.

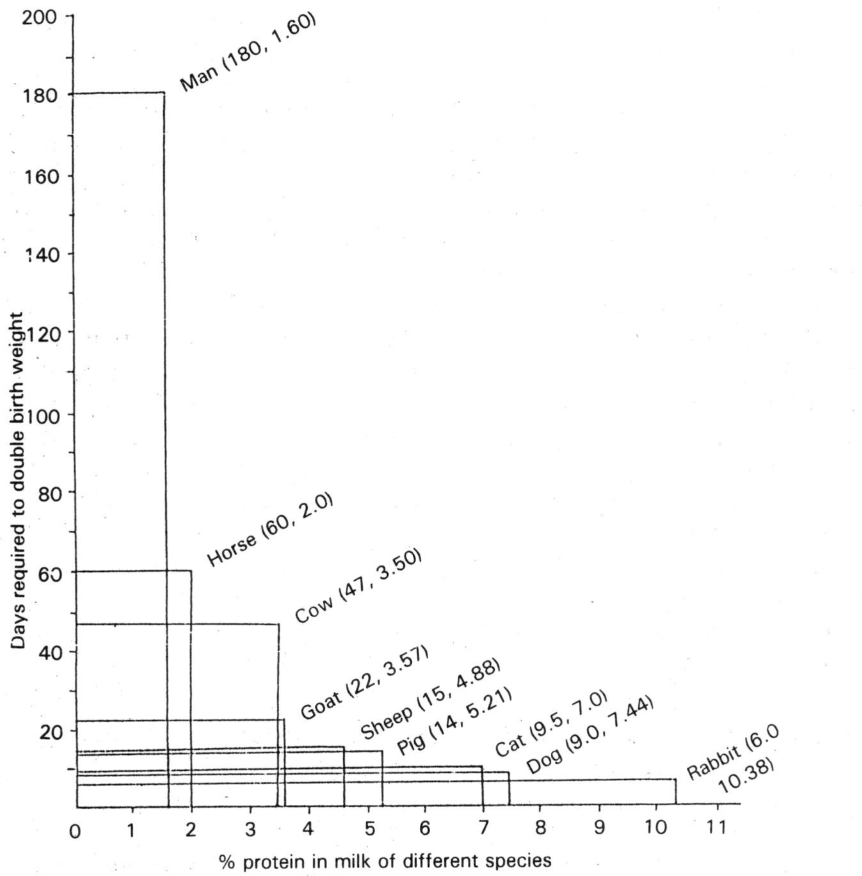

Fig. 10. Relationship between rate of growth of mammals with its milk protein percentage

Rabbit's milk contains 10.38 per cent milk protein and it takes 6.0 days to double its weight, whereas human milk contains only 1.60 per cent protein and an human infant takes 180 days to double its weight. Lactose content of human milk and monkey's milk are the highest. Human milk also contains lactose - containing milk oligosaccharides which were identified and isolated while attempting to identify the bifidus factor from the intestine of breast-fed babies. High lactose content of human milk coupled with high content of lactose containing milk oligosaccharides are practically responsible for high intake of galactose, fucose, (6-deoxy-L-galactose), other nitrogen containing mono-saccharides and the neuraminic acid derivatives. Galactose is a part and parcel of cerebrosides and myelin sheaths of nerves. 6-deoxy-L-galactose may undergo metabolic conversion to galactose in the brain and neuraminic acid derivatives may be responsible for the fertility of human brain. One notable finding is that monkey's milk also has a high lactose content. However, the lactose containing milk oligosaccharides of monkey's milk is not known. Also, the lactose content of dog's milk is uncertain. Had these values been known, then a correlation between the intake of lactose and lactose containing milk oligosaccharides and steady development of human brain during the evolutionary process could have been made. Anyhow, it is interesting to hypothesize that during the evolutionary process of human being, development of human brain was 'largely due to the intake of human milk by human infants'. Lactose content of reindeer milk is minimum whereas the fat percentage is highest in reindeer's milk. During the evolutionary process it is indeed interesting to conjecture that the composition of milk of different species changed with the change of the environments. Eskimos drink reindeer milk which contains the highest fat. Eskimos stay in the coldest part of the world and they need high fat diet for their maintenance and such diet is provided by the fat rich Reindeer's milk. Ewe's milk contains 4.8 to 6.7 percent protein, 6.3 to 8.6 percent fat and 4.9 percent lactose; one report gives its lactose content as 4.3 to 4.8 percent. Ewes milk is common with nomads who being nomadic take lot of milk and milk products which mainly supply their maintenance ration. Goat's milk is common to mountainous climate. In very high altitude of mountains yak's milk is common whereas in the desert areas camel's milk is very common. Buffalo milk is very common in the Indian subcontinents' Myanmar etc. As a matter of fact, buffalo milk contributes to 60 percent of the total milk production in India. The percentage composition of milk of different species are summarised in the Figure 11.

Compositional Differences of Buffalo Milk with Cow Milk

A brief summary of the significant differences in the level of various constitutents and physico-chemical properties is given below.

a. The fat, SNF, protein, lactose and calcium is higher in buffalo milk than that of cow milk.

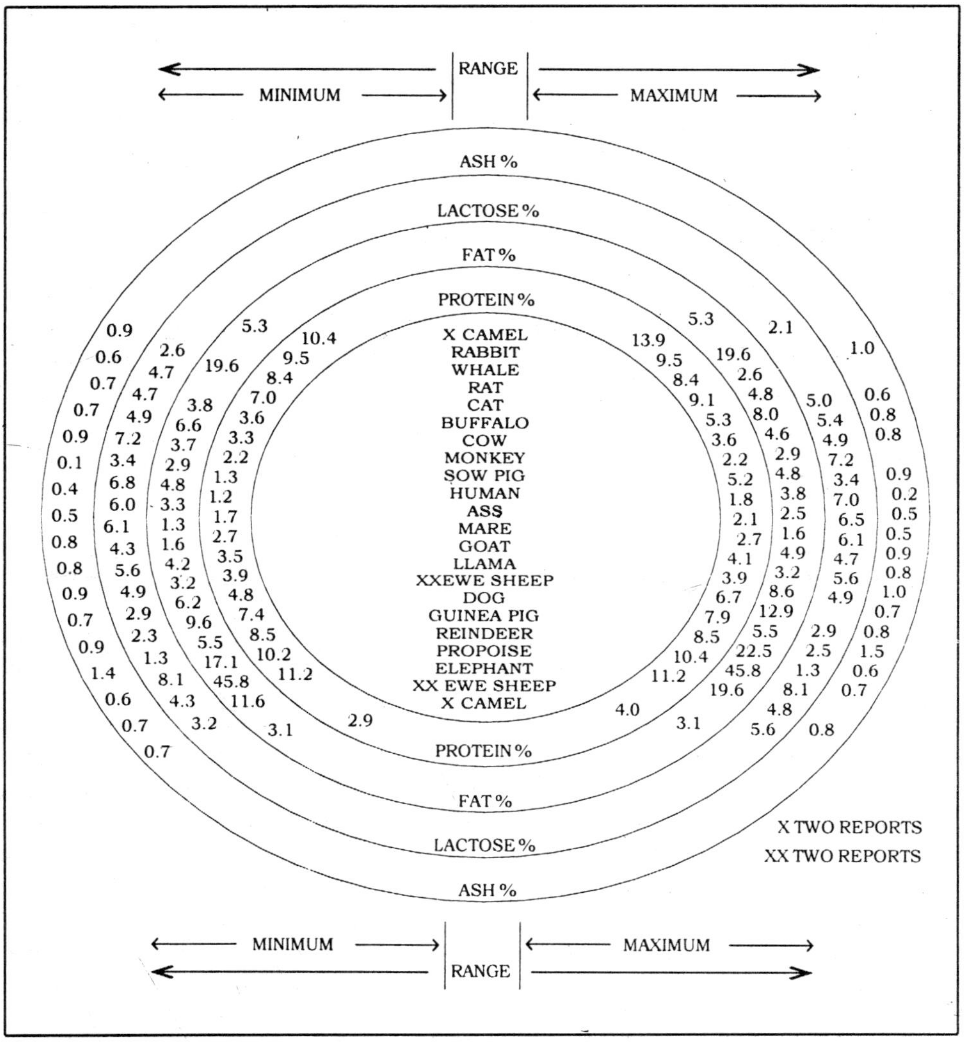

FIG. 11. Showing the approximate percentage of protein, fat, lactose and ash in milk of different species in general

b. The composition of buffalo milk is influenced more by stage of lactation, age and season.

c. Buffalo milk requires larger starting volume for determination of its fat content than cow milk.

d. Buffalo milk has higher pH, viscosity, curd tension and lower rennet clotting time and less heat stability due to higher salt balance ratio than cow milk.

e. Cow milk has higher soluble casein but buffalo casein micelles are bigger in size (135 μM buffalo and 90 μM cow), more opaque and aggregated than cow milk. The voluminosity and sialic acid content is lower in buffalo milk but it has more calcium and phosphorus. The primary phase of rennet action is slow in buffalo milk but secondary action is fast.

f. Buffalo milk has more ß-casein and K-casein fractions while the mobility of α s-casein fraction in case of buffalo milk is slower.

g. The β-variant of α-lactalbumin and ß-lactoglobulin of buffalo milk are synonymous with the α-lactalbumin and ß-lactoglobulin of cow milk. Buffalo milk contains less proteose-peptone as compared to cow milk.

h. The size of the fat globules in buffalo milk are bigger. The Reichert-Meissl, Krischner value and Melting Point values are higher whereas buffalo milk has lower Polenske Value, Iodine Value and Butyrorefractometer Reading index. Buffalo milk fat undergoes less oxidative rancidity. It contains more butyric acid and long chain fatty acids like palmitic, stearic, oleic. It is richer in tetraenoic acids and conjugated pentaenoic acid but lower in caproic, caprylic, capric, lauric and myristic acid. The phospholipid composition is similar.

i. Buffalo milk contains more of colloidal calcium (0.22 percent) and phosphate (0.13 percent) as compared to cow (0.12 percent) and (0.09 percent), respectively. It contains more cations (Ca^{++} and Mg^{++}) and less anions (Phosphate and Citrate) in comparison to cow milk. It has lower soluble calcium, magnesium and citrate.

Problems Encountered on Buffalo Milk Processing

i. In the manufacture of cheese from buffalo milk, the defects encountered and probable reasons and modifications necessary are given in Flow diagram 7.1.

ii. In the manufacture of condensed milk and evaporated milk the defects encountered, and the probable reasons and the modification necessary particularly for condensed milk are given in Flow diagram 7.2.

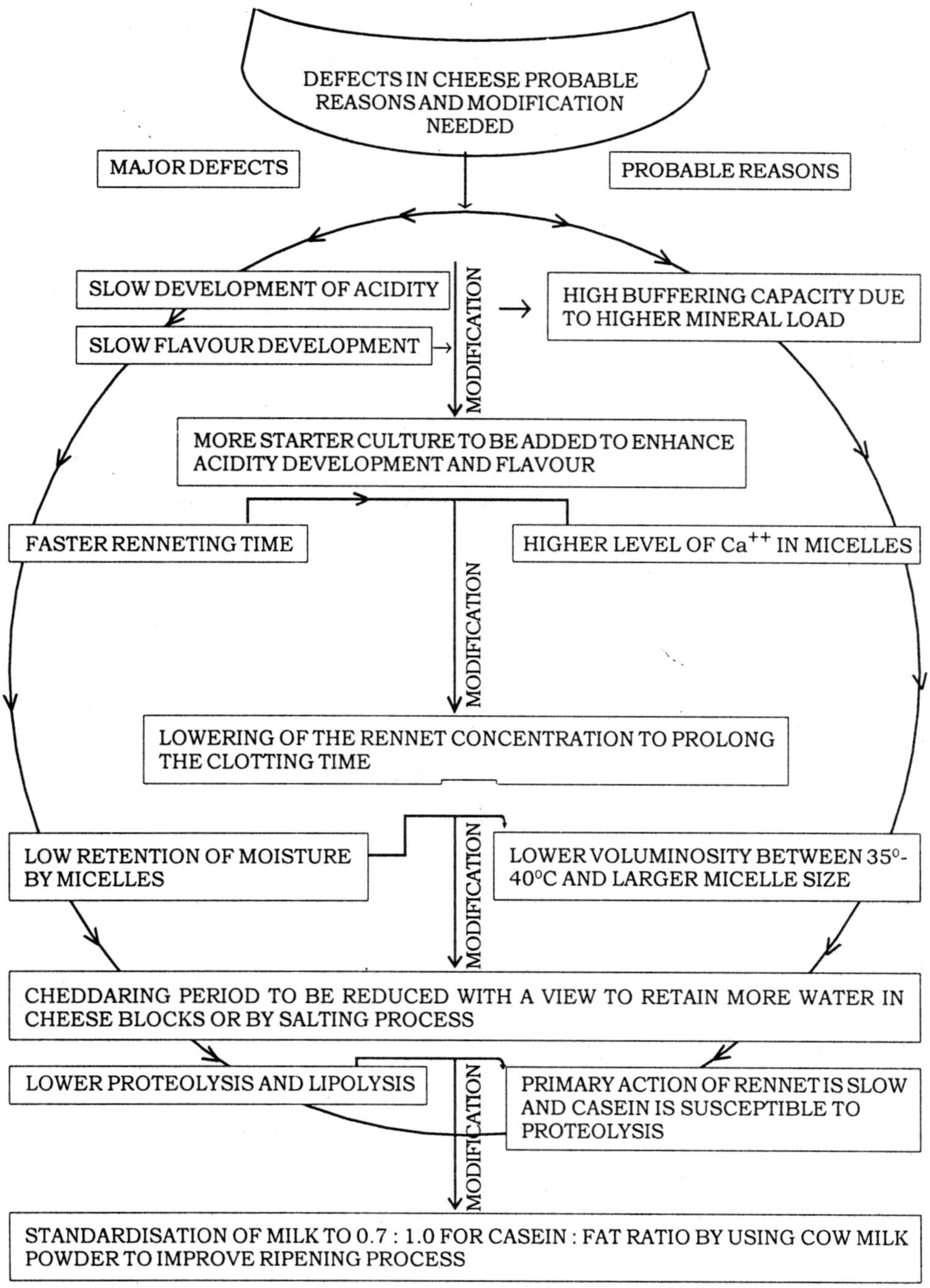

Flow Diagram 7.1 : Defects in cheese, probable reasons and modification needed.

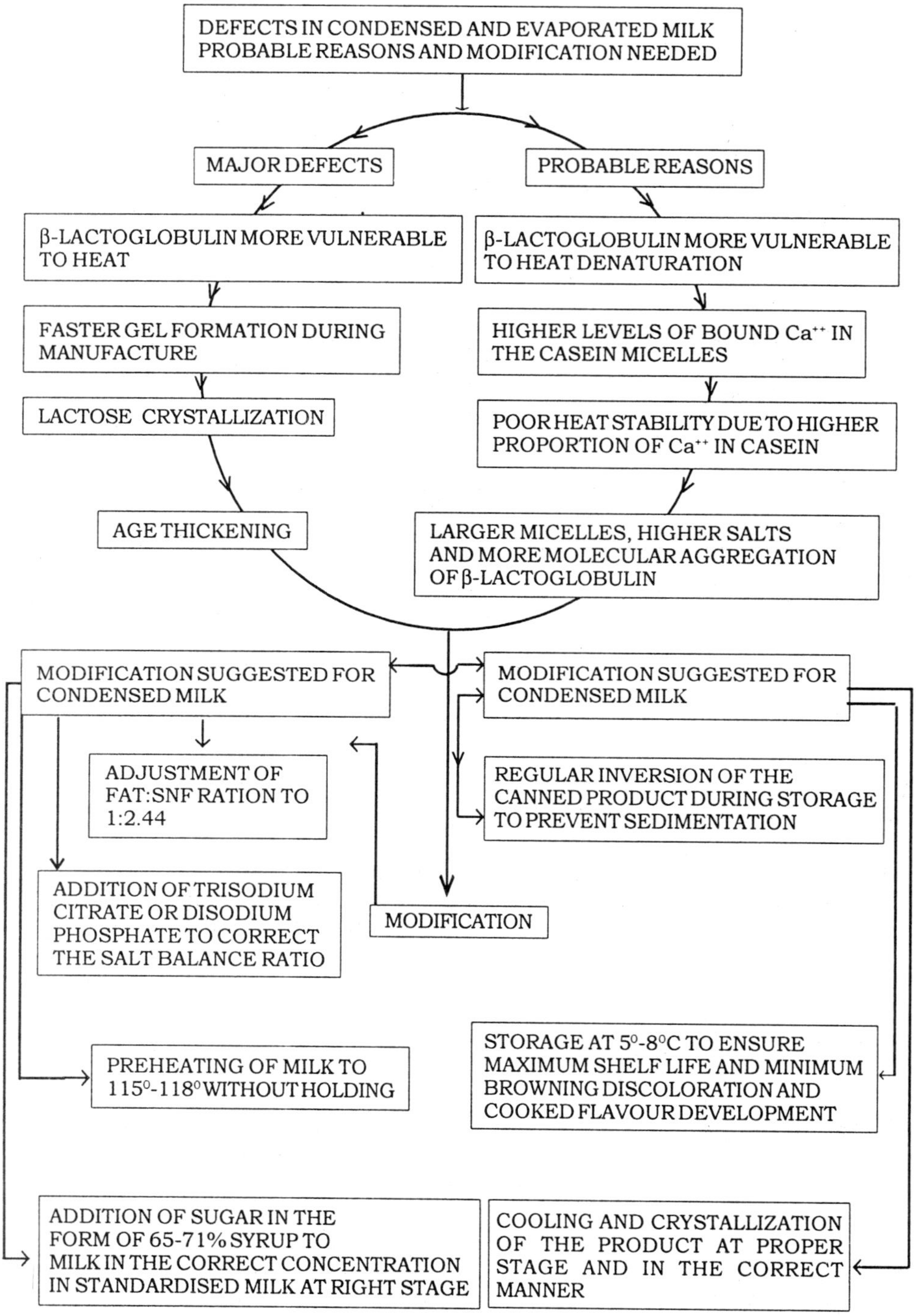

Flow Diagram 7.2 : Defects in condensed and evaporated milk, probable reason and modification needed

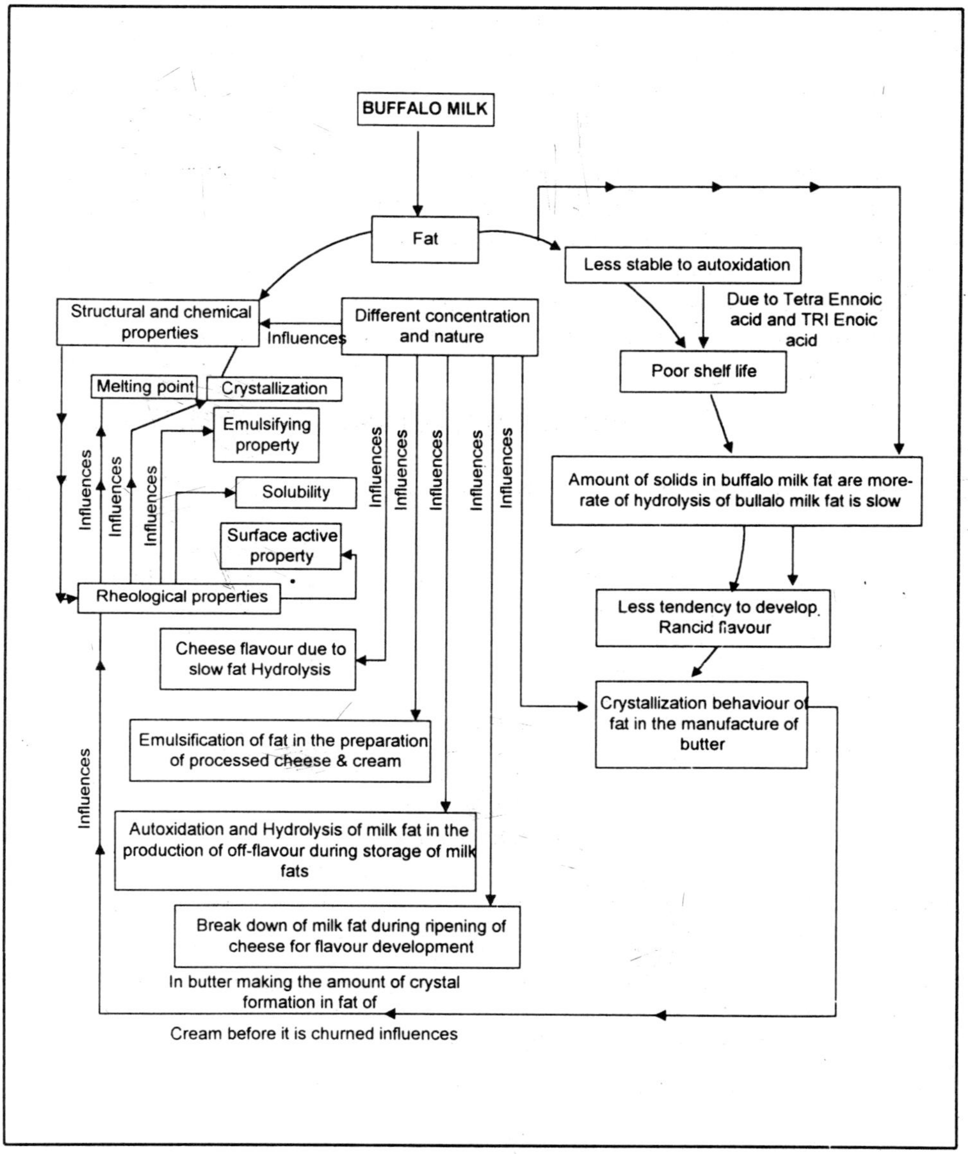

Flow Diagram 7.3 : Defects and problems due to fat in various products

iii. The difference in concentration and nature of buffalo milk fat in comparison to cow milk leads to many problems particularly in the manufacture of numerous products. Such problems are enunciated in Flow diagram 7.3.

iv. In the manufacture and storage of milk powder from buffalo milk, the main problem that stands out prominently is the lack of heat stability and low solubility of the milk powder. However, with the adjustment of fat/SNF ratio to that of cow, addition of stabilisers to correct salt balance ratio, use of correct preheating temperature (air inlet and outlet) and gas packing, the problem can be overcome.

REVIEW QUESTIONS

1. Give the average gross composition of cow, buffalo, goat and sheep milk.
2. Which species produce milk of highest protein content and which has highest lactose content?
3. How does buffalo milk compare with cow milk?
4. List some of the defects that are usually encountered during manufacture of cheese and condensed milk from buffalo milk.

REFERENCES

Ganguli, N.C. 1977. *Realities in Buffalo milk processing.* Paper presented at the XXIII Dairy Industry Conference, Haryana Agricultural University, Hisar.

Ganguli, N.C. 1984. *The chemical strength* and processing potential of buffalo milk. Lecture delivered under the joint auspices of the Indian Dairy Association (North zone) and Glaxo Laboratories (India) Ltd., Aligarh, July.

Ganguli, N.C. 1979. Buffalo milk technology. *World Animal Review* No. 30, p. 2.

8

PHYSICO-CHEMICAL, BIOCHEMICAL AND BIO-LOGICAL APPROACHES TO COMMON DAIRY PROCESSES

Microbe Induced Dairy Process

IN biological systems, the source of energy is Adenosine-triphosphate (ATP). The cleavage of the terminal phosphate of ATP molecule liberates energy. In case of prokaryotes and eukaryotes, catabolism or breakdown of glucose produces energy which is utilised in the anabolism or biosynthesis of multiple biological compounds. Therefore, most of the micro-organisms utilize carbon as a source for maintenance, growth and multiplication. Microorganisms in milk utilise lactose and citrate as a source of energy. Mainly, two types of micro-organisms are found in milk, viz. heterofermentative and homofermentative micro-organisms. The various metabolic pathways are elucidated later. In cultured milks, the production of diacetyl produces a sweet aroma and thereby enhances the flavour of certain cultured milk products. Citrate is the substrate for diacetyl production by certain specific micro-organisms in cultured dairy products. An important intermediate for diacetyl production is pyruvate. Though lactose is also the substrate for pyruvate production by many milk micro-organisms, such a pool of pyruvate cannot be utilized for diacetyl production. This is because the pyruvate pool produced from lactose fermentation produce 4 ATP's from each molecule of glucose, out of these 4 ATPs two are utilised earlier in the pathway. Also, the reduced NAD^+ produced during lactic acid production from pyruvate is also utilized earlier in the catabolic pathway. Therefore, to run the cycle, the pyruvate pool cannot be disturbed. Hence citrate is the only source of diacetyl production.

Starter micro-organisms under the main group of lactic acid bacteria, hydrolyzes lactose to D-glucose and D-galactose. The liberated D-glucose then enters the classical Embden-Meyerhoff (EMP) pathway. The liberated D-galactose can not enter the EMP pathway directly but follows a longer route and ultimately enters the EMP pathway directly via glucose-1-phosphate. The pathway for the conversion of D-galactose to glucose-1-phosphate.

1. $$\text{Galactose} + \text{ATP} \xrightarrow{\text{Galactokinase}} \text{Galactose-1-phosphate}$$

2. Galactose-1-phosphate + UDP–glucose —(Phospho–galactase / uridyl transferase)—> UDP–Galactose + Glucose–1–phosphate

3. UDP–Galactose —(UDP–Galactose–4–epimerase)—> UDP–glucose

Glucose-1-phosphate thus produced is converted to glucose-6-phosphate by the enzyme phosphoglucomutase. Thus, both the hexose moiety of lactose enters the EMP pathway and is degraded to lactic acid. The overall simplified scheme would be as follows :

Hexose $\rightarrow$ 2 triose-phosphate

2-triose phosphate $\rightarrow$ 2 pyruvic acid + 2 NADH + H^+

2-pyruvic acid + NADH + H^+ $\rightarrow$ + 2 lactic acid + 2 NAD^+
NAD + H^+

The proof that such a pathway exists in the fermentation of lactose by starter micro-organisms came through radioactive studies. Different carbon atoms of glucose were labelled and the labelling in lactic acid was found as follows :

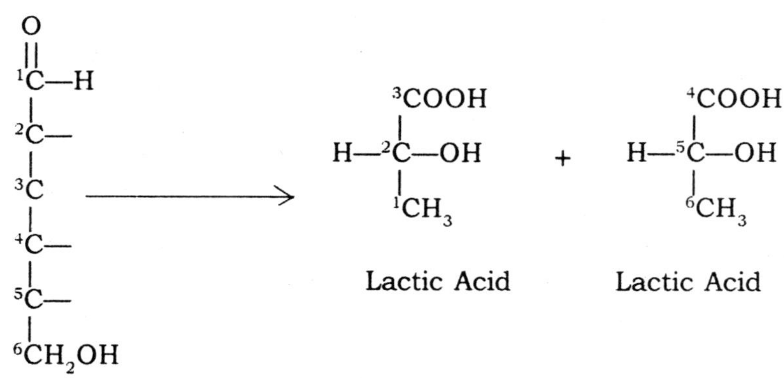

It therefore, becomes clear that lactose fermentation by homofermentative lactic streptococci follows the classical EMP pathway. However, heterofermentation bacteria cannot follow this pathway because they do not have the enzyme aldolase and hence catabolise glucose-6-phosphate by a different pathway. This pathway would be considered later. However, homofermentative bacteria also have the enzyme system to convert glucose-6-phosphate to ribulose-5-phosphate. It is, therefore, natural to surmise that homofermentative bacteria are more flexible at least as regard lactose fermentation than heterofementative bacteria. Any discussion on lactose fermentation by homofermentative bacteria would remain incomplete if the EMP pathway is not given.

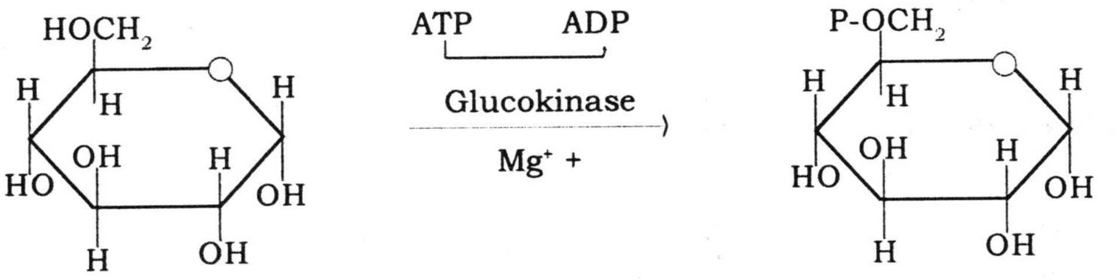

Glucose Glucose–6–P

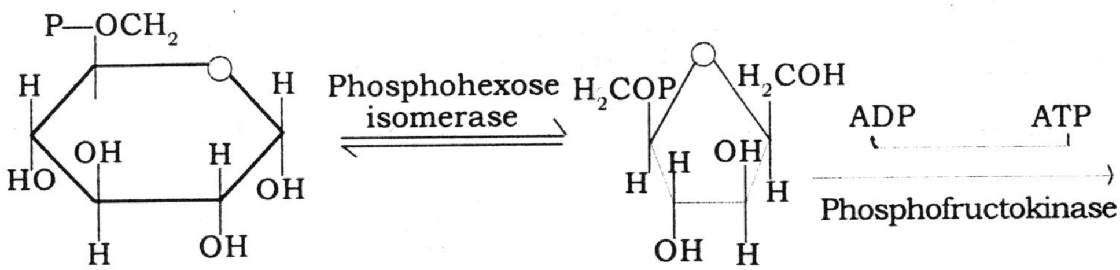

Glucose–6–P Fructose–6–P

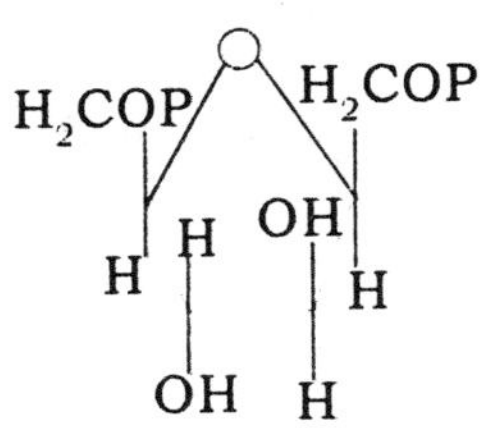

Fructose–1–6 diP

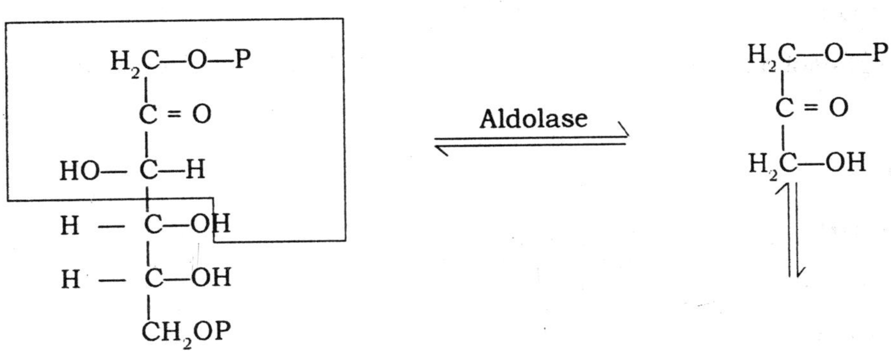

Fructose–1–6 diP

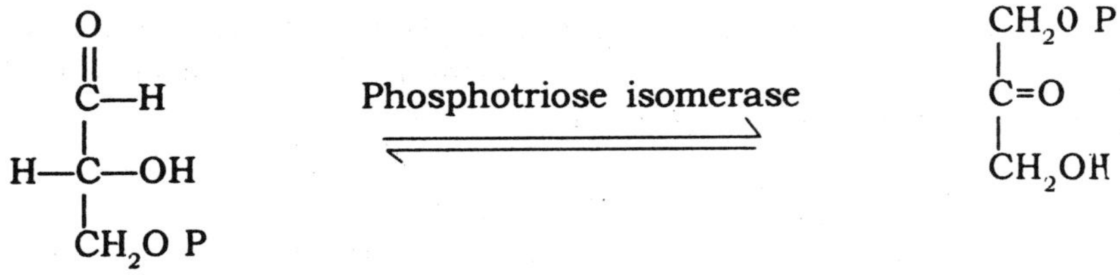

Glyceraldehyde–3–phosphate Dihydroxyacetone phosphate

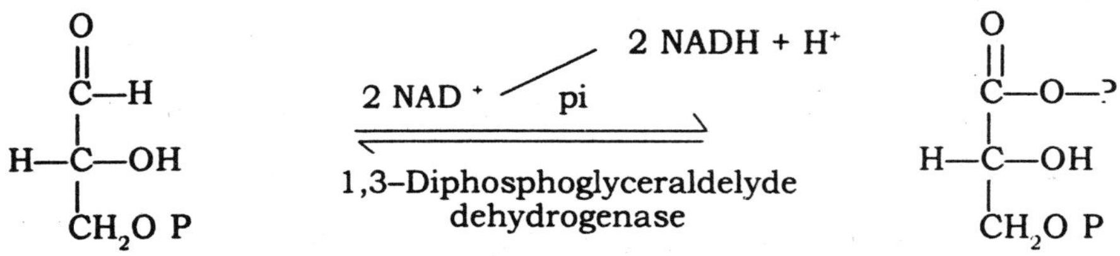

3–Phosphoglyceraldehyde 2–1–3–diphosphoglyceric acid

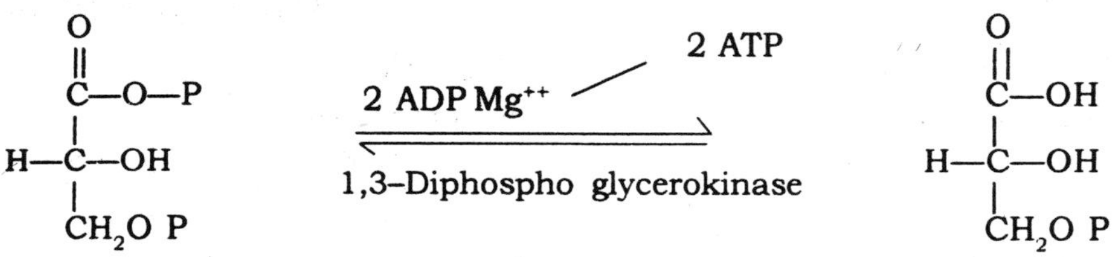

2–1–3–Diphosphoglyceric acid 2–3–phosphoglyceric acid

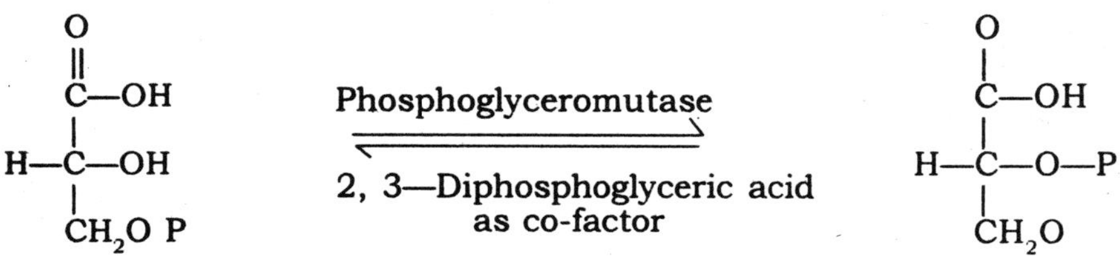

2–3–phosphoglyceric acid 2–2–phosphoglyceric acid

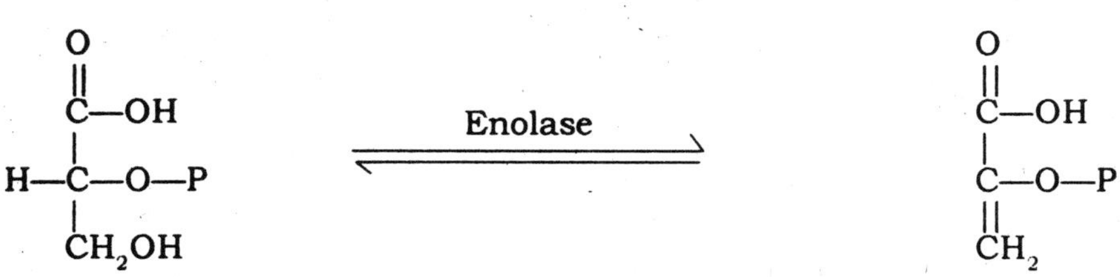

2–2–phosphoglyceric acid 2–phosphoenolpyruvic acid

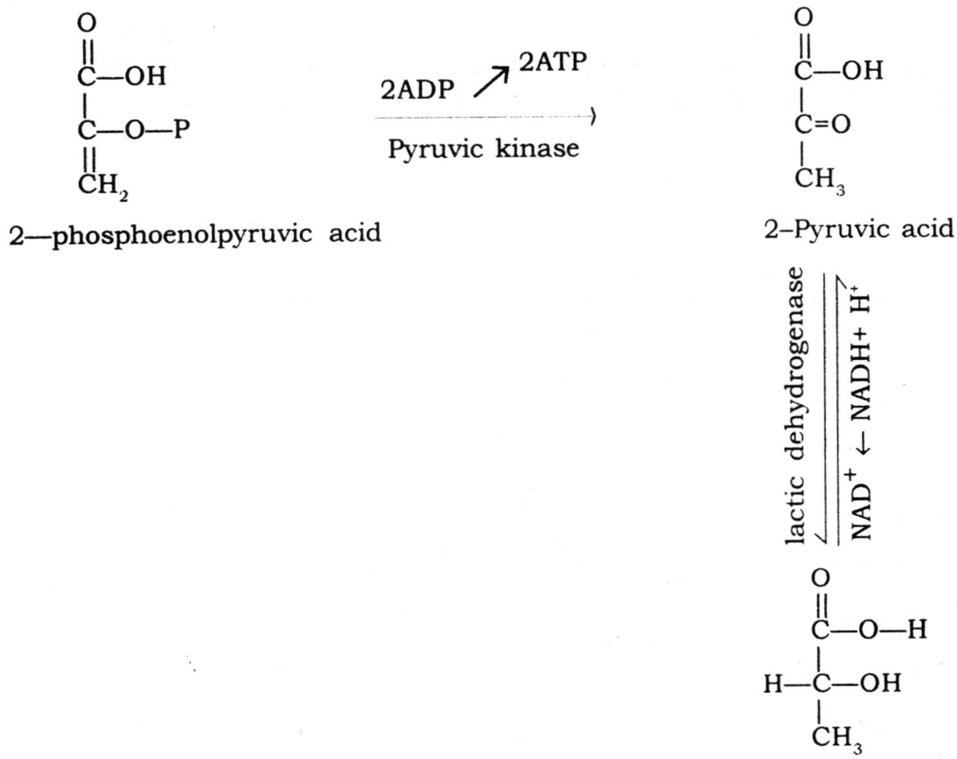

2—phosphoenolpyruvic acid → 2-Pyruvic acid → 2-lactic acid

The D-form of lactic acid is produced mainly by *Streptococcus lactis* whereas the L-form is produced predominantly by *Streptococcus citrovorus* and *Paracitrovorous.* In this pathway, the NAD^+ acts as an electron acceptor, i.e. as an oxidising agent while oxidising 2-triose-phosphate to 2-pyruvic acid. In this process it gets reduced to NADH + H^+. In total 2 NAD^+ are oxidised to NADH + H^+ . The reduced NADH + H^+ again get oxidised to NAD^+ while reducing 2-pyruvic acid to 2-lactic acid and thereby the cycle continues.

In the fermentation process, 2 ATPs are consumed while 4ATPs are genrtated, thereby producing two ATPs by substrate level phosphorylation. It should be remembered that fermentation of lactose by lactic acid bacteria takes place anaerobically, which means, in the absence of oxygen. Where aerobic catabolism (breakdown) occur wherein NAD^+s are generated, each NAD^+ through oxidative phosphorylation produces 3 ATPs. However, as fermentation of lactose by lactic acid bacteria is anaerobic, in total there is a gain of 2 ATPs in the process. Adenosine triphosphate is a high energy biological compound cleavage of whose terminal phosphate liberates biological energy which is so necessary in the fermentation process wherein such energy is utilized by the microbes while carrying out its anabolic (biosynthetic) processes.

Heterofermentative bacteria do not have the enzyme aldolase and, therefore, are unable to ferment lactose by the classical EMP pathway (as above). The route followed by heterolactic bacteria is as follows :

$$\text{Hexose} \xrightarrow[\text{Hexokinase}]{\text{ATP} \nearrow \text{ADP} \quad Mg^{++}} \text{Glucose–6–phosphate}$$

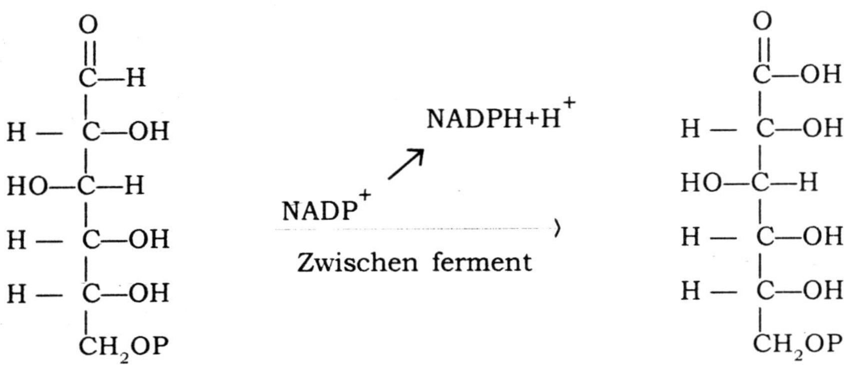

glucose—6—Phosphate 6—Phosphogluconate

This is oxidation reduction

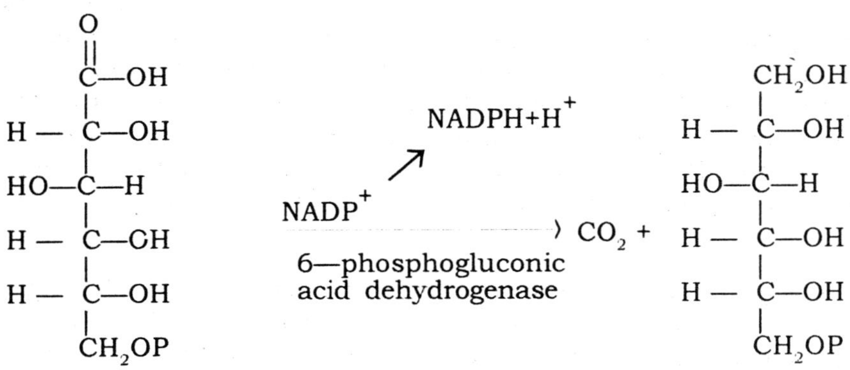

6—Phosphogluconate Ribulose—5—phosphate

Conversion of 6-phosphogluconate to CO_2 and Ribulose-5-phosphate is oxidative.

Decarboxylation Reaction

The preceding reaction and this reaction are carried out by *Leuconostocs* and altogether 2 $NADP^+$ are required to carry out the reactions to the right. The reduced $NADP^+$ produced thereby are regenerated later on in the pathway with the production of ethanol instead of acetic acid thereby causing a loss in the production of ATP while not producing acetic acid. But the Leuconostocs are compelled to do this in order to regenerate the $NADP^+$ required for conversion of hexose monophosphate to ribulose -5-phosphate.

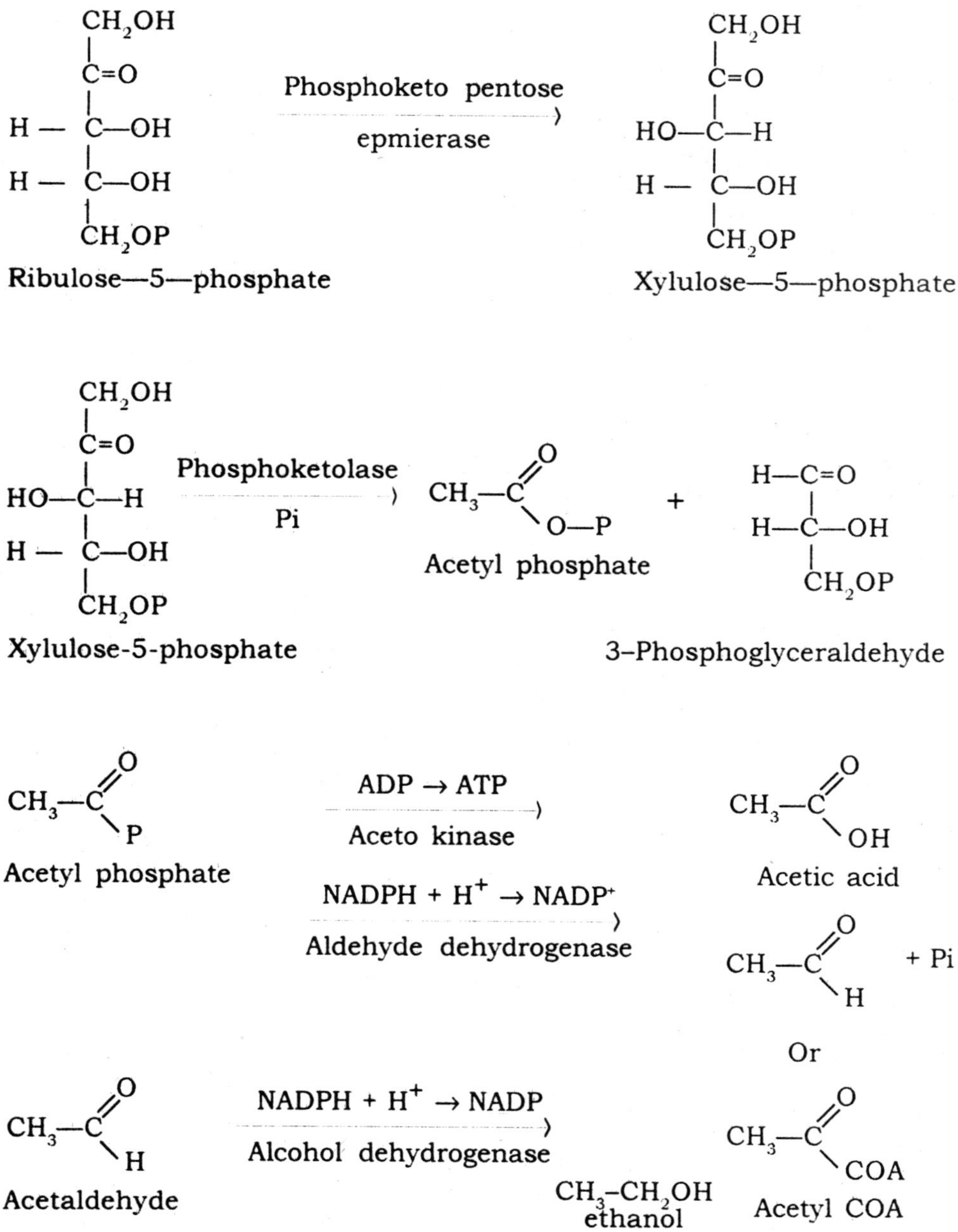

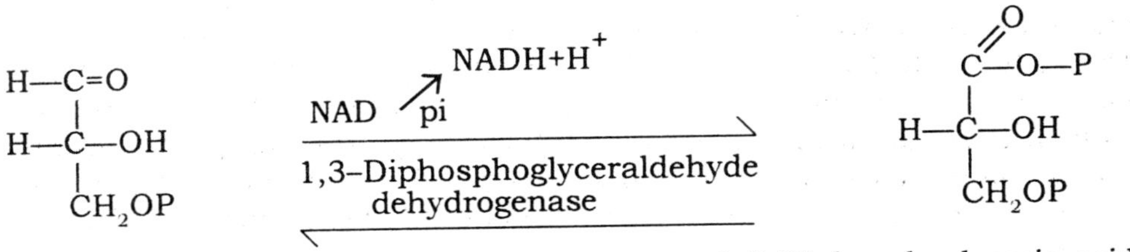

3-Phosphoglyceraldehyde → 1,3 Diphosphoglyceric acid

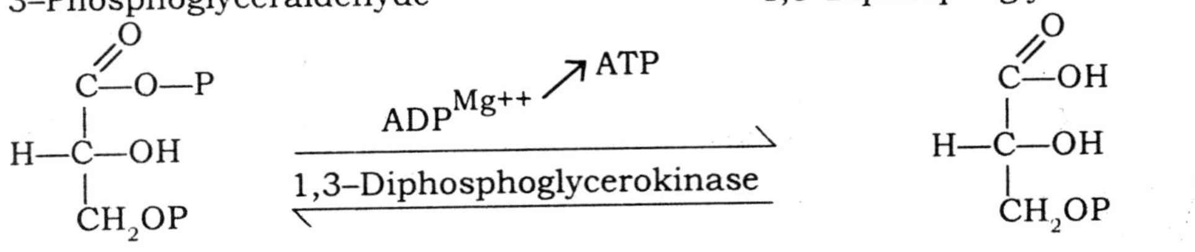

1,3 Diphosphoglyceric acid → 3-Phosphoglyceric acid

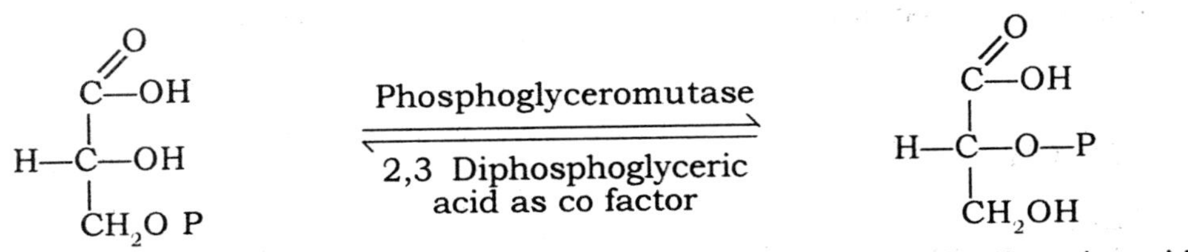

3-Phosphoglyceric acid → 2-Phosphoglyceric acid

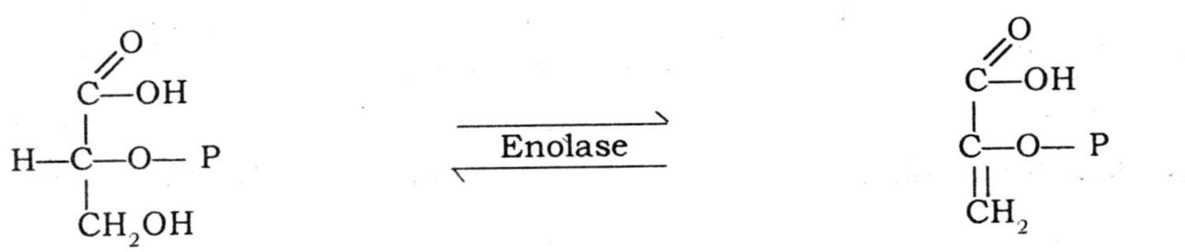

2-Phosphoglyceric acid → 2-Phosphoenol pyruvic acid

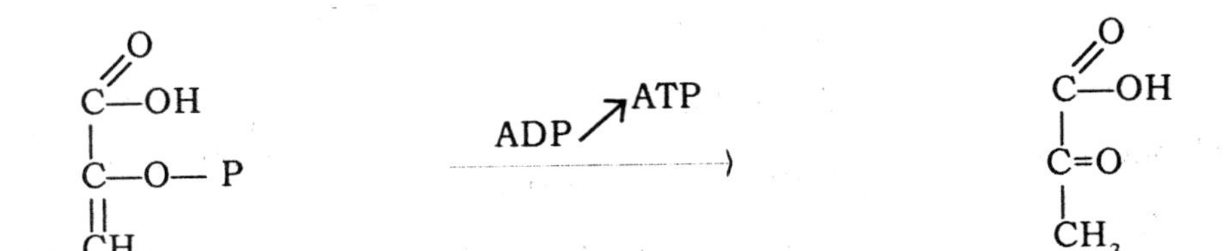

2-Phosphoenol pyruvic acid → Pyruvic acid

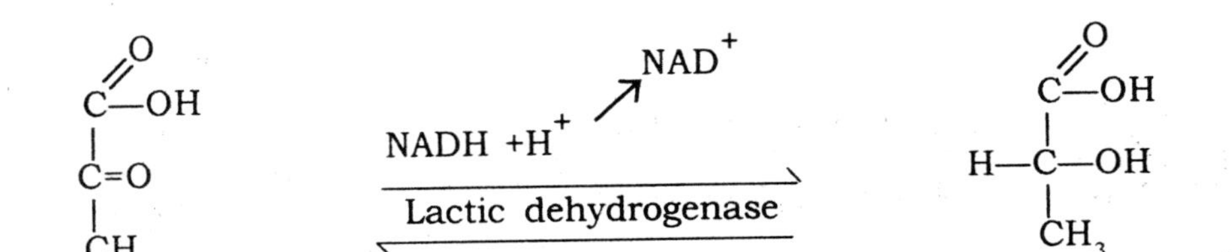

Pyruvic acid → Lactic acid

Conversion of ribulose-5-phosphate to xylulose-5-phosphate is a epimerisation reaction. This reaction is followed by a phosphoroclastic split reaction in which xylulose-5-phosphate is converted to acetyl phosphate and 3-phosphoglyceraldehyde by the reaction as depicted under.

The overall reaction for this pathway is :

Hexose——————————>CO_2 + ethanol or acetic acid + lactic acid, when a hexose (Glucose) was labelled with C^{14} at number 1, 2, 3, 4, 5 and 6 carbon atoms labelling number 1, 2, 3, 4, 5 and 6 carbon atoms appeared as follows.

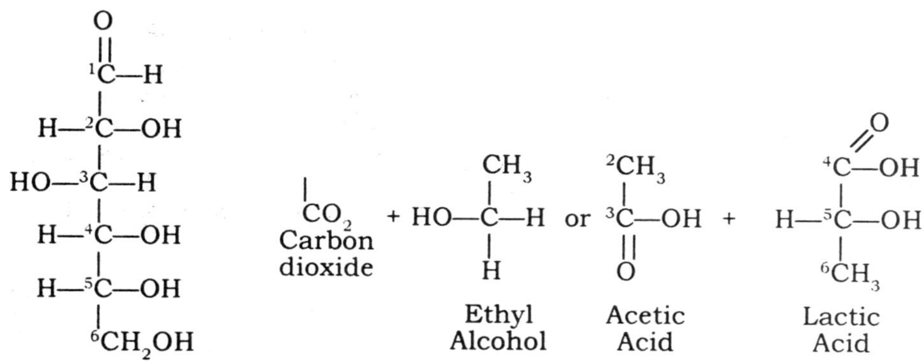

Apart from EMP pathway and phosphoketolase pathway, pentose phosphate pathway, Entner-Doudoroff Pathway are also the pathways of glucose metabolism. The pentose phosphate pathway is as on page 232.

In the previous paragraps references have been made of several bacteria which are classified as in Flow Diagram - 8.1.

In the preceding paragraphs the pathway of fermentation of glucose by homofermentative and heterofermentative bacteria by initially phosphorylating glucose by adenosine triphosphate has been discussed. A high energy biological compound, cleavage of whose terminal phosphate liberates or releases energy.

However, recent works clearly indicate that many micro-organisms have the ability to phosphorylate glucose without using ATP.

Glucose + phosphoenol pyruvate ——————————> Glucose-6-phosphate + pyruvate.

How did the story begin ? Way back in 1964 Kunding, Ghos and Hoseman described a novel bacterial p-transferase (PT) system which enzymatically transferred a phosphate from phosphoenol-pyruvate to various carbohydrates according to the scheme (As shown on page 235).

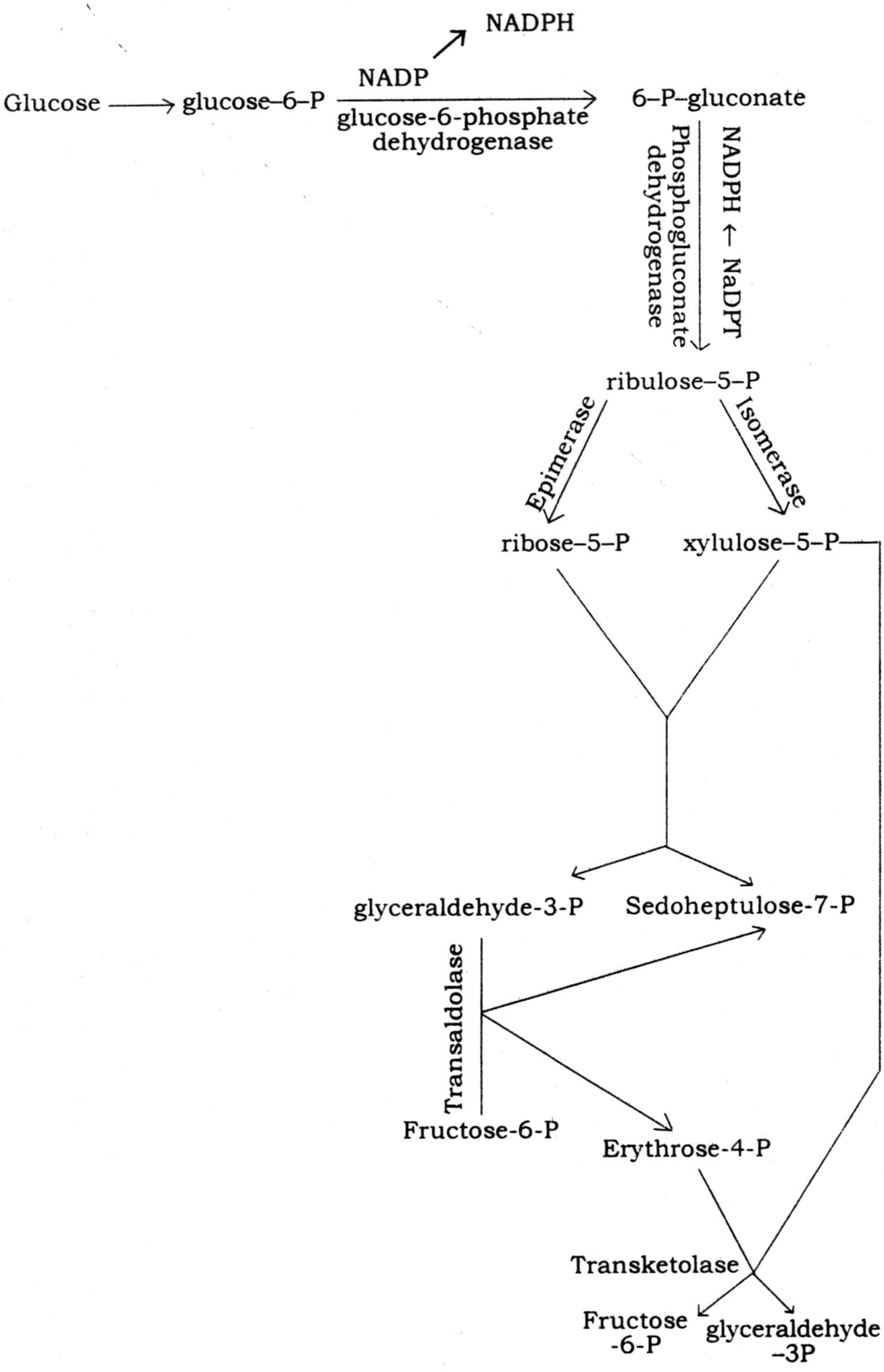
NADPH
NADP
Glucose
glucose-6-P
glucose-6-phosphate dehydrogenase
6-P-gluconate
NADPH ← NaDPT
Phosphogluconate dehydrogenase
ribulose-5-P
Epimerase
Isomerase
ribose-5-P
xylulose-5-P
glyceraldehyde-3-P
Sedoheptulose-7-P
Transaldolase
Fructose-6-P
Erythrose-4-P
Transketolase
Fructose -6-P
glyceraldehyde -3P

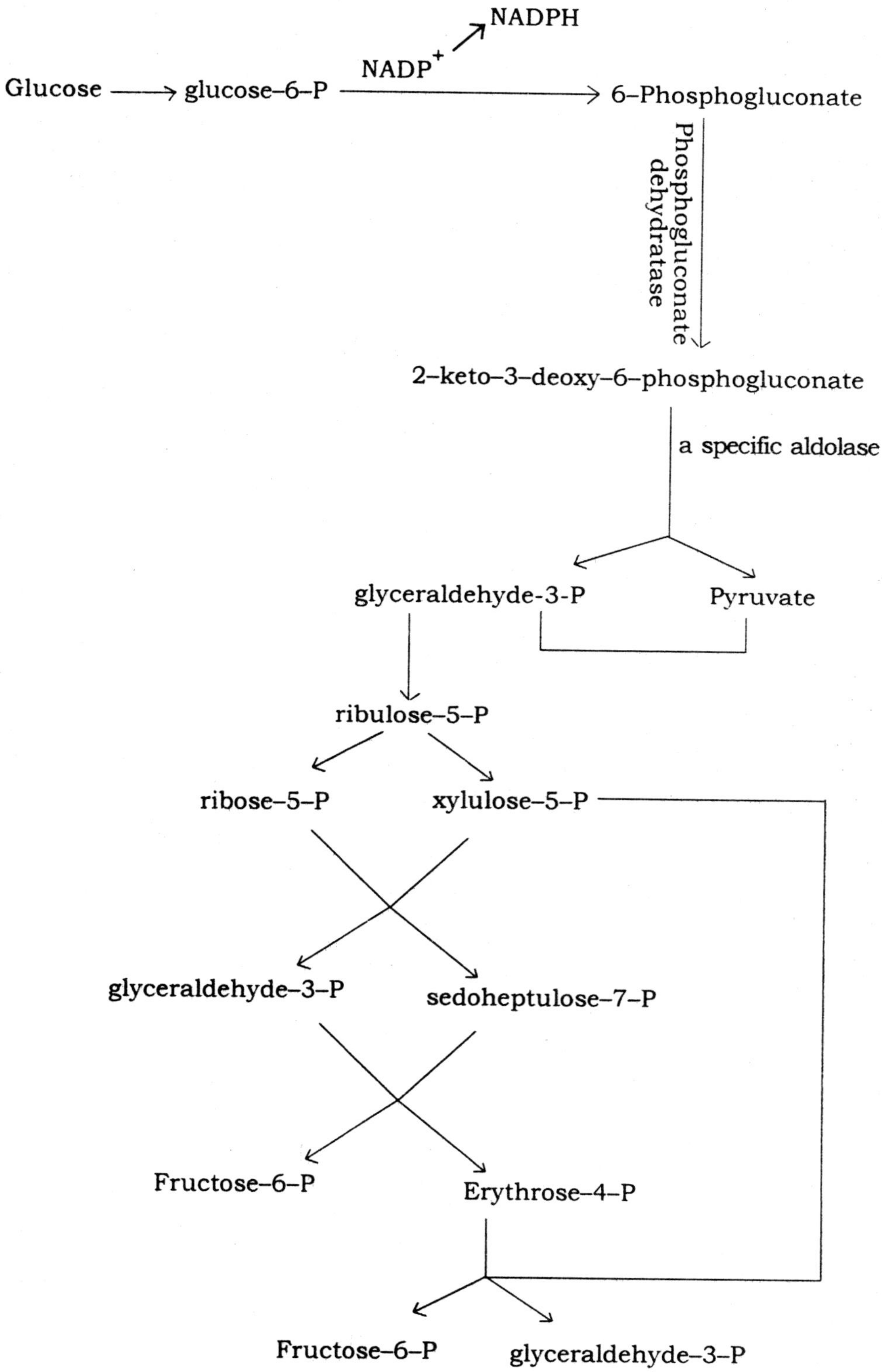
NADPH
NADP+
Glucose
glucose–6–P
6–Phosphogluconate
Phosphogluconate dehydratase
2–keto–3–deoxy–6–phosphogluconate
a specific aldolase
glyceraldehyde-3-P
Pyruvate
ribulose–5–P
ribose–5–P
xylulose–5–P
glyceraldehyde–3–P
sedoheptulose–7–P
Fructose–6–P
Erythrose–4–P
Fructose–6–P
glyceraldehyde–3–P

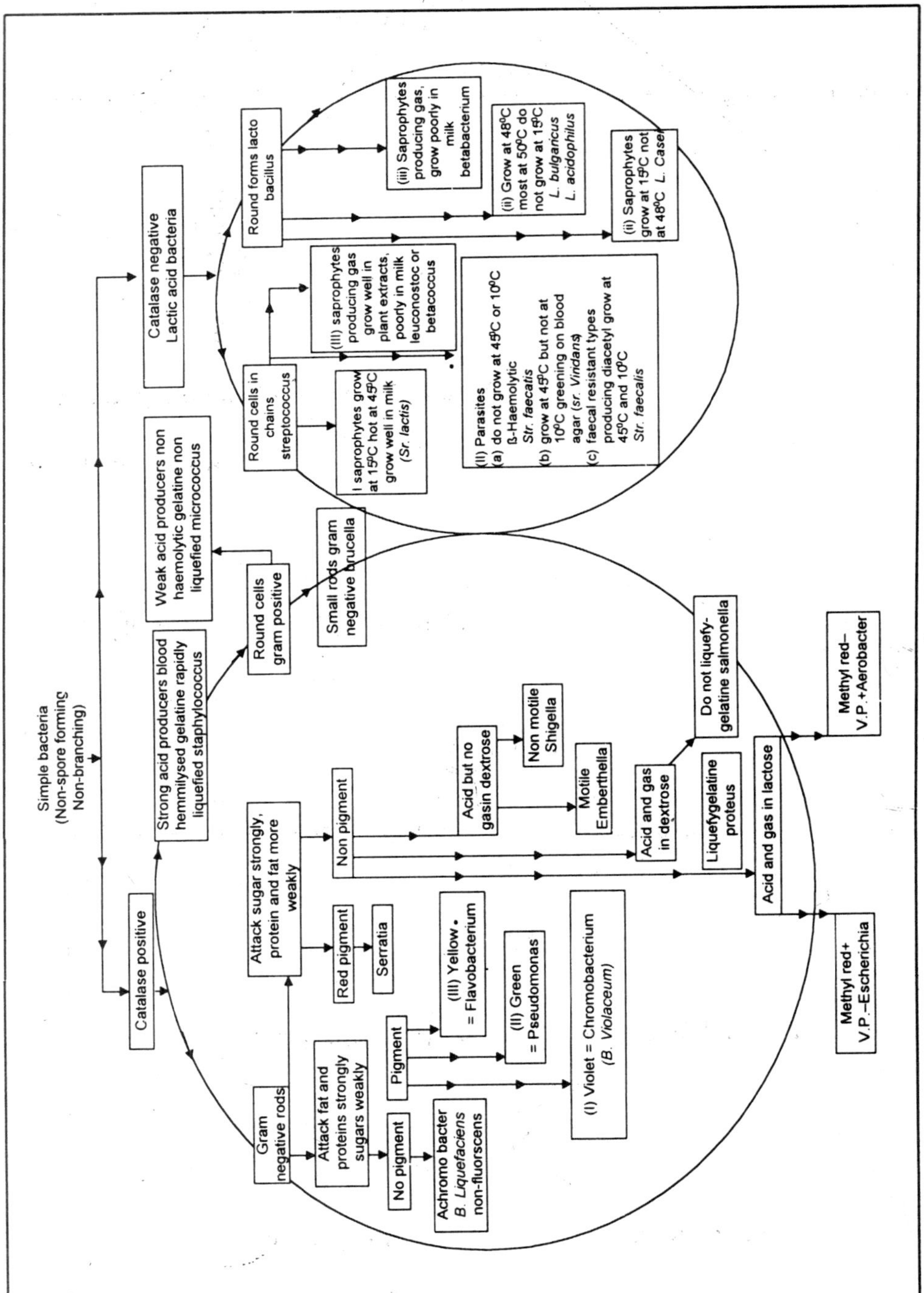

Flow Diagram 8.1 Classification of simple bacteria (non-spore forming non-branching)

Sugar + phosphoenol pyruvate ——————>Sugar - P + pyruvate.

Various sugars were phosphate acceptors while phosphoenol pyruvate was the phosphate donor.

The reactions of phosphoenol pyruvate system may be summed up as follows :

$$\text{p-enolpyruvate} + \text{HPr} \xrightleftharpoons{\text{Enzyme I, Mg}^{++}} \text{Pyruvate} + \text{P — HPr}$$

$$\text{P—HPr} + \text{Sugar} \xrightarrow[\text{Factor III}]{\text{Enzyme II, Mg}^{++}} \text{Sugar—P} + \text{HPr}$$

Sum:

$$\text{P-enolpyruvate} + \text{Sugar} \xrightarrow[\substack{\text{HPr +} \\ \text{My ++} \\ \text{Enzyme II} \\ \text{Factor IIII}}]{\text{Enzyme I}} \text{Sugar— + Pyruvate}$$

The various bacteria that can carry out these reactions are *Aerobacter aerogenes, Staphylococcus aureus, E. coli, Salmonella typhimurium.*

The phosphoenol pyruvate transferase system is abbreviated as (PEP:PTS) system.

The patthways of lactose, glucose and galactose fermentation by micro-organisms, PEP, PTS (phosphoenol pyruvate transferase system) is given in the Flow diagram 8.2.

Formation of Diacetyl in Cultured Dairy Products

The production of diacetyl in cultured dairy products is necessary to impart good aroma and flavour to the respective cultured dairy products. The precursor for diacetyl, production is citric acid and the micro-organisms responsible for its production are mainly *S-diacetylactics* and *L-citrovorum.*

For citric acid fermentation the following scheme is generally accepted. The enzyme citratase of aroma producing bacteria cleaves citric acid in the following manner.

$$\text{Citrate} \xrightarrow{\text{Citratase}} \text{acetic acid} + \text{oxaloacetic acid}$$

$$\text{Oxalo acetic acid} \xrightarrow{\text{oxaloacetic decarboxylase}} \text{Pyruvic acid} + CO_2$$

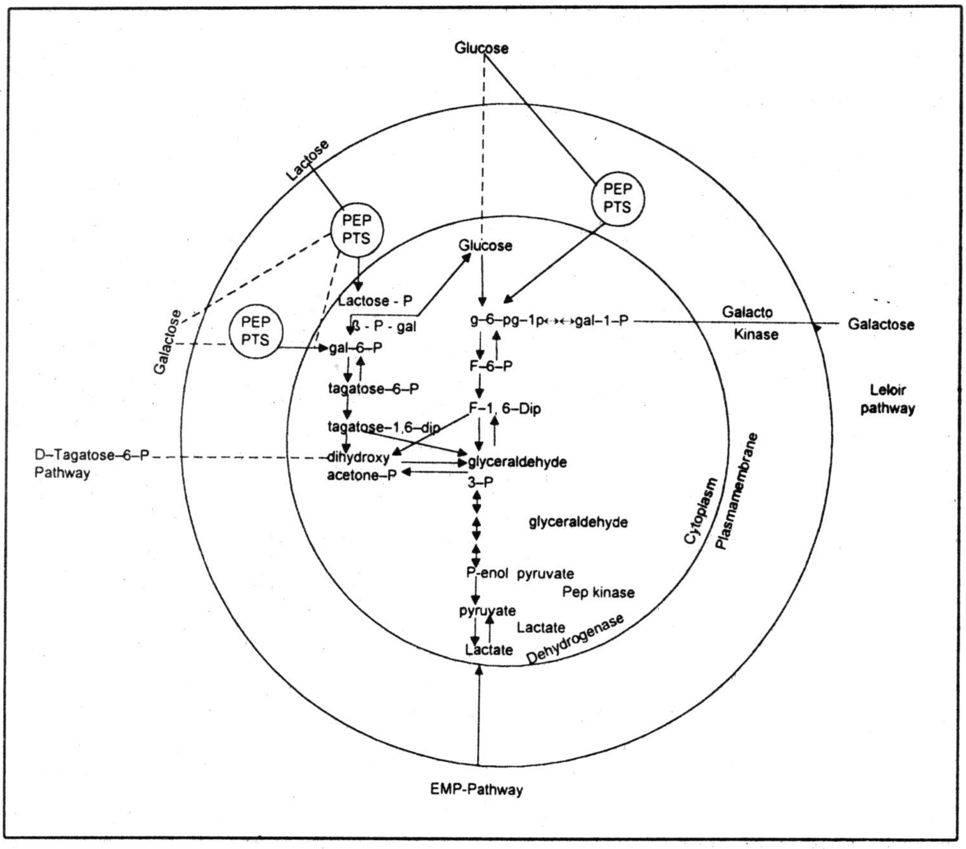

Flow Diagram 8.2 Pathways of Lactose, Glucose and Galactose fermentation by micro-organisms PEP, PTS (Phospho Enol Pyruvate transferase system)

Conversion of pyruvic acid to diacetyl or acetone by the aroma producing micro-organisms depends on the pH. However, at low pH acetoin is produced.

Is lactose or citrate the best substrate for acetoin and diacetyl production in starters?

Lactose alone cannot serve as a substrate for diacetyl production. Citrate alone or a mixture of citrate and lactose can act as a precursor for diacetyl and acetoin biosynthesis.

Pyruvic acid is an intermediate in both lactose and citric acid fermentation by bacteria. However, pyruvic acid generated during lactose fermentation is entirely reduced to lactic acid in the presence of the coenzyme, reduced NAD^+, thereby regenerating oxidised NAD^+ which is required spontaneously for catabolism of lactose to pyruvic acid. Therefore, pyruvic acid produced during lactose fermentation cannot be utilised for diacetyl and acetoin production. In the fermentation of citrate by aroma producing micro-organism, no reduced NAD^+ is generated and, therefore, there is no necessity of regeneration of oxidised NAD^+. This enables *Leuconostic cremoris* and *streptococcus diacetylactis* to metabolise the pyruvic acid produced on cleavage of citric acid completely and biosynthesise diacetyl and acetoin. When both citric acid and lactose are present, there is more pyruvic acid that can be converted to lactic acid and this excess pyruvate is then metabolised to aroma compounds at a low pH. This large pyruvate pool is fed by both lactose and citric acid of milk.

Until 22 years ago, the most prevalent hypothesis was that diacetyl was formed through spontaneous oxidation of acetoin (acetyl methyl carbinol) and decarboxylation of alpha-acetolactate.

However, it has been established that diacetyl is produced by the following pathway.

acetaldehyde-TPP + diacetyl + CoASH + TPP
acetyl CoA Reduced coenzyme

The Flow diagram 8.3 depicts the pathway for diacetyl production.

Controlled Fermentation

Controlled fermentation of milk produces butter milk, sour cream, yoghurt and cheese. In addition to carbohydrate, organic acid, proteins and fat are also fermentable if fermentation is an energy yielding oxidation-reduction process.

Abnormal Fermentation in Milk

Ropiness or sliminess is caused due to abnormal fermentation of milk

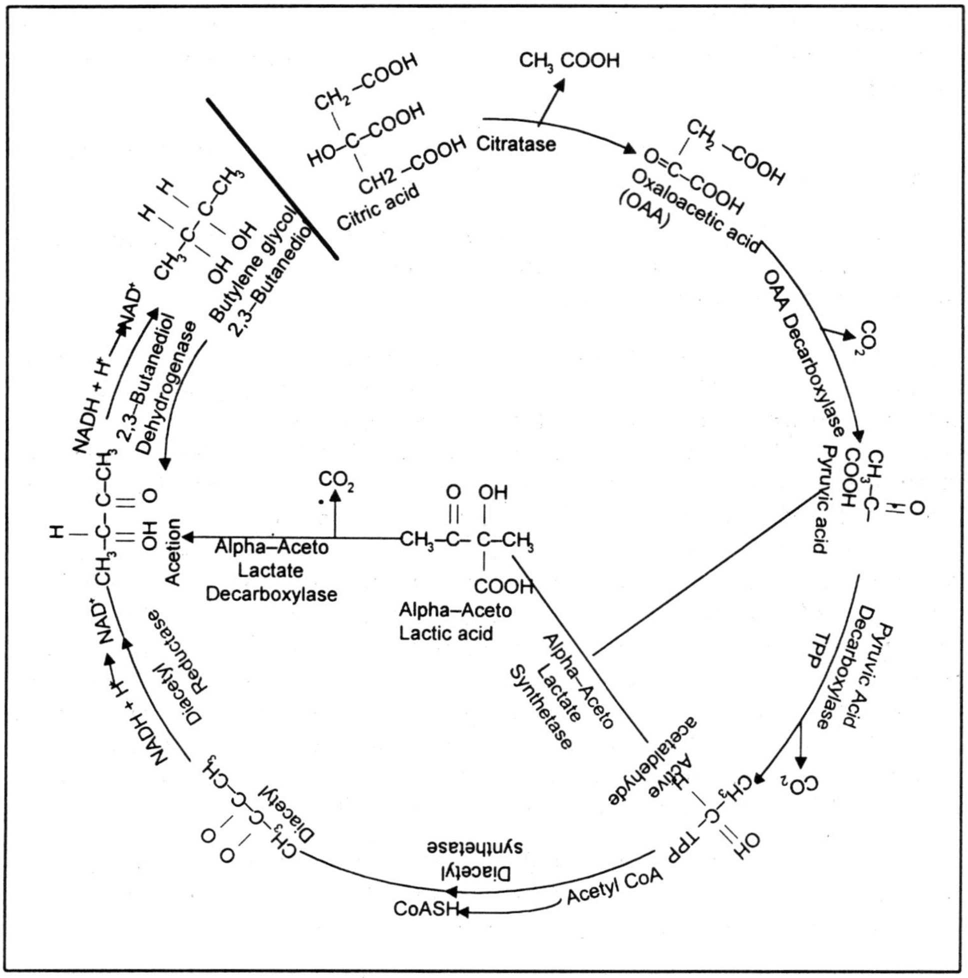

Flow Diagram 8.3 Citric acid fermentation by micro-organisms

in which mucins and polysaccharides are produced by several microogranisms requiring oxygen. The polysaccharides produced have an empirical formula of $(C_6H_{10}O_5)_6$. Ropiness is distinguished from stringy milk which is due to mastitis. Two pathways are present for polysaccharide synthesis. Wherever polysaccharides are synthesised, UDP-glucose is present. The microorganism producing ropiness in milk, therefore, can confidently be implicated with the biosynthesis of polysaccharides in ropy milk.

Ropiness may be produced by *micrococci, B. viscosum, Pseudomonas, B. clocae, B. aerogenes, Escherichia, Aerobacter* group, staphylococci, aerobic spore formers and lactic acid bacteria.

Cheese Processes

Milk clotting enzymes are used for the manufacture of cheese. All proteolytic enzymes clot milk under proper condition. Rennet is the name given to gastric extract of animals containing an enzyme rennin which is responsible for coagulation of milk. Proteolytic enzyme which also clot milk are produced by fungi, higher plants and various animal organs.

During ancient times nomadic men carried milk in the stomach of animals to preserve milk on distant routes. They observed that such transport of milk, coagulated the milk in the stomach of animals. Soon realisation came that it is an easy way to convert a big bulk of milk to smaller junk which is palatable, easily digestible, and transportable. No doubt this finding was accidental.

The variety of cheese prepared in the world today derived their names to geographical topography. Quality attribute of a variety of cheese beginning from soft, semi hard and ending with hard cheeses depends particularly on the use of rennet (chymosin) as a milk coagulant and this practice goes back to the days when mankind have just put their foot prints on civilisation or probably have begun to knock at the doors of civilisation.

Rennet induced and microbe-induced processing are the criterion of cheese manufacture. Chymosin and starter cultures both act in symphony in the processing and ultimate ripening of cheese. The main criteria behind the choice of milk coagulant for cheese making is the definite ratio of the clotting and proteolytic activities. Proteolytic activities should be such as to produce least bitter peptides in ripened cheese. Low formation of bitter peptides in ripened cheese makes sure of a better flavour in the finished product. As a matter of fact, chymosin (rennin) attacks primarily the phenylalanine-methionine bond which are respectively the 105th and 106th amino acid residues of the K-casein. The mechanism of clotting of milk by chymosin (rennin) is briefly described further :

Milk clotting occurs in two or three separate phases ie. a primary phase in which the enzyme attach K-casein to destroy its stabilizing capacity and secondary non-enzymic phase in which the destabilized system clots in the presence of calcium ions. In addition to this the tertiary phase involves the slow general hydrolysis components followed by clotting. This, however, is not essential to clotting process.

Phenyl alanine-methionine peptide bond is the primary site of rennin attack on K-casein.

Rennin

Pyr—Glu—Gln ——— Phen — Meth ——————— Val.OH
105 106 169

Pyrrolidine
Carboxylic Acid

N-term Para-K-Casein C-term | N-term Macropeptide part C-term

Phenylalamine or Leucine | Methionine | valine alanine theonine serine

The rennet action can be described in three main phases :

a. Breaking of protective colloids of casein by enzyme followed by release of non-protein nitrogen.

b. Formation of a coagulum.

c. All casein components are slowly hydrolysed under the protective action of casein.

The reactions are described as follows :

1. K-casein $\xrightarrow{\text{Rennin}}$ Para K casein + NPN (soluble in TCA, enzymatic reaction)
or casein or Para casein

2. Para casein + Ca^{++} $\rightarrow$ Insoluble clot (calciumpara caseinate with calcium phosphate)

3. Calcium para caseinate $\rightarrow$ Peptides (overall hydrolysis)

Reaction number 1 and 2 takes place in cheese vat while reaction number three takes place in ripening room. The clot formed, when coagulating enzyme is added to milk contains more mineral matter than produced by acid coagulation and is much more plastic. Most of the fat is mechanically "entangled" in the curd, where as most of the lactose, lactalbumin, lactoglobulin and a large proportion of the mineral matter remains in the whey.

The diagrammatic representation of cheese processes is given in Flow Diagram 8.4.

Bacteria from starter cultures act on lactose to produce lactic acid. Proteins (casein) and salts (calcium) in presence of rennet and heat give rise to Ca-para-K-caseinate containing fat also. Calcium-para-K-caseinate in presence of heat and lactic acid produces the coagulum in the aqueous phase whey that have a mixture of lactose and whey proteins. The coagulum when heated produces the soft curd. Heat synerresis and acid synerresis of the soft curd occur in presence of lactic acid, thereby producing the firm acid curd with the ultimate drainage of the whey. Acid curd when mixed with salt produces raw cheese. The major fermentation and ripening products in a variety of cheeses are given in the Table 8.1.

Table 8.1. Major fermentation and ripening products in a variety of cheeses

Type of cheese	Major fermentation product	Major ripening products
Bacterial ripened (soft and hard)	Lactate, diacetyl	Amino acid, volatile fatty acids, pentanone CH_3SH, H_2S, NH_3
Surface ripened (Brick, Kimburger)	Lactate, diacetyl	$CH_2(SCH_3)_2.CH_3SH$
Mould-ripened (Blue, Gorgonzola, camembert)	Lactate	Volatile fatty acids, amino acids, ketones, lactones, hydrocarbons
Cheese with eyes (Swiss, Emmental)	Lactate propionate acetate	Peptides, amino acids, proline, CH_3SH, thioesters, CH_3SCH_3

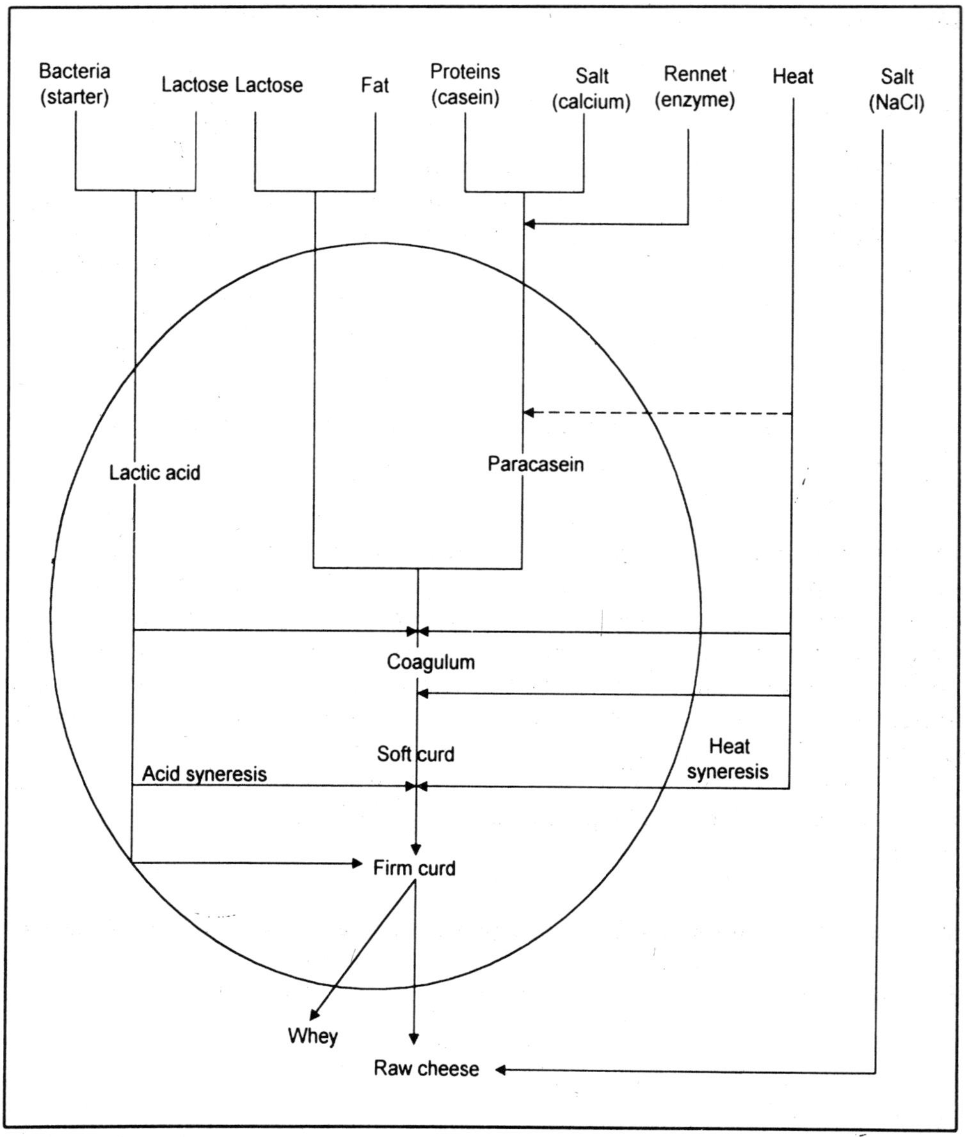

Flow Diagram 8.4 Diagrammatic representation of a cheese process

Lipolysis and Related Changes

In ripened cheese, volatile fatty acids are produced by starter streptococci. Milk fat is necessary for volatile fatty acid production (other than acetate). The pathways are briefly described in Flow diagram 8.5.

Formation of Propionate in Swiss Type of Cheese

Propionobacterium shermani though added in the starter does not play its role in the early stages of swiss-cheese ripening. Pathways leading to propionate acetate and CO_2 formation by *Propinobacterium shermani* is given in Flow diagram 8.6.

In mold-ripened cheese together with the mesophilic starter, spores of blue-green penicillium moulds are also inoculated. Mesophilic starters bring about the lipolysis and proteolysis accordingly. However, penicillium moulds are strongly proteolytic and lipolytic and they superimpose their action on the starters. The flavour that overrides and is distinct is that of fatty acids and methyl ketones (*P. roqueforti* produces both classes of compounds). The pathway for the formation of methyl ketones by *Penicillium roqueforti* is depicted in Flow diagram 8.7.

PROTEOLYSIS

Starter organisms secrete peptidases and proteinases. Starter organisms together with rennin, work on casein and break it down to peptides. These peptides are proteolysed to amino acids. In some cases, five amino acids are produced. Evidence is there for amino peptidase, tripeptidase and 2-dipeptidase activity in starter cultures of *Str. cremoris* and *Str. lactis.* Lipolysis and proteolysis produce small molecular weight compounds which has a direct relationship with the flavour of the ripened product. Though starters are mainly used as acidifying agents, they also produce flavours precursors. However, the flavour of each variety of cheese differs from the other. This is because of the distinct pathway followed by the micro-organisms during ripening of different varieties of cheeses. Flow diagram 8.8, depicts the pathway of casein breakdown by starter organisms, rennet and milk organism.

General mechanism of protein breakdown during cheese ripening is given in Flow diagram 8.9.

During the process of ripening of cheeses, chymosin exhibits proteolytic activities and produces bitter peptides by attacking α_{s1}-casein and ß-casein fractions. In α_{s1}-casein, the peptide bond between phenylalanine residue number 23 and phenylalanine residue 24 is very susceptible. The specific sites of chymosin attach on α_s-casein fraction are the amide linkages between phenylalanine residue numbers 23 and 24 or phenylalanine residue number 24 and valine residue number 25. Peptide bonds, 25 in

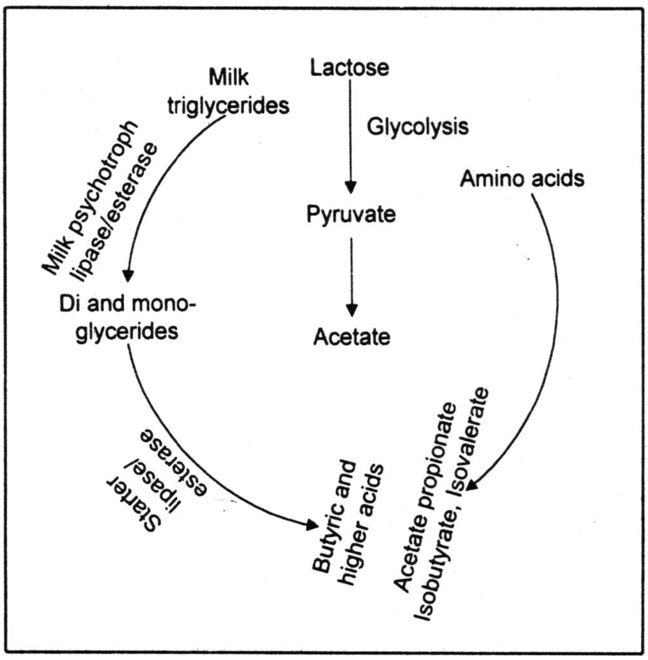

Flow Diagram 8.5 Possible pathways involved in the formation of volatile fatty acids by starters in cheese

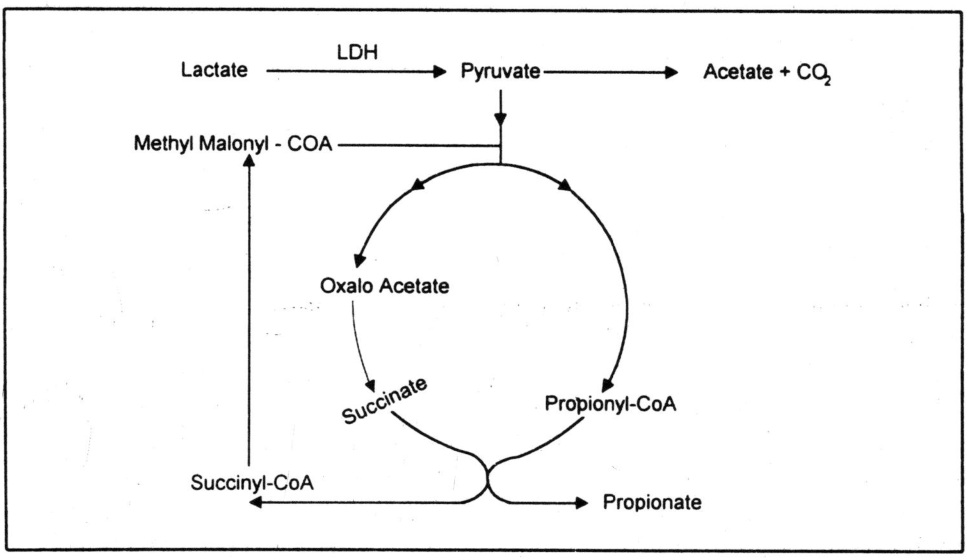

Flow Diagram 8.6 Pathways leading to Propionate acetate and CO_2 formation by *Propionobacterium shermani*

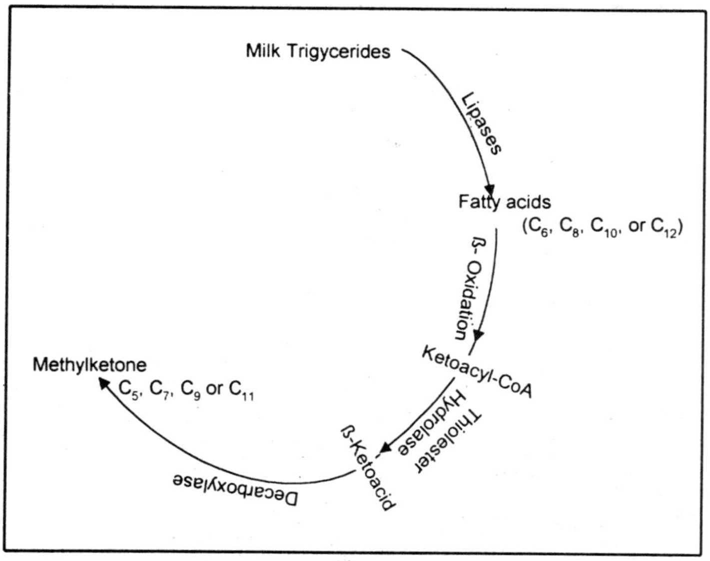

Flow Diagram 8.7 Pathway for formation of Methyl ketones by *Penicillium roqueforti* (mold ripened cheeses)

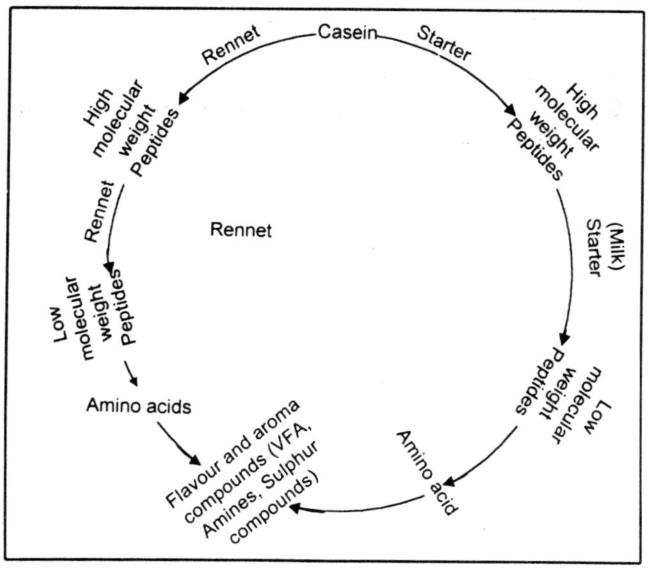

Flow Diagram 8.8 Pathway of casein breakdown by starter organisms, rennet and milk proteinase

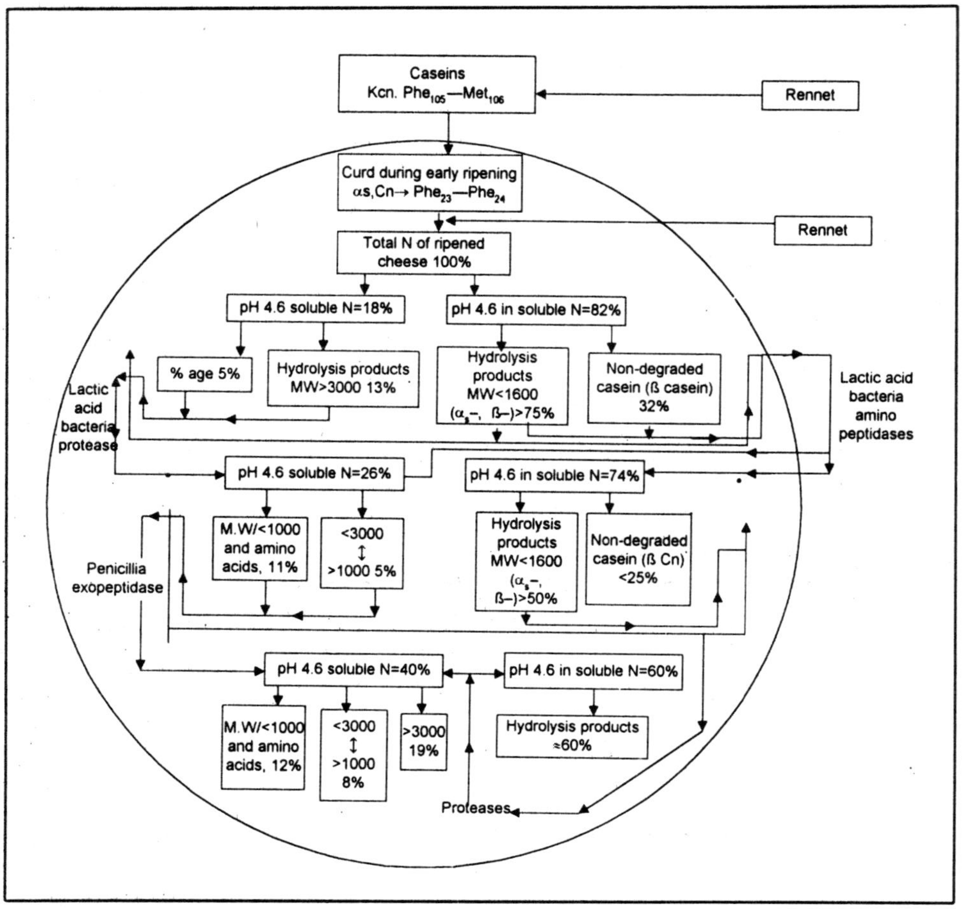

Flow Diagram 8.9 General mechanism of protein breakdown during cheese ripening

number are extensively hydrolysed by chymosin in α_{s1}-casein B. The carboxy groups of phenylalanine and leucine are preferentially attacked in these bonds. To date, eight bitter peptides have been isolated. Whereas ß-casein is hydrolysed by chymosin to three products, viz. B-1 (164-165), B-2 (189-190), and B-3 (192-193). So far, three bitter peptides have been isolated. Addition of salts inhibits production of bitter peptides from ß-casein fractions by chymosin. Therefore, monitoring of the salt level in cheese controls the flavour and hence the quality of the ripened cheese. However, amongst the milk coagulants so far used in the cheese industry chymosin produces least amount of bitter peptides in the ripened and finished products, thereby being preferred over most of the other milk coagulants for cheese making.

The amino acid sequence of the bitter peptides are an inherent part of the casein micelle. A question arises why then when paneer or any other milk product containing appreciable amount of casein is taken as such, bitter taste is not detected by the taste buds.

It has almost always been noticed that bitter peptides are caused by hydrophobic amino acids i.e. amino acids having non-polar hydrocarbon side chains. The hydrophobic amino acid residues of α_{s1}- and ß-casein fractions exist in the interior of the casein micelle and are so hidden that it does not reach the taste buds when casein or any other casein products are eaten. In ripened cheese which has undergone extreme proteolysis, peptide fractions are released to the outer sphere of the casein micelles. Those peptides which have hydrophobic amino acid residues now on the outer surface of the casein micelle very easily touch the taste buds. Hence the cheese so ripened gives a bitter taste. Nevertheless, attempts should be made by monitoring and manouvering the salt level of the cheese and also modifying the procedure by adjusting the pH of the milk to be coagulated to such a level as to decrease the attack of milk coagulants on α_s-and ß-casein fractions. Once the proteolytic attack on α_s- and ß-casein fractions is reduced, there will be a concommitant decrease in the production of bitter peptides. The pH chosen for control on the breakdown of α_s-casein and ß-casein fractions by a particular milk coagulant should be such as to warrant optimum milk clotting activity. The figure shows the enzymatic cheese ripening process which ultimately gives the characteristic of body texture, flavour and aroma. Flavour is dependant on the enzymatic breakdown of proteins, sugar and fats. Aroma is dependant on hydrogen sulphide produced from the enzymatic breakdown of proteins, on esters of C_2-C_{10} fatty acids and also on C_2-C_{10} fatty acids. Perfect body of cheese depends on breakdown products of proteins and $C_{12}C$-$_{22}$ fatty acids arising from the enzymatic breakdown of fats. As regards texture of ripened cheese, good texture is due to enzymatic breakdown of protein and C_{12}-C_{22} of fatty acids produced by enzymatic lipolysis of fats (Flow diagram 8.10).

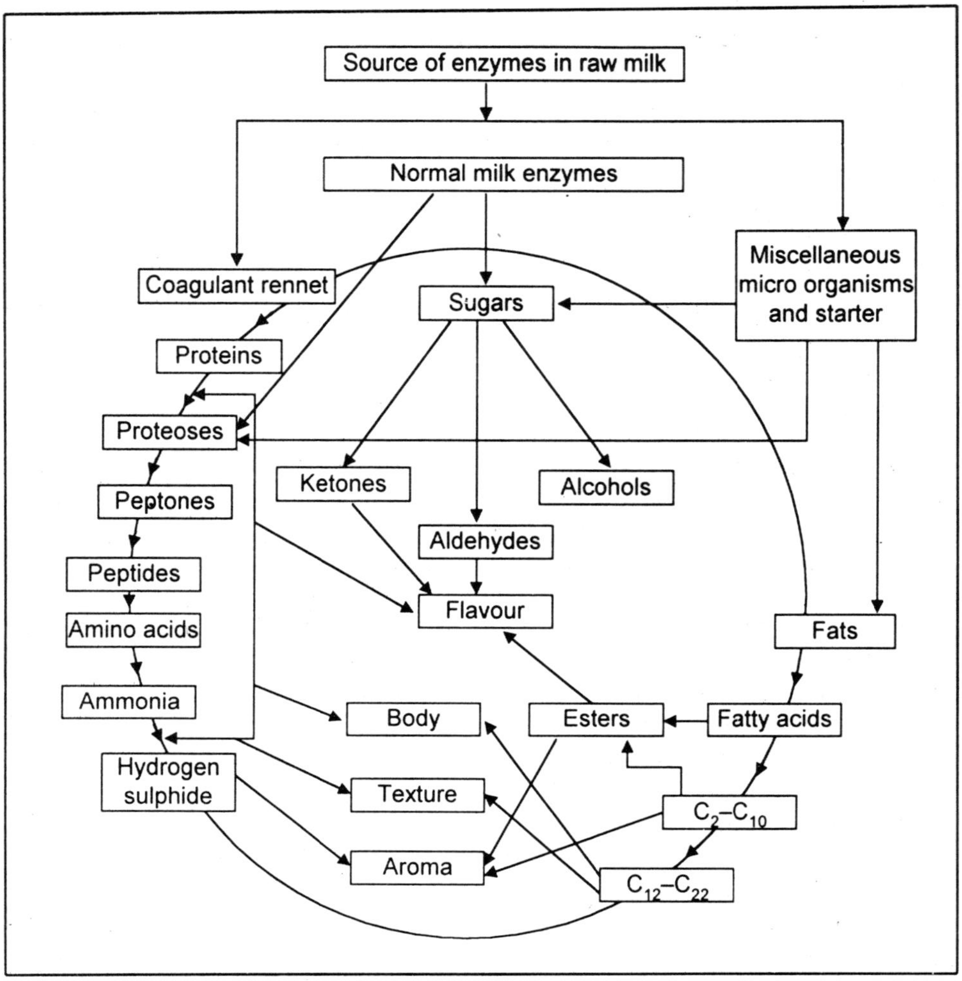

Flow Diagram 8.10 Diagrammatic representation of the enzyme sources and degradation of curd constituents to give the characteristics of body, texture, flavour and aroma

Milk coagulants other than rennet (chymosin) have been used from time to time in cheese making. This is primarily because of commercial rennet (Hansens rennet) becoming scarce and expensive. Moreover, religious sentiments of some vegetarians are safeguarded in consuming cheese not prepared by Hansen rennet. In addition to this, since the alternative source of chymosin are available it should be explored.

The Table below gives the detail of various sources of enzymes and their suitability for cheese making (Table 8.2).

Table 8.2. Summarised information on the research carried out for the substitution/supplementation of calf rennet for cheese making.

Milk coagulants	Mode of action, fractions formed	Suitability in cheese manufacture
1	2	3
Pig pepsin, chymotrypsin, microbial rennet	Acts on K-casein polypeptide fraction with N-terminal methionine + para K-casein produced having identical electrical electrophoretic mobility compared to that produced by chymosin.	Proteolytic activity being more than than that of chymosin, bitter peptides are more and the quality of cheese manufactured is inferior to that produced by chymosin.
P. polycephalum	Acts on K-casein polypeptide fraction with N-terminal methionine. Acts on the para-K-casein like carboxy-peptidase, release N-terminal amino acids.	Proteolytic activity more than chymosin and hence more bitter peptides are produced. Flavour of the cheese not acceptable.
Carboxy-peptidase A	Act on the para K-casein releases N-terminal amino acids.	Same as above
M. pusillus, Endothia parasitica	Hydrolyse chymosin sensitive bonds of K-casein	Proteolytic activity slightly more chymosin. However, cheese quality inferior.
Trypsin	Acts on K-casein, Hydrolyse 97, 111 and 112 amino acid residue to give two moieties of K-casein. Sufficiently similar in properties to the true glycomacropeptide and para-K-casein of chymosin hydrolysis, can attack on the 14 different sites on peptide chain.	Same as above
Chymotrypsin	Same as above	Same as above
Papain, ficin and bromelain (sulphydryl protease)	Milk clotting is similar to that of chymosin but extensive hydrolysis of all the casein fractions occur.	Proteolytic activity much higher than that of chymosin. Hence, the cheese manufactured is not suitable.

Acid proteases other than chymosin	Non-specific proteolysis occurs. Acid proteases from non-bovine sources are more generally proteolytic than chymosin. They are and also confined to hydrolyse bond of K-casein as compared to chymosin.	Due to non-specific proteolysis occurrence of bitter peptides in the ripened cheese is higher and hence not suitable for cheese manufacture.
B. polymyxa, E.parasitica	The proteases from these two organisms catalyses rapid hydrolysis of α_s and ß-casein fraction	Proteolytic rate higher than chymo sin. Bitter peptides are produced. Hence unsuitable.
M. meihei, Mucor pussillus	The proteases from these two sources hydrolyse α_{s1}- and ß-casein fractions more than rennet. Hydrolyse ß-casein fractions more than a_{s1} casein fraction.	Mucor coagulants have a greater acceptance and are more suitable for cheese manufacture.
Acid proteinases	These act on the presence of 2-carboxy group, one in the protonated and one in the de-protonated form; each molecule. Hydrolyse peptide bonds franked by bulky hydrophobic amino acids.	
Withania coagulans berries	–	Satisfactory cottage and soft cheese. Despite modification of the manufacturing technique defects in cheddar cheese.
Cynara carduncules		Soft cheese from sheep milk (portugal. Serra from sheep milk. Due to high proteolytic activity, defective flavour and texture for Edam and Roquefort cheese.
Plant Proteases	Highly proteolytic	Produces bitter peptides in the ripened cheese and hence unsuit able.
Cardoon flowers	More proteolytic than rennet	Superior to rennet in production of serra cheese from sheep milk.
Bacillus subtilis	Lower ratio of milk clotting to proteolytic activity than rennet. Extensive hydrolysis of ß-casein occurs.	Suitable for soft or semi-soft cheese.
B. mesentericas	Loss of fat and nitrogen in whey occurs when compared to rennet.	Fairly good for semi-hard cheese.
Irfexlacteas	Slightly more proteolytic than rennet	Not suitable for semi-hard and hard cheese.
Byesochlamys fulva	–do–	–do–

Rhizopus	Highly proteolytic	Too proteolytic for cheddar cheese.
Bacillus proteinases	Extensive hydrolysis of ß-casein occurs. Rennet acts on K-casein primarily cleaves the phenylalanine-methionine bond (105) and forms Ca-K-caseinate and glycomacropeptide.	Bitter peptides are produced and hence unsuitable.
Rennet:Pepsin 50:50	Pepsin acts on K-casein as K-casein as detailed earlier	Suitable
Rennet:*E. parasitica* (50:50)	Rennet and *E. parasitica* acts on K-casein as written earlier.	-do
Bovine pepsin: Pig pepsin 50:50	Both bovine pepsin and pig pepsin acts on K-casein as mentioned earlier.	-do-
Rennet:Bovine pepsin (94:6)	Greater action of rennet on K-casein than bovine pepsin on K-casein.	-do-
Rennet:Bovine pepsin (9:91)	Reverse is true	-do-
M. pussillus proteinases:pig pepsin (50:50)	Mode of action of *M. pussillus* is similar to that mentioned earlier and that of pepsin remains the same as before.	Suitable cheese produced.
Rennet from chicken entrails and chicken pepsin	Mode of action similar to that of Hansen rennet.	Quality of cheese produced is better than that of microbial rennet.
Rennilase	-	Good camembert, Tilsit, Edam, Italian cheese, St. paulin, Emmental, Yugoslavian cheese, acceptable cheddar cheese in UK.
Rennilase and Farmase	-	Acceptable Gouda cheese

Acid Induced Dairy Processes

Coagulation of the milk system is attributed to destablisation of the casein micelles and this phenomenon reveals summation of many changes in colloid system. The acid coagulation of milk is achieved at 35-50°C and pH 4.2.-4.6 (isoelectric point of casein at pH 4.6-4.7). The lowering of pH of milk can be achieved by adding normally acetic acid (10%), hydrochloric acid (10%) or lactic acid. Casein is present in milk in the form of rather large colloidal particles associated with calcium and phosphorus. These colloidal particles are known as calcium phosphocaseinate or calcium caseinate phosphate. When acid is added to milk and pH lowered calcium and phosphorus which is associated with casein progressively dissolves until the isoelectric point or casein (pH 4.6 to 4.7) is reached and casein is free of bound salts. The curd thus formed is mainly casein associated with fat which can be removed by washing with fat solvent. When only casein is desired skim milk is used. The whey which remains after removal of precipitate or curd contains α-lactalbumin, ß-lactoglobulin, several minor proteins, lactose, soluble minerals, salts and vitamins etc. A point needs to be mentioned that at elevated temperatures milk whey proteins get denatured and K-casein interacts with ß-lactoglobulin to form a complex which has an isolectric pH of 5.7 to 5.9 as against casein whose pI is 4.6. Also self aggregation of the complex leads to a typical spongy and close-knit texture.

Temperature Induced Dairy Process

The fluid milk is normally heated after clarification to improve its durability and to kill disease causing micro-organisms. There are mainly three types of heat treatment given to milk, pasteurisation (71-74°C for 15-40 sec., 62-65°C for 30 min.) UHT treatment (135-140°C for 6-10 sec. or 140-145°C for 2-4 sec.) and sterilisation (110-120°C for 10-20 min). The ability of milk to withstand high processing temperature is known as heat stability of milk. It is determined by heating a small sample of milk (1-1.5 ml) sealed in a narrow glass tube, in a thermostatically controlled oil bath usually at 140°C for milk of standard concentration or at 120°C for concentrated milks, until particles of coagulated proteins are observed in the flowing milk. Heat coagulation time (HCT) depends on several factors. Depending on the shape of the curve of HCT/pH, the milk sample is referred to as type A or B. Some mlk show a very pronounced stability meaximum and minimum and are known as type A milk while other show no maximum or minimum are known as type B milk. In type B milk stability increases with increase of pH.

Normally at pasteurisation, UHT or sterilisation temperature no coagulation takes place. However, such process affect several milk constitutents. Coagulation of milk takes place at high temperature-time combination and type of milk. Several factors affect heat coagulation of milk. The chemistry and reaction of changes that leads to heat coagulation are extremely difficult to envisage but several factors and changes are widely accepted which induces heat coagulation. These are outlined as follows :

a. During heating at high temperature, the pH of milk decreases to 5.5 to 6.0 from its normal pH. This decrease in pH is due to production of organic acid principally formic acid from lactose, precipitation of primary and secondary calcium phosphate as tertiary phosphate with concomitant release of H^+, and hydrolysis of organic (casein) phosphate and its subsequent precipitation of $Ca_3(PO_4)_2$ with release of H^+.

b. Calcium phosphate precipitates on heating, may be as hydroxyapatite, $Ca_{10}(PO_4)_6 (OH)_2$ which differs from indigenous colloidal calcium phosphate (CCP) which is a complex of calcium citrate and calcium phosphate.

c. During heating milk at 140°C, browning occurs. The maillard type of browning is due to the interaction of the carbonyl group of sugars with amino compounds.

d. Casein, the principle protein of milk is completely dephosphorylated with high temperature (5 hrs at 120°C).

e. With high heat temperature and prolonged periods peptide bonds are cleaved with the liberation of NPN. The ability of K-casein to stabilise α_s-casein against Ca^{2+} is impaired or destroyed.

f. Micellar state of casein in milk is altered by heat. Due to heat, colloidal calcium is removed which leads to aggregation and dissociation of casein micelles.

g. Different changes that occur in milk system on heating at high temperature also leads to a reduction of the zeta potential.

h. The physical changes that take place on heating of milk are; change in colour, viscosity, surface tension, residual oxygen

(Direct .... 0 = 2 ppm and indirect .... 0 = 8 ppm.

UHT Process

Recently UHT (Ultra High Temperature) milk have become very popular all over the world. In UHT plants milk is heated either by a direct or indirect process at 130°-150°C for a few seconds and it is then asceptically packaged, usually in cartons.

Coalescence, age - thicknening, decomposition and gelation of UHT milk are the changes that take place in the casein particles during storage of UHT-sterilised milk. It was known that in conventional sterilisation (13 minutes, 118°C), the general appearance of the casein particles did not vary for six months at 28°C. In UHT sterilised (15 econds, 135°C) concentrated milk, it was usually observed that immediately before the milk solidifies completely due to age thickening, the casein particles started to coalesce.

UHT sterilised evaporated skim milk was made from milk to which approximately 50 percent extra K-casein (milk B) or approximately 20 percent extra (α-$_{s1}$+ß) casein (milk C) had been added. The size of the casein micelles in these milks and their behaviour with respect to the process of age-thickening were compared with those of concentrated skim milk without any addition of casein (milk A). It was found that the casein micelles in milk B were smaller and those in milk C were bigger than the micelles in milk A. This meant that addition of K-casein caused the casein micelles to coalesce and had no effect on the gelation pattern of the three milk systems. The hypothesis is put forward that an increase in the ratio of K-casein to (a_{s1}+ß) casein permits the coating of a larger total surface area of micelles with K-casein. A decrease in this ratio reduces the amount of available K-casein, thus forcing the micelles to grow in order to decrease the total surface area.

The stability of the milk proteins to ethanol, calcium ions and rennet decreased with time of storage. However, preliminary autoclaving of the UHT milk induced a high degree of stability.

The extent of protein decomposition that had occurred in the UHT milk samples depended on the time and temperature of storage. There was negligible decomposition when the samples had been autoclaved before storage at 4° and 20°C. Storage at 37°C, led to significant decomposition. UHT milk samples gelled after being stored for 13 months at 4°C, 20°C and 30°C, but not at 37°C. The autoclaved milk samples was still fluid after being stored for 2 years at these temperatures. The time of onset of gelation did not depend on the degree of protein breakdown and it seemed, therefore, that proteolysis was not the primary cause of gelation on storage. Because of bulk tank collection, the storage time of raw milk in the farm has been prolonged. It is stored at 5°C but psychrotrophic bacteria develop at this temperature and produce extracellular lipases and proteases which have a negative effect on the quality of the raw product.

The gelation pattern in UHT sterilised milk inoculated prior to UHT-sterilisation with psychrotrophic bacteria produced heat resistant proteases. Uninnoculated twice UHT sterilised milk did not coagulate. However, innoculated twice UHT sterilised milk coagulated. Neither uninnoculated nor innoculated UHT sterilised whey or casein solution coagulated. However, a product that coagulated after being stored for a few days was obtained if calcium and phosphate were added to casein solution at the same concentration in milk.

During storage, sialic acid was released in the inoculated milk and the amount of sialic acid in the serum part of the milk increased. The conclusion may be drawn that pseudomonas cultures produce proteases that survive UHT-sterilisation treatment and these heat-resistant proteases will attack the casein and split off the lycomacropeptide from K-casein whereby sialic acid is released. Development of bitter flavour or gelation in UHT sterilised milk is due to bacterial proteases which survive the heat treatment. The heat-resistance of these proteases appear to preclude their destruction by heat during processing. A proteanase containing fraction from UHT-sterilised milk was added to UHT-sterilised milk and the proteolysis in the sample increased compared with that in control samples without added proteinase. The increase in proteolytic activity and deterioration in samples containing added proteinases indicated that an enzyme process occurred in UHT-sterilised milk which could influence gel-formation.

It has been observed that heat-treatment itself did not change the electrophoretic pattern of casein. During storage the increase in the whey protein concentration only partially is due to the restoration of denatured whey proteins in the native state. This increase, in fact, is particularly due to proteose-peptone and albumin fractions which progressively reach the concentration values higher than in milk before UHT sterilisation process. The increase in the proteose-peptone and albumin concentration is also due to the products of casein and probably, of some whey protein break-down.

The first step of gel formation behaviour, the dissociation of casein/ whey proteins complex agatinate into two former fractions is purely due to its chemical-physicl nature, when the total quantity of calories supplied is insufficient either for stable, physical entanglement or for sufficient inter-molecular non-covalent bonds.

On the other hand, the second step of gel formation behaviour, the slow coagulation of casein no longer protected, can be looked upon as a gentle; enzymatic process due to the action of residual proteases which survived the heat treatment or were reactivated during storage.

Reverse Osmosis (RO) and Ultra Filtration (UF)

Membrane separation process are based on the ability of semi perme-able membrane of the appropriate physical and chemical nature to dis-criminate between molecules primarily on the basis of size, and to a lesser extent, on shape and chemical composition. The distinction between three chief membrane separation process namely reverse osmosis, ultrafiltration and microfiltration are arbitarary. However, in its broader sense RO is essentially a dewatering technique, UF is a method for simultaneously purifying, concentrating and fractionating macromolecules or fine colloidal suspensions, and MF is a clarification process that separates molecules and particles on the basis of size and solubility. The three processes differ in their membrane pore-size (0.5–2nm for reverse osmosis, 1–50 nm for ultra filtration, 100-2000 nm for micro filtration), operating atmospheric pressure (15–75 for reverse osmosis, 115 for ultra filtration, 0.3–2 for micro filtration). The permeate in case of reverse osmosis is water, dissolved solutes in case of micro filteration. Retentate includes water + small molecules in case of ultrafiltration while water + ions and organics over 200 MW in reverse osmosis while over 1000 MW in ultrafiltration and large suspended particles and emulsion in microfiltration.

Membrane used in these processes are made of different organic materials and some of these are known as cellulose acetate membrane, polyamide membrane, polysulphon membrane, polyfurane membrane, polyacrilonitrile, polycarbonate, polyvinyl chloride and noncellulosic membrane made of ziroconium oxide. These different membranes have their particular characteristic and ability for separation.

Membrane processes can carry out concentration and fractionation of liquid dairy products without phase change and, therefore, the energy requirements are less than conventional concentration processes. Besides the risk of thermal damage (denaturation or degradation of heat sensitive constitutents such as proteins and vitamins) and aroma, loss is minimised.

Separation mechanism of reverse osmosis is based on the fact that solutions containing salts whose surface tension increases with concentration, such as inorganic salts will have a "negative excess" of the solute absorbed on the membrane surface i.e. there will be a layer of water molecules on the membrane surface. If the membrane contains pores of the appropriate size (twice the thickness of the water layer) and a pressure sufficient to overcome the osmotic pressure or chemical potential difference is applied, the absorbed solvent (water) layer will flow through these pores. Thus the control of the pore size and providing an appropriate membrane surface is critical to the success of RO.

The mechanism of separation of UF is largely a seiving effect, based on the size of molecules in the feed solution. The ability of the membrane to substantially retain a defined macromolecule of known molecular weight is generally used to specify the porosity of the membrane. The term used is "molecular weight" cut off (MWCO), which is the molecular weight of the smallest test macromolecule that is largely rejected by the membrane.

Some of the important applications of these processes in dairy industry are listed below :

1. Concentration of whey by RO - The production of cheese, paneer and casein create large volume of whey (approximately 80-90 percent of the initial volume of milk). Whey contains sufficient amount of protein, lactose, minerals which could be utilised. The RO process removes major part of the water from whey more economically than evaporation process. It may be used for concentration of whey for using as such or concentrated to whey powder, whey protein powder, lactose and demineralised whey powder. These may be used in food industry for example in bakery product, confections and candies.

2. Concentration of UF permeate by RO - RO play an important role in concentration of UF permeate with minimum energy requirements and least thermal degradation. This concentration UF permeate by RO can be used for manufacture of lactose.

3. Concentration of whole milk by RO - Whole milk concentration is important. Use of RO for concentration of whole milk is primarily for reducing the transportation costs, manufacture of dairy products and use in fluid products.

4. Whole milk powder from RO concentrate - Whole milk powder is prepared using RO as a first concentrate step followed by either evaporation and spray drying or spray drying alone.

5. Concentration of skim milk and skim milk powder. To be used for ice cream, ymer (Danish product) and yoghurt.

6. Cheddar cheese from RO concentrated milk. Use of RO concentrated milk for cheese has the advantage of increased yield, use of 50 per cent less starter and 60 per cent less rennet and increase in titratable acidity of curd during cheddaring is faster which reduces the cheddaring time. No appreciable loss of proteins, amino acid and vitamins and minerals occur in whey from manufacturing of cheddar cheese from RO concentrate.

7. Khoa from RO concentrate–The khoa prepared from RO concentrated milk has slightly higher moisture content and significantly higher fat and lower ash content on dry matter basis in comparison with a control khoa made by conventional method.

8. Ultrafiltration of milk on the farm. Ultrafiltration of milk on the farm have the following advantages, approximately 50 percent reduction in milk collection cost, saving in energy, labour and product treatment, reduction in feed costs, optimum protein and fat standardisation for cheese making, and improved microbiological and organoleptic quality of cheese.

9. Ultrafiltration can recover and concentrate the proteins from whey (over 99 percent of protein can be recovered). These proteins can be used in baked foods, frozen desserts confections, infant foods and cheese products. Lactose, fractionated from whey can also be used in these products or converted to alcohol, yeast, vitamins, lactic acid and other sugars.

10. Cheese making. The principle of making cheese with ultrafiltered milk is to include the whey proteins in the cheese and increase the yield of product. Numerous varieties of cheese are preapred from ultrafiltered process.

11. In cheese making 70 per cent of the coagulating enzyme is lost in the whey, only 30 per cent being retained in the cheese to degrade the protein for ripening. It is observed that the UF of whey could be recovered and reused. The whey proteins would of course, be concen-

trated at the same time. It is stated that adding a 20-fold UF whey concentrate while cheese-making would reduce the rennet requirement in cheese making by two-third of normal consumption.

12. Enzymes and biologicals can be purified and concentrated with UF processes with less labour and cost than the normal processes.

Reverse osmosis can be employed for concentration of whey, skim milk, UF permeate and whole milk. Concentration of whey and skim milk prior to evaporqation and spray drying are technologically and economically feasible for RO process. Substantial saving in costs and energy, particularly in the transport of whey and whole milk, may be achieved when using RO process. This process has potential application in the manufacture of many dairy products like ice cream, yoghurt, powders, khoa, fluid products and cheese.

Ultrafiltration is now an established unit operation for the purification, fractionation and concentration of many dairy liquids. Fractionation of cheese whey into protein rich and lactose rich streams, and their subsequent utilisation in manufacture of whey protein concentrates and lactose derivatives is one of the successful application of ultrafiltration. More economic benefits could be realised by concentrating milk on the farms or at collection centre prior to shipment to the dairy plant. This could provide substantial saving in cooling, heating and transportation costs.

Emulsifier and Stabilizer

Emulsifier

Emulsion is known as dispersion of tiny droplets of one liquid to another liquid. Water is usually one of the component and other is an oil or liquid insoluble in water. The emulsion composed of water and oil are not stable and need a substance, emulsifier to stabilise the system. Emulsifier, therefore, are substances which tend to concentrate in the interface between the fat and the plasma and reduce the surface tension of the system.

Emulsifier in Cheese

The emulsifier has very important roles to play in the processed cheese. These are (i) to prevent separation of fat from cheese during the heating operations, (ii) to cause specific soft and smooth characteristics in the body and texture of the finished product, (iii) to produce desirable melting and slicing properties in the finished cheese.

The dispersion of the fat globule in mix is due to the fact that one portion of the emulsifier molecule is soluble in fat, the other in water i.e. the lipophilic and hydrophilic group.

Emulsifier in Ice-cream

Emulsifier improves the whipping quality of the mix, the production of a smooth texture, giving a dry stiff product at the time, it is drawn from the freezer and a more exact control during the various manufacturing process.

Examples of Emulsifier

The following is a list of chemicals used as emulsifier. Sodium citrate, sodium tartrate, sodium potassium tartarate (Rochelle salt) Trisodium orthophosphate (Na_3PO_4), Disodium orthophosphate (Na_2HPO_4) sodium metaphosphate ($NaPO_3$). Tetra sodium phosphate ($Na_4P_2O_7$) sodium pyrophosphate ($Na_5P_3O_{10}$), sodium acid phosphate, sodium hexameta phosphate, potassium citrate, dipotassium phosphate, calcium citrate, ammonium tartarate, ammonium chloride, Lactate, Mucate, Malate and Gluconate.

Commonly used emulsifier for process cheese and related products . Sodium citrate ($2Na_3C_6H_5O_7.11H_2O$) - it is versatile, produces firm cheese with good melting and is inexpensive. It has best general all round qualities. It is used at the rate of 0.5 to 3.0 percent by weight of cheese.

Disodium Phosphate (Na_2HPO_4): It is least expensive. It has good firming and good melting properties. The amount of stabiliser varies from 0.5 to 3.0 per cent by weight of cheese.

Trisodium Phosphate (Na_3PO_4): It is highly alkaline. It is used in small amounts. It helps in producing good slicing in combination with other emulsifiers.

Sodium Hexametaphosphate ($(NaPO_3)_6$): It produces tart flavour and very firm body but not an easily melted product. It is least soluble.

Two types of emulsifiers are used in the manufacture of ice-cream. Mono- and di-glycerides derived by the chemical reaction of naturally occurring glycerides and polyoxyethylene derivaties of hexahydricalcohols, glycols and glycol esters. The mono-glycerides improve fat dispersion and whipping ability and have a moderate effect on stiffness and the melting rate. The poly-derivatives are effective in producing dryness, stiffness and increasing the melting time.

Stabilizers

The primary purpose of using stabilizer in ice cream are to produce smoothness in body and texture, retard or reduce ice crystals growth during storage, provide uniformity of the product and resistance to melting. Stabilizers function through their ability to form gel structure in water or their ability to combine with water as water of hydration.

Commonly Used Stabilizers in Ice-cream

Gelatin: It forms gel in the mix and prevents formation of large ice crystals. It contributes to smoothness in texture and firmness in body. Though gelatin is incomplete protein, yet amino acids contribute to food value and is a widely used stabilizer. Concentration of gelatin to be used in ice cream varies with the animal source from which gelatin is obtained.

Sodium Alginate: This is of vegetable origin. It improves whipping ability and leaves a slightly cleaner flavour in mouth. It dissolves at 68-71°C. It produces same stabilizing effect as gelatin.

Carageenan: This is extracted from carageen (Irish Moss), a seaweed. It is used along with gum stabilizers.

Agar-agar: This is extracted from red algae (sea weed). It is useful for sherbets and ices.

CMC: Carboxymethyl cellulose has a high water-holding capacity and can be easily dissolved in the mix. It does not form as firm gel as gelatin and other vegetable stabilizers but has a good use in ice cream. It is used at the rate of 0.45 per cent.

Pectin: This is a carbohydrate obtained from citrus fruits. It is suitable for sherbets and ices. It is used at the rate of 0.18 per cent.

Guar Gum: This carbohydrate is obtained from a legume. It is used at the rate of 0.20 per cent and often used in combination with carageenan.

Centrifugation

Cream is manufactured usually by the centrifugation process. The differences in the density or specific gravity between the fat globules and the rest is the principal criterion of cream separation. In a centrifuge, a centrifugal force is applied wherein the particles which have a higher density or specific gravity are expelled to the periphery while the fat globule which have less density or specific gravity remain at the centre. Based on this theory cream separators are designed which have one outlet near the centre and another outlet near the periphery. Thus, the outlet near the centre causes the exit of mainly fat globules dispersed in water and the outlet near the periphery skim milk containing negligible amount of fat, which comes out of the cream separator. In milk clarifiers the same principle is used.

REVIEW QUESTIONS

1. Describe controlled fermentation, abnormal fermentation and citric acid fermentation.

2. Briefly discuss the three phases of rennet action.
3. Diagramatically show how protein breakdown in cheese takes place.
4. List few selected enzyme from microbial source used for cheese making.
5. Compare the acid, heat and rennet coagulation of milk.
6. Define reverse osmosis and ultra filtration.
7. List some of the stabilizers used in dairy industry.
8. What is centrifugation?

REFERENCES

Deepa Pande and Mathur,. M.P. 1989. Enzymes in relation to UHT processing of milk. *Indian Dairyman* 41(5):235.

Nirmal Sinha, 1981. Reverse Osmosis (RO) and ultrafiltration (UF) processes - their differential features and application in the Dairy Industry. *Indian Dairyman*33(5):323.

Prajapati, P.S., Thakar, P.N. and Upadhyay, K.G. 1986. Reverse Osmosis — An important process for Dairy Industry. Dairy Guide 8(8):11.

Prasad, C., Prasad, S. and Murari, K. 1988. Interaction of milk proteins during UHT processing and some chemical changes. Indian Dairyman 40(4):209.

Sen, D.C. 1986. Advances of ultrafiltration in whey processing. Dairy Guide 8(4):35.

Singh, R.R.B. and Patil, G.R. 1987. Physico-chemical and sensory changes during processing and storage of UHT milk. Indian Dairyman 39(5):209.

Webb, B.H., Johnson, A. and Alford, J.A. 1978. Fundamentals of Dairy Chemistry. The AVI Publ. Co. Inc. West Port, Connecticut. Chapter-12. Milk-Clotting Enzymes and Cheese Chemistry.

9

CHEMISTRY OF MILK PRODUCT

Heat/Acid Coagulated Products

IN the Kusana and Saka Satavahana periods (75-300 AD) there are references of a product made from warm milk and curds. During that period the Sakas and Gujars migrated from Central Asia first to Persia then to India. These stock of people were nomads and were the first to develop a certain variety of cheese. This nomadic variety of cheese is known in Iran as Paneer Khiki (Paneer means container and Khiki means stomach lining from goat or sheet hence the name for the similar variety of cheese made from rennet). In Afghanisthan it is a staple food and known as **Paneer-Khom** (from raw milk) and **Paneer-e-Pokhta** (from boiled milk). March of Islam in India was synonymous with the arrival of Paneer particularly in North-west India and Kashmir and it is very popular now in North, West and Central India.

Paneer

Traditionally, buffalo milk in bulk, is boiled in a suitable iron vessel and the coagulant (usually sour whey) is added to the hot milk and stirred with a ladle till coagulation is complete. The curd is collected by filtering and draining the whey using a coarse cloth. The curd is washed with water and pressed to remove more whey and moisture. Paneer is also prepared at pilot plant level using buffalo milk (6 per cent fat) heated to 82°C for 5 min. The coagulation was brought about at 70°C by slowly adding 1 per cent citric acid solution with constant stirring. The curd was filled in a hoop and pressed to remove whey and water. Good quality paneer was obtained when standardized buffalo milk (fat:SNF, 1:1.65) heated to 90°C and coagulating the milk with 1 per cent citric acid at 70°C.

The quality of paneer is influenced by a number of factors as reviewed below :

Type of Milk: Buffalo milk is considered more suitable compared to cow milk in terms of body, texture and spongy characteristic of paneer. The cow and buffalo milk mixed in the ratio of 50:50 having a fat content of 5 per cent yields a superior product than cow milk alone.

Quality of Milk: Acidic milk having a titratable activity (TA) of 0.20-0.23 per cent yields a product with inferior quality. Homogenization of cow milk is recommended to bring about improvement in the yield and to improve the organoleptic score of paneer.

Heat Treatment of Milk: Heating milk to 82°C for 5 min and cooling to 70°C before coagulation affects the sensory and microbiological quality of paneer.

Type of Coagulant: Different coagulants such as citric acid solution, one day old whey, 48 hr old sour whey. Whey cultured with *L. acidophilus* have been used to prepare paneer. In addition to these, coagulants such as hydrochloric acid, phosphoric acid, tartaric acid and acetic acid have also been used in the manufacture of paneer.

Strength of Coagulant: Best quality paneer can be obtained using 1 per cent citric acid. Two per cent citric acid solution also yielded good quality paneer when cow milk is used. Satisfactory quality products are obtained when non-conventional coagulants like hydrochloric acid/ phosphoric acid and acidophilus whey (0.6 per cent) are used.

Amount of Coagulant: For making good quality paneer from cow milk 2.34 g of citric acid or lactic acid is required for 1 kg of milk while for buffalo milk 2.1 g of citric acid or lactic acid per kg is required.

Coagulation Temperature: Coagulation temperature of 70°C has been suggested to be most effective. However, good quality paneer can also be prepared at a coagulation of 80°C from both buffalo milk and cow milk.

pH of Coagulation: The optimum pH of coagulation is 5.35 for buffalo milk while for cow milk pH is 5.0.

The yield of paneer decreases with increase in the strength of coagulant, coagulation temperature and lowering the pH of coagulation in case of buffalo milk and it increases with increase in the coagulation temperature in cow milk paneer. The yield of paneer can also be increased by co-precipitation of casein and whey protein. Higher level of fat in paneer milk leads to lower yield of the product in case of buffalo and cow milk. Homogenization of paneer milk improved the paneer yield.

Wide variations in the chemical composition of paneer are observed by different workers (Table 9.1) which may be attributed to the difference in the initial composition of milk, method of manufacture and losses of milk solids in whey.

The coagulation process in paneer manufacture is regarded as a consequence of the chemical and physical changes in casein brought about by the combined influence of heat and acid treatment. This phenomena involves the formation of large structural aggregates of casein from the normal colloidal dispersion of discrete casein micelles, in which milk fat and coagulated serum proteins are entrapped together with whey. During

Table 9.1. Chemical composition of paneer

Sample Source	Moisture	Total solids	Fat	Protein	Lactose	Ash
	Constitutents (%)					
Buffalo milk (5% fat)	56.77	43.23	22.30	–	–	–
Buffalo milk (5.8% fat)	50.72	49.28	27.13	17.99	2.29	1.87
Market sample	51.91	48.09	26.59	17.65	2.05	1.78
Buffalo milk (5% fat)	56.43	43.57	22.50	–	–	–
Mixed milk (5% fat)	57.14	42.86	22.32	–	–	–
Cow milk (5% fat)	53.90	46.10	24.80	17.60	–	–
Buffalo milk (5.5% fat)	55.19	44.81	23.80	17.99	–	–
Buffalo milk (5.8% fat)	54.10	45.90	23.50	18.20	2.40	1.80
Buffalo milk (6% fat)	48.30	51.70	29.20	18.50	–	–
Cow milk (4% fat)	52.40	47.60	23.40	20.30	–	–
Mixed milk (5% fat)	49.00	51.00	26.10	20.80	–	–
Buffalo milk (5.9% fat)	51.12	48.88	26.86	17.38	–	2.00

Source : Reproduced from *Ind. J. Dairy Sci.* **45**:285, 1992.

the formation of coagulum, the major changes that take place include (i) the progressive removal of tricalcium phosphate from the surface of casein and its conversion into mono calcium phosphate and soluble calcium salt, and (ii) progressive removal of calcium from calcium hydrogen caseinate to form soluble calcium salt and free casein. When the pH of milk system drops, the colloidal particles become isoelectric, i.e. the net electric charge becomes zero with the formation of "Zwitter-ion". Under such circumstances the dispersion is no longer stable, the casein gets precipitated and forms a coagulum (Ling, 1956). Isoelectric precipitation may be induced by the addition of calcium as it has a key role in the formation of coagulum. It increases the curd tension by providing closer and more abundant linkage between casein micelles. Probably this might be one of the several reasons for good quality paneer obtained from buffalo milk. The development of typical rheological characteristics of paneer could be due to preponderant and intensive heat induced protein-protein interaction. It is believed that β-lactoglobulin and K-casein interact by sulphydryl disulphide interchange when heated together. Interaction between heated K-casein and β-lactoglobulin as evident by electrophoretic changes is initiated at approximately 65°C, increasing to a maximum of 83 per cent at 85°C and decreasing to 76 per cent at 99°C. However, there are reports that α-lactalbumin and β-lactoglobulin also interact and the complex so formed appears to be able to interact with K-casein.

The microbiological quality of paneer depends mainly upon the condition of post manufacture, handling, packaging and storage of the product.

When paneer samples were stored for 6 days under refrigerated condition, the loss of moisture was relatively less in polyethylene bags compared to parchment paper. A shelf-life of 18 days under refrigeration was observed for paneer when packed in parchment paper and paneer could also be stored up to 16 days when tightly wrapped heat shrunk film (cryovac) is used. Paneer prepared from mixed cow and buffalo milk of 5 per cent fat could retain its original characteristics only up to 3 days when stored at 15°C and 8 days at 5°C, without developing any fungus growth or foul smell. Storage of paneer at sub zero temperature, i.e., at –13°C and –32°C for 120 days did not affect its flavour and appearance but the body and texture of the product deteriorated and became crumbly and fluffy.

Paneer cubes packed in tins and sterilized in autoclave at 15 Psi for 15 min kept well over a period of 50 days at room temperature. A slight amount of browning accompanied with a cooked flavour affected the organoleptic quality of paneer. Paneer cubes fried prior to sterilization, however, spoiled earlier due to the development of pronounced oxidised flavour after 40 days of storage.

The shelf-life of paneer mainly depends on the surface spoilage. Various antifungal agents have been used for controlling the surface spoilage. These include chlorinated water (35 ppm), buffered water (pH 7.5), acidified water (pH 5.5 and 2.5), pottassium sorbate (0.1-2 per cent), delvocid, brine (5 per cent), hydrogen prooxide (0.2 per cent). Delvocid (Fungicide) when used in combination with hydrogen peroxide (germicide) have excellent results extending the shelf-life of paneer up to 32 days at 8-10°C.

No significant changes were observed in acidity and pH during storage of paneer at 10°C up to 6 days, while slow reduction in total nitrogen and appreciable increase in non-protein nitrogen content was recorded after the third day onwards. Progressive increase in acidity from 0.41 to 0.49 and rapid increase from 0.44 to 0.62 per cent was reported after 3 days of storage at 15°C and room temperature.

The increase in acidity, water soluble nitrogen and free fatty acid content in the stored paneer after 90 hrs of storage was rapid at room temperature, progressively slower at 15° and 5°C. When the product was stored at sub zero temperature, viz. - 13° and –32°C, no change in pH, moisture and total nitrogen content was observed up to 90 days of storage. However, a marked increase was noticed in non-protein nitrogen content which might be due to protein degradation. Paneer provides a favourable condition for the growth of a wide variety of organisms depending upon the temperature of storage. The total counts as well as yeast and mould counts increased during storage of paneer up to 10 days at 5°C. A stready increase in psychrotorophic, proteolytic and lipolytic count in market paneer samples during storage at 7°C up to 7 days were observed. On subsequent storage, the count increased tremendously. A reduction in mesophillic count was observed, initially at 5°C but rapid increase was noticed after 4 days in unwrapped and after 6 days in wrapped sorbic acid coated paper. No significant increase in proteolytic and lipolytic counts was observed in cheese prepared from milk containing sorbic acid and stored at 5°C, but at room temperature the count increased steadily upto 18 days. Using antimicrobial agents such as TBHQ and BHA, no significant increase in total and yeast and mould counts at 5°C up to 10 days were observed. But at 25°C rapid increase was noticed earlier during storage. Paneer is highly nutritious as it contains in concentrated form almost all proteins and fat alongwith essential minerals and vitamins. It retains about 90 per cent fat and protein, 50 per cent minerals and 10 per cent lactose of the original milk. The paneer from buffalo milk has high PER (3.4) and b.v. (86.56) as compared to paneer of cow milk with a count 2.3 and 81.88 while the mixed milk cheese has 2.3 and 80.38. Buffalo milk paneer has a higher NPU as compared to cow milk and mixed with paneer. About 76.8 per cent Ca and 68.3 per cent Mg of the original milk was retained in paneer.

Paneer manufactured using citric acid had maximum hardness but lower values for gumminess and chewiness. However, paneer prepared using citric acid and tartaric acid as coagulants showed maximum gumminess and chewiness. Buffalo milk paneer showed higher value of firmness as compared to cow milk paneer. Electron microscopy revealed that raw paneer samples had a granular structure consisting of protein particles having a core and lining ultrastructure irrespective of the type of milk used and is similar to other milk products obtained by coagulating hot milk with acid to a pH in the vicinity of 5.5.

Chhana

Chhana is a heat-acid coagulated milk product which forms the base of several popular Indian sweets, such as *rassogolla, rasmalai, chum-chum, sandesh,* etc. It has been estimated that about 4 per cent of the total milk produced in the country is converted into *channa.* It originated in the eastern states, but is now popular in north and north-western regions of India.

Cow milk is taken in a boiling pan and is coagulated at high temperature (70°C) using sour whey but some organic acids (lactic or citric) may also be used. Whey is then drained off by straining through a cloth. The coagulum is collected in a cloth bag, and free whey allowed to drain off. It is creamish white in appearance, has compact but spongy body, and has sweetish acidic flavour.

The process of coagulation of milk in *chhana*-making may be regarded due to the combined chemical and physical changes in the casein brought about by the action of acid aided by relatively higher temperature of coagulation. This involves the formation of large structural aggregates of casein, or curd from the normal colloidal dispersion of discrete casein micelles in which milk fat and coagulated serum proteins are entrained together with whey. In forming the coagulum, the dispersed phase separates entirely from the dispersion medium (the milk serum) as a flocculant precipitate. During the process of boiling and subsequent coagulation by addition of acid, the casein micelles lose a part of calcium and phosphate etc. originally bound to it.

The chemical composition of chhana depends mainly on the initial composition of milk, the conditions of coagulation, the technique of straining (which in turn determines the percentage moisture retained) and the losses of milk solids in the whey. An average chemical composition of fresh laboratory-made *chhana* is given in Table 9.2.

Table 9.2. Composition of *chhana*.

	Cow milk	Buffalo milk
Moisture %	53.4	51.6
Fat %	24.8	29.8
Protein %	17.6	14.5
Lactose %	2.1	2.3
Ash %	2.1	2.0

On an average 100 g *chhana* may contain mineral matter (2.6 g), calcium (208 mg), phosphorus (138 mg) vitamin A (366 I.U.), riboflavin (15 μg), thiamin (73 μg) and ascorbic acid (2.8 mg).

Some of the factors which affect the physical quality of *chhana* are :

Type of Milk: The *chhana* prepared from cow milk gives a good quality of *rossogolla* while that from buffalo milk produces *chhana* with hard body, coarse texture and excessive greasiness which render it unsuitable for *rossogolla*. However, undiluted buffalo milk gives good *chhana*. A mixture of 75 percent buffalo milk and 25 percent cow milk produces suitable *chhana* which offers good *rossogolla*.

Mineral Level of Milk: The hardness of *chhana* is directly related to the calcium content of milk. Addition of calcium or magnesium chloride produces hard *chhana* while sodium chloride, sodium acetate and sodium citrate did not affect the body and texture of *chhana*.

Fat Content: *Chhana* prepared from cow milk (4 to 5 percent fat) is most suitable for *rossogolla*. Lower fat content below 3 percent produces *chhana* unsuitable for *rossogolla*.

Presence of Colostrum: Presence of colostrum milk produces too pasty *chhana* with deeper yellow colour which is unsuitable for *rossogolla* preparation.

Freshness of Milk: Fresh milk is preferred for *chhana*. Acidic milk produces undesirable sour smell and bitter taste in *chhana*. When acidity is neutralised with sodium carbonate the *chhana* prepared can be used for *sandesh* only.

Adulteration of Milk with Water and Starch: Water up to 25 percent is effective. Addition of starch produces gummy mass or gelatinous structure which makes *chhana* unfit to be used for *rossogolla*.

Curd Tension of Milk: Curd tension influences the body and texture of *chhana*. Buffalo milk *chhana* has higher curd tension than that of cow milk. This is due to casein micelle (miceller size), voluminosity, compositional heterogeneity and mineral make up. The *chhana* prepared from buffalo milk is distinctly harder than cow milk because more of long chain saturated fatty acids are present in buffalo milk fat. It is also because of higher calcium, magnesium and SNF. Addition of calgon, glycerine monostearate and lecithin to milk and homogenisation of milk reduces curd tension.

Type and Strength of Coagulant: Citric acid does not impart flavour to *chhana* while lactic acid and sour *chhana* whey produce slightly sour and acidic taste of *chana*. Citric acid (1-2 per cent) and lactic acid (0.5-0.75 percent) produce most suitable *chhana*.

pH and Temperature of Coagulation: The optimum pH and temperature for good *chhana* is 5.7 and 70°C. Temperature below 60°C and above 80°C are unsuitable for *chhana*, to be used for *rossogolla*. At pH 5.3 smooth and soft *chhana* is produced.

Chhana is a concentrated form of protein and fat. The BV of *chhana* is 88, digestibility coefficient 92, PER value 3.1 (casein 2.8), NPU 71.5 (casein 83.5) and NPR 5.25 (casein 4.55). Superior nutritional quality of *chhana* compared to casein may be attributed to the association of whey proteins, which complexes with casein during the heating of milk in the manufacturing process.

Shelf-life of the product packaged in vegetable parchment was 2, 3 and 12 days at about 38°C, 24°C and 7°C, respectively. Compositional as well as quality standards for *chhana* have been developed by PFA and ISI.

Heat Desiccated Products

Vedic literature (Rigveda III, 52.1, VII 91.3) cite eloquently the manufacture and use of heat desiccated milk products. Buffalo milk was the principal source of such products (Taittiriya Samhita, VII 5.31). References are there of Samtanika (Malai) and Kurchika, a tonic prepared from inspissated milk during the Kusana and Satavahana period and used mainly for soldiers (Susruta. Samhita. 46.403). Post Gupta period literature is full of accounts of milk concentrated to one-sixth of the original quantity for use in the preparation of sweets. Milk concentrated to one-eighth of the original volume was named as Surkara and was fed to soldiers (Manas. III 1567). Khoa, Khurchan and Rabri have probably evolved to its present status as sweets, as rich descendants of an ancient and dignifiied cultural heritage of our country.

Khoa

Khoa (Syn. *Mawa. Kava, Palghoa*) refers to heat desicated milk product and occupies an important place among indigenous Indian milk products. About 7 percent of the total milk produced in India is converted into *khoa*. Typically buffalo milk is desiccated with vigorous stirring in shallow pan over direct fire, till a semi-solid mass is formed that begins to leave the surface on stirring. *Khoa* is an intermediate product and forms base for a large number of sweets such as *burfi, peda*, milk cake, *kalakand, gulabjamun*, etc. According to the PFA Rules (1976), khoa is the product obtained from cow or buffalo (or goat or sheep) milk, or a combination thereof, by rapid drying. The milk fat content should not be less than 20 per cent of the finished product.

Three types of khoa have been identified. These differ in quality and purpose for which it is prepared.

Danedar Khoa: This form of *khoa* is characterised by granular texture and uneven body. Usually, 0.1 percent citric acid (on product basis) is added during the heat desiccation process in shallow pan to obtain such texture. It has 20-25 percent fat and 35-40 percent moisture. It is mainly used for the preparation of *kalakand*, granular *burfi*, etc.

Dhap (Katccha *Kava*): It is characterised by loose and sticky body and smooth texture, and fine grain structure. It contains about 20-23 per cent fat and 37-44 per cent moisture. It is preferred for the preparation of *gulabjamum* as it permits formation of uniform balls suitable for frying and absorption of sugar syrup.

Pindi: It is characterised as a circular ball of hemi-spherical pat with smooth and homogenous body and texture. This posses characteristic cooked flavour and is preferred for the preparation of *burfi* and *peda*. It contains 19 per cent moisture; 37 per cent fat, 18 per cent protein, 22 per cent lactose and 3.6 per cent ash.

The chemical composition of *khoa* depends mainly on the initial composition of milk, the degree of concentration of milk solids and the losses or gains in handling. An average chemical composition of fresh, laboratory made whole milk *khoa* is given in Table 9.3.

Table 9.3. Average chemical composition of *khoa* (in per cent)

Content	Milk	
	Buffalo	Cow
Moisture	19.25	25.66
Fat	45.93	34.83
Protein (dry basis)	22.06	25.83
Lactose	27.42	34.33
Iron (ppm)	125	139

Source : De, S. and Ray, S.C. (1952). *Ind. J. Dairy Sci.* 5, 147.

During the conversion of milk into *khoa*, following changes take place:

Change of State: The liquid milk is converted to solid *khoa* due to considerable dehydration. The water is removed and an increase in concentration of milk solids by 5 to 6 times takes place. The water is also dispersed as fine droplets in the mass of *khoa* and, therefore, *khoa* do not appear to be wet in the final form.

Change in Intensity of Colour: Milk is white in colour. The colour of milk changes from 'light' to a more intense shade of colour (with a tinge of brown). This abrupt change of colour and also consistency is due to denaturation and coagulation of protein either during the process or towards the end of the process.

Change in Milk Fat: Due to the vigorous agitation of milk at high temperature, the fat globules are appreciably sub-divided. Considerable free fat is also produced due to the rupturing of the fat globule membrane by the vigorous scraping action of the stirrer. The vigorous agitation of hot milk has an appreciable homogenising action so that when the stage of coagulation is reached all the fat globules are entrained in the coagulum. Almost half of the globular fat is released as free fat - the extent of which depends upon the type and fat content of milk, and the manufacturing process. Usually, 44.8 to 62.8 per cent of fat appears as free fat in *khoa*.

Changes in some of the fat related constituents and fatty acid composition during *khoa* preparation are shown in Table 9.4 and Table 9.5.

Table 9.4. Variation in certain fat related constituents at various stages during *khoa*-making based on laboratory trials (Average value).

	Raw milk		Boiled milk		*Dhap*		*Khoa*	
	Cow	Buffalo	Cow	Buffalo	Cow	Buffalo	Cow	Buffalo
Total phospho-lipids %	0.440	0.288	0.434	0.322	0.432	0.389	0.464	0.368
Total cholesterol (in mg%)	360	271	355	262	366	255	364	274
Free cholesterol (in mg%)	312	214	300	210	301	215	304	210
Esterifier cholesterol (in mg%)	45	63	42	62	43	66	45	63
Total carbonyls (μM/g fat)	6.73	9.96	7.94	12.43	8.70	14.75	14.65	19.36
Free fatty acid (μM/g fat)	112	47	125	59	109	68	109	65
Iodine value	41	33.7	41.3	33.4	35.6	36.9	36.6	34.1

Source : Summarised from studies on lipids isolated at various stages during preparation of *khoa* from cow and buffalo milk. Jessa Ram Arora, M.Sc. Dissertation submitted to Kurukshetra University, Kurukshetra, 1978.

Table 9.5. Free acid composition of cow milk fat and buffalo milk fat obtained at various stages during manufacture of *khoa*.

Fatty	Raw milk		Boiled milk		*Dhap*		*Khoa*	
acids	Cow	Buffalo	Cow	Buffalo	Cow	Buffalo	Cow	Buffalo
Total saturated	3.47	4.48	3.84	4.43	4.54	2.68	4.58	2.58
Total unsaturated $C_{16} + C_{18}$ $C_{16:1} + C_{16:2} +$ $C_{18:1} + C_{18:2} +$ $C_{18:3}$	1.64	2.53	1.84	2.57	2.32	1.52	2.13	1.44

Source : Summarised from studies on lipids isolated at various stages during preparation of *khoa* from cow and buffalo milk. Jessa Ram Arora, M.Sc. Dissertation submitted to Kurukshetra University, Kurukshetra, 1978.

Heat Coagulation of Milk Protein: The main reaction in the preparation of *khoa* is the heat denaturation and coagulation of milk proteins. Most of the albumin and globulin are rapidly denatured, the protective properties of the other colloids are destroyed early in the boiling process, and the process is accelerated by the incorporation of air and the frothing during stirring. Total heat coagulation of the proteins occurs when the boiling mixture thickens to a buttery consistency in the pan. Coagulation of milk protein is brought about by heating to 132°C-136°C. Albumin and globulin are coagulated below 100°C, while casein is coagulated above 100°C. During desiccation, whey proteins are almost fully denatured while casein is also irreversibly denatured from colloidal state of non-dispersible state. The factors that influence the heat coagulation are temperature and time of holding, concentration of casein, acidity of milk, salt balance and precise heat treatment.

Change in Lactose: The lactose is present in milk in dilute solution. The lactose in *khoa* is present mostly as a super-saturated solution, since the available water is only sufficient to make roughly a 50 percent solution, while the solubility of 1-lactose hydrate at atmospheric temperature is only 16 per cent. In the hot *khoa* the lactose would be anhydrous but it is doubtful if any of this would crystallise out, since the drops of water are too small in size. In no case, however, are large crystals of lactose found in ordinary *khoa*. the viscosity of the product may come in the way of crystallisation.

Changes in protein, fat and lactose during conversion of milk to *khoa* are shown in Table 9.6.

Table 9.6. Proximate principles for milk, *dhap* and *khoa* (Average value expressed as percent)

	Protein	Fat	Moisture	Lactose
Milk	4.14	5.0	82.89	4.42
Dhap	16.55	21.62	48.57	12.63
Khoa	19.81	22.29	35.59	20.99

Precipitation of Milk Salt: A portion of the milk salts are precipitated by the action of heat.

Increase in Iron Content: From 2 to 4 ppm in milk, the iron content in *khoa* exceeds 100 ppm (due to the incorporation of additional quantities of iron from the pan-surface into the finished product by the vigorous scraping of the stirrer).

Changes in Vitamins: Different observations are available on the changes in vitamins during processing of *khoa*. A loss of vitamin A was reported in one report in *khoa* preparation, which could be madeup by fortification of Vitamin A in milk itself or by adding synthetic vitamin A to khoa itself. In another report a loss of ascorbic acid, thiamine, riboflavin and nicotinic acid as well as vitamin A was reported in the course of heating of milk for making *khoa*. Stability of different vitamins during *khoa* preparation are shown in Table 9.7. The percentage loss of different vitamins during *khoa* preparation are shown in Table 9.8.

Table 9.7. Stability of different vitamins during *khoa* preparation

	Vitamin A		Vitamin B_2		Vitamin B_6		Folic acid		Vitamin C	
	IU/100g sample	IU/100g of fat	Ug/100g sample	Ug/100g total solids	Ug/100g sample	Ug/100g total solids	Ug/100g	Ug/100g total solid	Ug/100g sample	Ug/10g total solid
Milk	160.97	32.19	213.38	1376.38	27.50	164.13	0.22	1.31	1.88	11.30
Dhap	598.25	27.66	520.03	1028.97	81.08	142.58	0.64	1.15	4.93	8.77
Khoa	581.72	26.07	622.24	979.63	85.10	132.27	0.68	1.06	5.43	8.46

Table 9.8. Percentage loss of different vitamins during *khoa* preparation.

	Vitamin A	Vitamin B_2	Vitamin B_6	Folic acid	Vitamin C
Milk	–	–	–	–	–
Dhap	14.30	25.25	13.11	12.07	22.31
Khoa	18.78	28.78	19.39	18.47	25.05

Change on Storage: *Khoa* has a low keeping-quality at room temperature and develops rancid flavour. This may be due to fat hydrolysis which is due to lipase action. Mouldy surface developed during storage are due to high moisture content of *khoa* and excessively high humidity of storage room. When *khoa* is kept for a long period at low temperature (5-10°C) staleness develops. In addition to these, several deteriorative factors cause off-flavour development in *khoa* during storage. There is increase in various types of micro-organisms during storage at room temperature. The keeping quality of *khoa* during storage depends on better storage condition, temperature of storage, quality of initial milk, initial moisture content, packaging material used and sanitary conditions observed during manufacture. The shelf-life of *khoa* is about 2-4 days under ambient conditions and 3 weeks under refrigerated conditions.

The PER value of *khoa* is about 3.82 (milk 4.32), BV 99.36 (milk 98.66) and NPU 85.93 per cent (milk 82.03).

Khoa Based Sweets

Though *khoa* based sweets have been known in India for centuries yet no systematic information is available on their preparation properties, quality, packaging and shelf-life. There are mainly four *khoa* based sweets – *Burfi, Gulabjamun, Kalakand* and *Peda.*

Burfi is white to light cream colour with firm body and smooth to granular texture. It is known all over India and prepared by mixing *khoa* and sugar (25-35%) in a shallow pan with concommitant heating and stirring. *Burfi* may contain shredded coconut, nuts, chocolate and special toppings.

Gulabjamun is round or cylindrical in shape, dark brown exterior, soaked in thick sugar syrup with firm body and smooth texture. It has origin in northern, western and central region of India. *Khoa* and *maida* or *suji* are mixed in about 3:1 ratio, with about 1 percent baking powder. The mixture is kneaded with water to form a dough. Ball or cylindrical shapes are prepared and fried in cooking medium till golden brown or dark brown colour develops. The balls are then soaked in sugar syrup (33 per cent). *Khoa* could also be replaced with roller dried/spray dried milk powder.

Kalakand is of light caramel colour, sweet having granular texture and firm body. It has its origin in northern regions of India. It is prepared by the method similar to that of *burfi* but some citric acid is also added to *khoa* during the heating process to permit formation of well-defined grains. The granular khoa is mixed with ground sugar in desired proportion and kneaded at room temperature. Flavours are added according to liking.

Peda is ball-shaped having compact body and smooth texture. It has its origin in northern, western and central regions of India as well as some parts of south India. *Khoa* is mixed with sugar cardamon, nuts, etc. and heated in a shallow pan with constant stirring till sufficiently caramelised. This is patted into a compact mass in the shape of balls with flattened top and bottom ends.

Rabri is a heat desiccated, thickened milk sweet, popularly prepared from buffalo milk in the northern and central parts of India. It is a favoured dessert both in urban and rural areas of central and north India. However, in the eastern part of India, milk is invariably from cow's milk. Buffalo milk is simmered in a shallow vessel with frequent scraping of the heating surface to resuspend clots into thickened milk. When clots have become well defined, sugar is added and stirring continued till clots become firm and a desirable consistency of thickened milk is obtained.

The product is pale yellow to light brown in colour having sweetish, nutty, caramel flavour. It has caramelised clots with creamy viscous body. The product has approximately the composition as : moisture 30-35 percent, fat 20 percent, protein 10-13 percent, lactose 16 percent, sugar 15 percent and ash 3.5 percent. The product has a shelf-life of 5-7 days under refrigerated conditions. Storage studies conducted on *Rabri* showed that samples kept in polystyrene cups and high density polyethylene bags remain in acceptable form upto a maximum of 18 hr at 30°C. Bulging and development of high acidity were the major causes of spoilage of the product at this temperature. At 5°C, *rabri* samples remained acceptable upto 20 days in both types of packaging materials. Lactose crystallisation coupled with loss of freshness were the main defects of *rabri* at refrigerated storage. After 20 days the product started developing sourness due to increase in acidity. The rate of chemical and microbiological changes in *rabri* packaged in polystyrene cups were slower as compared with that packaged in polyethylene bags.

Fat-Rich Product

In the records of Indian civilization, the earliest reference to clarified butter (ghee) goes back to the times of early Aryans. In the Rigveda period, use of ghee was very common (C. 1500 BC). Susruta called articles fried in clarified butter 'ghrta taila pakvap'. A cake made of rice or barkey meal cooked in clarified butter on slow fire was known as 'Apupa' (Hindi - Pua). 'Divalika' was saline small cake prepared with wheat flour fried in clarified butter (Angv. P. 182). 'Haiyamgavina' was the butter churned out of the curds prepared from previous day's milk (Susruta, Su. 45.94). In the beginning of the fourth century, the author of Angavijja names a number of articles of food such as 'Morendaka', 'Saskyl', 'Phenaka' and 'Utkarika' which are made by frying even to this day.

Butter

Butter is an important dairy product. It is available as table butter or as cooking butter. On an average table butter contains fat 80 per cent (min), salt 3.0 per cent (max.), curd 1.5 per cent (min) and diacetyl 1.4 per cent (ppm max). The minimum fat percentage in *desi* butter (cooking butter) is 76 per cent.

Butter is made from cream by churning. There are three views regarding churning of cream into butter. According to Fisher and Hookers phase reversal theory, churning involves phase reversal of cream. In cream fat is present as dispersed phase and milk serum the continuous (oil in water) while in butter, fat is continuous phase and droplets of skimmed or separated milk are dispersed in it (water in oil). This interchange of position between the two constitutents of an emulsion is termed phase inversion and in butter churning it is brought about mechanically. During churning process, agitation of cream causes coalescence and clumping of fat globules until eventually the ratio of the surface area to the volume of fat units become so small that it can no longer contain all the butter milk in stable form. The fat in water emulsion then suddenly breaks, yielding butter grains (consisting of an emulsion of water-in-butter fat) and free butter milk. The theory suffers from the fact that butter is not a true-water-in-fat emulsion. Microscopic studies reveal that a proportion of fat globules in butter are still intact in the worked butter.

According to Rahn's Foam theory, churning produces a foam with air bubbles dispersed in the cream. Fat globules collect in "clumps" round the edges of the air bubbles. The "clump" increase in size until the foam suddenly collapses leaving the butter "grains" suspended in the butter milk. The weakness in this hypothesis is that foam formation (i.e. the presence of air) is not required in some continuous butter-making process.

The King's modern theory is midway between the above two theories. The fat is present as clusters (clumps) of fat globules in cooled cream at churning temperature and within each globule it is present partly in solid and partly in liquid form. The clusters break during churning causing foam/froth formation. The fat globules get concentrated to the film around the air bubbles in the foam and brought to close contact with each other. The movement of fat globules over one another and violent shaking between them causes wearing away of the emulsion-protecting surface layer of the phospholipid-protein complex. The fat globules further adhere together to form larger and larger particle which finally become visible as butter grains or granules. Some of these granules enclose air from the foam. The fat in the granule is in globular form. The working of butter grain causes globules to move over another. Due to friction and pressure some of the release liquid fat while others are broken up. However, there remains enough free liquid fat which enclose water droplets, air bubbles and intact fat globules.

Though butter is stable in character, yet it undergoes spoilage and its quality deteriorates during storage. Various factors which are responsible for this chemical deterioration include storage temperature, sunlight, moisture and metallic contamination. The spoilage is chemically due to oxidation of the fatty acid esters. In addition to this, spoilage is also due to erratic enzymatic activities of the micro-organism which give rise to the development of hydrolytic rancidity and other defects in butter. The microbial activity is also responsible for undesirable changes such as decomposition of fats, proteins and lactose which take place generally in water phase. Several micro-organisms cause mould growth in butter. *Escherichia coli* is the chief among those bacteria. The shelf-life of butter can be improved by addition of potassium sorbate along with sodium chloride which reduces the release of free fatty acids and check the microbiological population.

Microstructure or Physical Structure of Butter

The structure of butter means the constituents of butter and their arrangement in the butter. Butter has the following composition with respect to major constitutents :

Constituents	Table Butter (%)	*Desi* Butter (%)
Fat	< 80	< 76
Moisture	>16	

Salt	>3
Curd	~1.5
Air	Small amount

All the constituents influence the quality of butter depending on the quantity and their arrangement in butter. Fat causes hardness and consistency texture is related to water and other constituents which mostly affect the butter colour e.g., yellow, white or moldy colour of butter. Yellow colour of butter is not always due to pigment, but due to the arrangement of constituents such as air volume and water keeping quality of butter.

Constituents of Butter

Fat in Butter: Fat in butter exists in two forms - (a) globular form nearly equal to 30 per cent and (b) free form nearly 70 per cent of fat in butter. These two forms vary depending on type of butter and extent of churning. As soon as butter is formed from the cream or break point takes place, the butter granules are collected at that time. Globular form is in higher percentage and the free form in lower concentration. Due to working of butter the globular form is reduced continuously and free form is increased as the breaking of fat globule membrane (FGM) during working also takes place. This free fat is the medium for taking uncrushed fat globules water droplets and air bubbles present in butter. It means free fat of butter forms the continuous phase whereas globular form of fat and other constitutents such as water, salt and air form the discontinuous phase. Globular and free fat also exist in two forms depending on the composition and melting point of the triglyceride, the two forms are – (a) crystalline form, (b) liquid form. In a butter at normal conditions, these two forms exist in equal concentrations (approximately). The object of tempering or ageing of cream at room temperature before churning is to make fat in two forms – crystalline and liquid fat. The proportion of solid and liquid fat depends upon the tempering, time and temperature. Hardners or consistency or spreadability of butter depends on ratio of globular to free and also crystalline to liquid fat. If shortage of liquid fat percent in these two forms i.e. globular and free fat, then butter is said to be hard, crumbly or gritty. Otherwise butter will be of open, brittle and crumbly texture.

If excess of liquid fat, say more than 70 per cent or even more, upto 90 per cent due to certain feeding effect or lactation period or room temperature. It will give opposite effect such as leakage of fat in butter or oiling of butter or soft butter. This is also considered to be quality or textural defect.

Other than hardness or softness, the compositional factors also influence the quality of butter. The crystalline butter exist in colloidal particles of average diameter 0.1μ in size. If the size of the crystalline fat is bigger than 0.1μ in free fat then it is called mealy texture which can be judged by cutting the butter and spreading it on the hand.

Aqueous Phase or Layer in Butter: Aqueous or water contents of butter contains other soluble and dispersed constitutents. Aqueous phase is also present in two forms –

(a) Internal moisture – means that moisture which is present inside the butter granules.

(b) External moisture – means that moisture is outside butter granules.

The size of moisture present in the form of water droplets is in the range 1-30 μ size and under normal working conditions water droplets are 1000-30000 droplets per cu. cm. of butter depending upon the working time. If there is large number of finely dipersed water droplets in the butter than it affects the butter appearance. It gives opaque or whitish colour to butter. If number of water droplets is less but of large diameter it gives slightly deeper yellow of butter due to light dispersion. In normal butter, if there is uneven distribution of small and big droplets in butter than it causes discolouration defects in butter.

Air in Butter: Butter contains 3-5 per cent air by volume i.e. no air at all. It is mostly dispersed and partly dissolved in water or liquid part of butter. It influences the butter in physical properties. It causes weakness in interior parts of butter and affects the hardness and spreadibility of butter. Most of the air present in the form of air bubbles depends upon the working time. The butter contains air in weak spots and forms an open texture when no air is incorporated then we have the closed texture of butter. When more air is incorporated then it may produce inner cracks, more weak spots and quality of butter with respect to body and texture is effected.

Salt: In butter common salt is added > 3 percent. Salt adds flavour to butter as well as checks the growth of micro organisms. Before working, butter salt should be added in the fine form. When salt is not properly mixed, it gives local salty taste. Salt is mostly present in soluble state in water droplets, as butter contains water about 16 per cent (max) and salt about 3 per cent (max), therefore, water droplets contain salt about 15.8 per cent (max). As micro organisms also grow in water phase but the water phase contain very high concentration of salt and due to very high osmotic pressure the growth of micro organisms is retarded.

Makkhan

Makkhan is known since Vedic times. It is used all over India and is an intermediate product to traditional ghee-making process. Makkhan is used for garnishing curry dishes and chapatis, ghee manufacture and direct consumption.

Traditionally, cold water is mixed with *dahi* (curd), and churning carried out by specially designed wooden devices. Grains of makkhan appears on the top of fluid butter milk, which are scooped out and patted into a smooth compact mass. Operations are confined to the domestic spheres and form part of unorganized marketing.

Buffalo milk yields, white makkhan, while cow milk yields yellowish product. The body is similar to that of creamy butter, but the product has typical diacetyl flavour, which originates from curd during fermentation process. Makkhan consists of 78.0-80.0 per cent fat, 1.5 to 2.0 per cent SNF and 15-20 per cent moisture.

Cream

Cream is one of the most important dairy products. It is used as such, or as a raw material for the manufacture of table butter, or for preparing ghee. Sterilised cream is cream containing 20 per cent or more of fat which has been homogenised, canned and sterilised. It may contain sodium salts of phosphoric and polyphosphoric acid, citric acid, carbonic acid and calcium chloride in quantities not exceeding 0.3 per cent. According to the PFA rules (1976) cream is the product of cow or buffalo milk or a combination thereof which contains not less than 25 per cent milk fat. Cream is distinguished into various types (depending on fat percentage) such as table cream, light cream, coffee cream, whipping cream, heavy cream and plastic cream. Cream contains all constituents such as water, fat, protein, lactose, ash, total solids and solids not fat. Cream is rich in fat soluble vitamin A, D, E and K.

Cream is separated from milk either by gravity or centrifugal method. This is based on the fact that milk fat is lighter than skim milk portion. The average density of milk fat and of skimmed milk at 16°C are 0.93 and 1.036, respectively. When milk (a mixture of milk fat and skim milk) is subjected to either gravity or centrifugal force, the two components, that is cream (fat) and skim milk, separate from one another by virtue of their different densities. Several factors affect separation of cream by both the above methods.

In Gravity Method milk is allowed to stand for sometimes so that fat level-rises. The velocity at which fat globule rise, is given by Stokes Law :

$$V = \frac{2G}{9}\left(\frac{ds - df}{n}\right) r^2$$

Where,

V	=	Velocity at which a single fat globule rises
G	=	acceleration due to gravity
ds	=	density of skim milk
df	=	density of fat
r	=	radius of fat globule
n	=	viscosity of skim milk

On the basis of Stoke's Law, theoretically velocity is increased by an increase in the radius of fat globule, increase in densities of skim milk and fat and decrease in viscosity of skim milk. In practice size of fat globule, temperature clumping and addition of adhesive affects the velocity of fat rise in cream in gravity method.

In centrifugal method milk is subjected to centrifugal force higher than gravitation force. Since densities of skim milk and cream are different, fat portion being lighter moves towards centre while skim milk is forced to the periphery and are separated through separate outlets. Varios types of cream separators are follows :

$$V = r^2 \frac{(ds - df)}{n} N^2 \; R.K.$$

Where, V = Velocity of movement of a single fat globule

r = radius of fat globule

ds = density of skim milk

df = density of fat

N = speed of bowl (r.p.m.)

R = distance of fat globule from the axis of rotation

K = Constant

n = Velocity of skim milk

The observation based on formula are that the speed of cream separation is increased by greater radius of fat globule, greater difference in density between skim milk and fat, greater speed of bowl (r.p.m.), greater size of bowl and lower viscosity of skim milk.

Ghee

Ghee dessicated butter fat occupies the most prominent status amongst the traditional milk products of India at present. Apart from its dietary usage, ghee is also used for performing religious and ceremonial rites. The composition of ghee varies from region to region and season to season depending on the type of animal and food given. Rangappa and Achaya (1974) estimated that about 39 per cent of the total milk production was converted into the highly aromatic form of milk fat. As buffalo milk is a rich source of fat, and more than 55 per cent of the total milk production comes from buffalo milk, ghee is mainly produced from buffalo milk in India. Ghee enjoys a supreme status in India because of its pleasing flavour and aroma. Out of the total production, 80 per cent of ghee is used for culinary purposes, the remaining 20 per cent is used for confectionery, including small amounts consumed on auspicious occasions like religious ceremonies.

Several methods for the manufacture of ghee are in vogue. Traditionally, ghee is prepared by the milk-*dahi-makkhan*-ghee method. In the organized sector, the milk -cream-ghee or milk-cream-butter-ghee, or milk-cream-butter-pre-stratification-ghee method is employed. The final stage of ghee preparation involves heat desiccation in steam jacketed vessel.

Chemically, ghee is a complex lipid of glycerides (usually mixed) 99 to 99.5 per cent; free fatty acid (per cent oleic acid 2.8), phospholipid, sterols, sterol esters, fat soluble vitamins (vitamin A I.U. 17-30),tocopherol (18-48 μg/g), carbonyls, hydrocarbons, carotenoids (cow ghee 3.2 - 7.4 ug/g),small amount charred casein and traces of calcium, phosphorus, iron copper etc. It contains moisture not, more than 0.5 percent. The sterols are mostly cholesterol.The structure of pattern of arrangement of fatty acids in the triglyceride molecule is an importnat aspect of compositional studies of ghee or milk fat. The glyceride structure greatly influences the physico-chemical properties of a fat. It has been seen that the glyceride composition of cow ghee is distinctly different from that of buffalo ghee. The proportion of high melting triglycerides is much higher in buffalo (av 8.7 per cent) than in cow ghee (av 4.9 per cent). This is because buffalo ghee has larger proportions of long chain saturated fatty acids-palmitic and stearic acids. For the same reason buffalo ghee is distinctly harder than cow ghee. The content of long chain triglycerides is higher in cow(av 62 per cent) than in buffalo *ghee* (av 55 per cent). On the other hand the proportion of short chain triglycerides is higher in buffalo (av 45 per cent) than in cow *ghee* (38 per cent). This is because buffalo *ghee* contains higher amounts of butyric acid than cow *ghee*. The levels of trisaturated glycerides are similar in cow and buffalo *ghee* (21 to 43 per cent). *Ghee* has a melting range of 28 to 44°C. Its butyrofractometer reading is from 40 to 45 at 40°C. The saponification number is not less than 220. The iodine number should vary from 26 to 38. The Reichert-Meissl number (RM) varies from 26 to 32 depending on *ghee* from cow, buffalo, goat or sheep milk. In cotton seed fed area, *ghee* has RM of 20 which is specific for the region. The Polenske (P) value should not be more than 2.

Ghee made from buffalo milk is whitish (lack of carotenoids) with greenish tinge and that made from cow milk is golden yellow. The characteristic colour of buffalo fat has been attributed to tetrapyrozole pigments-biliverdin and bilirubin (Chandravadana and Dastur, 1976). This pigment is conjugated to a protein in milk, but is released during the manufacturing process of *ghee* making. Thus, imparting yellowish-green colour to buffalo *ghee*. Another desirable feature much sought-often, by the connoisseurs of *ghee* is the graininess, temperature and rate of cooling after heat clarification are two important factors influencing the size and structure of grains in ghee. Other factors, such as fatty acid composition of ghee, temperature of clarification/method of preparation, presence of free fatty acids etc. also influence grain formation to a smaller extent. Seeding can favourably induce good grain formation in *ghee* (Rama Murthy and

Narayanan, 1975). Typically, *ghee* possesses a pleasant buttery, light caramelised sweet (rather than pungent) acidic aroma. Flavour development and its profile in *ghee* is quite complex and is a subject of active investigation. According to Jain and Bindal (1968), the flavour of ghee developed through the possible interaction during heating process among a protein (probably casein degradation product), a reducing sugar (lactose) and minerals. Some of the fermentation products formed during souring of milk, impart to ghee its characteristic flavour. When fermentation process is omitted the flavour of ghee tends to be flat oily. A wide range of chemical compounds have been identified as possible contributors to the typical aroma of ghee. These can be listed as follows in Table 9.9.

Table 9.9. Chemical compounds affecting aroma of ghee.

Propanone-ethanol n-butanol, butanone-2	Jain & Bindal (1968)
2-ketohydrocarbons of 3 to 9 carbon chain lengths	Singhal & Jain (1968)
Monocarbonyls	Bector and Narayanan (1974)
Ketoglycerides oct-2 enals dec-2 enals	Ganguli & Jain (1973)
delta-lactones with 10 C, 12 C, 14 C, 16 C, 18 C gamma-lactones with 6 C to 12 C, 15 C, 16 C and 18 C	Wadhwa & Jain (1982)

There is a general similarity in the gross patterns of volatile carbonyls isolated from differently produced ghee. Of the eleven carbonyles in most of the *ghee* produced, six have been tentatively identified as propanone, butanone-2, pentanone-2, heptanone-2, octanone-2 and nonanone-2. Small but significant differences in the quality and quantity of volatile carbonyls in different types of *ghee* have been reported. About 95 percent of the carbonyls in *ghee* are non-volatile. Cow *ghee* contains more volatile carbonyls. The total carbonyls of buffalo ghee is higher than that of cow *ghee* irrespective of the method of preparation and temperature of clarification. The ketoglycerides constitute 50 to 60 percent of the total carbonyls in *ghee*, buffalo *ghee* (4.4 μmoles/g) having a higher proportion than cow *ghee* (2.4 μmoles/g). Oct-2 enal and dec-2-enal are the main alk-2-enal in the volatile as well as monocarbonyls in *ghee*. The patterns of alknals from cow and buffalo *ghee* are similar and consist of ethanol, pentanol, hexanal, heptanal, octanal, nonanal, decanal, undecanal and dodecanal.

Layer formation is typical in *ghee* if stored above 20°C. The chemical properties of these layers are significantly different. Differences are evident in Richert-Meissal (RM) number of these layers. The liquid layers always has a higher RM than the semi-solid layers. *Ghee* undergoes physico-chemical changes dependent primarily on the temperature of storage. Crystallisation occurs with the formation of solid-semi-solid, and liquid layers of *ghee* (cow and buffalo), kept either in metal or glass containers at 20°C or below, solidifies uniformly with fine crystals. Above 20°C and below 30°C, solidification results in a loose stucture. The liquid portion had a singificantly higher flaky portion of the same ghee. A similar trend in the iodine number in these layers also occurs. Detailed investigations on layer formation suggested that it would be preferable to store *ghee* below 20°C to avoid layer formation. *Ghee* stored at higher temperature is also susceptible to oxidative deterioration, rancidity and off-flavour. Due to inadequate storage facilities in India, much *ghee* loses its market susceptibility.

Shelf-life of *ghee* is dependant on the method of manufacture. Direct boiling of cream for *ghee*-making is reported to provide greater resistance to oxidation during storage than *ghee* made from butter. This is believed to be caused by increased serum, solid content and greater production of sulphydryl compounds during heating.

When *ghee* is prepared, packed and stored under ideal conditions it can have shelf-life upto 12 months. The shelf-life in general is affected by increase in acidity and development of off-flavours. Some of the factors that affect shelf-life are temperature of storage, initial moisture content, initial acidity content, copper and iron content, method of packaging and exposure to light and sunlight.

The shelf-life of butter oil was increased upto one year by heating 8.5 per cent of milk solids not fat with butter oil at 200°C at 15 mm pressure for 15 min. which resulted in the formation of antioxidant substances. Higher amounts of SNF in cream enhanced the keeping quality of ghee. Off-flavours developed faster in ghee prepared by butter method. Butter *ghee* had the lowest induction period. Higher free fatty acid and peroxide value have been reported at the beginning and during storage. Curd processed *ghee* had a proper storage quality than cream *ghee*. Reduction of shelf-life of *ghee* during storage in warm weather in the presence of air is a result of autoxidation. Such a spoilage is accompanied by the loss of unsaturated fatty acids, production of objectionable flavour, destruction of thiamins and carotenes, formation of toxic products, loss of attractive colour and flavour and decreased nutritive values. The principal sites of attack are the unsaturated portions of the fatty acids in the glyceride structure accompanied with a slow oxidation of saturated fatty acyl groups. Spoilage of *ghee* can be further minimized by improving the method of processing, packaging, storage and transport. Antioxidants are substances which when present in small quantities (naturally or by addition) in fat or fat containing food are able to prevent or delay the rate of oxidation and thereby prolong the wholesomeness and palatability of the food. Antioxidant should have no harmful effect, no objectionable flavour, effective in low concentration, be fat soluble, provide protection to food, be readily available and be economical. Antioxidant may be natural or synthetic. Tocopherols, phospholipids, amino acids, peptides, protein hydrolyzates, extract of plants like cocoa shells, roasted coffee powder, cereals, herbs, spices are natural antioxidants or contain substances having antioxidant properties.

The phospholipids are considered as major antioxidative substances and their content in *ghee* depends on the method of preparation, time and temperature of heating, moisture content in *ghee* and levels of milk solids in butter. The oxidative spoilage of fat can be minimized by using synthetic antioxidants such as gallates (ethyl, propyl, octyl), butyl hydroxy toluene (BHT), butyl hydroxy anisole (BHA), ascorbic acid and alphatocopherol in small amount. The addition of an antioxidant is only effective if it is added before the oxidation sets in. Recently, it has been observed that among the five antioxidants namely topanol-354, trolox-c, tertiary-butyl-hydroxy quinone (TBHQ), butyl-hydroxy anisole (BHA) and butyl-hydrodxy toluene (BHT) studied for their efficiency to enhance the shelf-life of *ghee*, TBHQ appeared to provide better stability.

Butter oil has a bland flavour whereas *ghee* has a pleasing flavour. *Ghee* has less moisture, contains more protein solids and differs in fatty acid and phsopholipids as compared to butter oil. Butter oil is prepared by melting butter at a temperature not exceeding 80°C whereas *ghee* is manufactured at 100 to 140°C. Butter oil can be reconstituted with skim milk powder whereas *ghee* can not be. Intensive work has been done on the simulation of *ghee* flavour in butter oil and methods standardial for rancid ghee. The butter oil in which ghee flavour is introduced has less shelf-life than *ghee*. The shelf-life of flavoured butter oil enhanced with 0.02 percent BHA and BHT and compared well with *ghee*. Addition of TBHQ resulted in an increase of shelf-life two times more than that of *ghee*.

A combination of halphen test, methylene blue reduction test and opacity test can be used for the detection of tallow and lard in *ghee*.

Packing of ghee is permitted ordinarily in 15, 4, 2 and 1 kg tinned cans. Permission is also given to pack *ghee* in 1 kg and half kg returnable glass bottles. *Ghee* is also packed in plastic satchets. Flexible film polylaminates have emerged as an attractive alternative to costly tin packages. Packaging of *ghee* in polyaminate containers that can 'stand up' are now being used in the industry.

Ghee residue is charred (burnt) light to dark brown material obtained as residue when ghee is filtered. It is a by-product in *ghee* preparation. The composition of ghee residue depends on non-fatty constituents (in butter or cream), technique for production and method of separation of *ghee*-residue (hand or mechanical pressing). In general it contains heat-denatured milk proteins, caramelized lactose and varying amounts of entrapped fat alongwith minerals and water. The composition of *ghee* residue is given in Table 9.10.

Table 9.10. Chemical composition of hand pressed ghee-residue.

Type of residue	Chemical composition					Average
	Water	Fat	Protein	Lactose	Ash	yield per 100 kg butter/ cream kg
From *desi* butter						
Cow	14.4	32.4	36.0	12.0	5.2	1.6
Buffalo	13.4	33.4	32.8	15.4	5.2	1.6
From creamery butter (unsalted)						
Cow	9.7	61.4	24.8	Traces	4.1	1.7
Buffalo	5.7	65.0	25.5	Traces	3.8	1.2
From cream (sweet unwashed)						
Cow	3.0	62.5	24.1	8.2	2.2	12.0
Buffalo	4.1	63.2	18.0	12.3	2.4	7.7
From cream (sweet washed)						
Cow	1.4	70.2	27.3	Traces	1.1	3.6
Buffalo	1.7	80.8	16.2	Traces	1.3	3.5

Adapted from S.N. Prahlad, M.Sc. Thesis, University of Bombay (1954).

Fermented Milk Product

From Paleolithic and Neolithic times, fermented milk products in some forms was known to the inhabitants of Indian sub-continent. This is supported from the carvings and inscriptions in the walls of Harappa and Mohenjadoro. Ample references are there regarding the antiquity of fermented foods in the early vedic period, viz., Dadhisara, referring to the cream of curds (Mrcchakatika Act. 1). Dadhanvat, referring to a kind of cheese (Rv. VI: 48.18); Dadhyodana, referring to boiled rice mixed with curd (Brhad up. VI. 4.15); Mathita referring to curds churned without water and Takra referring to a drink by mixing three parts of curd with one part of water (Manasoltasa III, 1571) and many others. Manu laid down the curds and its preparation could be consumed even if they had turned sour (Manu V. 10). Some of the fermented milk products relevant to modern dietary regimen are described in this section.

Dahi (Curd)

Among fermented milk products, *dahi* (curd) is the most popular in Indian homes. It has been estimated that about 15 per cent of the milk produced is used for production of curd. It is an item of daily diet in almost all regions of India. A major portion of curd is prepared in villages, mostly for marketing in towns and for domestic consumption. It is also prepared in sweetmeat shops and hotels and restaurants for local sale. Curd also acts as a base material for the preparation of *lassi, chakka, shrikhand* and indigenous butter used for *ghee* manufacture. Some varieties of soft and unripened cheese like Surati, Bandel and Dacca cheese are also prepared from curd as base.

Dahi or curd is the product obtained from pasteurised or boiled milk by souring, natural or otherwise, by a harmless lactic acid or other bacterial culture. It may contain additional canesugar. It should have the same percentage of fat and solid-not-fat as the milk from which it is prepared.

Traditionally milk (cow or buffalo) is standardised/pasteurised or boiled and then lactic culture is inoculated at 22-25°C, and incubation carried out at the same temperature for 8 to 10 hours till acidity of about 0.8 per cent is developed.

Curd is white to creamish yellow colour. Well-formed curd can be sliced. It has creamy consistency, rich diacetyl aroma and mildly acidic flavour.

Sweet curd should not have more than 0.7 per cent acidity while in sour curd, this could go upto 1.0 per cent. Yeast and mould count per gram at maximum level should be 100 while coliform count per gram (maximum) 10 and this should not respond to phosphatase test. Cultures normally used for sweet curd are *Str. lactic, Str. diacetylactic* and *Str. creamoris* while for sour curd same alongwith *Lact. bulgaricus* or *Str. thermophillus* can be used. The general composition of curd is given in Table 9.11.

Table 9.11. General composition of curd

Product	Water %	Fat %	Protein %	Lactose %	L.A. %	Ash %	Calcium (mg) %	Phosphorus (mg) %
Curd whole milk	85-88	5—8	3.2-3.4	4.6-5.2	0.5-1.1	0.7-0.75	0.12-0.14	0.09-0.11
Curd skimmed milk	90-91	0.05-0.1	3.3-3.5	4.7-5.3	0.5-1.1	0.7-0.75	0.12-0.14	0.09-0.11

The following changes take place in milk during conversion of milk to curd.

a. During the formation of curd, both specific gravity and total solids are observed to decrease. The fall in specific gravity is due to slight expansion in volume on curd formation. Total solid decreases as a result of formation of decomposing products like ammonia, volatile acids and alcohol on fermentation. The changes in these constitutents are shown in the Table 9.12.

Table 9.12. Changes in some constitutents during formation of curd.

	Specific gravity	Total solid %	Titratable acidity %	Volatile acid (ml of 0.01N NaOH for 100 g)	Alcohol % by volume
Cow milk	1.0372	9.37	0.12	13.5	0.05
Curd (1 day)	1.0315	–	1.05	20.5	0.66
Curd (7 days)	1.0259	9.08	1.59	24.0	1.33
Buffalo milk	1.0420	11.07	0.11	7.8	0.04
Curd (1 day)	1.0351	10.29	1.71	57.0	0.73
Curd (7 days)	1.0300	9.37	1.95	–	1.47

b. Curd prepared from skim milk differ with that of whole milk, in that the former has little fat, a slightly higher water and lactose content.

c. The lactose in whole buffalo milk seems to ferment faster and to a larger extent than the cow milk. In a seven-day-old curd 71.5 per cent of sugar is converted into acid in case of buffalo milk while it is 47.3 percent in cow milk.

d. Considerable changes occur in the mineral contents of milk. During souring at pH 4.6, all the calcium is in ionic form and all the phosphorus, except that conjugated to casein, exists as ionic phosphate.

e. Changes in vitamin contents of milk on fermentation depends on the type of organism used. Different results have been reported by different workers. A mixture culture of *L. bulgaricus* and *S. paracitrovorus* is found to draw on the thiamin, riboflavin and nicotinic acid for their growth, pure cultures of *S. cremoris* and *S. lectis* could synthesise thiamin and raise its concentration in curd from 2-20% over that of milk. some workers have reported that riboflavin and nicotinic acid content decreases in curd as compared to milk, when cultures like *L. bulgaricus, L. acidophilus* and *L. casei* were used while others attribute increase of riboflavin to increase in fat content. Cultures like *S. lactic, S. thermophilus, S. citrophilus, L.bulgaricus,* L. *helveticus* and *L. acidophilus* increased the folic and folinic acid activities in curd making. A group of workers have observed that there is no fall in vitamin A content, however, it could be fortified with synthetic vitamin A, if necessary. Average vitamins content are reported in Table 9.13.

Table 9.13. Average vitamin content of milk and curd (per 100 g of milk)

	Milk	Dahi
Vitamin A (I.U.)	118	102
Vitamin B_1	55	49
Riboflavin	167	157
Nicotinic acid	96	86
Biotin (μg)	29	3.2
Panthothenic acid	20.2	183
Folic acid	161	178
Vitamin B_{12}	0.15	–
Ascorbic acid (mg)	1.4	1.3

f. Inconsistent observations are recorded on the changes of nitrogenous constituents because different fermentative organisms have been used for curd preparation. Table 9.14 gives an average value of nitrogenous constituents on conversion of milk to curd.

Table 9.14. Changes in nitrogenous constitutents in curd

	Total Nitrogen %	NPN %	Protein N %	Dialys-able N %	Ammo-nical N %
Cow milk (Raw)	0.56	0.048	0.512	0.017	0.24
Curd (1 day)	0.57	0.073	0.497	0.027	9.74
Curd (7 days)	0.55	0.144	0.406	0.060	–
Buffalo milk (Raw)	0.65	0.038	0.612	0.021	0.16
Curd (1 day)	0.65	0.052	0.598	0.022	12.20
Curd (7 days)	0.65	0.081	0.579	0.034	9.74

The digestibility coefficients of curd are superior to that of milk from which it is prepared. There is a definite increase in the vitamin B group curd as well as intestinal synthesis as a result of the ingestion of curd.

Curd has a therapeutic value and displays antagonistic acitivity against enteropathogenic bacteria. Western workers have reported the anticarcinogenic properties of yoghurt, the occidental version of curd.

Curd can be aseptically packed in packs made of plastics or cartons of 100g, 200g and 500g of conical shape. Asceptic packing machines for curd on commercial scale manufacture with laminar flow unit/U.V. irradiation unit/system are developed.

Chakka

It is an intermediate product for the preparation of Shrikhand. It is commonly produced in the western region of India. It is essentially an indigenous Indian Dairy product prepared under household conditions. Traditionally, *chakka* is prepared by draining the whey from lactic fermented curd obtained from buffalo milk. *Chakka* is marble white, closely knit, plastic mass having acidic diacetyl aroma.

The composition of *chakka* will depend on the initial composition of milk, the degree of fermentation (i.e. acidity developed) and the extent of whey removed. The average composition of *chakka* is given in Table 9.15.

Table 9.15. Composition of *chakka* (per cent)

Moisture	Fat	Protein	Total Sugar	Ash	Titratable acidity (as lactic acid)
59.55	22.4	10.30	4.40	1.03	2.32

Shrikhand

This is a popular fermented milk product in the western regions of India, and consumed as a dessert or snack food. It is also used to make *Shrikhand wadi.*

Shrikhand is a semi-soft, sweetish-sour, whole milk product prepared from lactic fermenting curd. The curd (*dahi*) is partially strained through a cloth to remove the whey and thus produce a solid mass called *chakka* (the basic ingredient for *Shrikhand*). This *chakka* is mixed with the required amount of sugar, saffron, cardamom, aromatic spices, dry fruits, nuts, etc. and kneaded to smooth consistency.

Shrikhand can be prepared by using cow milk or buffalo milk or skim milk with cream (40 per cent fat). The composition of *Shrikhand* varied widely with fat content. Other factors which influence the composition of *shrikhand* are initial composition of milk, degree of fermentation, extent of whey removal and amount of sugar added in *chakka*. An approximate composition of *shrikhand* is given in Table 9.16.

Table 9.16. Composition of Shrikhand (per cent).

Moisture	Total solid	Fat	Protein	Lactose	Sugar
34.48 to 35.66	64.34 to 66.52	1.93 to 5.56	5.53 to 6.13	1.56 to 2.18	52.55 to 53.76

Shrikhand is usually packed and served in earthern pots. It can also be packed in wide-mouthed glass bottles, tin cans, plastic pouches, etc. It could be stored at refrigerated temperature (4°C) for about 10-12 days without much deterioration in its quality. Longer storage may result in loss of body and texture and development of rancid flavour.

It has high calorific value in nutrition. It has higher shelf-life than milk and curd. It is popular because of its characteristic flavour, taste, palatable nature and possible therapeutic value during intestinal disorders.

Shrikhand Wadi

Shrikhand Wadi or *Shrikhand Vade* is a sweet made out of *Shirkhand* in western regions of India. *Shrikhand* is heat desiccated to a plastic mass in shallow vessel. Shrikhand is deep fried in ghee in an open pan with constant stirring. When the mass begins to get hard, a small quantity of it is taken and tested for stickiness. When it does not stick it is ready for addition of spices like cardamom, saffron, etc. After addition of spices, the product is rolled in the form of thick *chapati* and then cut into desired shape and size. An average composition of *shrikhand wadi* is given in Table 9.17.

Table 9.17. Composition of *Shrikhand Wadi* (per cent).

Moisture	Fat	Protein	Lactose	Ash	Sugar	Lactic acid
6.5	7.4	7.7	15.9	0.8	62.9	1.0

Lassi

It is a popular refreshing beverage in the northern and western states of India. Whole milk curd is stirred with about 10-20 per cent water and sugar (6-9 per cent) or salt. Required amount of essence may also be added. It is desirable to homogenise the product for improved body and texture. The consistency of *Lassi* should be such that it can be sipped through a straw from the bottle.

It is whitish, viscous fluid with creamy mouth feel, rich aroma predominantly of diacetyl and mildly to highly acidic. On an average the composition of *Lassi* may range from fat 3.0-3.5 per cent, total solid 16-18 per cent and acidity varying from 0.75-0.88 per cent.

Sometimes the name *lassi* is also given to fluid left over after churning of curd for butter preparation. It is then known as *Mattha* (Butter milk) and is common in all regions of India. It is highly flavourful and mildly to highly acidic. It is essentially a rural domestic product. It contains about 0.5 per cent fat and 6-7 per cent milk solids. The composition of *lassi* or butter milk varies considerably, depending upon the composition of curd, the extent at which the curd is diluted when churning and the efficiency of fat removal.

Yoghurt

Yoghurt is a coagulated milk product obtained by lactic acid fermentation through the action of *Lactobacillus bulgaricus* and *Streptococcus thermophilus*, from milk and milk product (pasteurised or concentrated milk) with or without optional additions (milk powder, skimmed milk powder, whey powder, etc.) The micro-organisms in the final product must be viable and abundant. However, in some places lactose fermenting yeast, *Leuconostoc strains, Streptococcus diacetilactic* and *Lact. acidophilus* are also added to improve the flavour of the yoghurt.

Yoghurt is the occidental counterpart of curd. From the early origins in the Balcans and Middle East, yoghurt has achieved world-wide acclaim. Yoghurt is an excellent food. It is the traditional form of fermented milk consumed in Bulgaria, Turkey and adjacent countries. The average composition of yoghurt is given in Table 9.18.

Table 9.18. Composition of Yoghurt (per cent)

Fat	Total solid	Total protein	Ash	Lactic acid	pH
1.7	11.0	3.5	0.8	–	4.3

The process of yoghurt manufacture involves standardisation and homogenisation of milk and then it is heated to 85-90°C, held for 15-30 min/cooled to 45°C and inoculated at the rate of 3 per cent with a mixture of *Lactobacillus bulgaricus* and *Streptococcus thermophilus* followed by incubation at 42°C for 3-4 hours. The final acidity of yoghurt ranges between 0.8 to 1.5 and pH between 4.4 and 4.9. The yoghurt is held at 5°C in a refrigerator until use. During the growth of yoghurt, lactic culture in milk, lactic acid, diacetyl, acetaldehyde, acetone acetoin are produced by fermentation of lactose and breakdown of proteins. All these substances contribute to the flavour of the product but the most important is acetaldehyde product. Viscosity of yoghurt plays an important part in its organoleptic quality. It depends on the ability of *L. bulgaricus* and *S. thermophilus* to produce a polysaccharide from galactose (lactose ——> glucose + galactose), which acts as a thickener.

Yoghurt is more easily digested than milk. These products have significantly higher amino acids, peptides, compared to milk. There is significant increase in folic acid and loss of riboflavin, pantothenic acid, thiamin and biotin in yoghurt than milk from which it is prepared. The biological value of yoghurt is higher than that of milk protein. The fermented yoghurt had higher concentration of anti cholesteremic milk factor than unfermented milk.

Frozen Milk Product

Kulfi: It is a frozen dessert from concentrated milk. It has a typical icy texture and caramel flavour. It is prepared mostly in the northern, central and western parts of India. Traditionally, it is prepared by concentrating milk in a shallow pan with constant agitation. Sugar, nuts, essence, colour, etc. are incorporated towards the finishing stages. Concentrated mass is poured into conical metallic moulds and frozen in a mixture of ice and salt.

Kulfi has whitish to slight brownish appearance, compact body ice texture, nutty and caramel flavour. It contains on an average fat – 7-8%, SNF 17-18% and sugar 13-15%.

Ice Cream: Ice cream, fruit ice cream, nut ice-cream, chocolate ice cream means the frozen food made from heat-treated mix made out of milk, cream and/or other milk products (derived from cow and buffalo milk) and with or without sweetening ingredient, eggs, water, fruits, nuts, chocolate, permitted stabilizer/permitted emulsifier, edible common salt, permissible flavouring and colouring matter.

Ice cream in general has the following composition as shown in Table 9.19.

Table 9.19. Composition of ice cream

Characteristic	Requirement
Weight in grams, per litre (Min)	525
Total solids, per cent by weight (Min)	36.0
Milk fat, per cent (Min)	19, 8 if contains fruits, nut, etc.
Acidity per cent (as lactic acid) (Max)	0.25
Sucrose, per cent by weight (Max)	15.0
Protein per cent by weight (Min)	3.5
Emulsifier/stabilizer, per cent by weight (Max)	0.5
Standard plate count (per g)	Not more than 250000
Coliform count (per g)	Not more than 90
Phosphatase test of the mix	Negative
Water per cent	70-90

Milk fat is an important constituent in ice cream. It contributes to flavour and texture. Milk fat softens the ice cream. If fat is even slightly off-flavoured, the ice cream will develop off flavour. Fat also gives stability to ice cream but impairs whipping ability.

Milk solids, not fat (MSNF) consist of milk proteins, milk sugar and mineral matter. They have not much to contribute towards off-flavour but improves the body and texture of ice cream. Milk sugar also adds to sweet taste while protein makes ice cream more compact and smooth. Excess of MSNF is desirable provided it does not produces sandiness. Sandiness encountered in ice cream is caused by the formation of lactose exists in two isomeric forms known as alpha-lactose and beta-lactose. In the freezer, the solution containing alpha and beta forms get concentrated because some water is removed as ice. This leaves the alpha lactose in super saturated condition and a tendency for the alpha form sets into crystallise. Beta lactose does not crystallise for the simple reason that it is more soluble than alpha lactose.

Sugar increases the acceptability of ice cream. Sweetening effect can only be produced by sucrose which is the cheapest source of total solids in the mix. It also enhances the flavour. Stabilizers are used in ice cream to prevent the formation of objectionably large ice crystals in ice cream, especially during storage. Stabilizers of vegetable origin should be preferred because no aging is required. Stabilizers generally have a negligible influence on the food value and flavour of ice cream. Emulsifiers are mainly used to improve upon and provide a uniform whipping quality to the mixture and gives the frozen ice cream a drier appearance. Flavour increases the acceptability of ice cream while colour improves its appearance and aids in identifying flavour. Various processing treatment such as preparation of ice-cream mix, homogenisation, pasteurisation, cooling, aging and freezing of ice-cream mix, filling, packaging, hardening and storage affect the quality of ice cream.

The freezing process involves lowering of temperature of mix, freezing a portion of the water in mix, incorporating air into the mix and cooling the ice cream mix. The temperature of the mix drops rapidly when mix is put in freezer without the ice crystals being formed. Rapid agitation reduces the viscosity by destroying the gel structure and breaking fat globule clusters.

The liquid water changes to ice crystals as the freezing point is reached. These ice crystals are pure water in solid form. The sugar and other salts become concentrated in the remaining liquid water. Increasing the concentration of these solutes depresses the freezing point of the liquid portion, so that the temperature of mix must be lowered before any more crystal will form. Thus, in freezing ice cream, the freezing point is continually being lowered by the formation of ice crystals; and while the temperature drops, more ice crystals are formed, increasing the concentration of sugar and other solutes in the remaining liquid water until the concentration is so great that further freezing will not occur. This is then put in the container and palced in the hardening room. The freezing process is continued without agitation until the temperature of ice cream reaches –

18°C (0°F) or below, quick hardening is desirable as slow hardening favours large crystals and coarseness. The ice cream is generally stored between –12 to –45°C. Shrinkage is one of the many defects in ice cream and is caused by temperature fluctuation during storage, excessive over run and rough transportation.

Physico-Chemical Nature of Ice Cream

Effect of Pasteurisation: Although the prime role of pasteurisation is to ensure the microbiological safety of the product, the heating also affects the physico-chemical structure of the mix. The emulsifier is melted and heat-activated stabilizers brought into colloidal solution. The whey proteins present in the SNF are partially denatured and uncoil, exposing the lipophilic portion of the molecule. As a consequence, the whey proteins begin to act as emulsifiers while at the same time, the water binding capacity is increased. Denaturation also increases the number of available binding sites for protein:hydrocolloid interactions and thus, enhances the action of stabilizers such as carrageenan. Pasteurisation is generally beneficial to ice cream quality, but excessive heat treatment leads to unacceptable organoleptic deterioration.

Role of Stabilizers and Emulsifiers: It has been commonly stated that stabilizers interfere with ice crystallisation and limit the growth of large crystals. Until recently this has not been studied systematically, although qualitative support is offered by the observation that the use of stabilizers results in an ice cream of smoother body. At a given rate of heat removal, the coarseness of a population of crystals depends on the relative rates of nucleation and growth and it has been postulated that stabilizers act either by increasing nucleation or reducing the rate of growth. Investigations into each of these possibilities showed that while stabilizers have no significant effect on rate of nucleation, the growth of ice, in a model system, is drastically retarded and the morphology altered by the presence of a stabilizer gel network.

The effect on ice growth rate and morphology has been explained qualitatively in terms of mechanical interference with the growth of ice. It is probable that the gel fibres cause the ice-interface to develop a radius of curvature of similar size to the mesh size of the network and thus to depress the freezing point. The gel fibres themselves are placed under stress and may rupture, but the significance of this is unclear. The effect of stabilizers is greater than expected and this may be due to the presence of sucrose as an additional solute.

The effect of stabilizers cannot be explained wholly in terms of the effect on ice growth. The three-dimensional network formed binds, restricts movement of water in the frozen product and thus minimizes the potential for formation of large crystal during temperature fluctuations. The viscosity of the mix is also increased as a result of interactions between gum-type stabilizers and partially denatured proteins.

Emulsifiers act as surface active agents and reduce the energy required to maintain the integrity of the fat globules. In an emulsifier such as glycerol monostearate, the lipophilic portion of the molecule, the fatty acid chain enters the fat phase, while the hydrophilic portion the glycerol portion enters the water phase. Added emulsifiers are more effective than milk proteins due to their smaller molecular size and greater mobility.

Paradoxically an additional beneficial function of emulsifiers is to destabilize the fat globule. This permits a degree of clustering and prevents rapid melting.

Effect of Ageing

During ageing a number of important changes occur. Skim milk powder and stabilizers become fully hydrated and the emulsifier-induced disruption of proteins continues. Substantial crystallisation of fat also occurs. Triacylglycerols of the highest melting point are the first to crystallise and are situated closest to the surface of the fat globules. The crystallisation process continues successively with triacylglycerols of progressively lower melting points, creating multiple-shelled fat globules with a core of liquid fat. The quantity of fat remaining liquid depends on the specific fat involved, but a balance between liquid and crystallised fat is essential for good quality. The poor quality associated with ice cream made from unsaturated fats results from the relatively low degree of fat crystallisation.

The Freezing Process and Structure of the Final Ice Cream

The final structure of ice cream is determined during the freezing and aeration of the mix. Ice cream is a complex physico-chemical system consisting of air cells dispersed in a continuous liquid phase in which ice crystals are embedded. The liquid phase also contains solidified fat particles, insoluble salts, milk proteins, lactose crystals, stabilizers of colloidal dimensions as well as sugars and salts in true solution.

Over the years many values have been quoted for the size of air cells, values for the mean diameter varying from 30 to 150 μm. These figures largely refer to ice cream made with batch freezers, and with continuous freezers, air cell diameters range from 5 to 300 μm, with a mean value of ~ 60 μm. Smaller cells are coated with a layer of fat crystals. Large air cells produce a smooth texture. A mean air cell diameter of 60-100 μm is satisfactory for most types of ice cream.

The interface between air cell and the mix is a relatively thin, continuous layer of unsaturated material coated with a layer of discrete fat globules which project into the interior of air cells. Air cells and ice crystals are separated by lamellae of variable thickness which contain both individual fat globules and casein micelles.

The amount of water existing as ice in ice cream varies with temperature, but is ~ 50 per cent at an extrusion temperature of –5°C. ~ 95 per cent at a hardening room temperature of –30°C and ~. 70 per cent at a storage temperature of –11°C. The results of investigations concerning the size of ice crystals have been variable, but the overall mean size would appear to be 35-40 μm, with crystals being 5-8 μm apart.

Fat globules in ice cream mix are of two types, large globules ranging in size from 0.5 to 4 μm but with most less than 2 μm diameter, the overall mean being ~ 1 μm. In addition very fine globules are also present which range in diameter from 0.04 to 3 μm with a mean diameter of 0.5 μm. Destabilisation of the fat emulsion, initiated by the action of emulsifiers, continues as a result of the combined effect of freezing and mechanical agitation during the freezing process. The fat globule membrane is ruptured and the liquid fat released to appear at the mix/air interface. Whipping properties are reduced and some coalescence occurs. A higher level of fat destabilisation occurs at low temperatures and when a high proportion of liquid fat is present.

Dried Milk Product

Whole milk powder is prepared by spray drying or roller drying of whole fresh milk or standardized milk obtained from cow or buffalo or a mixture thereof which does not contain any added preservative chemical or alkali. It may contain calcium chloride, citric acid, sodium citrate, sodium salt of orthophosphoric acid and polyphosphoric acids not exceeding 0.3 per cent. Butylated hydroxyanisole may be added to the extent of 0.01 per cent. The powder should not contain more than 5.0 per cent moisture and not less than 26.0 per cent fat. The total acidity of milk powder should not be more than 1.2 per cent. The standard plate count may not exceed 50,000/g and the coliform count may not exceed 90/g. The maximum solubility index should be 15.0 for roller dried and 2.0 for a spray-dried product.

Skim milk powder is prepared by spray drying or roller drying of whole fresh milk of cow or buffalo or mixture thereof, from which the fat has been removed substantially and which does not contain any added preservative, chemical or alkali. It may contain any added calcium chloride, citric acid, sodium citrate, sodium salts of orthophosphoric acid and polyphosphoric acid, not exceeding 0.3 per cent. It may contain 1.5 per cent milk fat and moisture not more than 5.0 per cent. The total acidity should not exceed 1.5 per cent. The standard plate count should not exceed 50000/g and coliform count must not exceed 90/g. The solubility index should be 15.0 for roller-dried and 2.0 for a spray-dried product. There are essentially three processes of drying milk. The basic principle upon which drying of milk by the use of cold has been developoed are, drying by freezing out the water and separating the ice crystals from the milk solids by centrifuging and freezing the milk and evaporating the water from the frozen product by sublimation. The system has become obsolete now. In roller drying concentrated milk is applied in a thin film upon the smooth surface of a continuously rotating steam heated metal drum, roller or cylinder, and the film of dried milk is continuously scraped off by a stationary knife/blade/scraper. The milk-film has to be ground to obtain the powder. The spray drying process consists in atomizing the preheated and concentrated milk to form a spray of very minute droplets which are directed into a large suitably designed drying chamber, where they mix intimately with a current of hot air. Owing to large surface area, the milk particles surrender their moisture practically instantaneously and dry to a fine powder, which is removed continuously.

In both types of drying of milk i.e. roller drying and spray drying, milk undergoes several processes like pasteurisation, pre-heating, clarification, condensing, etc. The pre-heating temperature time combination classify the power as low heat powder, medium heat powder, high heat powder. For low heat powder pre-heating is done at 185°F or 75-85°C/3 minutes and for higher heat powder, it is done at 220-230°F or 105-106°C/3 minutes. The important consideration with respect to pre-heat treatment is that low heat treatment means less denaturation of proteins, more solubility, less surface tension, more reconstitubility and dispersibility. In high heat powder, high heat treatment leads to generation of –SH group from amino acid. These –SH groups neutralise oxygen and restart fat oxidation. So, the shelf-life of high heat powder is more than low heat powder. High heat powder also removes some volatile substances responsible for off flavour including fat oxidation. Roller drying is done by two types – (i) atmosphere roller drying, (ii) vacuum roller drying. In atmosphere roller drying, the temperature is kept at 212-217°F and revolution is kept 15-20 revolutions/minute whereas under vacuum drying temperature is maintained upto 100-200°F and sometimes it can be 170-°F and steam pressure is maintained at 20-40 lbs. In roller drying concentrated milk is applied in a thin film upon the smooth surface of a continuously rotating steam heated metal drum, roller or cylinder, and the film of dried milk is continuously scraped off by a stationary knife/blade/scraper. The milk film has to be ground to obtain the powder. Spray drying is also done by two types – (i) nozzle type, (ii) disc type. In nozzle type process, milk is allowed to pass through a small orifice under pressure, so that it is converted to fog. In this way vast area of milk particle is obtained and in the presence of hot air, it surrenders moisture instantaneously in the drying chamber and dried to a fine powder, which is removed continuously.

Physico-chemical changes take place during manufacture of dried milks, sweetened condensed milk and evaporated milk.

During the process of drying, dried milks are susceptible to (a) Maillard reaction, that is reaction of lactose with amino compounds in milk, (b) lipid oxidation, (c) heat induced milk coagulation.

Maillard, while attempting to establish the mechanism of protein biosynthesis, accidentally struck upon the reaction between lactose and amino compounds in heated milk. Maillard reaction is enhanced in dried milk, condensed sweetened milk and evaporated milk due to the following factors, (a) high moisture content, (b) water activity, (c) head space of containers, (d) high temperature of processing, (e) high storage temperature, (f) longer storage period.

The ultimate product of Maillard reaction is melanoidin, a brown polymeric compound which has been found to be useful in small concentration to act as a preservative in refrigerated stored milk. The important intermediates on the non-enzymic pathway to malanodin formation are shift base of HMF, hydroxy methyl furfural (HMF), lactulose lysine, lactose lysine, furosine, pyridosine, aldoseamines, ketoseamines, mutagens, aldose condensation products, enediols, strecker degradation products, reductones, etc. Furosine has been found to be an index of early Maillard reaction.

Dried milks are susceptible to lipid oxidation. The rate of oxidation is affected by fatty acid composition, degree of unsaturation, partial pressure of oxygen, the nature of surface being exposed to oxygen and the storage temperature, light exposure, moisture content and head space of container. Lipid oxidation has direct effect on the flavour of dried milk products. Homogenisation of milk for dried milk increased the free fat content. Storage of dried milk increases the free fat content from 0.17 per cent of initial value to 2.16 per cent after two months of storage. Such storage period increases the solubility index from 1.33 to 2.83, dispersibility decreased gradually to 75.17 per cent from 78.13 per cent, wettability measured in minutes was 31.1 at the beginning and 42.8 after storage for two months in glass bottles. Moisture content increased by 0.08 per cent after storage for the same period. Per cent lactic acid increased from 0.1 to 0.12 with concomittant decrease in pH.

The ability of bovine milk to withstand high processing temperature is essential for many modern milk processing operations. Non casein proteins are relatively heat-labile. But casein is remarkably heat-stable due to lack of secondary and tertiary structure. Casein exists in milk mainly as colloidal tertiary calcium phosphate. Milk composition affects heat stability, e.g. milk salts, pH, species, stage of lactation, relative importance of serum and colloidal phase, whey proteins, detergents, etc.

Casein and whey proteins are the principal milk proteins of milk. The whey proteins comprise α-lactalbumin, ß-lactoglobulin, bovine serum albumin, euglobulin, immunoglobulin and lactoferrin. The whey proteins are also known as serum proteins which, as against earlier views do not influence heat stability. However, ß-lactoglobulin and K-casein interact during heat treatment. Apart from the principal milk protein casein, the minor milk proteins are γ-casein, and proteose-peptone fractions which arise from enzymatic degradation of milk stored at low temperatures for a period more than 24 hrs. The isoelectric point or pI of a protein is that pH at which it becomes less soluble or in case of casein it coagulates. The principal casein and fractions are $\alpha\text{-}_s$, K and ß-casein fractions. The isoelectric points or pI of α_{-s}, $\alpha\text{-}_{s1}$, K-casein and ß-casein are respectively 4.1, 4.1, 4.1 and 4.5.

Properties of Milk Powder

There are two main properties of milk powder - (i) physical property, (ii) chemical or biochemical property.

Physical Property

Physical property include shape, size, surface density and aligned properties like reconstibility, densities, flow-ability dispersibility, etc.

Shape of milk powder depends upon the way it is prepared (roller or spray dried) shape of roller dried powder is irregular shape, rough edges with wrinkles at surface. Whereas spray dried milk powder particles are smooth, slightly elongated to some extent. Appearance of wrinkles depends on temperature gradient of outlet and inlet air. More is the difference, more wrinkles are formed. The size of roller dried powder particle vary from 8-20 μ and spray dried milk powder particle size vary from 10-150 μ by method of preparation. Size of particles also depend on total solids during condensing of milk. More are the total solids bigger is the size. During spraying enough air cells are incorporated and are almost absent in roller dried. In spray dried very fine air cells are incorporated, they may be throughout the particles or inside the particle. Large bubbles are present as inter-particle whereas smaller bubbles are intra-particle.

Chemical Properties

Lactose: It exists in two forms as – α-monohydrate and ß-anhydrate depending on the temperature. When the temperature is below 93.5°C, it exists as α-monohydrate. Above 93.5°C it exists in amorphous state or lactose glass, because at this stage it absorbs moisture and thus leads to crystallisation of lactose. After drying the temperature is lower than 93.5°C, lactose exists in amorphous state and distributed throughout the particle. When moisture is about 5 percent and relative humidity of air is above 50 percent or higher than the lactose is considered to be present as supersaturated solution and it becomes hydroscopic. It absorbs moisture and becomes sticky (soeption) with various particles stuck to each other and thus leads to caking. So to avoid cake formation, powder should be stored in atmosphere where relative humidity <50 percent and moisture of powder is <5 percent. Crystal development in milk powder depends on total solids in condensed milk. If total solids are more, then crystallisation will be faster.

When lactose crystallisation takes place, the most apparent effect are the cracks at the surface and it thus becomes more permeable to gases and solvents. Crystals of lactose favour denaturation of protein, because immediately after drying casein micelle is loosely held. Lactose crystals rupture fat globule membrane and thus free fat amount is considerably increased and it leads to an oily layer formation, greasiness, etc. after reconstitution and faster oxidation. Thus, crystallisation of lactose affects the dispersibility, baking and sensory quality of the product.

Milk Fat: Milk fat, exists as fat globule or free fat. In roller dried powder about 90 percent of fat exists as free fat whereas in spray dried milk powder, most of the fat exists as globule with membrane intact. The free fat means that it is not emulsified and it leads to greasiness and gives oily layer due to free fat in reconstituted milk. In roller dried powder fat exists as a pool of fat present on the surface of particle. At ambient temperature, this fat is partly solid and partly liquid. In roller dried, more free fat is due to severe heat and knife action. When we increase fat of dried milk its tendency increases to act in free form. In spray dried milk powder, the action is similar to homogenisation, ghee fat globules subdivide, otherwise remain distributed throughout the particle in globular form. During spray, air is incorporated so fat tends to surround the air particles. Spray dried milk powder also contain some free fat and this free fat leads to oxidation.

The free fat can be reduced by –

1. Lowering the temperature-time combination of preheat treatment.
2. Condensing heat treatment should be minimum.
3. During drying, larger the orfice at low pressure, less is the free fat.
4. Homogenisation of milk before drying also reduces the free fat.

Air: In dried milk, sufficient gases are entrapped and present between and inside the particles. About 25-80 per cent dry whole milk powder particles contain air. This variation is due to the method of preparation. In foam spray drying method more air is incorporated than nozzle spraying. More is the total solid more is the air incorporated. Immediately after drying, air content is much less in both the methods of drying. Most of the air comes in contact after drying so it should be packaged under vaccum to minimise air content.

Moisture: High moisture, excessive absorption of moisture by lactose resulting in lactose crystallisation leads to lumping/caking of powder. High moisture powder is more prone to insolubility due to denaturation of protein. More is the moisture absorbed, more is the free fat formation and this favours browning reaction and auto-oxidation and thus decreases solubility. Some gases like carbon dioxide also increase in high moisture powder.

Evaporated Milk

Evaporated milk is made by the evaporation of water from whole milk under a vaccum. To improve its stability, low percentage of sodium phosphate, sodium citrate, calcium chloride, and/or carrageenan may be added. Vitamin D is also usually added. The product is homogenised and canned, then sterilised under pressure at 117°C for 15 minutes or at 127°C for 2 minutes. Evaporated milk is also known as unsweetened condensed milk.

Evaporated milk should have the following characteristics – the milk should be clean, sweet, free from off flavours and odours and reasonably free from extraneous material. Contamination by antibiotics, pesticides and other chemical residues and metals is highly undesirable. No abnormal milk should be accepted. Acid development is objectionable. Maximum acceptable range extend from 0.18 to 0.22 per cent. The milk should not coagulate with equal volume of 70 per cent alcohol indicating heat stability of milk which is very necessary during sterilisation. Milk with substandard stability may be improved with sodium citrate or disodium phosphate or calcium chloride. The milk sample should be tested for bacterial count, sensory characteristic, sediment test, acidity, alcohol test, alcoholalizrin test and clot on boiling test. After inspection of these attributes milk should be correctly standardised for fat and non-fat solids.

The evaporated milk should have the following score card (ADSA).

Item	Perfect Score
Flavour and odour	40
Body and texture	35
Colour	5
Fat content	10
Total solids	10
Adulterants and preservatives	–
	100

The evaporated milk should have water 31 per cent, T.S. 69 per cent, fat 9 per cent, SNF 22 per cent, protein 8.3 per cent, lactose 12.2 per cent, ash 1.5 per cent and total milk solids 31 per cent.

During manufacture of evaporated milk, certain physical and chemical changes occur. These include development of brown colour, changes in viscosity, sedimentation, decrease in pH and heat stability, leading to coagulation or gelation of milk protein during sterilisation process. These problems are further aggravated during storage to cause formation of fat-protein lumps leading to fat separation, age thickening or gelation and organoleptic changes including browning of the product and development of flavours.

Changes Resulting from Concentration

Effect of Concentration: The degree of concentration, the concentration factor (R) may be defined as the ratio of dry matter content (by weight) of the concentrated milk to that of the unconcentrated. Concentration has a number of effects which may be modified by other factors such as preheat temperature.

Changes Resulting from Concentration

1. Lowering of a_w level.
2. Alteration of physical properties (increase in density and refractive index, decrease in thermal conductivity).
3. Increase in viscosity liquid becoming non-Newtonian (shear thinning and viscolastic) and then fairly solid as R exceeds 9.
4. Diffusion coefficient decreases.
5. Hydroscopicity increases.
6. Osmotic pressure, freezing point depression and boiling point elevation (colligative properties) increase as does the electrical conductivity.
7. Thermo-dynamic activity of solutes increases.
8. Ionic strength increases with a corresponding decrease in activity coefficient of ionic species. This in turn causes an increase in ionisation and solubility of salts.
9. Dissolved substances become supersaturated and may precipitate.
10. Conformation of proteins may change. Tendency towards formation of a compact structure and association increases.

11. Casein micelles increases in size. This is due to covalescence rather than swelling and the increase is less if the milk is preheated sufficiently to form casein-whey protein complexes.

The consequence of these changes is to accelerate most of the reactions initiated during preheating. The effect is limited in evaporated milk where the value of R is rarely greater than 2.5 (a_w level ~ 0.98) but is more pronounced in sweetened and condensed milks where the value of R (before sucrose addition) usually approximates to 7.0 (a_w level ~ 0.83).

Other Changes During Evaporation

Considerable damage occurs to the fat globule during evaporation. Globules are split to form a number of small globules with changed surface layers and are highly prone to lipolysis unless inactivation of lipases is complete. Evaporation also removes dissolved gases and reduces off-flavours by removal of volatilizers.

Heat Stability of Concentrated Milk

Heating of concentrated and unconcentrated milk ultimately leads to coagulation and stability is usually defined in terms of the time needed to cause visible coagulation of milk at a given temperature – the heat coagulation time (HCT). Coagulation of evaporated milk during sterilisation has been recognised as a problem for many years but, until recently, the underlying cause has been poorly understood.

Early work was largely restricted to stuudies of the effects of calcium/magnesium and citrate/phophsate on the HCT and led to the "salt balance theory". This "theory" permitted the development of the procedures still used industrially to control coagulation, but did little to explain the mechanism involved. Subsequently a "working hypothesis" was developed, based on the colloidal caseinate phosphate and the changes induced by heating, reaction between β-lactoglobulin and the K-casein of the casein micelles being considered the most important factor in altering susceptibility to coagulation. More recently, model describing heat coagulation in unconcentrated milk has been developed and this has now been extended to concentrated milk.

In the case of milk which has not been preheated and which has an initial pH value below 6.5, coagulation occurs as the temperature increases during heating. Coagulation appears to be induced by aggregation of denatured serum proteins and depends on calcium and hydrogen ion activities. Coagulation of this type cannot occur in pre-heated milk (the commerical situation) since the serum proteins are denatured, although the stabilizing effect of pre-heating is determined by the manner and extent of aggregation of the denatured serum proteins rather than the denaturation itself. In all pre-heated concentrated milk coagulation is due to aggregation of casein micelles. Behaviour during coagulation is, however, strongly modified by pH value of the milk.

At pH values below ~ 6.5 casein micelles flocculate into voluminous irregularly shaped clusters, while in contrast to higher pH values the aggregate tends to fuse into compact particles of approximately shpherical shape. The compact particles become anisometric shortly before the HCT (Heat coagulation time). The aggregation rate is determined by the colloidal stability of particles and the rate constant (s) of bond forming reactions, both of which are significantly affected by pH value. Dissociation of K-casein from the casein micelles proceeds more rapidly at high initial pH value and has a profound effect on micellar stability. The formation of micellar calcium phosphate-like salt bridges is also highly dependant on high initial pH value, since the reaction only occurs significantly where casein micelles are K-casein depleted. Other types of salt bridge (ionic and Ca^{2+} mediated) are less affected by pH value. Covalent cross-linking of proteins may also be involved in heat coagulation, but requires close contact between specific protein residues. This would be promoted by a large heat-induced fall in pH value, or by depletion of K-casein from the micelles.

Heat coagulation time is primarily determined by the rate of particle aggregation and the geometry of the emerging particles, the latter being strongly influenced by dissociation of K-casein and the supersaturation of calcium phosphates. Protein content is a further major factor determining heat stability of low pH value concentrated milk, but not that of high pH value. In recently developed models, the relationship between HCT, and protein content is explained by assuming that flocks formed in low pH milk are of a fractal nature, with fractal dimensionality of ~ 2.1. Fractal flocks need not become large to cause gelation in a concentrated dispersion and only aggregation due to encounters resulting from Brownian motion can be important. In high pH value milk, it is likely that the HCT is determined primarily by the time after which aggregates become anisometric and, therefore, voluminous. This time is itself determined by the ratio of the time it takes for two particles to fuse to the time elapsing between subsequent aggregation steps.

The role of phosphate in stabilizing concentrated milk has also been examined. The basic effect of phosphate may be partly explained in terms of a pH increase. The overall effect is more complicated, however, and not fully understood. Phosphate addition disturbs the existing equilibrium is attained only slowly at 20°C. During heating at 120°C, a continuing association of calcium phosphates with the micelles occurs over a period of at least 10 min, while in concentrate without added phosphate the association is completed during heating up. The different behaviour in the presence or absence of added phosphate may have consequences affecting rates of serum protein association, K-casein dissociation and bond formation during the sterilisation process.

Age Thickening and Gelation of Concentrated Milk

Change resulting from concentration means that a tendency to age thickening and gelation is a property of all concentrated milk. Particular problems are encountered occasionally with canned, sterilised, evaporated milk. This is not due to protein breakdown, but due to the formation of a three-dimensional network in which aggregated casein micelles are connected by thread-like structures. Age-thickening and gelation is strongly promoted by cold storage of the concentrate before sterilisation and this is attributed to a physico-chemical transformation of the casein micelles. Age-thickening occurs in two distinct stages, the rate being lower at higher sterilisation heat loads.

1. Viscosity increases accompanied by "spotty lump" formation but the milk is still fluid after stirring.

2. Irreversible gelation accompanied by syneresis and flocculation after stirring.

Age-thickening is also a potential problem in sweetened, condensed milk and appears also to involve aggregation of casein micelles into a three-dimensional network. Heated whey proteins are not directly involved, but may promote the development of networks by acting as links between adjacent micelles. Whey proteins may also be involved in complex formation with fats in whole sweetened condensed milk, forming a net-like structure which enmeshes casein.

A decrease in pH value occurs during age-thickening which increases casein/casein interactions and reduced ionisation of seryl phosphate residues. Both reactions tend to increase gelation, although this may be reduced but not eliminated, by addition of polyphosphates which have an affinity for casein.

Defects in Evaporated Milk

Flavour: The flavour which may occur in evaporated milks are usually unlike those commonly occurring in bottled fresh milk. This is due to the removal of volatile off-flavour during the concentration of milk under vacuum. The most common flavour defect in evaporated milk is that which seems to be associated with progressive age, darkening or browning of the product. No one term seems to describe the off-flavour adequately. Such terms as slightly acidic, stale-coffee, old, sour and strong suggest the nature of the defect. Also, the term caramel though suggesting the nature of the defect also hints at the brownish colour of the evaporated milk. However, the caramel flavour has a pleasing appetising taste unlike the defect associated with the age, darkening of evaporated milk and subsequent gelation. The age, darkened flavour defect is easily detected. When the sample is taken in the mouth, the flavour sensation is not particularly different from normal evaporated milk, but soon the delayed slightly acidic reaction occurs. This taste remains even after the sample has been taken out for some time. The off-flavour is accompanied by only a slight odour suggesting a staleness. The underlying taste of age darkened milk is acidic because there is an increase in acidity in such products when stored at high temperature.

Body and Texture: Manufacture of evaporated milk is subjected to such laboratory controls that particularly no body + texture defect exists in milk. However, no long holding some body and texture defects may be encountered.

Physical Defects: The physical defects are (a) buttery fat separation, (b) curdiness, (c) feathering, (d) bloats of evaporated milk due to causes other than micro-organisms, (e) bloats of evaporated milk due to bacterial causes, (f) coagulated evaporated milk, (g) bitter evaporated milk, (h) fishy flavour in evaporated milk, (i) mineral deposits, (j) gassy, (k) grainy, (l) low viscosity and (m) sedimentation.

Curdiness: The causes and preventions of the curdiness due to heat coagulation are as follows - (i) curdy evaporated milk is due to too low heat stability in the presence of heat treatment used for sterilisation, (ii) the standard commercial sterilising process of holding at 240°F for 20 minutes, or other temperature time ratios with equal lethal effect, represents the minimum temperature time ratio that will insure a sterile finished product, (iii) where the cooking time approached the heat stability time, there is danger of permanent curdiness in the product, (iv) if the heat stability of the evaporated milk is too low, it may be increased and curdiness prevented by the use of higher forwarming temperatures or by the right use of casein stabilizers, (v) too old or otherwise inferior raw milk as well as holding the evaporated milk too long, or at too high temperature before sterilisation are very frequent causes of curdiness in evaporated milk, (vi) high homogenising such as pressure between 1364 kg and 1818 kg or higher, lowers the heat stability and increases the curdiness of evaporated milk.

Fat Separation in Storage: It has been shown that the adoption of homogenisation as an integral function of the manufacture of evaporated milk eliminated the defect of fat separation as a serious problem. Yet, even in the homogenised product, the distribution of the fat globules is not permanently uniform.

If objectionable intensity of fat separation is to be avoided, in addition to efficient homogenisation, it is necessary to aim not for maximum heat stability but for a stability, sufficiently moderate to give the evaporated milk a full body and a viscosity adequate to definitely retard fat separation.

Feathering: Feathering in hot drinks can not be foretold by examining the products. Feathering is more when the viscous evaporated milk is used. When such milk is added to coffee, there is formation of flakes.

Bloats of evaporated milk are due to causes other than micro-organisms. The present discussions deal with bloats due to freezing to change in altitude, and to chemical action. When freezing, the content expand sufficiently to cause the cans to bulge at their ends, giving the impression of bloats. However, on heating to temperatures at which the contents of the can melt, the bulging ends almost flip back, to normal.

Change to high altitudes may cause the tins to bulge. This occurs invariably and especially when the cans are sealed at low altitude and then transported to high altitudes. Chemical action of the milk acids upon the base metal of the container causes evolution of gas sufficient to cause the appearance of bloats.

Bacterial Causes: Bloats of evaporated milk caused by gaseous fermentation are commercially the most serious bacterial defect of evaporated milk. The responsible organisms are usually, though not always a spore forming anaerobe. *Clostridium welchi* is one of the main organism. Aerobic spore formers like *Bacillus* have been isolated from evaporated milk cans wherein bloats have occurred. *Streptococcus* has also been isolated. However, bloats are usually due to anaerobic bacteria belonging to the butyric acid group and invariably the putrefactive types prevail. However, exceptions are there. The responsible organism in most cases has been found to be *Plectudium foetidum*. It is a highly objectionable anaerobe. The milk first curdles and then the organism digests the curd and leaves a clear yellow fluid similar in appearance to butter oil. Such drastic fermentation is accompanied by the evolution of a penetrating foul putrifying odour resembling that of hydrogen sulfide gas and bloats occur.

Coagulated Evaporated Milk: Bacterial coagulation is probably the most common bacterial defect occurring in evaporated milk. The coagulum found in different outbreaks has varied from the sweet curdling type to a sour curd and an intensively bitter curd. The causative organism was identified as *Bacillus cereus.*

Bitter Evaporated Milk: In many instances bacterial coagulum showed intense bitterness. This suggested proteolysis of proteins by inherent proteases and also proteinases of bacterial origin. The peptides so released caused bitterness. The main microorganism responsible for such bitterness is a *Facultative anaerobe* with an optimum incubation temperature of 32°C.

Fishy Flavour in Evaporated Milk: Coagulated evaporated milk often produces fishy flavour. The micro-organism responsible is *Proteus ichthyjosnius*.

Mineral Deposits: Under certain conditions there is a tendency for a whitish gritty deposit to appear in the bottom of the container. The deposit is made up of calcium citrate. Citric acid exists in milk in combination of mainly calcium and to certain extent magnesium. The citric acid content of milk has been found to be greatest when the cows are on green pasture. The citrates are, therefore, highest in late sporing or early summer. In India, it is usually in July, August, September. The higher the concentration of the evaporated milk, greater is the tendency for the appearance of mineral deposit during storage. Storage temperature is a controlling factor. Higher the temperature, higher is the deposit. However, at 5°C there is no salt precipitation.

Browning of Evaporated Milk: It is due to the non-enzymatic reaction between the free amino groups of protein with the potentially free aldehyde group of the glucose moiety of lactose.

Browning of evaporated milk can be prevented by (a) lowering the moisture content, (b) controlling the water activity, i.e., the ratio between bound and free water, (c) by decreasing to the maximum extent of the head space of the container, (c) by lowering the storage temperature, (e) by shortening the storage period.

Grainy: A grainy evaporated milk is one lacking smoothness and uniformity throughout. In such defects transmission of light is uneven when a film of evaporated milk across a loop or an open bottom spoon so suspended that light may be allowed to pass. Grainy evaporated milk is often associated with an excessively heavy viscous body. There is presence of pin-point size curd particles where light is transmitted through a film of evaporated milk.

Condensed Milk

Condensed milk was first prepared in 1809 by Nicholas Apart, a French Food Scientist. The method was as follows :

The milk was condensed by heating it in an open vessel placed in a water bath over fire to about two-third of the original volume. The condensed milk was strained, cooled and poured into glass bottles, filling them to the brim and corking them tightly. The bottles so filled and sealed were then held in a water bath for two hours. The milk so treated was found to have a very good aesthetic, safety and keeping quality.

The gross chemical composition of market milk samples of condensed milk are given in Table 9.20.

Table 9.20. Gross chemical composition of condensed milk

Type of brand	Total No. of samples	Fat % range	T.S. % range	Sugar % range	Milk solids % range	Lactic acidity % range
A	25	8.10-9.92 (8.84)	71.27 to 76.72 (74.23)	35.5 to 47.0 (42.67)	26.6 to 33.3 (29.90)	0.20 to 0.45 (0.30)
B	10	7.50-9.60 (8.78)	74.31 to 77.40 (76.13)	39.6 to 46.4 (44.21)	30.3 to 35.85 (31.87)	0.22 to 0.40 (0.31)
C	5	9.82-9.84 (9.34)	74.47 to 76.75 (75.95)	39.86 to 42.25 (41.45)	32.85 to 36.62 (34.40)	0.20 to 0.26
Over all	40	7.50-9.92 (9.85)	71.27 to 77.40 (76.85)	35.5 to 47.0 (42.94)	26.6 to 35.8 (30.99)	0.20 to 0.45 (0.30)

The gross composition of condensed milk, sweetened or unsweetened whole or skim milk is given in Table 9.21.

Table 9.21. Gross composition of condensed milk (percent) in developed countries

	Fat		TMS	
Type of condensed milk	UK	USA	UK	USA
Condensed milk	9.0	7.9	31.0	25.9
Evaporated milk	9.0	8.5	31.0	28.0
Skim milk (sweetened)	–	–	20.0	20.0
Skim unsweetened	–	–	26.0	24.0

The detalied composition of condensed milk is given in Table 9.22.

Table 9.22. Detailed composition of condensed milk (per cent)

Type	Lactose	Water	T.S.	Fat	Milk solid not fat	Protein	Ash	TMS	Sucrose
Condensed milk	12.2	26.6	74.0	9.0	22.0	8.3	1.5	31.0	43.0
Evaporated milk	12.2	31.0	69.0	9.0	22.0	8.3	1.5	31.0	–
Skim sweetened	14.0	29.0	71.0	0.5	25.5	9.3	2.2	26.0	41.0

The Indian standard specifications for sweetened condensed milks, whole or skim, are given in Table 9.23.

Table 9.23. Indian Standard specification for sweetened condensed milk

Characteristic	Requirement for Condensed milk	Skim sweetened
Total milk solids (% wt.) min.	31.0	26.0
Fat (% wt.)	Not less than 9.0	Not less than 0.5
Sucrose (% wt.) min	40	40
Acidity (%) lactic acid Max.	0.35	0.35
Bacterial count (per g)	500	50
Coliform count (per g)	–ve	–ve
Yeast and mould count (per g) max.	10	10

Physico–Chemical Properties

In the manufacture of condensed milk, water is evaporated. Hence, condensed milk has a higher specific gravity or density. Heating product produces a dark colour due to Maillard reaction and develops a caramel like taste flavour and also a cooked flavour. The free sulphydryl groups of the amino acids, methionine, cysteine and cystine release the sulphydryl groups which produces the offensive cooked flavour.

The development of brown colour and formation of malanoidin have been discussed earlier. The principal factors for browning reactions are – (a) heat treatment, (b) total solid concentration, (c) pH, (d) storage temperature and time, (e) head space of container, (f) water activity, (g) moisture content, (h) oxygen content, (i) added lactose glucose, etc. (j) viscosity, (k) concentration, (l) temperature of forewarming/preheating.

Defects in Sweetened Condensed Milk

Condensed milk is obtained by evaporation of part of the water under vaccum. Indian housewives preferably use sweetened condensed milk to prepare various recipes or dress various delicious items of food. Hence defects in sweetened condensed milk are for the most part disliked by Indian housewives. The defects in sweetened condensed milk may be broadly classified into two groups, viz., (a) defects due to bacterial/ microbial origin, (b) defects arising out of chemical reactions due to processing.

Defects due to Bacterial/Microbial Origin

Buttons, Lumpy: Buttons are evident by the presence of round, firm, cherry curds on top of the condensed milk. They represent the mold growth. The species related are *Aspergillus, Rhizopus, Mucor* and *Penicillium.* These species release enzymes and cause deterioration in the product. Buttons may be loosely or firmly attached to the product.

Gas Formation

The gas developed in sweetened condensed milk causes bulging and blowing of cans and tins and cause leakage and bursting of barrel. The gas comes from fermentation of sucrose by the exogenous and indigenous enzymes present in micro-organisms present in sweetened condensed milk.

$$\text{Sucrose} \xrightarrow{\text{Sucrase}} \text{glucose and fructose}$$

↓ both enter the EMP pathways and produce

$$4\ \text{Ethanol} + 4CO_2$$

The sucrose fermenting yeasts are of the species of *Torulopsis* and *Saccharomyces.* As for lactose, it also enters the EMP pathway and catabolised to ethanol and carbondioxide. The lactose fermenting organisms are of the species, *candida* and *kluvesomyces.*

Thickening: Sweetened condensed milk (SCM) undergoes continuous growth of organisms during storage causing thickening and producing abnormal flavours. *Micrococci* have the ability to hydrolyse sucrose and produce acetic acid through heterofermentative fermentation. Also, micro-organisms present in SCM produce proteases which behave like chymosin and hence thickening or gelation of SCM occurs. This phenomenon is known as sweet curdling in SCM.

Rancid Flavour: It is a typical butryic acid flavour, the resultant hydrolysis of triglycerides by exogenous lipases of micro-organisms produce butyric acid.

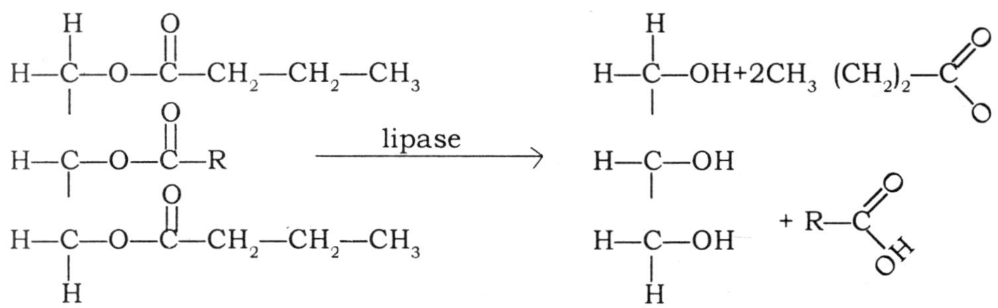

There are evidences of organisms surviving heat treatment. These organisms secrete lipases which hydrolyze fat, triesters of glycerol and fatty acid. Age of SCM determines the amount of rancid flavour in it. On opening such samples of SCM the rancid flavour is immediatley felt.

Fruity Flavour

This occurs in SCM due to micrococci.

Putrid Flavour

This is due to hydrolysis or proteolysis of proteins by proteinases of exogenous nature from micro-organisms surviving heat treatment. Proteolysis of proteins results in the production of bitter peptides.

Defects Arising out of Chemical Reactions

Flavour: Sweetened condensed milk (SCM) concentrated under vacuum will have none of the volatile flavours which occur in fresh milk. Being preserved by sugar instead of heat which does impart off-flavours in evaporated milk. SCM does show flavour defect. However, off flavours have been detected in SCM and some have been discussed under defects due to microbial origin. The rest are of chemical nature.

Metallic Flavour: The metallic flavour is associated with SCM and it is a chemical rather than a bacteriological defect. It is, usually related with copper as contaminant.

Strong: The term "Strong" describes the flavour accompanying progressive thickening and browning of condensed milk, while this flavour may be treated as a defect. However, it is not a serious one. Unfortunately the set, caked or jelled product with its deep brown colour suggests that the product might have undergone serious flavour changes. However, such milk usually develops a pleasant caramel like taste which is liked very much by children and adults. If the flavour is not accompanied by strong after taste, the off-flavour is not the most serious phase of this defect.

Tallowy Flavour: Tallowy flavour in SCM is much more prominent than rancidity and is less pronounced than in tallowy butter. The product gave a fishy smell which is very unacceptable. The flavour is pronounced and lingers for many days from the opened can.

Sandiness: The relative smoothness of texture of sweetened condensed milk (SCM) is determined mainly by size of lactose crystals it contains. Sandiness is due to presence of relatively large, coarse crystals, in milk of desired smooth, velvety texture crystals average 10 microns or less in length. Sandiness can be due to sucrose. When sucrose in water ratio of finished product exceed, 64.5%, it approaches the status of a saturated solution of sucrose. When such condensed milk is subjected to low temperature, some sucrose present in excess crystallises to give a coarse, sandy texture.

Sandiness due to rewarming of very cold condensed milk. When SCM is chilled to a very low temperature for packaging, is rewarmed in presence of agitation, the development of sandiness is unavoidable. Vigorous agitation while warming tends to cause smaller lactose crystals to redissolve. This increases supersaturation and promotes diffusion of dissolved lactose. The number of crystals is thus reduced. The remaining crystals have relatively much dissolved lactose to grow thereby forming sufficiently large aggregates to give SCM an objectionable sandiness. These findings are supported by the "Heat shoking theory" discussed by Dahle and by the findings of Roller and Williams that heat shocked ice-cream causes greater and more rapid growth of lactose crystals in ice-cream.

Sugar Sediments: Deposits of sugar sediments on bottom of can or barrel is common and consists predominantly of lactose crystals. The cause of such deposits lies in the difference in specific gravity between the crystallised lactose and remainder of condensed milk.

Effect of Crystal Size on Sugar Sediment: Condensed milk of normal viscosity in which lactose crystals do not exceed 10 μ in length remains free from objectionable sugar sediment. So, prevention of sugar sediment is largely a problem of crystal size and viscosity. This is in harmony and accordance with Stoke's law.

Viscosity: Greater the viscosity, greater is the resistance of force of gravity and hence less ability of lactose crystals to drop to the bottom. Concentration influences viscosity, greater the concentration, greater the viscosity.

Homogenisation: Homogenisation increases the viscosity of milk which then decreases the tendency of finished product to drop its lactose crystals and their effect increases with increasing homogenisation pressure.

Super Heating: The swelling of casein gives the product more body and to an extent tends to retard sugar sedimentation during subsequent storage.

Age Thickening (Chemical) : Immediate cause of viscosity of SCM and the tendency of product to thicken progressively with age is related to the phenomenon of colloidal swelling and hydration of proteins.

Reasons of Age Thickening : (a) SCM made during mid April to May/ June is least stable and shows greater tendency to suffer objectionable progressive thickening. From mid July onwards, it becomes increasingly stable.

(b) Forewarming Temperature: It is one of the major controlling factors of storage stability of SCM, its adjustment prevents age thickening.

(c) Sucrose Content: It has been observed that increase of sucrose in water ratio decreases the tendency to age thickening, as when sucrose is added to fresh milk used for SCM, it reduces viscosity of concentrated milk and retards age-thickening.

(d) Pan Temperature: Condensing at relatively high, pan temperature tends to unstabilise the finished product and increases the tendency to age thickening.

Solids-not-Fat: Milk proteins cause age thickening and their removal eliminates the probability of age thickening. However, K-casein adds to the stability of casein micelles, and its removal definitely destabilises the casein micelle. However, generally it is believed that increase in percentage of SNF, increases the tendency of age thickening.

Fat in SNF Ratio: Fat acts as a softener and diluent of colloidal portein of milk. Therefore, skim milk age thickens faster than whole milk.

Concentration: SCM of higher concentration has high initial viscosity and tends to hasten age thickening. However, too low concentration may cause fat separation and sugar deposits.

Storage Temperature: Storage temperature has a tremendous effect on age thickening. At or below 7.2°C temperature the SCM retains its initial viscosity rather permanently.

Bloats: These are formed due to thermal convection and expansion of constituents of tin and bloats of this character are possible only in case of cans, the seal of which is not entirely hermetical.

Dried Milk

Spray dried milk powders usually consist of particles which vary both in size and in particle density. The differences in the latter being due to vacuoles of different size present in the majority of the particles. Normally, the vacuoles are filled with air which has penetrated the particle walls by diffusion and they are, therefore, called air cells by many. The physico-chemical properties depend on a variety of factors like processing parameters and nature of ingredients used in the formulation of some physical characteristics. While bulk density determines the packaging and transportation costs of a powder, it also plays an important role in the reconstitution of the powder, its flowability, etc. Factors influencing the physico-chemical attributes of spray-dried powders are mainly bulk-density, wettability, sinkability, dispersibility, solubility index, flowability and free fat content.

Bulk Density: Bulk density is regarded as the weight per unit volume of the powder and is frequently expressed as gm/ml or gm/cm^3. It has a bearing on the volume of packed product (and, therefore, includes the packaging and transport costs of milk powders), reconstitution, internal surface area of the powder. Usually bulk density is expressed as packed and loose bulk density, former being determined by tapping or vibrating the sample to a constant volume while loose density for that weight per unit volume without tapping or vibrating.

Bulk density of a powder has been reported to be influenced by a number of processing conditions and also the powder structure, i.e., size and shape of the dried particles. Amongst the various factors responsible for controlling the bulk density, the degree of concentration, viscosity of the concentrate, drying air temperature, particle size distribution, the degree of particle agglomeration and handling of powders are the most important. Several other factors like particle density as given by particle material density and the content of occluded air within the particle were reported to be major factors governing the bulk density. Recently, inlet air temperature, concentrate flow rate and atomizing pressure also were shown to influence the bulk density of dried milks. It has been reported that an increase in the bulk density of spray-dried milk powders with the increase in dry matter content of concentrate which was possibly a result of decreased entrapped air and vacuole volume since amount of moisture to be dried was minimised by higher solid level. The viscosity of the concentrate was found to be directly proportional to the bulk density of dried material. With increase in viscosity of the concentrate, a significant increase in the bulk density of the resulting product was observed. The concentrate flow rate had a strong influence on bulk density since the bulk density significantly increased with higher flow rates. Inlet air temperature was found to be inversely proportional to the bulk density of milk, milk powders with higher temperatures giving rise to lower bulk densities. This decrease in bulk density could be attributable to rapid formation of the solid surface on particles. Thus, increasing the void size and quantity of vapours, which tended to cause ballooning of the particles. Amongst other drying conditions reduction in the speed of atomizer decreased the vacuole volume which resulted in an increased bulk density and mean particle density. The effect of increasing the atomizing pressure was generally to reduce the bulk density, moreover, it also increased the effect of inlet temperature and reduced that of flow rate which by themselves resulted in lower bulk densities of the product.

Enhanced bulk density of spray dried product could be obtained by increasing the dry matter content of the concentrate, increasing the viscosity of the concentrate, reducing the speed of rotation of the disc whose spray atomisation of concentrate is effected by means of a centrifugal disc or reducing the pressure of nozzle if spray drying under pressure or with compressed air, modifying the inlet air temperature of the drying air in the spray chamber.

Wettability: Reconstitutability is very important, yet extremely complicated and less understood phenomenon of the dried milks. Reconstitution of powder in water is not a single process by itself. It starts first with wetting of the dried particles. After wetting, the particles tend to sink in the aqueous medium which then disperse, on mixing and finally lead to the dissolution of powder. Thus, a process of reconstitution comprises settability, sinkability, dispersibility and solubility. Some authors hold the view that there is an intermediary process of penetrability after wettability and before sinkability, but normally wettability itself is considered as wetting of particles by penetration of water. Splitting of reconstitution process into above four steps is helpful for understanding the process. On the other hand, it was pointed out that there is no sharp border line between these individual reconstitution steps and they do not proceed one after another but simultaneously atleast two at a time, the individual steps influencing each other. For convenience of analytical purposes, these processes are determined separately and, therefore, are being delineated in the order of this classification. Wettability is defined as the case with which powder particles make contact with water surface. Wetting of particles takes place with water penetration by capillary force. It has been shown that the particle size influences the rate of wetting since water has to enter the mass of milk powder by capillary action. The larger and more irregular shaped is the mass particle, more readily it will wet. It was found that small particles of symmetrical shape enhanced close packing of the particles thus creating an interference to capillary rise of water in between powder particles.

Evaluation of processing conditions influencing the rate of wetting has been attempted. Wettability of the dried milk samples was increased by the addition of sufficient lactose to milk before drying to produce a concentration in the powder to 40 to 50 per cent. Further, addition of lactose was found to be detrimental for wetting of particles. Such was also the case when casein was added to milk before drying.

On an increase in wettability by the addition of lactose or sucrose to the concentrate before drying, a highly significant positive correlation between lactose content and wettability a negative correlation between fat content and wettability was established. Although preheating/treatment or spray drying temperature did not affect wettability, concentration of a solid content of 30 to 40 percent, homogenisation of the condensed milk just before spray drying, which was important for dried whole milk, yielding superior wettability. Studies on fractionation of fat in dried milks indicated that milk dried with low melting point fats had better wettability.

Studies on kinetics of reconstitution process indicated that as the concentration of reconstituted milk increased, the relative speed of solution dropped resulting in an impaired wettability of the milk powder. This behaviour was the result of reduction in diffusion rate of molecules of the dissolved substances, increase in viscosity of the solution, impairment of wettability of milk particles and slower moisture uptake. By having lactose in the crystallised form it seemed possible to get greater wettability of powder particles with less obstruction to capillary rise of water. It has been proved that larger particles in dry milk improves wettability. Threfore, it was seen necessary to form relatively larger droplets during atomisation of skim milk concentrate in the drying chamber. This could be accomplished by reducing the pressure by increasing the orifice size of spray nozzle and by increasing the viscosity of the concentrate either by increasing the relative amount of total solids or by crystallising the lactose. With decrease in the outlet temperature and speed of atomiser there was an increase in moisture content of the product. Also, with an increase in milk concentration, an increase in the wettability was observed that was attributable to the increase in the proportion of the large particles. Increased wettability of fried milk was obtained by increasing particle size and by decreasing rate of solution of proteins. Improvement in wettability by removing free-fat from the surface of particle by changing the surface tension between water and fat particles and by agglomeration of particles to 100 to 200 micro meter size has been recommended.

Sinkability: It is the ability of milk powder to sink down into the water after it has been wetted. A great deal of controversy exists on the study of sinkability patterns of dried milks.

The quick sinkability of powder is hampered if the dry product stays on the water surface for a longer period. This is more common in spray dried powder where particle surface is hydrophobic in nature resulting in clumping and difficulty in penetration of aqueous phase in the interior of dried particles. Any spray-dried product which has good sinkabiity would land itself to good reconstitution without excessive stirring. Sinkability of spray dried milk product is influenced by number of factors. Fractionation of fat in dried whole milk indicated that even as little as five percent of the high melting point fat decreased the sinkability of the product. It has been observed that even if higher bulk density improved the sinkability it should not be more than 0.4 gm per ml for the self-dispersion. Sinkability was closely related to wettability of the dried powders. The amount of occluded air has a pronounced influence on sinkability to the extent that a powder may be readily wettable but not readily sinkable because of the byoyancy of occluded air within powder particles. Agglomeration of regular spray-dried milks improved the sinkability chiefly by increasing the aggregate weight. Further effect of particle agglomeration on sinkability was found to be greater in non-fat dry milk than dry whole milk. The relation was thought to be indirect since fat has a negative influence on agglomeration.

However, sinkability from practical stand point is not an important consideration for reconstitution properties. However, some of the processing parameters that could enhance sinkability of dried milk have been identified, but are relatively less understood.

Dispersibility: This is the ability of milk powder to be distributed throughout the water when lumps or aggregates fall apart. Dispersibility is considered to be the best single criterion by the International Dairy Federation, for assessing the overall reconstitution process. Total heat treatment on casein during processing is an important consideration for good dispersibility.

The rate of dispersion of dried whole milk progressively increased with increasing powder temperature from 72° to 120°F (22.22° to 48.89°C). With dried skim milk, on the other hand, increase in powder temperature in the same range did not appear to have any measurable effect. Results on the particle size and dispersibility relationship indicated that a product with particle sizes between 30 to 50 microns had better wetting and dispersing properties than a product containing either larger or smaller sizes and it was considered to be a desirable feature to adjust the manufacturing techniques with the object of stimulating the smaller particles. Improved dispersibility was further associated with the larger particle size. However, very large particles of foam spray dried whole milk had less dispersibility than the smaller ones. Particles above 710 micron in average diameter showed a linear relationship with dispersibility. Nevertheless, very small particles were shown to retard dispersibility due to over drying taking place in the spray drier. These small particles had a tendency to flocculate and caused clumping. In absence of these small particles, water could readily penetrate the lump resulting in a good dispersibility. Higher level of fine particles (less than 10 microns) gave a low dispersibility. Dispersibility of milk powder was unaffected if the free-fat content of the powder is 20 percent or higher. Below this value there was a tendency for increased dispersibility with decreasing free-fat content. Dispersibility of powders of approximately same average particle size was considerably increased by homogenising the concentrated milk. Since both particle size and homogenisation possibly influenced the dispersibility, it was concluded that no direct effect could be found of the free-fat on dispersibility of dried whole milks. Powders with lower bulk density and having solubility index over 0.30 ml usually gave poor dispersibility. Dispersibility was also found to be adversely affected by the increased protein content of dried skim milk, indirectly heating lower whey protein nitrogen index, holding the concentrate at high temperature before drying, decreased particle size and pakaging at more than 76°C. Among various processing techniques, high heat treatment of liquid or condensed whole milk, recombination of condensed skim milk and cream or dried butter fat, milk treatment of condensed skim milk with a proteolytic enzyme before recombination with butter fat, reducing the fat content of the dried milk to 17 to 20 per cent, use of large orifice nozzles of relatively low nozzle pressures in the spray drier were found to enhance the self-dispersion of whole milk powder.

Solubility: This is considered to be the final step of the reconstitution process and measured by the level of insoluble fraction of the powder left after mixing, with water and centrifugation. It is expressed as the "solubility index" which is considered as a major defect in reconstitution of the milk powders. Solubility of the milk powder is controlled principally by the stability of proteins. Solubility is influenced by various factors like quality of raw milk, heat quality of raw milk, heat treatment before and after evaporation, homogenisation, concentrate flow rate, atomisation, inlet and outlet spray drying temperatures, moisture content and storage conditions of the powders.

In the case of skim milk powder, the solubility was influenced by the degree of pre-concentration. With progressively increasing total solids level of the concentrate solubility index also increased significantly. No similar influence was found in the case of whole milk powder. The viscosity of the concentrate rather than the actual total solids content was, therefore, considered to be the factor determining the changes in solubility of the powders. Increased viscosity of the concentrate resulted in a significant decrease in the solubility accompanied by a small increase in the particle density. Any increase in the total solids concentration caused only a very small increase in the solubility index. Single-stage homogenisation of concentrates with higher total solids content gave higher solubility index than non-homogenised samples. Two stage homogenisation enhanced the formation of fat-protein complex and clustering was largely prevented. In infant food, sucrose was found to increase the solubility as well as degree of particle agglomeration during drying.

Reduction in the outlet temperature from spray drier was thought to be the most effective method to manufacture whole milk powder with good solubility characteristics. Outlet temperature was found to be critical in controlling the solubility index. It should never exceed 90°C. The inlet temperature had also an effect on solubility index; with higher temperatures giving higher solubility index values. Higher concentrate flow rates and atomising pressures both reduced the solubility index. Several processing parameters, additives, drying and storage conditions play an important role in determining the solubility of spray-dried products. This provides useful guidelines in developing processes for new products.

Flowability: This is a desirable property in itself and is related with the loose and packed densities of the dried product. It is a rather complex phenomenon and several factors contribute to good flowability. The free flowing property, devoid of any stickiness and caking is one of the most important commercial requirement of dried milks.

The factors controlling the flowability were identified. They are (a) particle size, (b) shape, (c) surface morphology, (d) density and (e) electrostatic charge. Large particles tended to flow more easily than the fine ones. Uniformity in size was also advantageous. A wide variation in particle size was found to permit the fine particles to occupy the inner space between large particles. This resulted in closer packing with the greater friction, thus reducing the flowability. Spherical shaped and high density particles tend to possess a more desirable flowability. Increase in total solids content of the concentrate resulted in an improved flowability as measured from the angle of core found by the powder after free fall, the change in flowability was attributed to the probable variation in specific volume of the powder.

On the basis of available information, manufacturing parameters may be monitored to obtain a favourable impact on flowability of spray dried product.

Free Fat: The free fat content is of prime importance in determining the undesirable behaviour towards the disposition, wetting, flowability and oxidative stability of the powders. The term "free fat" was originally coined to designate the extractable fat not protected by a protein layer (or uncapsulated portion). The free fat comprised the fat occurring in the vacuoles and at the surface of the particle. It was later found that the free fat did not only comprise the surface fat but also the fat present in the drying cracks, the capillaries and inside of the powder particles accessible to the extraction medium.

Degree of preconcentration was found to influence the content of free fat in whole milk powder. Homogenisation of the concentrate before spray drying considerably reduced the free fat content. The disintegration of the fat globules during homogenisation of the concentrate meant that there were considerably fewer holes and cracks in the surface of the powder particles after atomisation. An increase in the total solids of concentrate resulted in reduction of free fat. Concentration and drying was found to increase the dispersion of fat phase and part of the destabilised fat becomes emulsified. Increasing the inlet and outlet temperatures results in a higher free fat content. Higher relative humidity and storage temperatures increased the free fat content and decreased the solubility. Higher fat content during storage was observed to increase the oxidation of fat and decrease the dispersibility.

The influence of processing parameters ingredients and handling conditions on the free fat content of dried milk is relatively less investigated. However, homogenisation seems to be the main determinant of free fat content or spray dried products.

There is a complex relationship between the processing conditions and physico-chemical characteristics of dried milks. A process which may improve properties like bulk density or flowability, need not necessarily improve the reconstitutability so also, the different properties themselves have a complex inter-relationship among each other. Therefore, a manufacturer has to select the optimum set of processing conditions so as to impart desirable properties in the final product.

Defect in Whole Milk Powder

Flavour: As the dry milk ages, it gradually loses the fine, appetizing pastry flavour. The flavour defects in dry whole milk are:

Stale and Old Storage: This flavour defect of dry whole milk appears to be the characteristic age defect. This is due to non-enzymatic browning reaction and especially in the absence of fat constituent. The chief reactants are protein and lactose. When the defect is intense, it may be accompanied by darkening of the product.

Rancid: It has a bitter, soapy unclean taste which is persistent after the sample has been expectorated. Sometimes it occurs due to improper preheat treatment to spray dried milk powder. This defect occurs due to lack of inactivation of milk lipases due to less preheat treatment.

Oxidised and Tallowy: This is the most troublesome flavour defect of dry whole milk. Not only is the flavour, which suggest tallow, the product is unpalatable. The occurrence of this defect is very frequent in whole milk powder.

Scorched: This flavour is produced in whole milk powder which has received excessive heat treatment during drying stage. If the product stays for a long period in the drying chamber, this defect definitely occurs. Concommitant with such defect, a dark colour usually occurs. The major two defects in body and texture are formation of lumps and caking.

Lumpiness: Lack of homogenity stands out in lumpy powder. This defect is prominent and stands out in spray dried powder. Whole milk powder on the other hand, due to high fat content has a softer lump. Insufficient drying, drainage from spray nozzle or exposure to moisture laden air are the principal cause of this defect, while cold products produce softer lumps which is due to agglomeration of particles. This type of defect should not be confused with a lumpy product, in which lumps are firm when passed between the fingers.

Caking: When dried whole milk powder becomes like rock, the defect is known as caked. This means the powder has lost its powdery consistency.

Colour: Light yellow in colour, but varies seasonally with fat colouration.

Browned or Darkened: This is due to longer, storage period and storage time and is due to non-enzymatic Maillard reaction. Usually in such defects, the products taste stale.

Scorched: This defect is usually manifested in roller dried powder. This is due to over heating and the product takes up a brown colour. Some particles in such defect takes up a dark almost burnt colour.

Lack of Uniformity: Maillard browning after packaging and scorched product during processing are the two principal causes of this defect. Browning is evident in small packages on the surface only.

Size Defect: Large-sized packages show this defect from the center of the product which appears like a core.

Defects in Skim Milk Powder

The defects in skim milk powder are usually those found in non-fat milk solids. These are:

Flavour: The chief defects are staleness, old storage, scorched, oxidised and tallowy.

Stale, Old Storage: In non-fat dry milks, this defect is more pronounced. The flavour of the product in such defect is scorched which directly points out to overheating of the product during processing. Also, appearance of brown colour in such defects suggest the initiation of Maillard reaction during processing and storage of the product. Due to overheating, caramelisation also takes place.

Scorched: This defect occurs due to over heating of the product during processing.

Oxidised + Tallowy: This defect is present in lesser extent when compared with whole milk powder because of its very low fat content.

Infant Food

Prior to the arrival of infant formulae in the world market, human babies were mostly breast fed. However, in case of lactation failure, insufficient milk secretion (due to malnutrition or faulty lactation management) and when the mother's milk was a carrier of pathogens or when the mother is suffering from breast cancer, breast milk substitutes were of cows, goats, sheep, camel, yak, reindeer and in our country that of mostly buffalo's milk. In many countries where other milk producing species existed, such milk was used for infants. The composition of milks used for infant feeding is given in Table 9.24.

Table 9.24. Composition of milk used for infant feeding (in percentage)

Species	Water	Fat	Protein	Lactose	Solids-not-fat	Calcium
Human	86.4	4.6	1.2	6.9	9.0	0.03
Cow	86.7	4.5	3.4	4.9	8.8	0.13
Buffalo	83.7	6.5	3.8	5.1	9.8	0.20
Goat	86.8	4.5	3.3	4.4	8.7	0.13
Sheep	81.6	7.5	5.6	4.4	10.9	0.20

However, such milks were usually a poor substitute of human milk. In the 20th century particularly after 1912, cultural, social and industrial renaissance took place. This resulted in a total metamorphosis of socio-economic standing of women who joined hands with man in many activities which were thought as man's kingdom and also, sharing in the economic responsibilities of family. Together with it there was a drastic and dramatic migration of rural population of cities. The patterns of infant feeding seen in the modern world are a sequel to the industrial growth and urbanisation, thereby giving birth to several infant formulae all over the world.

The compositional characteristics of human milk (HM), buffalo milk (BM), conventional infant formula (CIF) and maternalised infant formula (MIF) on dry matter basis is given in Table 9.25.

Table 9.25. Compositional characteristics of human milk (HM), buffalo milk (BM), conventional infant formula (CIF) and maternalised infant formula (MIF) on dry matter basis.

Constitutent	HM	BM	CIF	MIF
Casein (%)	3.67	20.36	17.80	4.96
Whey protein (%)	5.50	5.09	4.45	7.44
Total proteins (%)	9.17	25.45	22.25	12.40
Lipids (%)	31.67	41.12	19.57	26.24
Carbohydrates (%)	61.60	29.09	52.32	59.91
Ca (mg/100g)	249.90	1449.42	1014.30	998.11
P (mg/100g)	124.96	1124.64	786.80	495.42
K (mg/100g)	423.60	846.20	592.20	875.62
Na (mg/100g)	134.10	375.20	262.50	239.68
Fe (mg/100g)	1.21	0.67	5.00	5.01
Cu (μg/100g)	0.33	–	–	0.51
Mn (μg/100g)	0.38	–	–	0.51
Zn (mg/100g)	3.06	–	–	3.01
Total ash (%)	1.78	4.39	5.77	2.43

– Not available

Reproduced from Recent Advances in Manufacture of Infant Formulae.

It is evident from the table that the relative ratios of proteins: fat:carbohydrate:minerals of human milk differ considerably with that of buffalo or conventional infant formula. A marked difference is noticed in caseins and thereby total proteins, lipids, and carbohydrates. As for minerals, human milk is rich in calcium and low in most of the cases of most of the other minerals. High sodium in buffalo and CIF might be the root cause of hypertension in later part of any human subject. Human caseins produce a softer curd in the intestine compared to milk of other species of CIF.

The enzymes that hydrolyse lipids in the intestine of human infants are insufficient for high fat milk and, therefore, high lipid contents of buffalo milk are bound to cause physiological disorders in human infants. As for the carbohydrate make up of bovine milk, it is totally dissimilar to that of human milk. Human milk is a rice source of milk oligosaccharides. The nitrogenous derivatives of which have been identified and characterised as the "Bifidus Factor", the growth factor for *Lactobacillus bifidus* is very necessary for the maintenance of the intestinal digestive balance of human infants and is bacteriostatic against many pathogens harbouring the intestine of human infants.

In the table the composition of maternalised infant formulae exclusively at the National Dairy Research Institute is documented. This formula resembles human milk with regard to the whey protein, casein ratios, fatty acid profile calcium:phosphorus and sodium:potassium ratios as well as protein:fat:carbohydrate:mineral ratio.

Technology has been developed at the National Dairy Research Institute for the enzymic hydrolysis of lactose milk, which serves as the base for the manufacture of infant formula. The outstanding feature of the spray-dried product developed is that it contains 1.5×10^5 viable cells of *Bifidobacterium bifidum*.

The newly proposed compositional standards and existing ISI specifications for infant foods have been compared in Table 9.26.

Table 9.26. Newly proposed compositional standards of infant food

Constitutents	Existing ISI		Newly proposed standards	
	Min	Max	Min	Max
	per 100g		per 100g	
Protein (g)	20.0	–	12.0	–
Fat (g)	18.0	28.0	20.0*	–
Linoleate (mg)	–	–	210.0	–
Carbohydrate (g)	35.0	–	50.0	–
Total ash (g)	–	8.5	–	8.5
Iron (mg)	4.0	–	5.0	–

* In the event that vegetable oil is added, the milk fat content will not be less than 12 percent.

The Indian Standards Specifications for infant milk foods are given in Table 9.27.

Table 9.27. Indian Standards Specifications for infant milk foods (vide IS.1547.1968).

Characteristic	Requirement
Moisture (% wt) Max	4.5
Total milk protein (% wt) Min	20.0
Milk fat (% wt)	18.0 to 28.0
Total carbohydrates including sucrose, dextrose and dextrins, maltose and lactose (% wt) min.	35.0
Total ash (% wt) Max	8.5

Acid insoluble ash (% wt) Max.	0.01
Solubility (a) solubility index Max	15.0 (if rolled dried)
	2.0 (if spray dried)
(b) solubility (% wt) Min	85.0 (if rolled dried)
	98.5 (if spray dried)
Vit I.U. per 100 g (Min)	1500
Added Vit. D.I.U. per 100g	400 to 800
Iron mg per 100g (Min.)	4.0
Bacterial count per g (Max.)	50000
Coliform count per g (Max.)	10

Reproduced from *Outlines of Dairy Technology*, S. De.

The percentage composition of some infant foods manufactured in India are listed in Table 9.28.

Table 9.28. Composition of some infant foods (per cent)

Brand/Name	Moisture	Fat	Protein	Carbohydrate *	Ash
Amul	3.0	18.0	22.0	52.0	5.0
Glaxo	2.9	26.5	24.9	38.5	5.6
Lactogen	3.0	19.0	21.6	50.8	4.8
Nestogen	3.3	12.0	20.2	60.0	4.7
Vijaya spray	–	19.0	22.0	50.0	–
Parag	–	18.0	24.0	49.0	–

* Includes sucrose, dextrose or dextrins, maltose or lactose. Reproduced from Outlines of Dairy Technology, S. De.

Nutrient Status of Dried Milks

Nutrient levels in dried milks are affected by losses during production of the concentrate as well as by those which occur in the drying process. The amino acid composition of proteins is relatively unchanged by drying and instantizing. Losses of lysine are small in comparison with those which occur during preparation of the concentrate and usually amount to no more than 5 percent during spray drying.

Spray drying has relatively little effect on the vitamin content of milk, effects being most marked with vitamin B_{12} (20 to 30 per cent), vitamin C (~ 20 per cent) and vitamin B_1 (~ 10 per cent). Loss of other vitamins are minimal. The level of fat soluble vitamins in skim milk powder is, inevitably, low and supplementation is practised where such powder is used for infant nutrition.

Nutrient losses during storage are small provided that the relative humidity is low, the powders are not exposed to very high ambient temperatures and the length of storage does not exceed. Available lysine was reduced by ~ 8 per cent during storage of powder at 25 and 37°C for 1 year, the nutritive value of the proteins being reduced by the same order. Losses due to the Maillard reaction becomes significant when storage is prolonged even when storage temperatures are low.

Losses of vitamin B_1 and C are typically in the order of 10 per cent during storage for 2 years, although losses of upto 33 per cent have been reported in spray dried milk powder over 40 months storage. Vitamin C losses are reduced by minimizing the oxygen and water vapour permeability of the packaging. Light, impermeable packaging is also required to prevent loss of light sensitive vitamins, especially riboflavin.

Milk Sweets and Delicacies

Most of the indigenous milk products serve as the base for almost all the Indian sweets of milk origin. Honey, *guda* (jaggery) and *ghee* served as the basis for many sweets of ancient India. The pulse of Indian culture and the various religions, she nurtured since ages can be felt in the typical sweets of Indian origin where each species of Indian sweets let-off a fragrance of each typical culture and way of life preached by many prophets and saints. Indian sweets differ from zone to zone, state to state, district to district, taluk to taluk and uncoil and unfold their typicality from Kashmir to Kanyakumari.

Rossogolla: *Rossogolla* is one of the most popular *chhana* based sweet of Eastern part of India, being made from cow milk coagulam which is kneaded into small balls that are cooked in clarified sugar syrup having a characteristic spongy body and smooth texture. *Chhana* is kneaded with *maida, suji* and baking powder to form a dough. Balls are made of desired sizes, and cooked by boiling in sugar syrup. Cooked balls are soaked in sugar syrup and essence is added when cooled.

Efforts to make *rossogolla* from buffalo milk met with only limited success because it lacked sponginess due to constitutional difference between two milks.

According to ISI, it should have minimum 45 per cent total solid, 5 per cent fat, 5 per cent protein and 45 per cent sugar (maximum).

Table 9.29 gives average composition of market *rossogolla* from Bengal.

Table 9.29. Composition of *rossogolla*

	Ordinary variety (%)	Sponge variety (%)
Moisture	38.3	50
Fat	7.3	5.3
Sucrose	43.6	37.0
Total protein	6.4	5.5

As compared to this, on an average buffalo milk *rossogolla* contains moisture 37 per cent, sucrose 51.9 per cent, protein 6.3 per cent and fat 4.2 per cent.

Some of the factors affecting physico-chemical characteristics of *rossogolla* include syrup concentration (sugar concentration 60-80 per cent), time and temperature of cooking (20-30 minutes) and evaporation of water during cooking. During longer cooking concentrated sugar syrup causes caramelization imparting undesirable colour to the product. Major defects observed in *rossogolla* during storage are browning discolouration, loss of crispiness and flavour.

REVIEW QUESTIONS

1. Briefly describe the changes that take place during conversion of milk into *khoa*.
2. Discuss the factors which affect the physical quality of *chhana*.
3. How casein exists in milk and cheese.
4. Give the gross chemical composition of following products. *Rossogolla*, *Rabri* and *Cream*.
5. Discuss how ghee can be preserved for a longer period?
6. Briefly discuss the various theories for conversion of cream into butter.
7. What are the major changes that take place during conversion of milk into curd.
8. Give the gross chemical composition of the following : *Shrikhand*, *Chakka*, *Shrikhand Wadi*, *Lassi*, Yoghurt.
9. Discuss the physico-chemical changes that take place during manufacture of evaporated milk, sweetened, condensed milk and dried milk.
10. Discuss the physico-chemical properties of ice-cream mix.

11. What are the physico-chemical changes that take place during ripening of cheese.

12. List important milk clotting enzymes alongwith its source used for cheese preparation.

13. Why animal or microbial rennet is preferred over vegetable rennet?

REFERENCES

Bandyopadhyay, A.K. and Mathur, B.N. 1987. *Indian Milk Products : A Compendium of Dairy India.*

Bandyopadhyay, A.K., Soni, Kanwal and Ganguli, N.C. 1981. Production and properties of *chhana* and *Rasogolla* : A Review. Dairy Guide, March, 41.

Dairy Guide, 1982–1986. Publication Thyag Raj Nagar Market, Lodhi Colony, New Delhi–110 003.

De, S. 1980. *Outlines of Dairy Technology.* Oxford University Press, Delhi.

Ganguli, N.C. and Jain, M.K. 1973. *Ghee*, its chemistry, processing and technology, *J. Dairy Sci.* **56**, 19.

Indian Dairyman, 1980-90. Publications, Indian Dairy Association.

Ling, E.R. 1956. A Text book of Dairy Chemistry Volume I (Third edition revised) Chapinam & Hall Ltd., London).

Rai, M.M. 1964. Dairy Chemistry and Animal Nutrition, Ram Prasad and Sons, Agra-3.

Ramamurthy, M.K. and Narayanan, K.M. 1976. *Indian J. Dairy Sci.* 28, 163.

Rangappa, K.S. and Achaya, K.T. 1974. *Indian Dairy Products.* Asia Publishing House, Bombay.

Singhal, O.P. and Jain, M.K. 1968. *Annual Report*, National Dairy Research Institute, Karnal.

Wadhwa, B.K. and Jain, M.K. 1982. An Innovative process for making *ghee.* UNDP/ICAR National Symposium on "Present and Future Trends in *Ghee* and Fat Rich Dairy Products". Jan. 15–16, National Dairy Research Institute, Karnal.

Webb, E.H., Johnson, A. and Alford, J.A. 1978. *Fundamentals of Dairy Chemistry.* The AVI Publ. Co. Inc. West Port, Connecticut.

Webb, B.H. and Whittier, E.O. 1970. *Byproducts from milk.* The AVI Publ. Co. Inc. Westport, Connecticut.

10

ANALYTICAL ASPECTS OF DAIRY CHEMISTRY

THE analysis of any constituent of biorganic material is always based on theoretical principles. Here an attempt has been made to define the principles involved in the determination of milk constitutents. Students graduating in Dairy Science learn during their course of studies several biochemical techniques which find its application in the sophisticated analytical techniques of Dairy chemistry. Also care has been taken to outline the brief definition of various modern techniques.

Acidity

Freshly drawn milk does not contain lactic acid but is slightly acidic. The pH of fresh milk is usually 6.5. The natural acidity in milk is due to its constitutents such as casein, albumin, citrates, phosphates and carbondioxide. On standing, fresh milk at room temperature for sometime develops acidity due to bacterial action. This additional acidity through souring can be described as developed acidity. Thus, the total acidity of milk is made up partly of the "natural" and partly of the "developed" acidity. Total acidity could be measured by titrating a known volume of milk with a standard alkali to the end point of an indicator like phenolphthalein.

Churning of acid cream, for butter making, results in high fat loss. It is necessary to neutralise the cream to an optimum extent before it could be utilised in the manufacture of butter. For this purpose acidity of cream is must for butter making. During storage of *ghee* free fatty acids are produced. Similarly, compounds like proteins and fats on enzymic activity develop acids. The acidity thus produced could also be titrated with standard sodium hydroxide solution and expressed in terms of lactic acid. This can be explained as follows :

$CH_3CH(OH)\ COOH + NaOH$ $= CH_3CH(OH)\ COONa + H_2O$
Lactic acid Sod. lactate
Mol. wt. (90) g + 40 g

That is, 1 molecular weight of sodium hydroxide neutralises one molecular weight of lactic acid.

That is 40 g of sodium hydroxide neutralises 90 g of lactic acid. One litre of 1N sodium hydroxide contains 40 g of it.

Therefore, 1 litre N NaOH neutralises 90 g of lactic acid

or 1 litre of N/9 NaOH 10 g of lactic acid

or 1 ml of N/9 NaOH 0.01 g of lactic acid

0.09 g Of lactic acid

or 1 ml of N/10 NaOH N/10 NaOH is used).

If V ml of N/9 NaOH were required by 10 ml of milk/curd or 10 g of milk product, assuming that the acidity is due to lactic acid, then V ml of N/9 NaOH neutralises V × 0.01 g lactic acid. Therefore, 100 ml or 100 g would require 10 × V ml of N/9 NaOH corresponding to 10 × 0.01 × V g (0.1 × V g) of lactic acid equivalent. Thus, the percentage of lactic acid equivalent is obtained by dividing the volume of N/9 NaOH required by 10. Based on the above steps the acidity in the given sample is calculated as lactic acid equivalent. Normally either N/10 NaOH or N/9 NaOH is used for titration.

pH

pH is defined as the negative logarithum of the hydrogen ion concentration. The pH of milk or milk products is closely related to the chemical changes in milk proteins, salts, etc. Measurement of pH in milk or milk products can be done by pH meter, pH paper or pH solution (comparator method). The most convenient and reliable method for measuring pH is by the use of a pH meter. The instrument measures the EMF of a concentration cell using a reference electrode and a glass electrode reversible to hydrogen ion or alternatively a single electrode having both in it. The structure and construction of the electrodes can be studied elsewhere. For measuring pH, the pH meter is calibrated before use by means of a standard solution of known pH. An approximate idea of the pH of a solution can be obtained using indicators. These are organic compounds of natural or synthetic origin whose colour is dependent upon the pH of the solution. Indicators are usually weak acids which dissociate in solution.

$$\text{Indicator} = \text{Indicator}^- + H^+$$

Applying the Henderson-Hassel Balch equation,

$$pH = PK\ In + \log 10\ \text{Indicator}^-/\text{Indicator}$$

The two forms of the indicator have different colours and the actual colour of the solution depends on the PK, In and the pH. The colour change occurs over a wide range of pH, so indicators cannot be used to determine the pH of a solution to any degree of accuracy. Only a small quantity of indicator should be added to the solution being examined otherwise the acid-base equilibrium will be displaced. Indicators are affected by oxidising and reducing agents, salt concentration and protein so these facts must be borne in mind when using them. The indicators most suitable for determining the pH of milk colorimetrically are bromothymol blue (pH 6.0 to 7.6) and bromocresol purple (pH 5.2 to 6.8). Various models of comparator and disc are available. The colour obtained by adding indicator to milk is matched with colour in the disc and the pH noted. Several types of pH paper are available having the colour printed on the cover of paper strip alongwith the pH. The pH paper strip is dipped on the milk or solution to be tested. The colour of strip is then matched with the colour printed on the paper strip cover and the pH is noted. The principle involved in this is same as using liquid indicator. Both the methods using pH indicator and pH paper gives an approximate pH of the solution.

Fat

Fat in milk and milk product can be determined either by gravimetric method (Rose-Gottlieb) or volumetric method (Gerber).

In order to effect a quantitative separation of fat from milk or milk product it is necessary to break up the protective film surrounding the fat globule by using suitable agents such as ammonia and alcohol. Ammonia brings about break up of the protective layer and addition of alcohol facilitates the passage of fat globules from the aqueous phase to the solvents. The mixed solvent, petroleum ether and diethyl ether being non-miscible with water, effects a good extraction of fat from the non-fatty solids in the solution. Apparatus like Rose-Gottlieb tube and Mojonnier's tube are used to remove the fat from the solution. The extracted fat alongwith solvent ethers is dried and the weight of fat is then calculated.

When fat is to be determined in milk or some milk products, Gerber butyrometer is used. A known quantity of milk or diluted milk product is taken in butyrometer. To this, definite quantity of sulphuric acid and amyl alcohol are added. The proteins are dissolved and the fat globules thus set free and remain in liquid state due to the heat produced by the acid. On centrifugation fat being lighter separates on the top of the solution in butyrometer. The butyrometer is graduated in such a way so as to give fat percentage directly which is ready by levelling the fat layer meniscus.

SNF

Measurement of the specific gravity or density by a lactometer is based on the Archimedes principle. A floating object sinks till it has displaced a weight of fluid equal to its own weight. The greater the volume of displaced fluid, smaller is the density of the fluid and lower is the lactometer reading. The total solids and SNF content of milk are related to its fat percent and specific gravity by the Richmonds formula

% SNF = 0.25 D + 0.21 F + 0.66

where, D = 1000 (d-1) d = density of milk at 20°C

F = Percentage of fat.

Moisture

Moisture in products like *ghee*, cheese and skim milk powder is calculated by heating a known sample in dry flat bottom aluminium dish or porcelain dish for a fixed period at particular temperature. Moisture in *ghee* is obtained on heating the sample at 100°C for 1½ hrs, while for cheese it is estimated by drying of water at 100°C for 4 hrs and for skim milk powder by heating the sample in an oven at about 102°C for 3 hours.

Moisture in milk or product which contains higher percentage of water like ice cream or condensed milk is calculated by the difference. Total solid is first calculated and then moisture percentage is obtained by difference.

Protein

In Kjeldhal's method of protein estimation milk or milk product containing organic substance like protein is heated with concentrated sulphuric acid, all organic matter is charred on prolong heating. Carbon, hydrogen sulphur and phosphorus present in the material are converted into their respective oxides. All these escape into the atmosphere. However, the organic nitrogen present is converted into ammonium sulphate (inorganic nitrogen). The ammonium sulphate after the addition of sodium hydroxide is further converted to ammonia. This ammonia, on distillation can be trapped in boric acid to form ammonium borate and this is then titrated with a standard HCl using mixed indicator (Methylene blue + Methyl red). The colour change on addition to boric acid is violet which on absorption of ammonia becomes green. This on titration with standard HCl gives first bluish grey (which is the end point) and then turns to violet colour which is index of over titration. Alternatively, the ammonia liberated is abosrbed in a known excess of standard acid. The unused acid is titrated against a standard alkali. During the initial digestion copper sulphate is added to act as a catalyst, potassium sulphate is added to raise the temperature of the digestion mixture to hasten the digestion.

The nitrogen content of the sample is calculated on multiplying by a suitable factor (6.38) which gives the protein content.

$$\text{Percentage of Nitrogen} = \frac{\text{(A-B)} \times 0.014 \times 0.02\ \text{N (Normality of acid used} \times 100}{\text{wt. of milk taken}}$$

where, A is volume of HCl used for sample
B is volume of HCl used for blank

Percentage of Nitrogen

$$\frac{\left(\text{Volume of standard alkali used in blank} - \text{Volume of alkali used in sample}\right) \times .0014 \times 100}{\text{wt. of milk taken}}$$

If ammonia is absorbed in standard acid instead of boric acid.

Protein can also be determined by volumeteric method (formal titration of Pyne's method). Addition of excess formaldehyde to milk which is previously neutralised with standard alkali, would result in release of protons (H^+ ions) from the free amino groups of amino acids as follows :

$$^+NH_3CHR\ COO^- \longrightarrow NH_2CHR\ COO^- + H^+$$
(Zwitter ion form) → (anionic proton form)

$$NH_2CHR\ COO^- + 2\ HCHO \longrightarrow (HOCH)_2\ NCHR\ COO^- + 2H^+$$
(anionic form) → (complex anionic form)

The proton so liberated is titrated with standard NaCl to the end point of phenolphthalein (pH 8.0). The amount of alkali used for titration is a measure of the free amino groups of amino acids which in turn is roughly proportional to the amount of protein present in milk.

Another important method of protein estimation is by Follin's method. This method of estimation of protein in milk is less time consuming. The accuracy of the method is not to the same extent as of Micro-Kheldhal method. Folin's reagent is prepared with a mixture of 100 grams of sodium tungstate, 25 gram of sodium molybdate, 700 ml of water, 50 ml of 85 per cent phosphoric acid and 100 ml of concentrated hydrochloric acid is refluxed gently for 10 hrs in 1.5-2 litre flask. Now 150 gram of lithium sulphate and 50 ml of water are added followed by few drops of bromine water. This mixture is again boiled for 15 minutes without condenser to remove excess bromine. It is then cooled to litre and filter. The reagent should have no greenish tint. Acid concentration of the reagent is determined by titration with 1N sodium hydroxide to phenolphthalein end point.

A requisite quantity of diluted milk samples (0.5 ml) is taken. To this 5 ml of alkaline reagent is added. The alkaline reagent is first prepared by taking 100 ml of 2 percent sodium carbonate in 0.1N sodium hydroxide containing 0.5 ml of 1 percent copper sulphate and 0.5 ml of 2 percent sodium potassium tartarate. After 10 minutes of adding alkaline reagent 0.5 ml of Folin's reagent is added. This gives a blue colour. The blue colour is read in colorimeter at 660 mu or red filter after 30 minutes. A blank using 0.5 ml distilled water is also carried out. The readings are then converted to protein concentration from a standard curve in graph of a known standard protein prepared similarly. The final blue colour developed by adding Folin's reagent to protein solution which has been made alkaline previously is the result of (i) biuret reaction of protein with copper ions in alkali and (ii) reduction of the phosphomolybdic-phosphotungstic acid reagent by the tyrosine and tryptophan present in the treated protein.

Pro-milk, MK 11 the automatic electronic equipment marketed by A/ S N Foss Electric, Hillerod, Denmark for determining the protein content in milk has been in use for several years. The Pro-Milk method is based on the principle that basic amino acid residues of the protein form insoluble complexes with dye. The complex is removed by filtration and the excess dye is measured colorimetrically. The colour intensity of the filtrate is inversely proportional to the protein contents i.e. higher light transmission corresponds to a higher protein content. The colorimeter is directly calibrated for protein contents over the range 2.0-5.5 per cent.

Non-Protein Nitrogen

Non-protein nitrogen is generally present in 12 per cent or 15 per cent tirchloro acetic acid filtrate. This include urea, uric acid, ammonia, creatine, creatinine, α-amino acid and unaccounted nitrogen like hippuric acid, orotic acid and indican. This does not include casein, α-lact, ß-lact, and proteose-peptone. It does not have any biological values as protein, can not be utilised by body to perform functions characteristic of protein. Its content is highly variable in milk. In case of cheese it does not add to cheese yield.

Lactose

Lactose is the chief milk carbohydrate. Though trace amount of glucose, galactose and other sugars are present in milk, lactose is present in significant quantities. It generally ranges from 4.5 to 4.9 per cent. Lactose is one of the most important constituent of milk and it is correlated with many physico-chemical and biochemical processess. It is, therefore, necessary to measure the concentration of lactose in milk. Various methods have been used to determine the lactose content. The principles of some of the methods are outlined below :

Lane-Eynon Method: This method is based on the principle that lactose solution reduces quantitatively the alkaline cupric salt solution on boiling to red cuprous oxide; from the amount of copper salt reduced, the quantity of lactose is calculated. Milk filtrate is first prepared by taking a known quantity of milk, diluted with distilled water and then treated with acetic acid and making the volume to a known mark with distilled water quantitatively and then filtering the solution. Separately a known equal quantity of standard Fehling A (Copper sulphate solution in water) and Fehling B (sodium potassium tartarate and sodium hydroxide in water) are mixed. The mixture is heated to boiling and filtrate containing lactose is added to it. When the titration is about to complete (known from a preliminary titration) few drops of methylene blue indicator is added. The end point is indicated by the change of blue colour to colourless supernatant.

The percentage of lactose is then calculated by knowing the lactose in mg equivalent to a known amount of Fehling solution from the standard table.

Munson-Walker Method: The basis of this method concerns the reduction of a cupric salt complex of Cu_2O by lactose. When heated together under specified conditions the amount of oxide that precipitates from solution is proportional to the lactose present. A weighed sample of milk is freed of protein and fat by treating with a precipitating agent and filtering. An aliquot of the clear filtrate is first reacted with a known volume of Fehling A solution and then known volume of standard sodium hydroxide solution. After making up to a known volume it is filtered. To a known volume of mix (Fehling A + Fehling B), a known volume of filtrate is added and then heated. The amount of Cu_2O precipitated is filtered, washed gravimetrically in a Gooch crucible, dried and weighed. The lactose equivalent to the quantity of oxide is read from Munson-Walker tables. The percentage of lactose can then be calculated from the weight of lactose found in the aliquot of sample.

An alternate to this method consists of dissolving the wet precipitate of Cu_2O quantitatively in known quantity of saturated solution of ferric sulphate. This solution is then titrated with the standard solution of potassium permanganate. The weight of cuprous oxide in mg is then determined in known amount of the filtrate used in the precipitate on the basis that 1 ml of 0.1N potassium permanganate is equivalent to 0.00635 g of copper. Lactose equivalent of cuprous oxide is then determined from the table.

Polarimetric Method: Sugars contain an asymmetric carbon atom which makes them optically active or they are capable of rotating in the vibrational plane of polarised light. The degree and direction of this rotation is a property of the particular sugar. For measuring this property specially designed instrument known as polarimeter is used.

In measuring the lactose content of milk, the medium must first be freed of proteins and fat and a clear solution is obtained. This is usually accomplished by treating with acid mercuric nitrate reagent and filtering. The filtrate is polarised in a polarimeter tube, yielding the value a (angular rotation in degree). Specific rotation may be expressed as follows

$[a]_D 20^\circ = 100\ a/lc$ where $[a]_D 20^\circ$ = specific rotation at 20°C

referred to the D-line of spectrum, a = observed angular rotation in degree, l = length of column of solution in decimeters, C = concentration in grams per 100 ml. This can be simplified to $55.4 = 100\ a/2\ c$ by inserting the constant for lactose and the length of the polarimeter tube (200 nm). The grams of lactose per 100 ml of filtrate $c = a/1.108$. This value of lactose can then be converted to per cent lactose.

Chromatographic Method: Lactose is separated by using descending paper partition chromatography. The solvent system employed consist of ethyl acetate, pyridine and water or n-butanol-pyridine and water. The lactose spots are then detected by spraiying 2 per cent triphenyl tetrazolium chloride prepared in 1N alcoholic KOH and the paper is warmed in an oven at 75-80°C. The lactose spots appear as brown colour. The spots can be quantitatively estimated by cutting the spots and eluting the colour in alcohol and estimating the intensity of colour in photoelectric colorimeter. Another method also consists in spraying with silver nitrate reagent and eluting the brown colour with alkaline ferricyanide reagent, and measuring the intensity of colour in colorimeter. The amount of lactose is derived from a calibration curve, developed with known levels of lactose plotted against optical density of developed colour.

Picric Acid Method: The method is based on the fact that reducing sugars in alkaline solution reduce yellow picric acid to mahogany red picramic acid. A known quantity of milking is diluted to a specified volume with saturated picric acid. The well mixed solution is filtered and measured volume of the clear filterate is mixed with sodium carbonate reagent in a sugar tube and placed in a boiling water bath for development of colour. Optical density of developed colour at 520 mu is proportional to the lactose content. The concentration of lactose is then calculated by applying a known formula.

Chloramine-T Method: The protein and fat are removed from the milk with a precipitating agent, and to a measured amount of clear filtrate, a known quantity of chloramine-T reagent is added, together with KI solution. After standing for a specified period of time, the unreacted chloramine-T is measured by titration of liberated iodine with sodium thiosulphate solution. The difference in the amount of chloramine-T in a blank solution and that remaining in the sample analysed is equivalent to the lactose in sample. The principle of this method is based on the reducing property of lactose. The reagent N-chloro-p-toluene sulfonamide (chloramine-T) acts as an oxidising agent.

Sucrose: The polarimetric method is based on the principle of the clerget inversion. The method involves a milk treatment of the material with acid which produces complete hydrolysis of sucrose but almost none of lactose of other sugars. The sucrose content is derived from the change in rotating power of the solution. A clear filtrate of the sample, without mutarotation by lactose, is prepared by treatment of the solution with ammonia followed by neutralisation and clearing by the successive addition of zinc acetate and potassium ferrocyanide solution. In a part of the filtrate the sucrose is hydrolysed in a way which meets the special case of this milk filtrate. From the rotation of the filtrate before and after inversion, the sucrose content is calculated.

The volumetric method is based on the reducing property of sugar because of the presence of free aldehydic and free ketonic group. All the reducing sugars have the capacity of reducing cupric salts to cuprous salt which gives a brick red colour. This reduction is quantitative and sugar gets oxidised to corresponding acid.

Sucrose unlike lactose is non-reducing sugar because in lactose, glucose has free aldehyde group at 1^- and galactose after hydrolysis of ß-1^- ——>4 galactosidic linkage has free aldehyde group at 1-position as there is 1-4 linkage. While in sucrose due to 1,2 linkage, no free ketone or aldehyde group is available. The extract filtrate is treated with HCl which hydrolyse (invert) sucrose to monosaccharides glucose and fructose. After this the contents are estimated by usual Lane-Eynon method using Fehling solution and calculations are then made. Fructose becomes a reducing sugar through Enolization, (Enediol, Keto-enol tautomerism) formation.

Major Minerals

Ash

Milk contains salts like phosphates, citrates, sulphates, chlorides and bicarbonates of calcium, magnesium, potassium and sodium, etc. Heating milk at higher temperature decomposes organic matter and inorganic salts are behind in the form of ash.

Phosphorus

Because of the possibility of losses of phosphorus in the ashing process, the more satisfactory procedure to estimate phosphorus in milk is to oxidise the milk with oxidising agent, such as H_2O_2, $HClO_4$, HNO_3, etc. and digest it with H_2SO_4 which will affect complete conversion of all the phosphorus to H_3PO_4 (phosphoric acid). The acidic solution (digested material) is then treated with excess solution of ammonium molybdate which gives yellow coloured precipitate of ammonium phosphomolybdate. These precipitate on addition of reducing agent (1-amino-2-napthol-4-sulphonic acid or ANSA), forming blue colour complex which is unstable and uncertain in composition. The intensity of the blue colour is proportional to the amount of phosphorus present in sample and can be estimated colorimetrically using red filter. Amount of phosphorus in the same can be calculated from a standard curve.

$$P + (NH_4)_6\ MO_7O_{24}\ 4H_2O \longrightarrow (NH_4)_3\ PO_4.12MoO_3 \text{ (yellow ppt)}$$

$(NH_4)_3PO_4 12\ MoO_4$ +

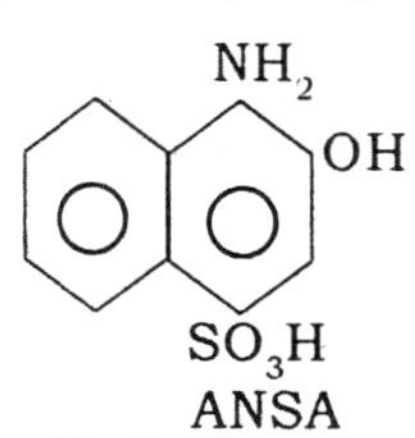

ANSA
(Reducing Agent)

$\longrightarrow$ Blue colour complex Molybdenum blue. probably oxides of molybdenum

Calcium

EDTA (ethylene-diamine-tetra acetic acid) is a dibasic acid which reacts with different types of metals and is a powerful sequestering/ chelating agent for divalent and polyvalent cations. Sodium salt of EDTA is used for practical purposes because this acid is unstable. The sodium salt when in solution supplies anions H_2V^{--} which can react with calcium as follows :

(1) 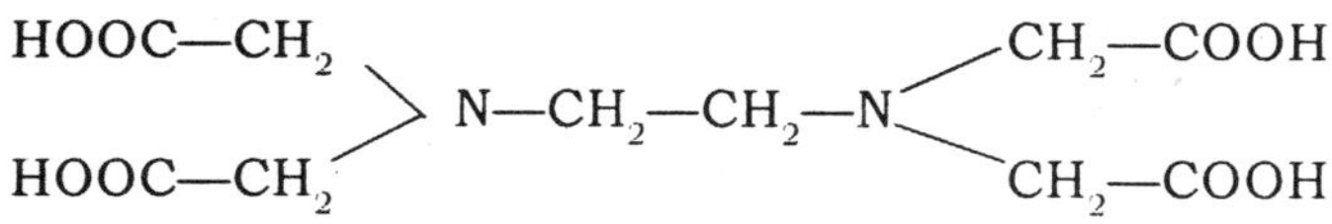 Dibasic acid

(2) 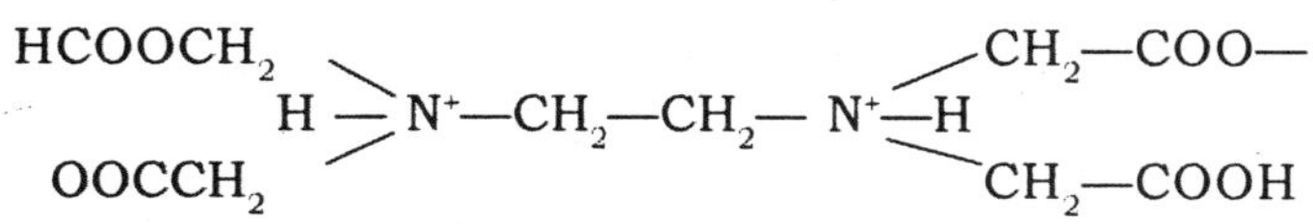 Zwitter ionform

(3) $Ca^{++} + H_2V^{--} \longrightarrow CaV^{--} + 2H^{+}$

$Mg^{++} + H_2V^{--} \longrightarrow MgV^{--} + 2H^{+}$

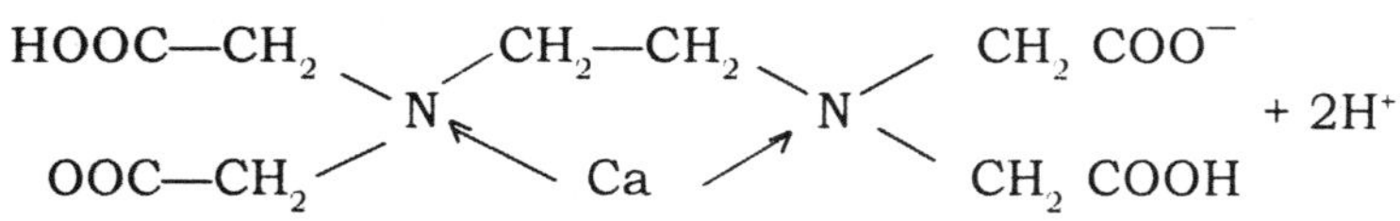

Since 'Ca' is bound to major milk proteins, i.e. casein, so proteins have to be precipitated to release 'Ca' from casein. Usually milk proteins are precipitated with TCA (Trichloroacetic acid) and casein free filtrate is used for calcium estimation. Calcium in above filtrate is first precipitated as calcium oxalate with the help of ammonium oxalate at low pH and then the precipitate is separated by centrifugation at 1400 rpm for 10 min. Titration is conducted at high pH due to phosphorus interference during titration by precipitating calcium as calcium phosphate. Excess EDTA solution at low pH is added in order to complete the reaction and then the pH is raised. Remaining EDTA is titrated against standard magnesium acetate.

Gel Filtration

Separation of molecules can be carried out on the basis of differences in size by passing them through a column containing swollen particles of the Sephadex gel. The mixture of large and small molecules is passed through the column. As they pass down the column, the small molecules become trapped within the gel and follow a longer path then the large molecules which are completely or partially excluded from the gel particles. Eventually complete separation occurs when large molecules leave the column first and the smaller ones in the last.

Cross linked dextrans are the basic materials used in gel filtration under the trade name of Sephadex. These dextrans supplied in the form of macroscopic beads swell in contact with aqueous solutions due to high content of hydroxyl groups. Sephadex is a stable material and is unaffected by bases and weak acid, but concentrated solutions of strong acids should be avoided since these hydrolyse the glycosidic links.

Electrophoresis

Electrophoresis is an analytical tool to separate charged particles or molecules in the electric field. The separation of molecules is done on a supporting medium such as paper, starch and polyacrylamide gel.

The paper electrophoresis has become an obsolete technique and not used to a greater extent. In the starch gel electrophoresis technique, hydrolysed starch is the supporting medium. The pore size of the gel depends on how it is prepared, the particular starch used and the pH and the nature of the buffer solution. The diameter of the pores in starch gel is such that molecules are separated on the basis of size as well as charge. A fine resolution of protein is achieved into different bands. It is difficult to elute compounds from this material and the zones are usually located by staining the gel.

Use of polyacrylamide gel as a supporting medium has distinct advantages over starch gel because (i) it is transparent, (ii) the resolution of components is better, (iii) it has lower electrical resistance, so a small column of gel can be used for separation, (iv) the pore size of the gel can be controlled and hence the separation of the proteins can be achieved on the basis of molecular weight also.

Chromatography

By chromatography is meant those processes which allow the resolution of mixture by affecting separation of some or all their components in concentrated zones or in phases different from those in which they are originally present, irrespective of the nature of the force or forces causing the substances to remove from one phase to another.

Paper Chromatography

Cellulose in the form of sheets of filter paper serves a useful support phase in the technique of paper chromatography. The mixture is spotted on the paper and dried. Then the solvent flows along the sheet either by gravity (descending chromatography) or capillary action (ascending and radial chromatography) or capillary action (ascending and radial chromatography). The solvent front is marked and, after drying the paper, the positions of the compound present in the mixture are visualised by suitably staining reaction. The ratio of the distance moved by a compound to that moved by the solvent is known as the Rf value and is more or less constant for a particular compound for particular solvent system. It depends on several factors like composition of solvent system, type of paper, direction of paper, manner or development (descending, ascending or radial) length of paper used, distance of starting line from solvent concentration of solute, presence of other substances, and temperature of development.

The resolution of mixture of solutes on filter paper may depend on surface adsorption, ion exchange or on partition between solvents. Cellulose fibres have strong affinity for water present in the solvent phase but little for organic phase or liquid. The paper is thought of as an inert support holding a stationary aqueous phase. As solvent flows through a section of the paper containing the solute, partition of this compound occurs between the mobile organic phase and the stationary water occurs between the mobile organic phase and the stationary water phase. Thus, some of the solute leaves the paper and enter the organic phase. When mobile liquid reaches a section of the paper containing no solute, partition again occurs. This time solute is transferred from the organic phase to paper phase. With continuous flow of solvent, the effect of this partition between the two phases is the transfer of a solute from the point of its application to the paper to a point some distance along the paper in the direction of solvent flow.

Ion Exchange Chromatography

Ion exchange chromatography can be defined as the separation of compounds on an insoluble matrix containing labile ions capable of exchanging with ions in the surrounding media. Several commercial resins are available which serve as insoluble matrix. Depending on the charge group, the ion exchange resins may be anionic or cationic according to the affinity for either negative or positive ions. Some of the important resins are CM-cellulose or Sepharose, DEAE cellulose, Amberlite and Dowex, etc. A column of one of these resins is prepared in a suitable buffer and protein is separated using this column. Ion exchange resins are able to separate mixtures of protein on the basis of differing charge distribution on the protein surface. The method will handle large quantities of protein and has a light resolving power, but requires the salt concentration of the protein sample to be extremely low.

A protein in solution will carry either a (+) or (–) charge, overall if it is not at its pI (Isoelectric point). Then it will form electrostatic bonds (salt link) to a –

1. Cationic exchange medium if it has an overall (+) charge.

2. Anionic exchange medium if it has an overall (–) charge.

3. Proteins can be given (+) or (–) charges if they are placed in a buffer which has a charge either below or above its pI.

Proteins above their isoelectric point will carry a net (–) charge and will bind to an anionic exchanger. As that protein is brought closer to the pI the greater (+) charges will develop and it will release from the exchanger. Proteins below their isoelectric point carry a net (+) charge and will bind to a cationic exchanger. It will release as it approaches the isoelectric point.

Partition Chromatography

The separation of compounds that are readily soluble in organic liquids but sparingly soluble in water is carried out by adsorption chromatography, while ionizable water soluble compounds are best separated by ion exchange chromatography. Partition chromatography is intermediate between adsorption and ion exchange chromatography, and compounds that are soluble in both water and organic solvents are readily separated by partition methods.

When a compound is shaken with two immiscible solvents it will distribute itself unevenly between the two phases and at equilibrium the ratio of the concentration of the compound in the two solvents is constant and is known as the partition coefficient (a).

$$\alpha = \frac{\text{Concentration of X in solvent 1}}{\text{Concentration of X in solvent 2}}$$

In partition chromatography one solvent, usually water, is held on the stationary supporting phase wich is in the form of a column or film of inert material. The other phase consists of a mobile water saturated organic-liquid that flows past the stationary phase. The components of a mixture are separated if their partition coefficients between the solvents are sufficiently different.

Affinity Chromatography

Affinity chromatography is a type of adsorption chromatography in which the bed material has biological affinity for the substance to be isolated. The specific adsorptive properties of the bed material are obtained by covalently coupling an appropriate binding ligand to an insoluble matrix. The binding ligand is then able to adsorb from solution the substance to be isolated, desorption being subsequently carried out by changing the experimental conditions after unbound substances have been washed away. Separation is rapid since in most cases only small bed volumes are required. Affinity chromatography provides opportunities for the isolation of substances according to their biological function, and thus differs radically from conventional chromatographic techniques in which separation depends on gross physical and chemical differences between the substances.

Dialysis

Dialysis is used for the gross separation of large and small molecules and depends on the fact that small molecules pass freely through a membrane while colloidal molecule do not. The charge and specific groups of the molecule also affect the rate of diffusion particularly with molecules in the intermediate size range. The rate of dialysis is also governed by the temperature so that the higher the temperature, the higher the dialysis rate. At an elevated temperature, the viscosity of the solvent is less and the rate of diffusion is increased. Many macromolecules are sensitive to temperature, so dialysis of proteins, etc. is usually carried out in the cold room. The solvent affects the rate of dialysis in several ways and needs to be selected with care. The rate of dialysis is greatest with distilled water but careful control of the ionic strength and pH is often needed to stabilise the molecules under investigation. The presence of salts reduces the rate of dialysis. The pH also controls the dialysis. The reagent present in solvent which tends to denature the molecule also reduces the rate of dialysis.

Some of the membranes used for dialysis are pigs bladder, parchment or colloidion, however, cellophane is the most preferred material. Dialysis tubing made of cellophane are commercially available.

Ultra Filtration

Ultra filtration is concerned with the separation of high molecular weight large molecules from low molecular weight small molecules in a system using semi-permeable membranes under hydraulic pressure. Several types of membranes are available depending on the system, such as those made of cellulose acetate, polyamides, polysulphones and zirconium oxide. A low pressure as low as 10-100 psi is usually applied to effect such separation. The important application of ultra filtration in dairying is in manufacture of fermented dairy product such as yoghurt, quart, soft and hard cheese and protein-enriched milk. In such processes concentration of desirable product is easily done without loss and energy saving as in conventional concentration processes.

Thin Layer Chromatography

Thin layer chromatography is a very useful technique in the separation and identification of the different components present in lipids and other organic and inorganic compounds. The glass plates coated with thin layers of silica gel G or aluminium oxide G are used for separation which is achieved in about 30 minutes. The equipment required is very inexpensive and simple to operate. This method is very sensitive and hundreths of a microgram of a sample can be used for analysis. It is also possible to resolve many milligrams of the mixture into its component for preparatory work. It is adsorption chromatography in which thin layer of adsorbent material coated on a glass plate serves as a stationary phase. The separation of a sample applied on the glass plate is very similar to that on filter paper used in paper chromatography. Glucosides, alkaloids, steroids, lipids, fats and fatty acids can be separated using thin layer chromatographic techniques.

GLC (Gas-liquid Chromatography)

Chromatography is a tool to separate mixtures of solutes based on the differences in physical properties. Separation is brought about by differential migration of solutes through a porous medium under the influence of a moving phase.

Gas being the moving phase, it is called gas chromatography. If the stationary phase is a solid it is called gas-solid chromatography (GSC) and if liquid it is called gas-liquid chromatography (GLC). Basically a gas chromatograph consists of (i) the pressure regulator and flow meter for the carrier has supply (N, He, H_2), (ii) sample injection system, (iii) the separation column, (iv) the thermal compartment, (v) the detection system, (vi) strip-chart recorder.

A liquid sample is injected into the carrier gas in the injection system which is made to vaporise before it reaches column. The vaporised sample mixed with carrier gas enters a pre-heated column contained in an oven. The column contains the stationary phase generally a partitioning liquid, coated on some inert material. After the sample is separated on column, it enters the pre-heated detector which senses the emergence of individual component. The signal from the detector is amplified several hundred folds before feeding to a recorder. The amplified signals are recorded as peaks called chromatograms.

HPLC (High Performance Liquid Chromatography)

High speed liquid chromatography or high performance liquid chromatography is a smilar technique as gas chromatography. Rapid separation and improved resolution and improved resolution of biological material such as protein peptides, nucleic acids, carbohydrates and organic acids can be achieved. In this chromatography the mobile phase is pumped at pressure upto 8000 psi. A pressurised delivery of the mobile phase is necessary to give a reasonable flow rate through a column packed with particles smaller than 100 μm in diameter. This technique is quite comparable to gas chromatography for speed, efficiency and resolution and it is inherently more versatile. It is not limited to volatile and thermally stable samples and the choice of stationary phase includes solid adsorbents, liquids, ion exchange and gel permeation materials. The assembly consists of pumps designed to operate at constant pressure and flow, device to flow mobile phase, column, injector, thermostated oven, detector, recorder and receiver.

Colorimetry

Use of colour as an index of concentration has been called colorimetric analysis or colorimetry and the instrument is called a colorimeter.

$$\text{Light energy} \xrightarrow{\text{Photo cell}} \text{Electrical energy}$$

The concentration of a biochemical compound can be determined by measuring the light absorbed by the solution using colorimeter. Many compounds are not coloured but can be made to absorb in the visible region by reaction with suitable reagents. When a ray of light of initial intensity Io passes through a solution in a transparent vessel, some of the light is absorbed, so the intensity of the transmitted light I is less than Io. The ratio of I to Io is known as the transmittance T and depends on the path length of the light through the solution and the concentration of the absorbing material. Two laws are considered in this connection. When a ray of monochromatic light passes through an absorbing medium, its intensity decreases exponentially as the length of the medium increases (Lambert's Law) and as the concentration of the absorbing medium increases (Beer's Law). These two laws are conveniently combined in the Beer-Lambert law, so that if l is the path length and C is the concentration of the solution, then

$$T = I/Io = {}_e\text{-}K'Cl$$

where, K' is a constant depending upon the nature of the absorbing material.

Taking logarithms,

$$\text{or} \quad \log_e I_o/I = K'Cl$$

$$\log_{10} I_o/I = KCl$$

The expression of $\log_{10} I_o/I$ is known as the extinction or absorbance. The extinction is sometimes called the optical density, but this name is not recommended. If the Beer-Lambert law is obeyed, then a plot of extinction against concentration gives a straight line passing through the origin. With the aid of a standard curve, the concentration of an unknown solution can then be readily determined. If l is 1 cm and C expressed as g mol/l then the abosrbance is equal to k and is known as the molecular extinction coefficient, which is characteristic for a compound. The Beer-Lambert law only applies for monochromatic light provided there is no change in ionization, association, dissociation or solvation of the solute with concentration.

One of the simplest colorimeter is Klett-Summerson colorimeter. In operation, with a suitable filter in place, the reference solution contained in a test tube is placed in the path of light striking one of the two photo cells, which is arranged in a potentiometric circuit so that the current from one cell is opposed to that from the other through a null point instrument (low sensitivity galvanometer). With the photometer set at zero (corresponding to zero optical density), the current output from the second photocell is adjusted so that it exactly balances that coming from the photocell which is subjected to the light emerging from the solution. This balance is indicated by a zero reading on the galvanometer. The reference solution is then removed and replaced by the solution under examination. Any light absorption by this solution will throw the two photocells out of electrical balance, balance is then restored by turning the potentiometer dial until the galvanometer again reads zero. The reading on the potentiometer scale at this point is the measure of the light absorption of the solution. The reading can be converted into optical density as follows :

$$\frac{100 \times D}{2} = R$$

Where, D is optical density and R is the reading on potentiometer.

Ultra Centrifugation

Ultracentrifugation means centrifugation at high speed normally at 20,000 rpm or above. The instrument is known as ultracentrifuge. It can measure certain properties of a molecule, such as molecular weight, shape, size and density, as well as the number of components in a protein solution. The ultracentrifuge subjects a small volume of protein solution to a carefully controlled centrifugal force, and records, by means of optical and photographic system, the movement of the macromolecules in the centrifugal field. The solute molecules which are initially uniformly distributed throughout the solution in centrifuge tube, are forced towards the bottom of the tube by the centrifugal field. This migration leaves a region at the top of the centrifuge tube that is devoid of solute and contains only solvent molecules. The migration also leaves a region in the centrifuge tube where the solute concentration is uniform. A boundary is set up in the centrifuge tube between solvent and solution in which concentration varies with distance from the axis of rotation. The measurement of the boundary's movement, which represents the movement of the protein molecules, is the basis of the analytic method. By the data obtained, namely, the sedimentation rate, the Svedberg unit (S) can be calculated. A Svedberg unit, named in the honour of T. Svedberg, the Swedish pioneer in the field, is defined as the velocity of the sedimenting molecule per unit of gravitational field or 1 × 10^{-13}cm/sec/dyne/g. Typical S values are 4.4 for bovine serum albumin, 1.83 for cytochrome C, and 185 for tobacco mosaic virus. With a knowledge of the diffusion coefficient, molecular weights can be readily calculated. The basic equation relates S and molecular weight :

$$\text{Mol. wt.} = \frac{RTS}{D\,(1 - vp)}$$

Where, R is the gas constant, T the absolute temperature, S the Svedberg unit, D is the diffusion constant, v the partial specific volume, and p the density of the solution.

To determine the number of components in a solution, a simple centrifugation can be readily made and the number of boundaries based on concentration gradient peaks can be determined. Diffusion coefficient measurements need not be made.

REVIEW QUESTIONS

1. Differentiate between natural and developed acidity.
2. What is the normal pH of cow and buffalo milk?
3. What is the principle of Kjeldahl's, Pyne's and Follin's method of protein estimation.
4. Briefly explain Lane-Eynone method of lactose estimation.
5. In phosphorous estimation using ANSA what is blue colour complex due to?

REFERENCES

Dairy Guide, 1982–1986. Publication 20, Thyag Raj Nagar Market, Lodhi Colony, New Delhi1100 003.

David, T. Plummer, 1971. An introduction to practical biochemistry. Tata MacGraw-Hill Publishing Company Ltd.

Indian Dairyman 1980-1990. Publication, Indian Dairy Association.

A Laboratory Manual of Dairy Chemistry, 1963. Board of Dairy Education, Govt. of India.

Indian Standard Specifications, Indian Standard Institution, Manak Bhawan, New Delhi.

11

LEGAL STANDARDS AND QUALITY CONTROL

Legal Standards of Milk

MILK, by its inherent nature is very susceptible to adulteration. Due to lack of a definite composition, infinite variations influenced by various factors make definition of pure milk difficult and impractical. Further, milk is indispensable in day to day life and demand for milk invariably outstrips the available sources of supply. It is no wonder then that milk is probably the most tampered or abused food item today.

To safeguard the interests of the consumers and also the producer and distributors of milk, in India, the Prevention of Food Adulteration Act was passed in 1954 by the Parliament and the PFA Rules were framed which came into force on June 1, 1955. Different food articles have been defined accurately and the limits of the various analytical parameters specified. Any food falling outside the permissible ranges is considered substandard and is deemed to be adulterated. Convictions made under the PFA Act are viewed seriously and invites severe legal punishment including heavy fines and imprisonment.

The PFA Act is implemented by a countrywide network of Food Inspectors who collect at random samples suspected of having been adulterated. The sample is divided into three parts, and the specified quantity of preservative is added. The sample bottles are then sealed and labelled. One sample is given to the vendor from whom it is drawn, one is retained as a preference and the third one is sent for analysis to the Public Health Laboratory. The Public Analyst performs the analysis according to specified procedures and gives his report. If the sample is found to be substandard with respect to any of the specified parameters, legal action is initiated against the vendor. Food adulteration cases are viewed seriously and if convicted the guilty is severely punished with both fines as well as non-bailable imprisonment. With an increase in consumer awareness and organization of consumers protection societies the number of cases being brought to light are on the increase. Extreme caution must, therefore, be observed before certifying milk or dairy products as fit for sale. Conformity to the delineated P.F.A. standards should be strictly adhered to and proper records should be maintained.

It is essential that milk designated as cow, buffalo or goat milk has to meet different minimum standards depending on the region in which it is produced or sold. Most of the milk packaged and sold through the organized sector is normally designated as the standardised or toned or double toned milk for which standards are uniform throughout India.

P.F.A. Rules – Definitions

Milk: It is the secretion derived from complete milking of healthy milk animals. it shall be free from colostrum.

Pasteurisation: Refers to the process of heating of every particle of milk or milk product to atleast 63°C and holding at such temperature continuously for atleast 30 min or heating to at least 71.5°C and holding at such temperature continuously for atleast 15 sec. or an approved temperature time combination that will serve to give a negative phosphatase test.

Sterilisation: means heating milk to a temperature of 115°C for 15 minutes or 145°C for 3 secs or equivalent to approved temperature time combination, for enzyme preservation at room temperature for a period of not less than 15 days from the date of manufacture. Sterilised milk shall show negative turbidity test and shall be sold in the container in which it was sterilised.

Boiled Milk: means milk which has been brought to boil.

Recombined Milk: homogenised product prepared from milk fat non-fat milk solids and water. It shall be pasteurised and show a negative phosphatase test.

Flavoured Milk: Called by whatever name, it may contain chocolate, coffee, or any other edible flavour, food colour and cane sugar. It shall be pasteurised, sterilised or boiled.

Standardised Milk: Means cow milk or buffalo milk or a combination thereof that has been standardised to a fat percentage 4.5 by the extraction and/or addition of milk fat or by the addition of skimmed milk or by addition of recombined or reconstituted skim milk. Standardised milk shall be pasteurised and shall show a negative phosphatase test.

Toned Milk: Means the product prepared by admixture of cow or buffalo milk or both with fresh skimmed milk or by admixture of cow or buffalo milk or both with non-fat milk solids or milk powder and water or by partial abstraction/addition of fat from/to milk. It shall be pasteurised and shall show a negative phosphatase test. When fat or dry non-fat milk solids are used it shall be ensured that the product remains homogenous and no deposition of solids takes place on standing. Double toned milk is also the same as toned milk.

Skimmed milk

It is the product prepared from milk from which almost all the milk fat has been removed mechanically.

Preservative

It is a substance which when added to food, is capable of inhibiting, retarding or averting the process of fermentation, acidification or other decomposition of food.

Rule 19

Any person taking a sample of any food for the purpose of analysis under the Act may add a preservative as may be prescribed from time to time, to the sample for the purpose of maintaining it in a condition suitable for analysis.

Rule 20

The preservative used in the case of sample of any milk in liquid or semi-liquid form shall be the liquid commonly known as formalin, that is to say a liquid containing about 40 per cent of formaldehyde in aqueous solution in the proportion of 0.1 ml (two drops) for 25 ml or 25 gm of sample.

Agmark Grading

Agricultural commodities do not normally conform to any specific standards but vary considerably and a strict control over their quality has to be maintained. In the case of products like *ghee* and butter there is a vast variation in quality and analytical parameter between the different regions of the country. These variations could arise due to the differences in crops, oils seeds, etc. used for feeding cattle. For example the Reichert Meissl (R.M.) value of ghee produced in cotton tract areas like Saurashtra, etc. is much lower than the minimum permissible limits specified for normal *ghee*. In such cases there is a possibility of even pure *ghee* being rejected as adulterated. To overcome these problems and to ensure that the consumer gets a pure, unadulterated product, the Directorate of Marketing and Inspections introduced the quality control and standardisation of *ghee* and butter in 1938. The grading of *ghee* under Agmark was introduced under the provisions of the Agricultural Produce (Grading and Marketing) Act, 1937.

The Directorate of Marketing and Inspections (DMI) issues a certificate of authority to the packer after confirming that the product conform to the rigid quality specification of the Agmark. The testing is done either at the producer's own established laboratory or at the State Grading Laboratory by a panel of chemists approved by the Directorate of Marketing and Inspection. After the *ghee* is packed, labelled and marketed, samples may be drawn occasionally from the market and tested at the Regional Analytical Laboratory.

Ghee is graded either as Special or General separately for cow and buffalo based on butyro-Refractometer reading at 40°C, moisture content, saponification value, reichert meissel value, polenske value, kirschner value and free fatty acids. To overcome the difficulty experienced in distinguishing between cow and buffalo *ghee*, a set of common standards for buffalo and cow, or mixed *ghee* was set up (Table 11.1).

Table 11.1. Agmark standards for ghee

Parameter	Special grade	General grade
Baudouin test	Negative	Negative
Butyro-refractometer reading at 40°C	40.0–43.0	40.0–43.0
Reichert-Meissel value	Not less than 28.0	Not less than 28.0
Polenske value	1.0–2.0	1.0–2.0
Moisture content	Not more than 0.3%	Not more than 0.2%
Percentage of free fatty acids	Not more than 1.4	Not more than 2.5

Similarly, two grades of butter viz. pasteurised table butter and unpasteurised table butter have been prescribed on the basis of flavour aroma, body and texture, colour, moisture content, acidity and curd, etc.

Ghee produced in cotton tract areas where cotton seed cake is fed extensively to the animals has the R.M. value standard of not less than 23.0. Thus, taking into consideration various factors, the Agmark standards have been realistically set, so that all pure ghee samples do not fail these standards. However, these standards do not take the rancidity factor into account. Though aroma, flavour, and free fatty acids are tested, rancidity test like peroxide value and kreis test, etc. are not performed.

Another factor which requires looking into, is the awareness among the public regarding the Agmark Standards. The general consumer has only a vague idea about these standards and is not fully aware of the different grades (Special, General) or labels (Yellow, Blue, Red and Green) specified by Agmark. Educating the public and consumer on these aspects is very essential.

Preservatives in Milk

Milk, apart from being a complete food for human beings also provides an ideal growth medium for the micro-organisms. Freshly drawn milk contains a heavy lot of microorganisms which proliferate and multiply very rapidly in raw or unprocessed milk. Consequently, there is an increase in acidity leading to souring and finally to spoilage. Under Indian conditions, the initial bacteriological quality of raw milk is poor and to add to it there is usually a long time lapse before the milk reaches the consumers or the dairy for processing. In such cases there is a tendency to use preservatives to prevent microbial proliferation and spoilage of milk.

Preservative is a substance which when added to food is capable of inhibiting, retarding or averting the process of fermentation, acidification or other decomposition of food.

Legally no preservative of any sort is permitted in milk meant for human consumption. However, unscrupulous producers take recourse to addition of preservatives to milk. The use of following preservatives has been observed to be very common :

Formaldehdye: Formaldehyde is strong preservatie affecting in even small doses. Commercially a solution called formalin is available which contains about 37 percent by weight of formaldehyde and 10-15 percent methanol to prevent polymerisation. Food samples taken for analysis are preserved with 0.1 ml (two drops) of formalin for 25 gms of the sample. However, in food meant for consumption this is not allowed.

Formaldeyde gives an intense violet colouration with concentrated sulphuric acid (Hehner's test) or concentrated hydrochloric acid (Leech test) in the presence of an oxidising agent. Its detection is very easy and even small quantities can be identified.

Hydrogen Peroxide: Hydrogen peroxide is a very strong oxidising agent, and is an efficient preservative in small quantities also. The catalase enzyme breaks up the hydrogen peroxide to water and oxygen. Hydrogen peroxide in combination with thiocynate activates the native lactoperoxidase system of milk and is gaining acceptance for the preservation of raw milk. The use of hydrogen peroxide as such is prohibited but it is used quite extensively and surreptitiously and has to be checked. Hydrogen peroxide gives a dark blue colouration with para phenylene diamine hydrochloride commonly referred to as 'ppdh'. Hydrogen peroxide present even upto 1:40,000 proportion can be detected by this method.

Boric Acid or Borates: These are commonly used preservatives and are available commercially. Their detection in milk is based on the fact that they turn turmeric paper red in an acidified medium.

Benzoic Acid and Benzoates: These are food grade preservatives commonly used in jams, sauces, etc. but are not permitted in milk. Milk, acidified with hydrochloric acid is extracted with ether and residue obtained after evaporation of ether is dissolved in ammonia and evaporated to dryness on a water bath. On addition of water and ferric chloride a flesh coloured precipitate of ferric benzoate is obtained. Benzoic acid is, however, not very commonly used in milk as a preservative.

Neutralisers in Milk

Milk freshly drawn from the udder has an acidity of 0.12–0.16 per cent expressed as per cent lactic acid. With passage of time this acidity increases steadily till the milk sours or curdles and becomes unfit for consumption or processing. The increase in acidity is due to the action of microorganisms on the milk constitutents. Mainly the milk carbohydrates-lactose is converted to lactic acid causing an increase in acidity, called the developed acidity. Often this developed acidity is neutralised with alkali and the milk is used for processing and consumption. Neutralisation of milk is not permitted by P.F.A. and is not desirable and has to be avoided as far as possible. Farmers and milk producers tend to neutralise milk to avoid rejection of sour milk at the collection centres. The common neutralisers used are sodium carbonate, sodium bicarbonate, sodium hydroxide, hydrated lime, etc. as discussed below :

Carbonate and Bicarbonate: These neutralizers are soluble in water and are odourless but have the disadvantage that they produce carbondioxide on decomposition. They have good neutralising capacity. They are cheap and readily available commercially as washing soda or cooking soda.

Their presence in milk can be readily detected by adding an equal quantity of ethyl alcohol and a few drops of Rosalic acid solution. Positive samples give a distinct rose-red colour and can be readily identified. Sodium hydroxide or caustic is a very strong alkali and has to be used with extreme caution to avoid over neutralisation. Sodium hydroxide cannot be detected by any simple means or platform test.

Hydrated Lime: is not very soluble in water and its neutralising capacity is only about 80 per cent of the theoretical value.

The presence of neutralisers can generally be detected by determining the alkalinity of the ash. A known quantity of milk is heated to ash and the alkalinity of the ash is estimated by titration against a decinormal standard hydrochloric acid. Values abnormally high would indicate neutralisation of milk.

pH: The normal range of pH is 6.6–6.8 and below 7.0. Values higher than these levels are abnormal and indicate the presence of neutralizer.

Developed Lactic Acid/Lactate Content

Freshly drawn milk contains very little lactic acid and its titratable acidity is mainly due to the milk constituents. Therefore, the estimation of developed lactic acid or lactate content in milk alongwith the titratable acidity will give a reliable indication of the neutralisation of milk. This is because the neutralisation of milk will decrease only the titratable acidity and not the lactate content/lactic acid.

Adulterants in Milk

Milk is probably the easiest food to be adulterated. Since the demand for milk in the country is far greater than the available supply, there is great temptation to adulterate milk with a motive to increase ready profits. Quite a significant amount of milk is supplied directly to the consumer from the producer through vendors without any sort of processing, packaging or quality control. Such milk is most prone to adulteration since the consumer is not aware or bothered about the adulteration and continues to believe that he is getting 'fresh' milk. A large quantity of such milk is adulterated and escapes unnoticed. But, however, in the organised sector suspicious samples at random are picked up and analysed for adulterants.

The most common pattern of adulterating milk is to add water to increase the total volume of the milk while keeping the fat percentage within permissible limits. The addition of water, however, decreases the SNF content and this loss is made good by the addition of cheaper solids like sugar, starch, salt or urea.

Detection of Added Water in Milk

Milk has drawn from the udder could contain upto 85 percent water and varies widely in composition. As such it is normally difficult to find any parameter that is constant. The freezing point of milk is fairly constant and has a value of –0.530 to –0.555°C. Freezing point is a colligative property and depends on the number of solute particles dissolved in the solvent. Water is the solvent and its freezing point depression is constant. Therefore, if water is added to milk there is a decrease in solute concentration resulting in an increase in the freezing point. The freezing point of milk is determined by the milk cryoscope and the added water calculated as

$$\text{Percent added water} = \frac{0.55 - \Delta T}{0.55} \times 100$$

Where, ΔT is the freezing point depression of the suspected sample. A Tolerance of 3 percent is allowed. The weight percentage of added water is calculated as

$$\text{Percent added water} = \frac{0.55 - \Delta T}{0.55} \times (100 - \text{T.S.})$$

This is necessary since the freezing point depression varies inversely with the amount of water (100–T.S.) per unit of solute. This increase in freezing point can be nullified by the addition of other solutes to the added water. It is seen that adding a solution in water of 9.62 per cent cane sugar; 5.05 per cent glucose; 1.74 per cent urea or 0.86 per cent sodium chloride alone or in combination did not change the freezing point of milk.

Detection of Cane Sugar

Cane sugar is a commonly used adulterant as it is cheaper than milk solids and is readily available. However, it can be easily detected by the sweet taste it imparts even in small quantities. Chemically it forms a red precipitate with resorcinol in an acidic medium on heating and can be easily detected.

Starch

Starch is another cheap and easily available material commonly used for adulterating milk. It has the advantage that it is tasteless, colourless and odourless and is difficult to detect organoleptically in small quantities. Starch forms an unstable blue starch–iodo complex when adulterated milk is heated, cooled and added with one percent iodine solution.

Glucose

Often glucose is added extraneously to milk to increase the total solid content. It is odourless, colourless and not as sweet as sugar. It is not a very common adulterant. It is detected by using the modified Barfoed's reagent either directly from milk or in the protein free filtrate from milk.

Urea and Ammonium Sulphate

These chemicals are readily available with the farmers as fertilizers and are often used to boost the total solid of milk after dilution with water. Their detection is not very easy or rapid and is not normally done alongwith the other platform test. Simple chemical test using protein free filtrate are available for their detection.

Skim Milk Powder

Though skim milk powder is not strictly adulterant its presence can be detected in milk. After centrifuging the milk, the residue is dissolved in concentrated nitric acid, diluted and added with liquor ammonia. Skim milk powder being highly proteinaceous in nature gives an orange colour while unadulterated milk gives a yellow colour.

Detergents

The primary pre-requisite for the production of good quality milk is the cleanliness and sterility of the equipment used. Cleaning and sterilisation are complementary process used in random to clean the residue of milk and to reduce the bacterial numbers as much as possible.

Cleaning processes in the dairy industry normally comprise the following stages :

a. A pre-rinse with cold or tepid water to remove most of the loose soil and to wet the surface.

b. Removal of soil from the surface by solution, emulsification, saponification or mechanical action or by a combination of these methods.

c. Dispersion of the undissolved soiling matter.

d. Removal of the used detergent solution together with the suspended and dissolved soil.

e. Final rinsing to remove the last traces of detergent.

Different types of detergents are available but the type of detergent to be used and its concentration depends on the specific requirement and method of washing. Broadly, all detergents are either organic or inorganic in nature.

The type and concentration of the detergent used depends largely on the nature of the soil to be removed and the method of cleaning. The soil deposited on the surface of the equipment varies with the processing operation involved. For example, in the dairy industry the soil could be in the form of an air dried milk film or heat precipitated film or hardened film or in the form of milk stones and foreign matter.

The soils essentially consist of the milk constituents and could be rich in fat, protein or minerals in varying amounts depending on the processing operations involved and the equipment used (Table 11.2).

Table 11.2. Chemical composition of typical milk soils.

Component	Cold milk	Type of soil Hot milk Swept' surface	Tubular heater	Milk stone
Lactose	38.11	–	–	–
Fat	29.9	43.0	23.0	3.6-17.7
Protein	26.6	41.1	30.3	4.1-43.8
Ash	5.3	11.9	46.6	42.3-67.3

The strength and type of detergent used is determined by the solubility and nature of the soils. Sugars are water soluble and easily cleaned unless they undergo caramelisation on heating. Fats are water insoluble but can be cleaned with alkali detergents. Proteins are insoluble in water and require alkali or acid cleaning but are difficult to remove after they get denatured on heating. The proportion of these constitutents in the soil varies and, therefore, the type of detergent used should also be changed according to the specific need.

Method of Cleaning

In the Dairy Industry, three methods of cleaning are generally employed, viz.

i. Manual or hand cleaning.

ii. Semi-mechanical.

iii. Mechanical cleaning.

In manual cleaning, removal of soil is achieved by scrubbing the surface with a suitable detergent solution. Normally a detergent concentration of 0.25 to 1.0 percent at 40-50°C is used.

In the semi-mechanical cleaning process, bottle, cans, crates, etc. are placed in the water containing a system of detergents, revolving brushes water rinsing, etc.

Cleaning-in-place (CIP) or In-place cleaning is a completely mechanical method where all parts of the equipment can be thoroughly cleaned without dismantling. Since no human contact is involved more severe conditions like stronger detergent and higher temperature can be employed. High pressure jets or spray can be used to effect removal of soils.

The concentration/strength of the detergent and the temperature of cleaning, etc. are determined by the method adopted for cleaning and the specific need of the equipment or situation.

Classification of Detergents

Detergents are generally classified as alkalies, acids, sequestering agents and surface active agents. These are the main class of detergents generally used and comprise caustic soda, soda ash, trisodium phosphate and sodium meta silicate. Most of these detergenets have specific characteristics and are blended in various proportions according to the requirements.

Caustic soda is a strong alkali with excellent dissolving, saponifying and peptizing property. It is a poor deflocculating, emulsifying and wetting agent, it is highly corrosive to metals and is avoided in cases of hand cleaning.

Sodium carbonate or washing soda is the most common and inexpensive ingredient used in dairy detergents. It has fair, deflocculating and emulsifying action, and is a good saponifier. It maintains its pH through good buffering action and is employed extensively in solution used over long period like bottle washer, etc. It is not harsh on the skin or hand and can be safely handled.

When sodium carbonate is used with hard water, it forms precipitate of calcium carbonate which promotes formation of stone and soils. This drawback can be overcome by using soda ash in combination with sequestering agents which bind up the hardness in water.

Trisodium phosphate (Na_3PO_4) is readily soluble and has deflocculating and emulsifying power. It is a fair water softener and forms insoluble phosphates with calcium and magnesium. It is corrosive towards tin but this can be countered by the presence of sodium meta silicate. It is relatively inexpensive and is widely used in detergent formulations.

Sodium meta silicate is strongly alkaline but not as corrosive as caustic soda. It also protects metals from the corrosive action of other alkalies. It has excellent deflocculating and emulsifying properties and is a fairly good water softener. The silicate formed with calcium and magnesium in hard water are insoluble and flocculent and can be easily removed. The deflocculating property of metasilicate keeps the soil in suspension during washing and then facilitates proper cleaning.

Acids

Several organic acid (inorganic) mineral acids are used as detergents. These include tartaric, citric, gluconic, acetic, phosphoric and nitric acid at approximately 0.1 to 0.5 percent concentration. Acids are used mainly in high temperature heat exchange equipment. Acid dissolves deposits of mineral salts and denatured or heat coagulated proteins which are difficult to remove with regular alkaline detergents. They also prevent precipitation of water hardening salts like carbonates by maintaining a pH on the acidic side. For optimum efficiency a pH of 2.5 or lower for a 0.1 percent solution is desirable, and should not be corrosive on the metals of the equipment.

Sequestering/Chelating Agents

The main function of sequestering agent is to soften water by combining with calcium and magnesium to form soluble complexes instead of insoluble precipitate. Sequestering agents are usually inorganic linear poly-phosphate like sodium-tetra phosphate, sodium-hexametaphosphate, sodium tripolyphosphate, etc. Their efficiency in binding calcium or magnesium could be selective and varying (Table 11.3).

Table 11.3. Relative merits of common sequestering agents

Chemical	Formula	Sequestering activity towards Calcium	Magnesium	Remarks
Tetra sodium pyrophosphate	$Na_4P_2O_7$	Poor	Very Good	Cheap, stable at high temperature and alkalinity, slow to dissolve
Sodium tri polyphosphite	$Na_5P_3O_{10}$	Good	Good	Soluble in warm water, unstable at high temperature
Sodium tetra	$Na_6P_4O_{13}$	Good	Good	and inert to lower forms in hot alkaline solutions.
Sodium hexa meta phosphate	$(NaPO_3)_6$	Excellent	Poor	Capacity to seque-ster calcium is decreased in presence of Magnesium. expensive, unst-able under high temperature and alkaline conditions

Organic sequestering agents which form ring type structures are termed chelating agents. The sequestering efficiency of organic chelating agent in hard water is much more than that of polyphosphate. Most of the commonly used chelating agents are sodium salts of ethylene diamine tetra acetate. Unlike polyphosphates, these are stable in solution and their efficiency increases with pH.

Surface Active/Wetting Agents

These are capable of influencing the interfacial conditions and possess varying degrees of emulsifying, dispersing or wetting property. They improve wetting and penetration of soil particles and surfaces. They are chemically stable and are effective under acid as well as alkaline conditions, and in hard water also.

The water soluble portion of the surfactant attaches itself to water molecules and oil soluble portion attaches to the fat molecules thus providing a link between two immiscible compounds. Surfactants are mainly of three types, anionic/cationic and non-ionic.

The hydrophilic property of anionic surfactants is derived from the acid, sulphate or sulphonate groups while those of the cationic is provided by the amine groups and is regulated by the length of the carbon chain. Non-ionics surfactants depend upon the terminal hydroxyl group and hydrogen bonding to water molecules for their hydrophilic property.

Another group of surfactants are the amphoteric agents which have multifunctional groups which are cationic in acidic solution and anionic in alkaline solution.

Normally a balance between hydrophilic and lypophilic values called HLB, is maintained. The HLB represents the varying proportions of water soluble and alcohol soluble groups on a numerical basis. Surfactants are considered more effective in removing fatty soils than protein deposits.

Properties of Detergent

A detergent is expected to possess some or all of the following properties.

i. Wetting or establishment of intimate contact between the surface and cleaning agent.

ii. Deflocculation or holding in suspension, removed soils down to micro particles size and preventing them from clumping or coagulating.

iii. Emulsification – holding unsaponified fat and oil in solution.

iv. Hydrolysis – The break up of fats, fatty acids, glycerides and soaps in manageable globules and simpler compounds.

Apart from these properties good rinsability and water softening property is also essential.

The Indian Standard Institute has proposed that a good cleaning or washing agent should have the following properties.

i. Saponifying power, i.e. making soap out of the fatty acid of oils and fats.

ii. Quick and complete solubility.

iii. Non corrosiveness to metal surfaces.

iv. Stable during storage.

v. Mild on hands if used for hand washing.

vi. Germicidal action

No single detergent has all the desirable properties and hence a blend of several compounds is often required.

Apart from these factors the easy availability and cost of the cleaning compounds also has to be considered. Depending upon the specific requirements, different detergent combinations are blended or formulated for the various cleaning operations in the dairy industry (Table 11.4).

Table 11.4. Properties of detergents.

Property	Strong alkali	Mild alkali	Mild acid	Strong acid	Surfactants
Corrosion	++++	++++	++	++++	0
Sequestering	0	+	0	0	0
Wetting	+	++	+	0	++++
Emulsifying	+	++	0	0	++++
Dissolving	++++	+++	+++	++++	+
Saponifying	++++	+++	0	0	+
Peptizing	+++	+++	++	+++	0
Dispersion	++	+++	+	0	+++
Rinsing	++	+++	+	0	++++

Key : Degree of Acidity :		
	Extreme	++++
	High	+++
	Medium	++
	Low	+
	None	0

Sanitizers

Destruction of all the pathogens and most of the non-pathogenic micro-organisms is commonly referred to as sanitisation. This destruction of micro-organisms can be brought about either by physical means or by the use of chemicals. Some of the commonly used methods of sanitisation are discussed below :

Heat: Heat is the most effective and reliable method of sanitisation when both temperature and time are controlled. Heat can be applied either through steam or hot water or hot air.

Sanitisation by steam or scalding water is effectively done at 15 psi for 5 min or 0 psi for 15 min. Steam is readily available in the dairy plants and widely used for sanitising the equipments. However, it suffers from certain disadvantages that heat stresses may cause cracks in soldered seams and welds, and cause rapid and extensive damage to rubber parts and hoses. It consumes a lot of energy since steam is dissipated into the atmosphere. It is also quite noisy and increases humidity. Unlike with chemical sanitizers and detergents steam does not require extensive rinsing before use of the equipment.

Hot Water: In small scale establishment where steam generation is not possible, hot water is often used for sanitisation. The equipment is either dipped in scalding water or the water is poured over the equipment. The temperature of the water should be as high as possible and should not be less than 85°C for sanitisation to be effective.

Hot Air: Hot air is an effective sanitising agent but its use is limited due to high cost and difficulty in application. Hot air cabinets are sometimes used. As a sanitizer, hot air is very effective since it has a high penetrating power and can destroy bacterial spores also. The effect of heat on micro-organism is principally on the heat labile proteins within the cells.

Irradiation: Ultraviolet light is very effective in destroying micro-organisms and can be used for sanitisation. But it is not easy to use and hence finds limited application. However, packaging material, etc. can be exposed to U.V. light before use.

Chemical Sanitisation: A large number of chemicals exhibit bactericidal action and are widely used as sanitizers. Prominent among these are the halogen bearing compounds like chlorine, iodine or bromine compounds; quarternary ammonium compounds and others.

Chlorine compounds: Free or available chlorine is the active agent responsible for the bactericidal action of chlorine compounds. The free chlorine forms hypochlorous acid HOCl in solution which destroys micro-organism by protein denaturation and enzyme inactivation.

For smooth metal surfaces a 1 minute contact with a solution containing 200 ppm of available chlorine or a 7 minute contact with a 25 ppm chlorine solution is sufficient for sanitisation. Sufficient contact time should be allowed and proper concentration of solution should be used for effective sanitisation. Increasing the contact time or the concentration beyond reasonable limits serves no practical purpose.

The properties of commonly available chlorine sanitizers are summarised in the Table 11.5.

Chlorine sanitizers offer a number of advantages which make them popular and useful. They are effective on a wide variety of bacteria, bacterial spores and phages. Their performance is not affected by hard water and they are relatively inexpensive and easily available.

On the obverse side, chlorine compounds have a number of disadvantages also. They are corrosive and irritating to the skin. The formation of hypochlorous acid (HOCl) which is the active agent of all chlorine sanitizers is pH dependant. With an increase in pH the HOCl formation decreases and consequently the effectiveness of the sanitizer also decreases. Chlorine sanitizers lose their strength on storage due to

Table 11.5. Chlorine sanitizers.

Compounds	Structure	Solubility	Percentage available chlorine	Remarks
Inorganic				
Gaseous chlorine	Cl:Cl	0.716g/100g	100	Difficult to handle, regulate or use. Limited application
Calcium hypochlorite	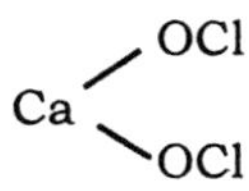	6.9g	100	Forms calcium precipitate easily available as bleaching powder, widely used.
Sodium hypochlorite	NaOCl	15g	100	Liquid, easy to use directly, leaves no sediment, widely used, likely to loose strength on extended storage.
Organic				
Chloramine T	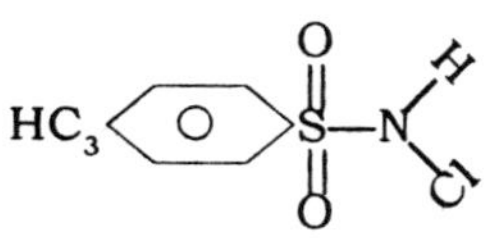	1.2	25	Organic sanitizers, lower solubility with varying percentage of available chlorine
Dichloro cyanucic acid	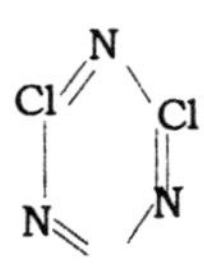	2.6	70	
Trichloro cyanuric acid	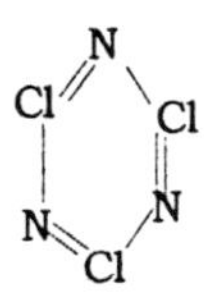	1.2	90	Slower rate of chlorine release through hydrolysis
Dichloro dimethyl hydration	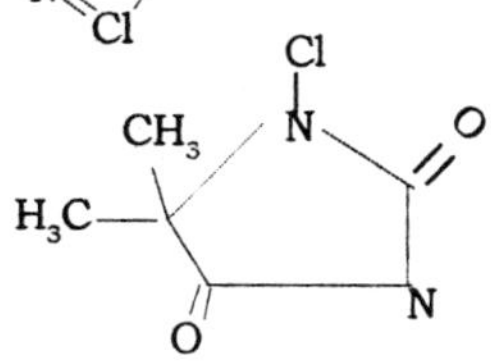		66	Not very freely available nor as widely utilised

rapid dissipation of chlorine. The use of chlorine sanitizers may be deterimental to certain equipment especially rubber parts.

Mechanism of Chlorine Sanitisation

Free chlorine forms hypochlorous acid HOCl with water

$Cl_2 + H_2O \longrightarrow HOCl + H^+ + Cl^-$

$HOCl \longrightarrow H^+ + OCl^-$ (pH dependent)

HOCl is the active agent, is nearly 80 times more effective than OCl. At pH 5.0 it is entirely present as HOCl but at pH 10.0 it is almost entirely OCl^-.

Organic compounds hydrolyse to produce one or more molecules of HOCl. The rate of hydrolysis is pH dependent being slower at high pH values. The release of Cl_2 from organic compounds is generally slower than from inorganic sanitizers.

Iodine Compounds

Iodine is an excellent bactericide but its corrosiveness, toxicity and low solubility makes it unsuitable for use in the food industry. It is less effective against bacterial spores and phages than chlorine and may also cause off-flavours apart from discolouring the equipment.

These difficulties have been overcome in the recent products called iodophores in which the iodine is loosely combined with a suitable non-ionic wetting agent which acts as carrier. Commonly, phosphoric acid is also added to increase the germicidal activity probably by the decrease in pH, to increase stability and to prevent formation of milk stone. The carrier or wetting agent acts as a solubilising medium for iodine. In this combination the germicidal activity of iodine is enhanced and the decreased vapour pressure increases the stability. Iodophores reduce the characteristic iodine odour, prevents permanent staining and corrosive action. It also retains its germicidal activity in hard water. As in the case of chlorine, the sanitising action of iodine compounds is related to the free iodine which forms HIO_3 which further produces IO^-_3 and H^+ ions.

Quarternary Ammonium Compounds (QAC)

These are cationic compounds which are finding increasing applications. Out of the hundreds of QAC available, only a few have disinfectant properties. The most common QAC used is cetyl trimethyl ammonium bromide (CTAB).

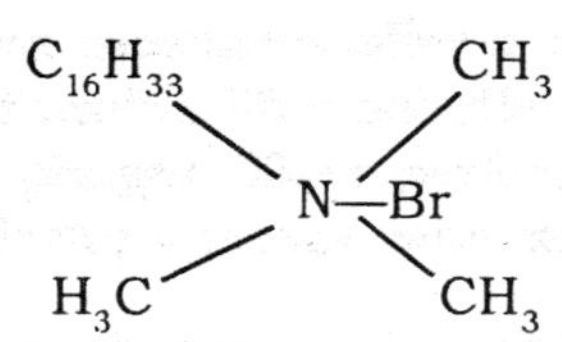

The QAC are colourless, odourless, tasteless and non-corrosive, making them ideal for use as disinfectants. They are highly surface active and reduce surface and interfacial tension considerably.

Action of QAC is different from that of the halogen compounds. The cationic part of the molecule is hydrophobic and tends to form a layer on the surface of the bacterial cell wall and ultimately leads to its destruction. The cationic portion concentrates the disinfectant at the required points. The action of QAC is strong and lasts for a long time after treatment. Since the QAC are cationic, it is natural that they are inhibited by certain anionic detergents. Therefore, the detergent used should be compatible with the QAC action. Soda ash is usually compatible while the common wetting agent inhibit QAC action. QAC are more effective against gram positive than gram negative bacteria especially in low concentration.

Water for Dairy Industry

A large volume of water is used in the dairy industry for such diverse purposes as chilling (refrigeration), heating (steam generation), reconstitution of dried milk, cleaning equipment, premises, etc. for sanitising and rinsing and finally for effluent treatment and disposal. In fact, easy availability of sufficient water is one of the prime considerations while choosing a dairy site.

The water used in a dairy should be of a very good quality both bacteriologically and chemically. It should be free from organic and suspended matter and should not contain any toxic or corrosive substances or harmful microflora. To ensure that these requirements are met, water supply to a dairy plant is subjected to various treatments.

Coagulation

Timely suspended material in the form of a colloidal suspension does not settle on standing and has to precipitate with a flocculating agent like alum. Aluminium sulphate alone or along with lime sodium aluminate, alumino, ferric or iron salts are commonly used. Alum or aluminium sulphate reacts with calcium bicarbonate.

$$Al_2(SO_4)_3 + 3Ca(HCO_3)_2 \longrightarrow 2\,Al(OH)_3 + 3CaSO_4 + 6CO_2$$

Along with soda ash or lime the reactions are :

$$Al_2(SO_4)_3 + 3Ca(OH)_2 \longrightarrow 2\,Al(OH)_3 + 3CaSO_4$$

$$Al_2(SO_4)_3 + 3Na_2CO_3 + 3H_2O \longrightarrow 2\,Al(OH)_3 + 3Na_2SO_4 + 3CO_2$$

The flocculent precipitate of aluminium hydroxide entrains or adsorbs the suspended matter of water. Silica sols when used in combination with other coagulants increases the rate of coagulation.

The coagulated impurities in the water can be removed in a settling tank as sludge which is removed from time to time. The water is then allowed to pass through sand filter during which most of the suspended impurities are removed.

Chlorination

The water supply is normally disinfected by the addition of regulated quantities of chlorine. The efficiency of disinfection depends on a number of factors like concentration of chlorine, exposure time, temperature, pH of the water, amount of organic matter and the type of micro-organism present. After treatment a residual concentration of 0.2-0.4 ppm is normal and sufficient.

The chlorine demand of water is the difference between the amount of chlorine applied and the amount of chlorine (free or combined as chloramines or total available) remaining at the end of the contact period. It is measured easily by treating a series of samples of the water with varying known dosages of chlorine or hypochlorite. After the desired contact period, determination of residual chlorine in the sample will demonstrate which dosage satisfied the requirement of chlorine demand.

The residual chlorine is determined by the orthotoludine (O-toludine) method. This aromatic compound is oxidised by chlorine, chloramines in an acidic medium to a yellow coloured holoquinone. The intensity of the yellow colour is measured in a Lovibond Comparator.

Hardness of Water

The capability of soap to form stable, copious foam or lather with water varies considerably from place to place depending on the type of water. This characteristic of water is called hardness. Hardness is caused by divalent metallic cations which react with soap to form precipitate and with certain anions to form scales. The cations calcium and magnesium which are responsible for the hardness must first be precipitated by sodium oleate in the soap before a lather can be obtained. This requires more soap and also leaves a ring of deposit or precipitate in the container. Most of the hardness encountered normally is due to the prescence calcium and magnesium, and other cations like strontium, ferrous, and manganese ions.

Hard water reduced the cleaning efficiency and economy of detergents. It also forms scales on boilers which reduces their efficiency considerably, and in some cases water may be corrosive also. Hence the hardness of water must be determined and steps taken to soften it to an acceptable degree before use.

The hardness of water does not pose any health hazard and hard water has not been implicated in any problem related to health.

Water is often classified according to its degree of hardness. For uniformity of expression hardness of water is always represented as milligram per litre (mg/l) or parts per million (ppm) of calcium carbonate, whatever be the cation or anion responsible for hardness. Generally accepted classification of water according to its hardness is as follows :

Soft	:	Hardness less than 50 ppm
Moderately hard	:	50-150 ppm
Hard	:	150-300 ppm
Very hard	:	Hardness more than 300 ppm

The salts causing the hardness determine the steps adopted for softening or removing the hardness. Based on this consideration water hardness is categorised as either temporary or permanent hardness.

Temporary Hardness

Temporary hardness in water arises mainly due to the presence of bicarbonate of calcium and magnesium. The alkalinity of natural water is due to hydroxides, carbonate or bicarbonates. When the pH of water is less than 8.3, the alkalinity consists mainly of bicarbonates. If the total alkalinity is more than or equal to the amount of calcium and magnesium present than the hardness is temporary, because the bicarbonates decompose to CO_2 on boiling.

$$Ca(HCO_3)_2 \longrightarrow CaCO_3 + H_2O + CO_2$$

$$Mg(HCO_3)_2 \longrightarrow Mg(OH)_2 + 2CO_2$$

The solubility of $CaCO_3$ is 14 mg/litre in cold and 19 mg/litre in hot water. Similarly $Mg(OH)_2$ is soluble upto 9 mg/litre in cold or 40 mg/litre in hot water. Therefore, when the water is boiled and cooled the $CaCO_3$ and $Mg(OH)_2$ are precipitated and the temporary hardness is removed. The carbondioxide released is capable of forming carbonic acid with water.

$$CO_2 + H_2O \longrightarrow H_2CO_3$$

Although carbonic acid is a very weak acid, and is present in a very dilute form, it is capable of causing corrosion due to continuous slow action.

Permanent Hardness

Hardness which cannot be removed by boiling is called permanent hardness. It is caused by the presence of chlorides or sulphates and other radicals of calcium and magnesium. Here the total alkalinity is less than the calcium and magnesium in terms of equivalent.

When hard water is used in a boiler the salts causing hardness mainly the sulphates get concentrated and eventually crystallise out to form a thick hard scale on the boiler surface. This is called scale formation and the boilers have to be descaled periodically.

Estimation of Hardness

It is essential to know the amount of calcium and magnesium in water and take measures to remove it. The hardness due to divalent ions like Ca^{++}, Mg^{++}, etc. is determined by making use of a strong chelating agent like EDTA, which forms stable complexes. In an alkaline buffer medium these cations form weak complexes with the indicator Erochrome Black T giving a wine red colour. On titration EDTA forms much stronger complexes with the cations and at the end point reverts the indicator to the original blue colour.

Hardness is always expressed in terms of parts per million of $CaCO_3$ irrespective of the salts causing hardness.

Treatment of Hard Water

The process of removing hardness from water is called softening. This step is essential to increase the efficiency and economy of dairy processer. Scale formation in boiler is a persistent serious problem leading to loss of efficiency in heat exchange and requires costly descaling operations from time to time. Similarly, the foaming capability of detergents is very much reduced, due to the formation of flocculation. More soap or detergent is required to give the same cleaning efficiency. These problems can be removed by softening the water before use.

Removing Temporary Hardness

Lime Soda Process: When hydrated lime and sodium carbonate are added in sufficient amounts, the calcium salts are precipitated as calcium carbonate and magnesium salts as magnesium hydroxide. This process is more efficient in the hot than the cold. At high temperature 70°C, the hardness can be reduced to very low levels. The amount of reagents to be added is determined by the hardness and the volume of water to be treated. The reagents are suspended in water as milk and a period of 2-3 hrs is allowed for the precipitate to settle before filtration. Due to this treatment the calcium and magnesium in water are replaced by equivalent quantities of sodium. Hence, the amount of sodium present in the water after treatment will depend on its initial hardness. The process of precipitation can be facilitated or speeded up by the addition of performed slurry or seed.

Treatment with Ion Exchange Resins: Zeolites are ion-exchange resins which are capable of inter changing the cations or anions with the salts in the water. Commercially used zeolites comprise of natural glauconite, treated clays, or synthetic resins prepared from sodium aluminate or sodium silicate. Base exchange zeolites substitute the calcium and magnesium from water with sodium. The zeolite is packed in a column or bed through which water is allowed to pass at a controlled rate. The hardness of the water is almost completely removed and the treated water can be used for all purposes. In the course of time all the sodium in the resin is completely depleted and substituted by calcium and magnesium from the water. At this stage the zeolite in the column can be regenerated for further use by treatment with sodium chloride or brine solution. The calcium and magnesium in the zeolite are again replaced with sodium and the column becomes ready for use again.

The ion exchange process offer several advantages. The plant occupies comparatively less space and the softened water with not more than 10 ppm hardness can be obtained. Since this involves an cation exchange process there is no sedimentation and hence no sludge removal involved. An optimum size column can be installed to deliver sufficient softened water at a steady flow rate. The zeolites can be regenerated and used repeatedly making the process economical.

The incoming water should be free from colloidal and suspended matter as these adversely affect the efficiency of the zeolite.

The zeolite process can be represented as follows :

$$Na_2 + Z + \left\{\begin{matrix} Ca^{++} \\ Mg^{++} \end{matrix}\right. \longrightarrow \left.\begin{matrix} Ca^{++} \\ Mg^{++} \end{matrix}\right\} Z + 2\,Na$$

or specifically

$$\left.\begin{matrix} Ca^{++} \\ Mg^{++} \end{matrix}\right\} + \left\{\begin{matrix} (HCO_3)_2 \\ SO_4 \\ Cl_2 \end{matrix}\right. \quad Na_2Z \longrightarrow \left.\begin{matrix} Ca \\ Mg \end{matrix}\right\} Z + \left\{\begin{matrix} 2\,NaHCO_3 \\ Na_2SO_4 \\ 2Nacl \end{matrix}\right.$$

Importance of Quality Control (Q.C.)

The concept of quality control varies when viewed at from different perspectives. The consumer views quality control as a watchdog which ensures that he gets a good product for the money that he is paying. The processing/production people look upon quality control as a necessary evil which they have to live with. The management views quality control as watch dog that keeps the consumers and the legal standards satisified while at the same time ensuring that the dairy does not incur losses by exceeding the prescribed standards for the different products.

Quality control thus plays a very important role, interacting with different agencies, department and the public.

The function of quality control in a dairy industry are manifold.

Quality Control of Incoming Raw Material

All the raw material entering the dairy has to undergo stringent testing and be approved before use. It has to be very clearly understood and emphasised that good quality finished product cannot be produced from poor quality raw material. All ingredients used in milk/product manufacture have to be vigorously inspected before purchase. These include milk, water, skim milk powder, butter/butter oil, sugar, salt, etc.

Milk supplies procured by the dairy have to be first tested for their acceptibility by the organoleptic tests and the platform tests. These tests include titratable acidity, the methylene blue reduction test, clot on boiling test, detection of additives like adulterants, preservatives or neutralisers and so on. After the milk is found acceptable, a small representative sample is taken for chemical analysis.

Milk found unsuitable should be summarily rejected and not allowed to enter the dairy. Even if a small quantity of poor quality milk is accepted and blended with the good quality milk, it will lead to a lot of subsequent problems in processing, storage and marketing.

Similarly, all other consumable ingredients have to be tested and approved before procurement.

Quality Control During Processing/Production

In this aspect, quality control plays an active supporting role to the processing or manufacture of dairy products. Once the raw material has been accepted and stored, quality control check gives regular report on its composition and quality. This helps the production manager in deciding how it should be used for product mix and diversification. The quality control reports are essential for standardising mixes and controlling the various parameters during production. An efficient quality control laboratory should give quick and accurate reports so that production can be monitored. This will help in avoiding rejection of product during the final quality control checking.

Quality control also should keep an eye on the cleaniness of the plant, equipment, premises, etc. This will help to a large extent in controlling post pasteurisation contamination and in maintaining a good microbiological quality of the finished product.

An efficient quality control department can advise the management on the different avenues which lead to losses and can help check losses.

Quality Control of Finished Products

Before milk or any product is marketed it has to be exhaustively analysed for its worthiness and for meeting the minimum legal requirements. Any product suspected to be substandard either by way of its composition, keeping quality, or packaging, etc. should be strictly rejected and not allowed to be put up for sale. This will prevent economic losses, avoid costly legal wrangles and also ensure that the customer is satisfied and keep up the good will of the dairy plant. This aspect of quality control is perhaps the most important one.

Apart from the above essential duties the quality control also has a few other important functions to perform. Standardisation of testing equipment has to be done by quality control to ensure that the test results reported are accurate. Milk production centres, cooperatives, and producers have their own milk testing equipment, and evaluate their milk quality before supplying the milk. Quality control has to help them in standardising their equipment and also train the personnel to ensure that there is co-ordination between their results and those of the dairy.

Quality control plays a major role in pollution control. The effluent water leaving the dairy premises has to be tested to ensure that it meets with the minimum legal requirements. Otherwise this could lead to trouble with the pollution control board and other agencies. Periodic maintenance of the effluent treatment plant helps in checking these problems. Liaision with the P.F.A. authorities and other standard agencies like the Agmark and I.S.I., is done by the quality control department. The P.F.A. nominations, I.S.I. and Agmark forms, undertakings, etc. have to be signed and maintained by the quality control department.

Thus, quality control in a dairy plant has to perform multifarious activities and keep track of different aspects while maintaining a strict vigil on the quality of every packet of every product originating from the Dairy.

Setting-up a Quality Control Laboratory

The quality control laboratory in a dairy plant should essentially be set up in a clean well-lighted and ventilated place. It should preferably be located in a corner place away from the hustle and bustle of the plant activities. The laboratory personnel will then be able to work without distraction and the laboratory will not become a thoroughfare.

In most modern dairies the quality control laboratory is located on the first floor overlooking the processing hall below. This has the advantage that quality control personnel can keep an eye on the processing activities. Another advantage is that the milk storage tanks are normally situated on the first floor and it will be convenient for drawing samples regularly during standardization and storage. At the beginning and end of each shift all the storage tanks, etc. should be sampled and a report maintaining the temperature, acidity, chemical quality processing history, etc. should be submitted to the incoming shift.

While setting up a quality control laboratory sufficient care and thought should go into organiing the equipment and floor space. As far as possible the microbiology section should be separated from the chemical laboratory. The microbiology room should be at the back so that only qualified and essential people can enter the room. Providing an air curtain will help in controlling contamination. The microbiology room should be self contained with all necessary equipment, glassware, U.V. chamber, counters, incubators, autocalves, etc.

The chemical laboratory room should be set up and organized properly. Routine tests like fat, acidity, S.N.F., etc. are done on hundreds of sample daily and these tests should be set up to handle large number of samples conveniently. Proper use should be made of the available wash basin, sink, etc. All samples should be properly labelled and after the sequential analysis the report should be recorded and dispatched as soon as possible. Normally, a few samples from each batch are retained to monitor its keeping quality and also for checking in case there are any market failure or complaints.

All solutions used for analysis in the quality control laboratory should be prepared carefully and standardised properly before use. Similarly, all equipment and glassware should be tested to verify that they confirm to the standards. One set of standardised glassware like milk pipette, butyrometer, lactometer, thermometer, etc. should be kept on the reference against which all other glassware can be standardised.

A proper record maintenance system should be established. A copy of all reports should be maintained and properly filed. An efficient filing system is very essential for a quality control laboratory since it has to keep track of so many diverse/different aspects.

The quality control laboratory is the nerve centre of the dairy and due importance should be given for setting it up properly.

Sampling of Milk for Quality Control

Milk received at a dairy or chilling plant is always tested for its quality, cleanliness and acceptability. Quality control personnel visit the raw milk reception dock and examine the milk for its acceptability before it is allowed to be unloaded and processed. For this the quality control personnel have to be adequately experienced and trained. By mere organoleptic examination they have to decide if the milk is acceptable or not. Doubtful samples are tested in the laboratory before deciding. Samples which are doubtful should be summarily rejected.

For the testing to be accurate and efficient, the sampling should be proper. The sample drawn should be representative of the bulk milk in all respects. A sufficient quantity of the sample has to be drawn for the various chemical and microbiological test. The quality and comparison of the sample should be identical with the rest of the milk, and for this an efficient system of sampling has to be employed. Different methods of sampling are adopted to draw samples from different types of containers.

Sampling of milk from a single container is easily achieved after proper mixing or if possible shaking or inverting the container. Milk in small containers can be poured from one vessel to another for mixing and a small portion can be drawn as a representative sample.

Often, a single composite sample has to be drawn from different containers. Wherever possible milk from different cans can be poured into a single large container like the dump tank or weighing bowl and mixed properly with a plunger. From this pooled milk a single sample can be drawn.

The level of milk in cans of the same height and diameter can be equalised by pouring from one to another. The same quantity of sample from each of these cans can be drawn after mixing, and pooled to get a single representative sample. This will ensure that a proper sample is obtained.

If different cans of the same diameter contain different levels of milk, then the tube method of sampling can be employed. A sample tube of uniform bore which is open at one end and having provision for closing the other end is used. After proper mixing of milk the tube is introduced in a vertical position. A column of milk equal in height of the height of milk in the can enters the tube. The top end of the tube is closed and withdrawn to deliver a proportionate quantity of milk. Similar samples are drawn from each can and pooled to obtain a single representative sample. This metod ensures that even if the different cans contain different quantity/quality of milk, a proportionate sample is obtained which is representative of the bulk or pooled milk. Milk is often received at the dairy in road tankers. Also samples have to be drawn from storage tank at periodic intervals. If the tank is equipped with a mechanical agitator, the milk should be agitated for not less than 15 minutes before drawing a sample. In road tankers, the man-hole at the top is opened and after reading the temperature, the milk should be agitated thoroughly with a plunger before sampling.

After drawing the sample it should be stored in clean sample bottles. The bottle should have a wide mouth to facilitate proper cleaning. It should have stopping side and well fitting caps. There should be no head space left after the sample is drawn. Lastly, the sample bottle should be properly labelled for its contents.

REVIEW QUESTIONS

1. What do you understand by the term P.F.A. and Agmark?
2. Name the common adulterant, neutraliser and preservatives used in milk and milk product.
3. Write the structure of different chlorine sanitizers.
4. Name the main classes of chemical sanitizers.
5. What is temporary and permanent hardness due to?
6. Why is milk stone a problem on dairy equipments.
7. Why is the water supply important in dairy farm sanitation?
8. What is quality control?

REFERENCES

Atherton, H.V. and Newlander, J.A. 1977. Chemistry and Testing of Dairy Products (Fourth Edition). The AVI Pub. Co. Inc. USA, reprinted by CBS Publishers & Distributors, Delhi.

Dairy Hand Book (Processing), 1980. National Dairy Research Institute, Karnal.

Dairy Hand Book (Processing), 1986. National Dairy Research Institute, Karnal.

Sud, S.K. 1984. Role of standardisation in microbiological quality of milk and milk product. Lecture delivered at the National Seminar on Safety of Milk and Milk Products, UNDP Centre of Excellence in Dairy Processing, NDRI, Karnal, March 9.

12

NUTRITIONAL IMPORTANCE OF MILK

ALL living beings require food for growth, supply of energy, maintenance and recovery from diseases, and for reproduction. Apart from fulfilling these basic functions foods also serve to titillate the sense of taste and appetite which is developed to a remarkable degree in the case of human beings. These requirements are met from varied sources all of which contribute to the diet depending on their nutritive value. In this context milk plays an indispensable role in the nutrition of human beings. From birth to weaning, milk is the only food consumed by the infant and has to provide all the nutrients in an assimilable form. After weaning, throughout the growth phase and into the adult stage milk continues to hold its place in the human diet either as fluid milk or in the form of a variety of products. During sickness or convalescence when most other forms of food are unacceptable, dairy products are depended upon to provide the required nutriton. The reasons for this exalted position of milk and milk products can be analysed in terms of the role played by each of the major constitutents of milk.

Milk Proteins

The proteins supply the amino acids, the building blocks from which tissue proteins are synthesised. This assists in the growth of children and makes possible the replacement of protein, used in metabolism or lost by injury. Out of the twenty amino acids required for synthesis of proteins, ten can be synthesised in the human body from available materials. The remaining amino acids cannot be synthesised and have to be supplied through the diet. These are known as Essential Amino Acids. Milk is a rich and ready source for the essential amino acids as shown in Table 12.1.

Milk protein compares very well with the standard reference protein proposed by FAO/WHO and even surpasses it to a great extent. Milk protein thus provides the human organism with amino acids in the proper concentrations and proportion required for optimum utilisation. Milk protein has an excellent supplementary value for blending with other foods to increase their nutritive value. Milk protein has an excellent buffering effect and consequently is a very good remedy for excessive stomach acidity. Dairy products are, therefore, used in the dietetic therapy of patients suffering from gastritis and stomach ulcers. On the adverse side milk proteins are also known to produce allergic responses in sensitive individuals. This reaction is usually species specific and it is best to avoid the milk of such species from the diet.

Table 12.1. Percentage of essential amino acids in the ideal reference protein and in milk protein.

Essential amino acids	In reference protein (FAO/WHO)	Grams in 100 grams of protein in milk protein
Isoleucine	4.0	6.3
Leucine	7.0	10.0
Lysine	5.5	8.1
Methionine + cystine	3.5	3.5
Phenylalanine + tyrosine	6.0	10.3
Threonine	4.0	4.9
Tryptophane	1.0	1.4
Valine	5.0	6.9
Total	36.0	51.4

Milk is probably the only source of animal protein for vegetarians who form a significant portion of the Indian population. On the whole, the role played by milk and milk products cannot be adequately substituted by any other source of protein. The biological value of milk protein is 90 (cow's milk). Higher biological value of milk protein is obtained when combined with potato and rice or maize protein.

Estimation of protein value is important to dieticians as also to those serving the profession of dairying. A dairyman must know the methods of estimation of protein value of milk, egg and soya protein.

The term "biological value" was first used in 1909 to indicate the nutritive value of proteins. The amount of nitrogen in protein absorbed by human body was the basis of calculation. The only source of nitrogen our body can utilise is from proteins. There is always a balance in our body between the anabolism (biosynthesis) and catabolism (breakdown of protein). Biological value (B.V.) may be defined as the quantity of nitrogen or protein in grams, absorbed from 100 gm of nitrogen or protein, that may be considered as having participated in tissue construction or renewal of protein or the percentage of nitrogen absorbed that is made use of by the body.

$$\text{B.V.} = \frac{\text{the utilised nitrogen}}{\text{the nitrogen absorbed from the food}} \times 100$$

Based on this method, the B.V. of hen's egg protein is 96, cow milk is 90, goat meat 76 and raw soya 57. When milk protein is supplemented with cereals the B.V. of milk protein increases. This is known as "The Supplementary Value" of milk protein.

Another method for evaluating the protein value is by calculating the net protein utilisation (N.P.U.).

The food protein ingested cannot be digested completely by the body. Assuming that 96 per cent of the cow milk protein is digested, N.P.U. can be calculated by the following formula

$$\text{B.V.} \times \text{Digestibility} = \text{N.P.U.}$$

$$90 \times 0.96 = 86.4 \text{ or } 86$$

The respective N.P.U. of cow milk protein, goat milk and raw soya are 86, 76 and 48.

Another method for estimation of protein value is the protein efficiency ratio (P.E.R.). This ratio gives the increase in weight of the body to grams of protein eaten.

$$\text{P.E.R.} = \frac{\text{increase in weight (in grams)}}{\text{grams of protein eaten}}$$

However, increase in body weight is not solely due to protein but water, lipid, and other sources also contribute. If increase in weight is calculated on the basis of newly synthesised tissue which has protein only and 80 per cent water, then,

$$\text{P.E.R.} = \frac{80 + 20}{20} = 5.0$$

However, practically this high value is never obtained. The P.E.R. of cow's milk is 3.2 and that of hen's egg protein is 3.8.

Milk Lipids

Milk fat is a very rich source of energy, providing up to 9 kilocalories per gram. Fat consumed in excess of the immediate requirement of the body gets deposited in the adipose tissue as body fat. This aspect is not always desirable and hence there is a trend towards consumption of low fat milk and dairy products. During periods of extreme starvation the stored body fat is broken down and used for meeting the energy needs of the body. The deposited fat thus acts as a storehouse of energy for contingencies.

Milk fat acts as a carrier for the fat soluble vitamins viz, A,D, E and K. This important function is underscored by the fact that fortification of these vitamins in dairy products invariably use milk fat as the medium for introducing the vitamins.

As in the case of essential amino acids, the body requires certain essential fatty acids which it cannot synthesise by itself. Requirements of linoleic and arachidonic acids are met through milk fat which contains these essential fatty acids in significant quantities.

Nearly 99 per cent of milk fat is in the form of triglycerides and this includes a good amount of esterified lower chain saturated fatty acids particularly butyric acid. These contribute significantly to the pleasant and acceptable flavour of milk fat. Phospholipids particularly lecithin, cephalin and sphingomyelin form upto 1 per cent of total milk fat and possess emulsifying properties. The presence of upto 0.40 per cent of sterols, mainly cholesterol is a discouraging factor in the consumption of excess milk fat. Cholesterol is implicated in promoting artherosclerosis and this awareness has led to people avoiding fats containing cholesterol.

Apart from directly contributing to nutrition, milk fat plays a secondary role which is no less important. The pleasant flavour imparted to foods containing milk fat greatly enhances their palatability and acceptance leading to increased consumption. Milk fat, therefore, is widely used in variety of foods and as a frying medium. Consequently it is the most highly priced constitutent of milk.

Carbohydrate

The predominant milk sugar is lactose which is present normally to the extent of about 5 per cent. Apart from providing a ready source of energy lactose adds to the taste of milk – the sweetness it imparts to milk greatly enhances its palatability. Carbohydrates are providers of heat and energy to the body. Complex carbohydrates ingested as food are broken down to their elementary monosaccharide structure.

Lactose takes a longer time than other sugars for ingestion in the stomach. This increased residence time of lactose is beneficial since it favours the growth of the lactobacillus species, in the stomach. Thus, a favourable microflora is maintained and the growth of undesirable putrifying organism is prevented. Lactose is essential for the absorption of calcium from the gastro intestinal tract. The sugar is believed to chelate calcium and delay its precipitation during transport. Lactose is broken down to its constitutent monosaccharides viz. glucose and galactose in the small intestine by the enzyme ß-galactosidase, commonly known as lactase. This enzyme could be destroyed due to illness, surgery etc. or in certain individuals there could be a congenital absence of this enzyme. Such persons are not able to utilise lactose in their diet. Intolerance to lactose need not necessarily mean intolerance to milk since other foods are ingested alongwith milk and the small amount of lactose consumed does not cause any adverse effects.

The intact lactose in the intestine is then utilized by purtrefactive bacteria for production of gas, alcohol and other undesirable fermentations. Such a condition is known as lactose intolerance and the affected individuals can not consume lactose in their diet.

Human milk is unique because it contains oligosaccharides including sialic acid or N-acetyl neuraminic acid. The intestinal flora of breast fed babies significantly differ by the presence of *Lactobacillus bifidus*. The 'bifidus' factor causing this difference is traced to the oligosaccharides of human milk. The growth of favourable microflora like lactobacillus, prevents the growth of harmful pathogens, etc. in the intestinal tract.

Minerals

The human body requires various inorganic minerals like calcium, phosphorus, iron, copper, manganese, sodium, potassium, chloride, iodine, cobalt, zinc, and so on for proper growth and functioning. Some of these are required in significant quantities while others are required only in trace quantities.

Milk and dairy products are a good source of most of these minerals, but particularly of calcium and phosphorus which are important in the formation of teeth and bones. For optimum utilisation by the body, calcium and phosphorus must be present in a ratio of 1.3:1. Milk contains these minerals in this ideal ratio and can meet almost the entire requirements of a growing child, and for most other age groups. Milk is said to be deficient in iron since it cannot adequately meet the requirements of these minerals.

Apart from the major minerals like calcium, magnesium, phosphorus, potassium, sodium, chloride and sulphur present in relatively high proportion milk also contains a plethora of trace elements present in ppm or μg quantities. The role played by these trace elements should not be underestimated, since nutritionally, they are essential in microquantities. Cobalt forms the centre of vitamin B_{12} molecules and is essential for its synthesis. Iron is essential for the formation of hemoglobin and hence its deficiency could cause anaemia. Iodine is a constitutent of thyroxine and plays an important physiological role. Iodine deficiency causes enlargement of the thyroid gland resulting in goiter. Iron is constitutent of heamoglobin and some other respiratory enzymes. Milk is a poor source of iron and has to be fortified by adding extraneous iron. Selenium is required in trace quantities but high concentrations are toxic. Milk is a good source for zinc, nickel, molybednum, etc. which are essential for certain specific enzyme activities.

By and large milk contains most of the minerals required by a human being but is considered a poor source of iron and manganese.

Vitamin

Milk contains most of the vitamins required by the body in various quantities. For some vitamins like Vitamin A, Riboflavin, Thiamin, etc. milk is considered a good source while for other vitamins like vitamin D, Niacin, vitamin C, etc. milk is a poor source. Deficiency of vitamins leads to various diseases and disabilities, like night-blindness (Vitamin A), Beri-beri (VItamin B complex), Scurvy (Vitamin C), Rickets (Vitamin D) and so on. Attempts are made to fortify the vitamin like vitamin D to enhance the nutritive value of milk. The fat soluble vitamin A, D, E and K can be introduced by blending with the milk fat. Some of the vitamins may be destroyed on heat processing or storage while others are more stable and resist deterioration. Therefore, the amount of vitamins available to the consumers could be significantly different from those estimated in raw milk.

Some data on composition and nutritive value of milk and milk products are given in Table 12.2.

Table 12.2. Quantities of certain nutritive components in milk and milk products (100 g).

Product	Calories	Water g	Protein g	Fat g	Ash g	Total carbo-hydrate g	Calcium mg	Phos-phorus mg	Iron mg
Whole cow milk	68	87.0	3.5	3.9	0.7	4.9	118	93	0.1
Skim cow milk	36	90.5	3.5	0.4	0.8	5.1	123	97	0.1
Evaporated milk	138	73.7	7.0	7.9	1.5	9.9	243	195	0.2
Condensed milk	320	27.0	8.1	8.4	1.7	54.8	273	228	0.2
Skim milk (dry)	362	3.5	35.6	1.0	7.9	52.0	1300	1030	0.6
Butter	716	15.5	0.6	31.0	2.5	0.4	20	16	0
Butter milk	36	90.5	3.5	0.1	0.8	5.1	118	93	0.1
Cheddar cheese	398	37	25.0	32.2	3.7	2.1	725	495	1.0
Cheddar cheese processed	370	40	23.2	29.9	4.9	2.0	673	787	0.9
Cottage cheese	95	76.5	19.5	0.5	1.5	2.0	96	189	0.3
Cream	204	72.5	2.9	20.0	0.6	4.0	97	77	0.1
Ice cream	207	62.1	4.0	12.5	0.8	20.6	123	9	0.1
Whey dried	344	6.2	12.5	1.2	7.7	72.4	679	576	-

Product	Sodium mg	Potassium mg	Vitamin A mg	Vitamin B_1 mg	Vitamin B_2 mg	Nico-tinic acid	Vitamin C mg	Niacin mg
Whole cow milk	50	140	160	0.04	0.17	0.1	1.0	0.1
Skim cow milk	52	150	Traces	0.04	0.18	0.1	1.0	0.1
Evaporated milk	100	270	400	0.05	0.36	0.2	1.0	0.2
Condensed milk	-	-	430	0.05	0.39	-	1.0	0.2
Skim milk (dry)	77	1130	40	0.35	1.96	1.1	7.0	1.1
Butter	980	23	2300	Traces	0.01	0.1	0	0.1
Butter milk	130	140	Traces	0.04	0.18	0.1	1.0	-
Cheddar cheese	700	92	1400	0.02	0.42	Traces	0	Traces
Cheddar cheese processed	1500	80	1300	0.02	0.41	Traces	0	-
Cottage cheese	290	72	20	0.02	0.31	0.1	0	-
Cream	-	-	830	0.03	0.14	0.1	1	-
Ice cream	100	90	520	0.04	0.19	0.1	1	0.1
Whey dried	-	-	50	0.49	2.5	0.8	-	-

Modified Milks to Suit Special Nutritional Purposes

Humanisation of Buffalo Milk

Often it so happens that infants even before weaning have to be fed other milks if mother's milk is insufficient. In such cases the alternative source of milk, e.g. from buffalo milk or cow may not be suitable or acceptable to the infant since it differs widely from human milk in the composition. To overcome this drawback, buffalo milk can be modified or humanised to make it more acceptable. This can be brought about by the following changes or modifications.

i) Removal of α_{s1} casein (by controlling tryptic proteolysis).

ii) Replacement of milk fat with vegetable fat

iii) Addition of lactose/polysaccharides.

iv) Enrichment of vitamins.

v) Removal of calcium by electrodialysis.

The product thus produced can be spray dried and used for infant feeding when required.

Soft Curd Milk

The curd formed in the stomach due to peptic action may be hard in the case of cow or buffalo milk as compared to mother's milk. The curd tension of milk can be decreased to bring it closer to that of human milk and thus facilitate better utilization. The curd tension of milk can be reduced by the base-exchange process. Milk acidified with 0.3 per cent citric acid is percolated through a zeolite bed (sodium aluminium silicate containing potassium). During this process upto 20 per cent of its calcium and phosphorus is replaced by sodium and potassium. By proper regulation it is possible to maintain the pH and the calcium:phosphorus ratio while appreciably reducing the curd tension.

Homogenisation, high heat treatment, ultrasonic sound and use of a pancreatic protease to partially digest the protein are some of the other methods that can be used to reduce the curd tension of milk.

Vitamin D Fortified Milk

Since milk is not a very good source of vitamin D, it was deemed fit to fortify milk to increase its nutritive value. Vitamin D regulates the absorption and metabolism of calcium and phosphorus in bone building. Consequently lack of vitamin D could lead to Rickets. Milk can be enriched with vitamin D by any of the following processes.

i) Addition of vitamin D concentrate to milk.

ii) Irradiation of milk with UV light to convert the precursor 7-dehydrocholesterol to vitamin D.

iii) Feeding irradiated yeast to animal.

Low Sodium Milk

People with heart and kidney trouble are advised to take a low sodium diet. Since milk contains approximately 500 mg sodium per litre, it is necessary to remove some of the sodium to make it acceptable by the low sodium diet. This can be done by ion exchange resins where sodium is replaced by potassium from the resin. The sodium concentration is reduced drastically while that of the potassium is increased considerably.

Lactose Hydrolysed Milk

Some individuals lack the enzyme ß-galactosidase or lactase as it is commonly known. This enzyme is essential for the breakdown of lactose, the principal milk sugar, into the component glucose and glactose moieties. Such people are not able to digest lactose and are called lactose intolerant.

Intolerance to lactose need not necessarily mean intolerance to milk since the concentration of lactose in milk may not be sufficient to cause clinical allergic reactions. Moreover, alongwith milk other foods are usually consumed and the effect of lactose may not be very drastic. But in the case of infants dependant on milk it could be a great problem. The problem of lactose intolerance can be overcome by hydrolysing the lactose enzymatically. Commercially, the lactase enzyme is immobilised on a column through which milk passes. The outcoming milk contains hydrolysed lactose and is acceptable by intolerant individuals. Alternatively the ß-galactosidase enzyme is available in small sachets which can be added to the milk before consumption. Hydrolysed lactose milk are available commercially to fulfill this specific need.

REVIEW QUESTIONS

1. Briefly mention the importance of proteins as food nutrient.

2. What do you understand by the term modified milk?

3. Give comparative values of milk protein (g/ 100 g) in milk and selected milk products.

4. What is the importance of carbohydrates and fat in the human body?

5. What are vitamins? Name the diseases which are caused by the deficiencies of different vitamins.

6. Give an account of the role of calcium and phosphorous in human nutrition.

REFERENCES

Atherton, H.V. and Newlander, J.A. 1977. Chemistry and Testing of Dairy Products (Fourth Edition). The AVI Pub. Co. Inc. USA, reprinted by CBS Publishers & Distributors, Delhi.

Dairy Hand Book (Processing), 1980. National Dairy Research Institute, Karnal,

Dairy Hand Book (Processing), 1986. National Dairy Research Institute, Karnal,

De, S. 1980. Outlines of Dairy Technology, Oxford University Press, Delhi.

Ganguli, N.C. 1974. Milk protein, ICAR, New Delhi. Chapter-10.

Renner, E. Nutritional Physiological Aspects of Milk Protein. Milk Proteins – Published by the Dutch Dairy Bureau. The Netherlands.

13

CHEESE AND ITS CHEMISTRY

Cheese

CHEESE is one of the oldest foods of mankind. In olden times nomads began carrying milk in the stomach of animals. This converted a large volume of milk into a small junk which was received by slaughtering the animals. The main reason of such transport was that nomads usually used to traverse long distances in search of food with change of weather and land climate. When they found it hard to travel much long distances with great bulk of milk, they accidentally discovered that feeding of milk to animals convert it to a small junk and it is not easily perishable as milk. Therefore, cheese was first accidentally produced by carrying of milk in the stomachs of animals where the milk clotting enzymes of the stomach converted the sour milk into a solid mass junket. With the advancement of time, cheese has been denied as a product made from the curd obtained from milk by coagulating the main protein casein with the help of rennet or similar enzymes in the presence of lactic acid produced by added or adventitious micro-organisms and from which part of the moisture has been removed by cutting, cooking and/or pressing, which has been shaped in a mould, then it is ripened by holding it for sometime at suitable temperatures and humidities.

In modern times over 400 varieties of cheese are available. These are named on the basis of their characteristic features. However, all the varieties are grouped into soft, semi-hard, hard and very hard type, based on the moisture content of cheese. In India varieties like cheddar, cottage, Gouda and Brick are popular. The reason for slow development of the cheese industry in India are :

i) Cheese is adopted from western food habits and not popular with the masses as yet.

ii) Indians in general prefer sweets and tend to dislike food with bitter tastes.

iii) Cost of cheese is very high when compared with other foods.

iv) Prejudice against calf rennet.

v) High cost of refrigeration, storage and transportation make cheese manufacturing and marketing very costly under tropical climatic conditions.

vi) Lack of technical know-how among industrialists.

Cheddar cheese is one of the most important variety of cheeses. Cheddar cheese originated in the town of Cheddar located in the country of Somerset in south-western England. It is probably the best-known cheese in the world; many variants are now recognised such as American, Australian, New Zealand, Canadian and Indian Cheddar. Starting from receiving the milk, cheddar cheese manufacturing involves almost twenty stages. In the final stage, the 'fresh cheese' is dipped in a liquid paraffin at 104–121°C to give on it a coating and allowed to remain in a cold room (ripening room) at 0-4°C for 3 to 6 months.

When the cheese remains in cold room the process is known as curing/ripening/souring/maturing. During this period the cheese undergoes variety of physico-chemical and biochemical changes. The term 'green cheese' is applied to hard pressed cheese in the early stages of ripening. The various changes that take place are as follows:

Flavour Development: The mildly acid taste and aroma of green cheese is converted to the characteristic flavours of ripened cheese. During ripening amino acids of paracasein are liberated as a water soluble extract which possess characteristic cheese flavour. In addition to common amino acids and peptides, compounds like α-aminobutyric acid and tyramine are also responsible for flavour. Diacetyl, a compound produced during citric acid fermentation, produces acid flavour. Acids like acetic, priopionic, butyric, caproic, caprylic and capric also contributes to cheese flavour. The action of lipolytic enzymes liberating fatty acids by hydrolysis of fat is important in ripening of cheese particularly with respect to flavour. In addition to fatty acids mentioned earlier which are identified as product of hydrolysis, keto acids and neutral carbonyl compounds also produce characteristic flavours. Starter bacteria which produces acids like lactic, acetic and propionic indirectly contributes towards flavour. Important bacterial proteinases, lipases and decarboxylases are active at the pH of hard rennet cheese and their end products like amino acids, amines and fatty acids are important as flavour compounds.

Body: There is a gradual change from rubbery body in green cheese to a mellow and waxy body in ripened cheese. This is due to loss of moisture during ripening which is controlled by the temperature of ripening room and its humidity.

Texture: Cured cheese tends to acquire a fairly compact texture.

An increase in acidity and decrease in pH takes place for the first few days of ripening. The pH is lowest in cheese on about the third or fourth day after pressing, and normally remains 5.10 to 5.05. It then decreases slowly and steadily during the curing period. The lactose present usually disappears in one to two weeks. Gas production is normal in certain types of cheese. During the first few months organism of starter culture like *S. lactis* predominates, later *Lactobacilli* and other form also dominates.

Chemical Changes

Protein: The insoluble nitrogen of protein is converted to soluble nitrogen. Cheese contains about 4 per cent of total proteins as soluble protein in the initial stages which increases to about one-third of total in well-ripened hard cheese. The insoluble proteins are converted to products like proteose, peptone, polypeptides, amino acids and ammonia. This proteolysis is induced chiefly by enzymes derived from micro-organisms added as starter, micro-organism present in milk, enzymes in milk itself and rennet. The rate of proteolysis increases with the increase in temperature of curing, rennet quality, moisture content, less salt, large size of cheese and moderate amount of acid. By coating the cheese with a layer of paraffin wax, moisture loss is controlled. Higher the moisture content of the green cheese more are the changes in protein content. Ripening is measured in terms of percentage of water soluble nitrogen. Increase in amino acids, amines and volatile acid correlates generally with the gradual appearance of typical cheese flavour.

Lactose: Lactose is converted into lactic acid which is further converted to propionic acid and acetic acid. These acids prevent other micro-organisms from growing. There are six major types of fermentation reaction in milk like lactic acid, propionic acid, citric acid, alcohol, butyric acid and coliform-gassy fermentation.

Lactose fermenting bacteria, yeast molds contain an enzyme lactose (ß-D-galactosidase) which hydrolyses lactose to glucose and galactose. This hydrolysis can also take place by action of mineral acids such as hydrochloric acid, lactase enzyme present in milk itself, and ion exchange resins i.e. sulfonated polystyrene resins.

The galactose is converted to lactic acid as follows :

i) D-galactose + ATP ——> Glucose-1-phosphate + ADP

The following steps are involved in the above reaction.

$$\text{D-galactose} + \text{ATP} \xrightarrow{\text{Galactokinase}} \text{Galactose-1-}PO_4 + \text{ADP}$$

$$\text{Galactose-1-}PO_4 + \text{UDP-glucose} \underset{\text{Uridyltransferase}}{\overset{\text{Phosphogalactose}}{\rightleftharpoons}} \text{UDP–Galactose} + \text{Glucose–1–}PO_4$$

$$\text{UDP-galactose} \xrightarrow{\text{UDP-galactose-4-epimerase}} \text{UDP-glucose}$$

Sum-

$$\text{D-galactose} + \text{ATP} \longrightarrow \text{Glucose-1-}PO_4 + \text{ADP}$$

ii) Glucose-1-phosphate —— Phosphoglucomutase (1,6 diphospho glucose as cofactor) ——> Glucose-6-Phosphate

iii) Glucose-6-phosphate (originating from galactose) and also glucose liberated by hydrolysis of lactose by lactase enters the Embden-Meyerhoff Glycolytic Pathway. The EMP pathway has been elucidated in the chapter chemical approach to dairy processes.

iv) The overall simplified scheme would be as follows :

Hexose ————> 2 triose phosphate

2 triose phosphate + 2 NAD^+ ————> 2 pyruvic acid + 2 NADH + H^+

2 pyruvic acid + 2 NADH + H^+ ————> 2 lactic acid + 2 NAD^+

v) Different bacteria produce different stereoisomers of lactic acid.

Streptococcuslactis-)	D-forms of lactic acid)	Following EMP pathway
S. citrovorus)	-L-form of lactic acid)	
S. *Paracitrovosous*)		
L. mesenteriodes)	Lactic acid do not follow EMP pathway as there is no aldolase enzyme in these bacteria which is required in EMP pathway, and hence ferment lactose through the heterofermentative pathway. This pathway has been described fully in the chapter chemical approach to dairy processes.	
L. dextranicu)		

Lipids: The lipid fraction of cheese contributes more to the development of flavour than any other component. The lipolytic enzymes liberate fatty acids by hydrolysis of the fat during ripening. Amount and type of fatty acid produced determines the flavour and off flavour of cheese. There are three sources of lipases. Those present inherently in normal milk, micro-organisms involved in the ripening process and some lipases present in the rennet preparations.

Lipases are activated by homogenisation and to some extent by agitation, foaming and even changes in temperature. Due to agitation lipolysis increases in cheese milk, producing unclean flavour, bitterness and rancidity. Rancid-flavour can also be produced by inoculating the milk with lipolytic bacteria isolated from silate and low grade milk. Milk lipases are inactivated at pasteurisation temperature, thus pasteurised milk produces a better quality of cheese.

Milk Clotting Enzymes

Nearly all proteolytic enzymes (proteases) clot milk under proper conditions. Milk clotting enzymes include sources like plant, microbial (bacterial), fungi and animal.

Plant Enzymes: Several parts of the plants like leaves, seeds fruits, flowers, shrubs and latex of the plant possess the ability to coagulate milk. These plant proteases are strongly proteolytic and cause extensive digestion of curd which results in reduced yields, bitter flavour and pasty bodied cheese. The following are the important plants used for coagulating milk.

Table 13.1. Different plant milk coagulants alongwith parts of plant used

Name	Part of plant used
Withania coagulans	Berries extract
Ficus carica and other plant of *Ficus* genus	Berries extract
Streblus asper	Leaves
Carica papaya	Latex of plant (enzyme papain)
Pineapple plant	Milk secretion (enzyme bromelain)
Cynara cardunculus	Extract of flower petals (enzyme cardoon)
Thistle	Flower of the plants
Castor oil plant	Seeds (enzyme ricin).

Microbial Enzymes

Cheese of acceptable quality have been made with microbial coagulant. At optimum concentration these enzymes do not produce bitterness. However, when used at higher concentration they produce bitterness. The following are important sources of enzyme.

i) *Bacillus* - All members of bacillus genus are good source of milk clotting enzyme, e.g. *B. cereus, B. subtilis, B. megatherium,* etc.

ii) *Micrococcus caseolyticus*

iii) *Fusarium* species

iii) *Streptococcus liquifaciens*

Fungi Enzyme: Enzymes from yeast, mould and fungi have been used from time to time but usually not with conspicuous success except few. Some important sources are :

i) *Syncephalastrum racemosum*

ii) *Cladosporium* herbarium

iii) *Bassochlamysfulva*

iv) *Rhizophus oligosporus*

v) *Basidiomycetes*

vi) *Aspergillus* group

vii) *Endothia parasitica*– it has been used with varying success for cheese making.

viii) Mucor – The chief protease in this genus are *Mucor pusillus* varlindt (M.P. protease) *Mucor miehei, Mucor rouxii* – enzymes from these species give sufficiently good quality cheese.

Animal Rennet: The abomasum (4th stomach) of milk suckling animals like cow, buffalo, goat, sheep and pig secrete the enzymes rennin. The extract "rennet" containing the enzyme rennin is prepared from the abomasum of these animals. Animal rennet is the best among all types of rennet and cow rennet (calf rennet), commercially known as hansen rennet, is the supreme among rennets which is due to its high ratio of clotting to proteolytic power. Hansen rennet factory was first established by Christopher Hansen (Danish) in 1874 in Denmark. In addition to this, porcine pepsin and bovine pepsin is also used for cheese making. Enzymes like pancreatin, trypsin and certain lipases can also be used for preparation of certain types of cheese though not very frequently.

Rennet: It is the crude separation or extract from the abomasum. It contains two principal enzymes viz., rennin and pepsin. These two enzymes though very similar, yet differ in their optimum pH (1.8–2.7 pepsin, 5.4 rennin), end product (pepsin-protease-peptone and peptides, rennin-paracasein) and ratio of clotting to proteolytic activity. Rennin is characteristic gastric digestive enzyme of the suckling animals while pepsin of the weaned animals. In addition to these two, trypsin and cathepsin may also be present which comes in crude extract by movement from other parts of the alimentary tract or by bacterial action. Rennin is an extremely powerful clotting enzyme, which cause rapid clotting without much proteolysis. On the other hand pepsin induces proteolysis, leading to bitterness in cheese.

The enzyme rennin has a variety of physiological function. One of its main properties being that of curdling or precipitating milk (casein and fat), so that the material doles out to the intestine for further digestion in small quantities (keep the milk in curd condition for further digestion). It also encourages and stimulates the stomach volume, strengthens the stomach muscles and gastric secretion. The enzyme rennin is present in an inactive form or zymogen form as "prorennin". This activates an acid pH by the release of peptides or the process is also autocatalytic conversion.

The abomasum is distinguished into two portions-pylorus or fundus and the outer wall of stomach consists of muscles and lining inside the mucosa. Mucosa of fundus portion is rich in rennin concentration. Rennin was first isolated by Heinz in 1851. It has molecular weight of approximately 40,000 dalton. It is globulin type of protein and contains about 19 amino acids.

The extraction of rennet from abomasa consists in chopping the cleaned stomach alongwith mucosa into small pieces and suspending it in 5 per cent sodium chloride solution containing 2.5 per cent boric acid added as preservative. The mixture is allowed to remain at room temperature for two to three days with occasional stirring. The whole suspension is then homogenised in a waring blender and insoluble tissue is discarded by centrifugation. The clear liquid is preserved in cold and activity of this preparation is checked. If necessary it can be further purified.

Alternatively the rennet can also be extracted through fistulation technique where animal is not to be slaughtered and in which a fistula is made through abomasum wall fitted with an elbonite cannula. The rennet secretion is washed with diluted whey, empty stomach and tapped through cannula. The collection can be done for a period of three to four months. During the period of collection, the animal is maintained solely on milk diet.

With higher concentration of rennet, the coagulation becomes faster. Maximum coagulation occurs between 40 to 42° C. There is no coagulation above 65°C or below 10°C. Heating the milk, delays coagulation because Ca^{++} ions are removed. Other reason of delayed coagulation are precipitation of whey protein on casein and rise of pH due to expulsion of CO_2. Acidic milk is more susceptible to coagulation. Higher the concentration of $CaCl_2$, faster is the coagulation. Homogenisation of milk accelerates clotting of milk which may be due to an increase in the degree of aggregation of casein micelles but decrease in curd tension. Inhibitors like albumin and globulin retard coagulation while peptone prevents clotting. Various other factors such as composition of milk, aging of milk, use of antiseptic and pH of milk also play a role in coagulation.

Various methods have been used to determine the rennet activity. These include viscosity change method (Marshall Cup test), visual precipitation method (Berridge's method), and methods based on proteolysis and NPN release.

Rennet Action : The action of rennet on milk can be studied in three phases. The first phase includes breaking of protective colloids of casein by enzyme followed by release of nonprotein nitrogen. In the second phase, a coagulum is formed. The third phase involves slow, over all hydrolysis of all casein components under the protective action of casein.

The above reaction can also be represented diagramatically as in Fig. 12 and also explained earlier in Chapter 8

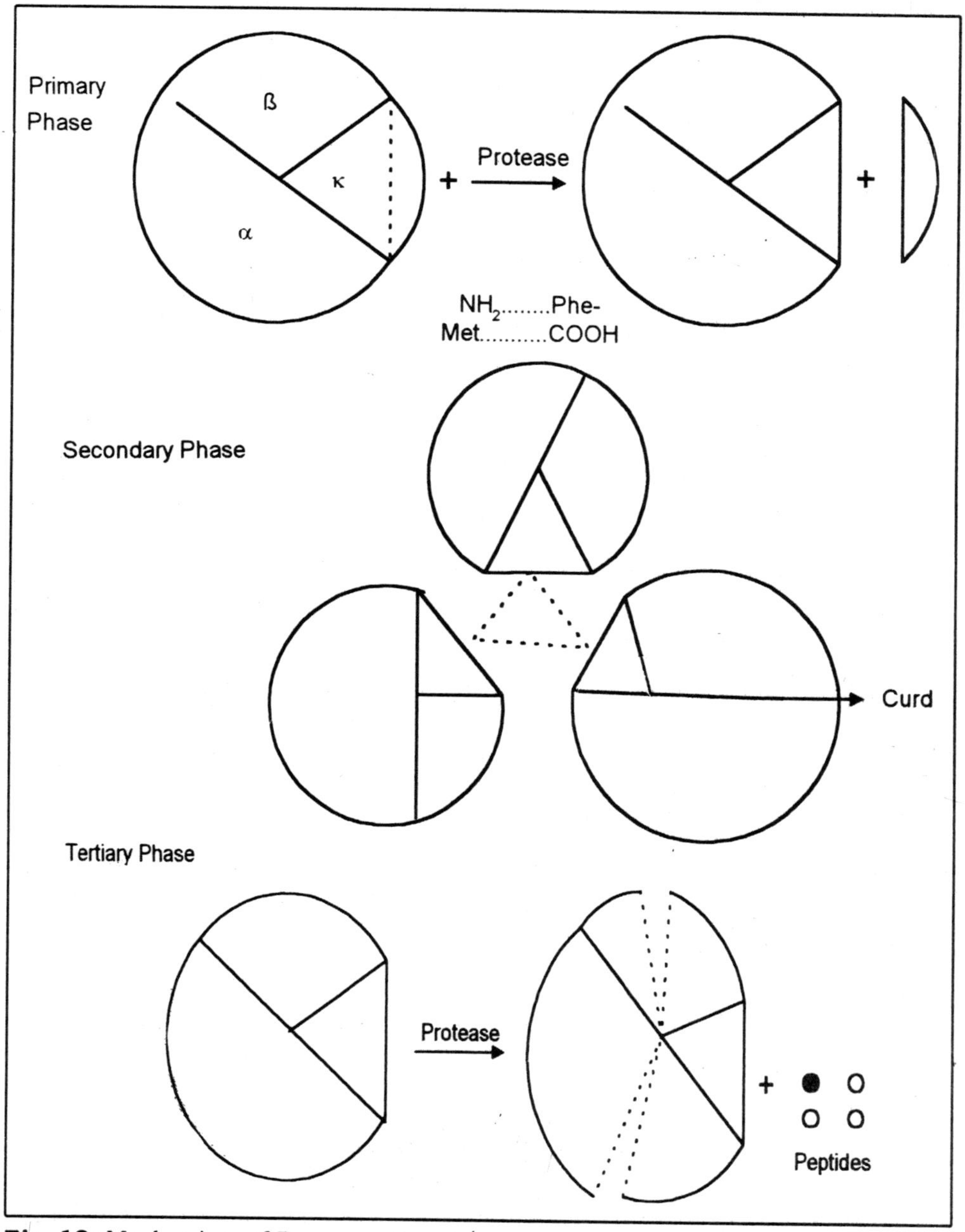

Fig. 12: Mechanism of Rennet action

Requirement for suitable cheese making coagulant :

i It must be produced from a raw material available in sufficient quantities and at acceptable price.

ii It must be toxicologically safe. There should not be any contaminating enzymes.

iii It must be suitable for the production of as large number of different cheese types as possible and it must not demand technological changes in the traditional cheese making process.

iv It must not cause any reduction in the quality and yield of cheese and should have desired firmness.

v It must be similar to the traditional calf rennet, particularly as far as chemical composition is concerned.

vi Extraction and manufacture on industrial scale must be possible in such a way that exclusion of unwanted enzymes is ensured and a high bacteriological standard obtained.

vii It should have a good strength. Minimum quantity should have to be used to coagulate the required amount of milk.

viii The amount of cheese that is obtained from a given quantity of milk of certain characteristic should be appreciable.

ix The qualitative and quantitative proteolytic characteristics of the rennet/coagulant should ensure a first class product in regard to taste and flavour.

x The relation of price/strength in the specific cheese factory especially considering coagulation activity. The price of rennet should be cheap as far as possible.

REVIEW QUESTIONS

1. What are the physico-chemical changes that take place during ripening of cheese.

2. List important milk clotting enzymes alongwith its source used for cheese preparation.

3. Why animal or microbial rennet is preferred over vegetable rennet?

REFERENCES

Berridge, N.J. 1952. An improved method of observing the clotting of milk containing rennin. J. Dairy Res. 19:328.

Datta Roy, D. 1981. Dairy Guide, Biosynthesis of Lactose 3, (10):9.

De, S. 1980. Outlines of Dairy Technology. Oxford University Press, Delhi, Chapter-7, Cheese.

Jennes, R. and Patton, S. 1959. Principles of Dairy Chemistry, John Wiley & Sons, Inc. New York.

Ling, E.R. 1956. A text book of Dairy Chemistry. Volume I (Third edition revised). Chapman & Hall Ltd., London. Chapter-6.

Rai, M.M. 1964. Dairy Chemistry and Animal Nutrition. Ram Prasad & Sons, Agra-3.

Tofte Jespersen, N.J. and Dinesen, V. 1974. Milk coagulants *Journal of the Society of Dairy Tech.* **32**(4):194.

Webb, B.H., Johnson, A. and Alford, J.A. 1978. Fundamentals of Dairy Chemistry. The AVI Publ. Co. Inc. West Port Connecticut. Chapter-12. Milk clotting enzymes and cheese chemistry.

14

CURRENT TOPICS

Opioids

THE word opiate means a drug containing opium used to induce sleep. Opioid is derived from opiate which means morphine-like. Peptides with morphine like activities are called opioid peptides or exorphins. These are isolated from some food proteins including milk proteins by treating proteins with enzyme proteinase, pepsin, etc. certain opioid peptides have also been detected in brain, pitutiary and other tissues and were named enkephalins and some time are called endorphins.

The opioid peptides have several physiological functions. These include analgesia, sedation and torpor, respiratory depression, hypotension, regulation of body temperature suppression of gastric secretion, regulation of sexual behavior and suppression of intestinal motility.

To function as opioid peptide in the central nervous system *in vivo* the exorphin must (a) be produced in the gastrointestinal tract, (b) survive degradation by the intestinal proteases, (c) be absorbed, without degradation, into the blood stream, (d) cross the blood brain barrier and thereby reach central opiate receptors, (e) interact as opiate with these receptors.

Opioids peptides produced from β-casein are called β casomorphins. Bovine as well as bullalo milk β-casein have been shown to release a β-casomorphin precursors upon *in vitro* incubation with gastrointestinal enzymes. β-casomorphin (human and bovine) and fragments (bovine) are commercially available. K-casein is also reported to release opioid peptides. Some of the casein peptides also have the opioid activity and are also commercially available.

The isloation, assay method and sequence determination of opioid peptides are subject of current interest, however, these are beyond the scope of this book.

Radionuclides

Radioactivity is a fundamental feature of the physical world and is due to naturally unstable atoms rearranging their inner parts in an attempt to become stable. It is the process of moving to a more stable state. Radioactivity occurs in our environment, within our homes, in the food we eat and even our bodies are radioactive.

Radionuclides are radioactive isotopes of some element. Such isotopes are always present in milk but in minute quantities. The naturally occurring radionuclides are those that have been deposited as a result of atomic energy in medical, industrial and scientific use. The artificial radionuclides are created by fission, burning of coal and oil, ground testing of fission as fusion devices, the operation of nuclear reactor and accident at reactors. These processes produce radionuclides like ^{40}K, ^{235}U, ^{238}U, ^{232}Th, ^{7}Be, ^{3}H, ^{87}Rb, ^{115}In, ^{214}Bi, ^{226}Rn, ^{90}Sr, ^{134}Cs, ^{137}Cs, ^{131}I and several others. From the atmosphere these get into the human body by different ways. These can be deposited on plants to be consumed directly by man, deposited on leaf surfaces to be consumed by animals whose products are then consumed by man or deposited on soil and then taken into plants to be consumed by man and animals.

The potentially hazardous radionuclides in milk are ^{90}Sr, ^{89}Sr, ^{131}I,^{133}I, ^{137}Cs, ^{140}Ba, and ^{40}K. The physical half life (period needed to reduce the radioactive emission by an isotope to half of its original level) and biological half life (time taken until half of the amount of an element ingested by a person has been excreted) of these nuclides are given in Table 14.1

Table 14.1 Important radionuclides of milk

Radoionuclide	Physical half life	Biological half life	Location
^{89}Sr	52 days	1-50years	80% in casein micelles
^{90}Sr	28 years	1-50years	80% in casein micelles
^{131}I	8 days	100 days	Serum
^{133}I	21hours		
^{137}Cs	33 years	230 Days	Serum
^{140}Ba	13 days	<3years	Plasma

Source : Reproduced from Dairy Chemistery and Physics by Walstra and Janness.

Strontium accumulates in bone and remains there for a long time. ^{90}Sr uptake causes a prolonged exposure to harmful radiation as it has a long physical half life. It is distributed in milk in much the same way as calcium, however, strontium phosphate is less soluble than calcium phosphate and a greater proportion of it is in the casein micelles.

Barium behaves more nearly like calcium, iodine accumulates in the thyroid gland where it attains high concentration and radioactive isotopes of ^{131}I is considered hazardous as it produces intense radiation, however, for short period, after a serious fall out. Iodine is found in milk serum, a small portion being strongly bound to proteins; about 2 percent is in the fat globules. Cesmium chemically behaves like Na^+ and K^+.

Atomic Energy Regulatory Board (AERB) standards of radioactivity are based on an annual exposure of 10 mrem when all items of food are contaminated to the recommended levels and intake is throughout the year. From milk and milk products alone exposure would be 1.6 MR/year. This is quite small compared to the natural and other radiation exposure that is received during day to day activities.

Radionuclides in milk and milk product observed in different countries are invariably lower than that of recommended limit in those countries. Presence of radionuclides in milk and milk product varies from country to country.

^{40}K is somewhat more hazardous due to higher energy of radiation compared to ^{137}Cs. Natural radioactivity due to ^{40}K present in milk powder is in the range of 600-800 Bq/l kg while liquid milk, in the range of 30-40 Bq/l in Indian milk and 50-80 Bq/l in foregin milk. It has also been reported that cow milk contains 44 Bq/l and buffalo milk contains 38 Bq/l. Since potassium salts are very soluble ^{40}K being retained in liquid phase in preparation of butter and it contains only 1.3 Bq/l kg. Cheddar cheese is reported to have 41 Bq/l kg of ^{40}K.

The ratio of $^{90}Sr/^{137}Cs$ in European milk is 1.8 while it is 0.5-1.58 in Indian milk. The mean value for ^{90}Sr is reported to be 3.9 pCi/l.

In India ^{137}Cs have been reported to be 0.02 Bq/l in milk and 0.6 Bq/l kg in milk powder which is far below than the accepted level of 30 Bq/l in milk and 210 Bq/l kg in milk powder. Most of the dairy products, in general contains ^{137}Cs to the extent of 40 Bq/kg. ^{131}I in milk is present to the extent of 20 Bq/l.

Pesticide Residues in Milk and Milk Products

Pesticides are toxic chemicals used to control the parasites on crops and animals. As many as 126 pesticides have been cleared for use in the country by the Registration Committee, out of which currently 57 are manufactured in our country. Nearly 40,000 tonnes of hazardous pesticides which are either banned or highly restricted for use in developed countries are still used every year in India. These pesticides which a study says are used "indiscriminately" include DDT, BHC, Aldrin, Chlordane, Heptachlor, Sodium Cyanide, 2,4-D and Lindane. They have been found to cause severe health hazards and to pollute the environment. The study was conducted by the National Institute of Science and Technology and Development Studies (NISTADS), a wing of the Council for Scientific and Industrial Research (CSIR). The highly toxic or condemned pesticides carry no warning or caution. Besides, the study says the use of pesticides is concentrated in a few areas only.

The source of contamination of pesticide residue in milk and milk product are contamination of utensils or milk, misuse of spray material and contamination of feed. It has been found through exhaustive analysis that the main channels of pesticide flow to milk are fodder, feed management practices and bad-farming husbandry. Fodder derive the pesticides from soil. Commonly used fodders are lucerne, berseem, maize green and wheat straw. Intensive agriculture makes the farmer treat the soils liberally with DDT to kill white ants. The quantity of DDT will increase each year as the farmers are becoming increasingly unjudicious about the application of pesticides to crops. The crops are liberally overdosed with the idea to completely exterminate the insects. To push up the sales pesticide dealers wrongly advise the farmers to over dose the pesticide four times. They are unmindful of the resultant effect. The fodder used has excess of pesticide and food obtained using such fodder will have residual pesticide which is very harmful.

Feeds are apparently the by-products of agriculture produce, such as wheat bran, rice polish, gram churi, gram husk, sarsoan cake, cotton-seed cake, groundnut cake, maize grain and molasses. Rice polish, cotton-seed cake, groundnut cake, maize gram and molasses are supposed to carry the residual effect from field i.e. preharvest time. Wheatbran, gram (chuni and husk) sarson cake carry more post harvest practices as these stored grains are mixed with heavy application of BHC and DDT and at the time of processing, hardly anyone heeds to wash the gram before milling. Thus, wheat bran, gram chuni and gram husk have been seen with exceptionally high amounts of BHC/DDT. It is apparently clear that pesticides residue would be a more serious problem where milk is solely produced from concentrate.

Direct spraying of insecticides in barns and stables for killing flies, ticks or direct application of insecticides on the animal body can be another source of pesticide residue in milk. This problem is far too less in rural areas except the urban stables and dairies. Over dozing of the crops, over looking the safe period, i.e. between last spray and the harvest time, direct mixing of pesticides in cereals and pulses are bad farming husbandry.

Milk is consumed by the young and old alike, it is of paramount importance, that the milk they consume should be free from any chemical contamination. The defence mechanism of the body in the new born are not powerful enough to withstand the threat posed by the ingestion of these harmful pesticide residues. Therefore, regular analysis of fluid milk in different milk processing centres would help to assess the level of pollution. Scores of children are being fed by infant formula manufactured by different companies and barring a few, none of them have facilities to check the level of the persistent pesticides in the infant formula they manufacture. The effects of these pesticide residues on the new born are not yet clearly defined. Of the various types of pesticides, only the organochlorine (non-polar) pesticides, such as Aldrin, BHC, DDE, DDD (TDE), O, P' and P' P-DDT, Heptachlorepoxide, Lindane, Methoxychlor, Dieldrin and Endrin find their way into milk through the animal's feed. These compounds are fat soluble and not readily excreted by the animal except into milk. Other pesticides such as organophosphates and carbamates are not fat soluble and they are broken down during metabolism and excreted by the animal.

Limited studies made on samples of fluid milk and milk products (curd, ghee) collected from different regions of India show heavy contamination with DDT, BHC, DDE and DDD in certain regions. The observed levels of these pesticides are above the WHO recommended levels for safety. Therefore, there is a great need to monitor milk and milk products for pesticide residues. In addition to these above pesticides milk and milk products are also found to contain Lindane, Dieldrin, Heptachlor Aldrin and HCH in this country.

The Table 14.2 shows an average value of pesticide detected in milk in particular area, while Table 14.3 shows value of pesticides in butter and ghee and differs region to region.

Table 14.2. Residue of pesticides in milk samples

Pesticides	Range (ppm)
DDT	Traces - 216 Traces - 513 (Human milk)
BHC	Traces - 13
Endrin	0.02-1.5
Heptachlor	Traces-0.60
Aldrin	0.03-0.52 0-44 (Human milk)
HCH	0.01-1.01 115-325 μg/kg (Human milk) 0.01-0.70 20-125 μg/kg (Human milk)
Dieldrin	0.12-2.5

Table 14.3 Residue of pesticides in butter and ghee.

Pesticide	Range (ppm)
DDT (Butter)	0.30-8.10
DDT (Ghee)	2.0-13.0
BHC (Butter)	0.61-5.50
BHC (Ghee)	2.23-6.40
HCH (Butter)	2.85
HCH (Ghee)	2.01

Pesticide residue analysis in dairy products usually consists of following five steps – (a) Sampling : The aim of sampling is to provide a reproduction of a portion of the environment on a scale that enables the sample to be handled in the laboratory. Analytical results are meaningful only if the collected samples are truly representatives, (b) Extraction of residues from the sample matrix : The residues are extracted with appropriate solvents from the sample substrate, the aim being to combine the maximum efficiency of extraction of the residue with the minimum co-extraction of any sample constituents which may interfere with the subsequent procedure, (c) Removal of interferring co-extractives (clean up) : The clean up stage aims at removing interfering materials from the extract to obtain a clean solution of the extracted residue in a solvent suitable for quantitative examination by the chosen method of determination, (d) Identification and estimation of the quantity of residues in the cleaned up extract usually at very low levels in the range of 10^{-9} to 10^{-12} grams. Estimation of the quantity of residues is normally done by Gas-Liquid Chromatography with electron capture detector, (e) Confirmation of the presence and identity of the residues. Chromatographic method is by far the most widely used method for the determination of pesticide residue, followed by spectrophotometry. Spectrophotometric methods are generally less sensitive and less selective than GLC. Organochlorine pesticides present in milk and dairy products are usually analysed with a tritium or ^{63}N electron capture detector. The directly coupled gas chromatograph-mass spectrometer is a powerful tool for positive identification of pesticides. HPLC has also been found increasingly useful for the analysis of organochloropesticides as it requires less extensive clear cut than required with GLC.

Immobilisation of Enzymes

Immobilisation of enzyme means insolubilisation of enzyme. An immobilised enzyme preparation can be defined as an enzyme which is modified or confined in such a manner that it can be recovered after usage or it can be used continuously and is not lost in the product. An immobilised enzyme system, therefore, consists of an enzyme, a support material and a method of immobilisation.

Advantages offered by immobilised enzyme over that of the soluble enzyme are :

1. Enzyme reusable.
2. Reaction easily terminated by separating substrate from enzyme.
3. More precise control.
4. Less product inhibition.

5. Greater pH and temperature stability.
6. Can use enzyme presently unusable for various reasons.
7. Continuous or batch use.
8. Greater reactor design flexibility.

Disadvantages :

1. Lower specific activity.
2. Inactivation in continued operation.
3. Cost of support and immobilisation method.

Use of Immobilised Enzyme

The various uses of immobilised enzyme include number of industrial process, biochemical research, biomedical research, chemotherapeutic application and use of commercially available enzyme in analytical laboratory.

Preparation of Immobilised Enzymes

There are five methods in general for preparation of immobilised enzymes. They are :

Covalent binding of the enzyme to an insoluble support.

Cross linking of the enzyme.

Micro-encapsulation of the enzyme in a porous membrane.

Inclusion in the structure of a polymer.

Non covalent adsorption on to an insoluble support.

Covalent Binding of the Enzyme to an Insoluble Support: It is frequently used and is best technique for immobilisation of enzymes. A covalent bond is formed between one or more of the enzyme amino acid residues and a functional group on the insoluble support. Some of the groups of enzymes that take part in covalent bond formation are amino group of protein, lysine, arginine, α-amino group of polypeptide, hydroxyl group of serine, threonine and aromatic ring of the tyrosine. The insoluble support may be inorganic, organic or biological material. It is necessary to "activate" the carrier material for immobilisation of enzyme. It is to synthesise groups on the carrier material to which the enzyme will be attached. For example when enzyme is immobilised on glass beads it is first cleaned with nitric acid and then treated with aminosilane (α-amino propyl triethoxy silane). This is to provide covalently linked amino derivative on glass surface. Enzymes are then attached directly to these amino groups using such agents as glutarldehyde. Covalent bonding are of five types – (a) carbodimide coupling, (b) acyl chloride coupling, (c) glutaraldehyde coupling, (d) coupling by way of diazonium salt, (e) cyanogen bromide coupling.

Cross Linking of the Enzyme: Immobilization of enzyme is achieved by cross linking the enzyme molecules to each other or to some functional group on a carrier matrix. Cross linking is achieved with bifunctional reagent which may either possess two identical functional groups such as glutaraldehyde, bis diazobenzidene-2-2' disulphonic acid or two different functional groups such as toluene-2-isocyanate-4-isothiocynate. Cross linking results in coupling of one enzyme molecule to another, thus forming large matrice of the enzyme molecule. If support material with functional group is used, some of the enzyme molecules may be covalently bound to the carrier. Cross-linking an enzyme to itself is both expensive and inefficient as some of the protein material will inevitably be acting mainly as a support, resulting in relatively low enzymic activity.

Microencapsulation: Enzymes may be encapsulated inside microcapsules made of nylon or cellulose nitrate. Depending on the method of preparation, diameter of the microcapsule varies 10 μm to 1 mm. Membrane of microcapsule is semi-permeable so that substrate and other products from the surrounding medium can readily diffuse into and out of the microcapsule whereas the enzymes remain trapped inside. By this method no chemical modification of protein and enzyme is done hence no change in kinetic properties of the microencapsulated enzyme with that of soluble enzyme. It has few applications and the stability of enzyme is less.

Enzyme Inclusion Within a Matrix: Enzymes may be immobilised by confining them within the lattices of polymerised gels. The lattices are made with a small enough pore size that free diffusion of low molecular weight substances is not impaired but the higher molecular weight proteins are entrapped within the lattice. Enzyme and a solution of monomer (e.g. acrylamide) initiate the polymerisation on addition of cross linking agent. The polymer may subsequently be broken up into the desired particle size. The major problem with this type of immobilization is that enzyme may slowly leak out of the gel matrix. Leakage is more in gel with high water content. The leakage can be inhibited by glutaraldehyde which provides additional cross linking. Occurrence of enzyme leakage makes this system unsuitable for analytical use.

Adsorption: Physical adsorption of an enzyme on to an insoluble matrix is simplest way of preparing immobilised enzyme. The method relies on non-specific physical interaction between the enzyme protein and the surface of the matrix brought about by mixing a concentrated solution of enzyme with the solid. Solids for adsorption are usually charcoal, alumina, cellulose, calcium phosphate, hydroxapatite, Dowex-50, and DEAE-sephadex. No activation step is necessary in this. It is a cheap method and easily carried out. The binding forces involved in this type of immobilization are hydrogen bond, multiple salt linkage and Vanderwaals forces. Because of weak nature of bonds involved, desorption of protein, due to changes in temperature, pH, ionic strength or even the mere presence of substrate is often observed. Another disadvantage is that the adsorption surface is non specific and will, therefore, adsorb further protein or other substances to which it is exposed in the course of its use.

Immobilized enzymes are used in special type of apparatus. These are called enzyme reactors and are of four types – (a) continuous stirred tank reactor, (b) tubular reactor, (c) packed bed reactor, and (d) fluidized bed reactor.

Properties of Immobilized Enzyme: There is a change in stability of an enzyme which is due to the native structure of enzyme. Immobilization affects enzyme's kinetic constant. There is decrease in specific activity of the enzyme, which may be due to inactivation of part of the enzyme or alteration in the catalytic mechanism of the enzyme. Upon immobilization, there is shift in pH optima. Proteolytic enzymes, when immobilized on polycationic carrier, the pH optima of the enzyme is shifted to the acid side, whereas a shift to the alkaline side was observed when polyanionic carrier was used. Charges on the carrier matrix may affect the Km of immobilized enzyme. Many of the substrate, temperature and pH values used are ideal for growth of micro-organisms. However, in some cases the temperature or pH conditions do not encourage growth. Certain support materials used for immobilization are resistant to microbial attack such as inorganic carriers and polyacrylamide, etc. but microbes can still attack the enzymes attached to the carriers. Mould and yeast contamination and growth have been reported during continuous treatment of whey with lactase immobilized on glass beads. Several types of bactericidal agents have been employed to control microbial growth in the reactors.

Applications

Detection and Determination of Biological Compounds: Glucose oxidase immobilised on nylon tubes or poly styrene tube was used to assay D-glucose. Upto 25000 assay can be performed per tube. UV absorption was taken as criteria of estimation and detection.

Structural Analysis of Macromolecules: Trypsin immobilized to maleic anhydride – ethylene copolymer cleaves 10 peptides bond in pepsinogen. Native trypsin cleaves maximally 15 peptide linkage. This suggests selective degradation nature of immobilized trypsin in its catalytic action and used for preparation of unique polypeptides of immunoglobulin and heavy meromycin.

Environmental Application: Nitrate ions in water can be detected using immobilized nitrate reductase.

Preparative Use: Trypsin inhibitor or trypsin plasmin inhibitor can be isolated by passing crude extracts of inhibitor through a column of immobilized enzyme and eluting the inhibitor by rupturing the immobilized enzyme – inhibitor complex with acidic buffer.

Food Industry: Starch can be fabricated with immobilized enzyme to change its texture (introduce sweetness and conversion to fructose). Glucose syrups can be prepared from starch using immobilized amylases and glucoamylase on DEAE cellulose.

Medical Use: Microcapsule containing L-asparaginase for suppressing asparagine dependent tumors in mice. When soluble enzyme was injected into the mice, tumors were not suppressed because of the organisms immunological rejection of foreign protein.

Membrane Bound Enzyme as Model System: For studying mode of action of membrane bound enzymes in relation to their physiological significance.

Functional Properties of Milk Proteins and Casein-Coprecipitate

Milk proteins, by virtue of their heterogenity, their range of physical properties and amphipolar characteristics are premier proteins used for multiple functions such as emulsion, stabilisation, water holding, coagulation, adhesion, etc.

Milk contains 3-4 per cent protein. Major protein is 80 per cent as casein while remaining is whey protein and others are 20 per cent. Coagulation of milk proteins by acid, heat and enzyme is an essential step in the production of various products such as cheese and paneer. Aside from being coagulatable, milk protein contributes to flavour and to the formation and stabilization of small air-cells in ice-cream and whipped cream. In foams, the proteins function at the air-liquid interface. Various milk powders (fat and water removed) and various milk protein fractions (casein and sodium caseinate, whey protein, etc.) are used as ingredients in other foods. Non fat dry milk (NFDM) is used by baking industry to improve the water absorption capacity of flour, to improve baking quality of weak flour, to control the rate of gas emission, to strengthen structure and texture, to retard moisture loss, to enhance crust colour and flavour development, to delay the salting process.

Casein promotes low-temperature stability of some frozen foods. Whey protein (especially lactalbumin) have good whipping characteristic and are used in dessert toppings for stabilisation of foam structure. NFDM is added to comminuted meat as a stabiliser for the emulsion and to increase moisture retention. Whey proteins are included in Candy and confection formulations (Caramels, fudges, toffies, fondants, icing formulations) because of their structure strengthening and moisture retaining properties.

Changes take place in structure of milk proteins when subjected to different processing treatment. Pasteurisation causes inactivation of degradative enzymes partially or totally. There is a slight denaturation of some whey protein (ß-lactoglobulin and proteins of the fat globule membrane) which leads to sulfide formation and a flavour change. In the native state, ß-lactoglobulin has a conformation that shields and renders the sulphydryl groups of cysteine unreactive. As ß-lactoglobulin unfolds upon heating, these sulfhydryl groups become more accessible and reactive, yielding H_2S. Important nutrients in milk are not altered substantially by pasteurisation. Complicated changes occur in milk during heat sterilisation. Temperatures exceeding that for pasteurisation (72°C) cause instability and coagulation of proteins. Depending on the time temperature relationship, heat treatments may partially dephosphorylate casein, denatured whey proteins initiate protein-protein and protein-lactose-interaction. Pasteurised homogenised, fluid milk can be frozen and held satisfactorrily at –24°C for as long as 4 months, but frozen concentrated milk can be satisfactorily stored for only short periods. In both instances, storage stability is limited primarily by the tendency of casein to form a precipitate which fails to disperse upon thawing. Spray drying is used to produce milk products. In spray drying, the liquid is finely atomised into a stream of rapidly moving hot air. Although heat is the principal agent causing protein alteration during spray drying, surface forces cannot be ignored. During spray drying of milk and milk products under properly controlled conditions, changes in milk protein structures and solubilities are relatively minor. In drum or roller drying, the product is dried in a thin film on a rotating drum that is internally heated with steam. The dried product has a scorched flavour and reduced protein solubility. Two new amino acid, furosine and pyridosine have been identified in scorched drum-dried milk powders. They arise from lysine. Furosine and pyridosine are found in relatively large amounts in acid hydrolyzate of drum-dried milk powders, in trace amounts in hydrolyzates of spray-dried milk products and not at all in freeze-dried milk.

Co-Precipitates

Combination of casein and whey protein precipiated together by acidification of heated milk is termed as co-precipitate or casein co-precipitate. The heat induced interaction between ß-lactoglobulin (whey protein) and K-casein which occurs at temperature above 65°C is of significance in the manufactre of co-precipitate. The reasons for commercial development of casein co-precipitate were to increase the yield of protein, to increase the nutritive value and the range of functional properties of milk proteins for use in foods. The curd formed by co-precipitate of casein and whey protein has characteristic difference from that of casein curd. Recovery of about 96 per cent milk proteins can be achieved under optimum processing conditions. In the Russian process, the co-precipitation of protein from heated skim milk was done by addition of $CaCl_2$. Co-precipitate are classified according to their Ca (and hence ash) contents. The terms high-, medium- and low-calcium coprecipitate were used to define co-precipitate with calcium contents in the ranges 2.5-3.0 per cent, 1.0-2.0 per cent and 0.5-0.8 per cent, respectively. Control of calcium level was achieved by varying the amount of calcium chloride added to it, changing the length of time at which the milk was held at about 90°C and, for low and medium-calcium co-precipitate, varying the pH of precipitate by acid.

A considerable amount of data is available on the properties of co-precipitation in relation to their solubility in various solvents and to their applications in food. The general physical properties of co-precipitate may be regarded as the physical properties of these products when examined alone or in water (dispensed or dissolved). These include pH, solubility (in water, alkali or acid) and rate of solution, density, particle size, viscosity of solutions and colour of powder and solutions.

The co-precipitate of casein contributes excellent nutritional value to food products but their principle value in most foods is their functional contribution. Functional properties provided by casein co-precipitate are as follows :

Functional property	Food application
Emulsification	Coffee whitener, soup, whipped topping, meat products
Stabilisation	Soups, meats, desserts
Fat absorption Meat products	
Water absorption or retention	Meat products, bread, cakes confections
Viscosity	Soups, gravies, chilli
Gelation	Simulated meats, cheese
Fibre/Texture	Simulated meats
Dough formation	Baked foods
Adhesion/Cohesion	Meat products, baked foods
Aeration	Whipped toppings, desserts, confections

The combination of wide range of physical and functional properties and superior nutritive properties allows co-precipitate to be used as ingredient in a variety of food products either by contributing a necessary functional property to the ingredient mixer or for nutritional purpose. Some of the most important uses of co-precipitate are in baked products, meat industry and dairy products.

Sanitation and Milk Hygiene

Introduction

Milk is a lacteal secretion produced by all mammals and triggered by lactogenic hormones. Prior to the processing of milk, milk from ruminants, e.g. cow and buffaloes, is drawn from the udders of milch animals either through machine milking or by hand-milking. In India, the organized Dairy Farms are very few and in most cases very little importance is given to sanitation and milk hygiene. In most of the milk processed in organized dairy sectors are by collection through chilling centres from individual farmers who ship milk in cans to chilling centres or directly to the reception dock of the dairy factory premises. Nowadays there is a practice of collecting milk from individual farmers in chilling stations freezing such milk in refrigerated road tankers and transporting the frozen milk to distant places. Decades back Aarey Milk Colony in Goregaon, Bombay received milk from Kaira District Co-operative Union, Anand, Gujarat in refrigerated Rail tankers.

The numerous ways of transporting milk to dairy factories all over India result in milk reaching the industry in quality which is sub-standard. Also there is lack of sanitation or hygiene prevalent in most of the villages in India which is a cause of grave concern in the Indian Dairy Industry. In villages particularly, mastitis is rampant. Entry in the dairy factory premises should be restricted as sanitation is essential to ensure better quality milk. At village level, better sanitary practices are possible through intensive and extensive extension work and also through revolutionary educative system.

The principal sources of contamination of milk that reaches the Dairy factory premises are - (a) contamination of milk in the bulk tanks at the milk reception docks due to poor cleaning and sanitization, (b) due to poor handling of milk at the reception dock, and (c) poor sanitary condition and milk hygiene at the reception dock and processing units.

The post processing contaminations are due to - (a) poor sanitary habits of the handlers at the processing units, (b) poor C.I.P. of the processing units, (c) lack of sanitary conditions present in the entire dairy factory premises and above all, (d) air.

The diseases that may arise through ingestion of such processed milk products are :

a. Diphtheria	*Corynebacterium diphtheriae*
b. Scarlet fever	*Streptococcus scarletinae*
c. Tuberculosis	*Mycobacterium tuberculosis*
d. Pneumonia	*Diplococcus pneumonia*
e. Whooping cough	*Bordella perlussis*
f. Food poisioning	*Staphylococcus aureus*
g. Dysentry, Diarrohea	*E. coli, A. aerogenous*
h. Small pox	Virus
i. Chicken pox	Virus
j. Measles	Virus
k. Mumps	Virus
l. Influenza	Virus
m. Common cold	Virus
n. Poliomyelitis	Virus

All these diseases are by faecal contamination and air.

Control of Faecal Contamination

Post-processing faecal contamination to milk and milk products can be controlled by intensive and extensive education to milk and milk product handlers and by devoted and dedicated cleaning in place of all dairy equipments.

Control of Air-Borne Contamination

Sunlight is the most patent weapon against air-borne infection of milk and milk products. However, there are absolutely no ways to clear the air entirely from the microbes, but certain steps could be taken to reduce the likelihood of contamination.

Air-borne contamination in the dairy premises is due to - (a) Quality ventilation, (b) dust and dirt which reaches freightening heights during dust storms, and (c) walking of people inside the dairy factory and stirring up the dust gathered in their shoes and floor.

The remedy is to have closed drains and regular disinfection of the floors. Also, all ventilators should have glass panes which should remain closed in areas where dust storms are rampant. The floors should be intermittently washed. Air filters may be used to remove micro-organisms as well as dirt from the air. In developed countries electronic precipitators are used to clean and purify air.

How to Ensure Milk Hygiene at the Farm House

In a farm house cleaniness of the milkers, milking machines, pipelines chilling tanks, road tankers, etc. are essential to ensure milk hygiene.

Observation on milking at National Dairy Research Institute, Karnal has shown that the cows are very quite and their handlers appreciate the need for a calm environment. The care these men take of the cows and buffaloes and their obvious "Feeling" for the animals is very impressive.

The practice of feeding a concentrate to cows and buffaloes prior to milking is good and undoubtedly assists in stimulating milk let-down.

The washing down of each animal using a hose directed initially at the teat area, followed by the hind legs and finally the udder may be less desirable since the technique of spreading a thin layer of wet feaces over the udder probably increases rather than decreases bacterial contamination of the teats.

Following the wash the udder is briefly rubbed with a rag soaked in a solution of the disinfectant kept in a central bucket. Such an udder cloth is an excellent factor for spreading infection, like mastitis.

It is suggested that at every milking bale, a hose be made available since running water is the best available sanitiser in a cow shed. The hose which should hang from an overhead supply could be shared by two milkers and be fitted with a simple valve. After washing excess dirt from the udder with water, the udder should be cleaned with a suitable soap. Since germicides add to the cost, a suitable cream soap can be prepared from available soap and water (equal weights) blended in a mixer with 10 per cent of coconut oil. This mixture sets to a soft consistency and would be convenient to use with each milker having a supply in a tin. After washing with soap, a final water rinse is necessary. In the case of machine milking this would assist in the stimulation necessary for adequate let down.

Post Milking Wash

Currently, at the end of milking no cleaning techniques appear to be used, since mastitis control requires the streak canal to be sealed following hand or machine milking. It is an excellent idea to treat each teat with an effective antiseptic. In developed countries, the use of iodine containing preparations has come to be widely practised, but since these are relatively expensive, they may not be suitable in our country.

The most suitable alternative would be a solution of sodium hypochlorite which can be generated by bubbling chlorine gas into a solution of caustic soda. It is used at a concentration of 0.5 per cent as a teat dip.

Sodium hypochlorite solutions are also ideal for cleaning all milking receptacles. Calcium hypochlorite seems to be widely used but because of the likelihood of building up milk stone deposits, the sodium solution would be preferable.

Sanitation and milk hygiene are pre-requisites at - (a) village level, (b) the level of middle-man who collects the milk from individual farmer and transports it to dairy plants, (c) the farm level and, (d) at the dairy plant. It goes without saying that cleaner the milk produced, handled, processed, packaged and stored, better it is for the entire humanity.

Protein Lipid Interactions

Interaction between proteins and lipids is a major field of biochemical research. Most of the work has concentrated on cellular and subcellular structures (especially membranes) on blood constitutents, and on model system interactions in aqueous solution. The status of this research has been the subject of numerous conferences, reviews and books.

In this chapter we will discuss mainly the following points :

- Current ideas about mechanisms of lipid protein interactions in model system, in tissues and in selected food systems.
- The role of protein lipid interactions in changes during processing of Dairy Food.

Nature of Lipid Protein Interaction

Types of bonding between protein and phospholipids -

There is no single type of bonding by which lipids interact with proteins. In most isolated natural lipid protein complexes, the lipoproteins contained phospholipids in their lipid constituents. Phospholipid interactions with proteins may involve electrostatic forces. Electrostatic bonding can arise from attraction between the negatively charged phosphate group (HO-P-O-OH) and positively charged protein group such as lysyl residue.
||
O

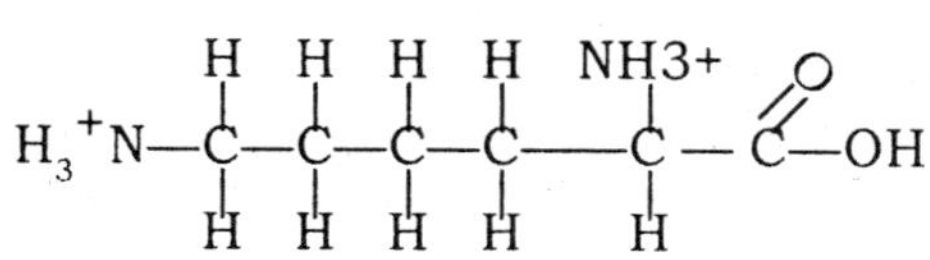

It may also involve a positively charged group in the choline residue

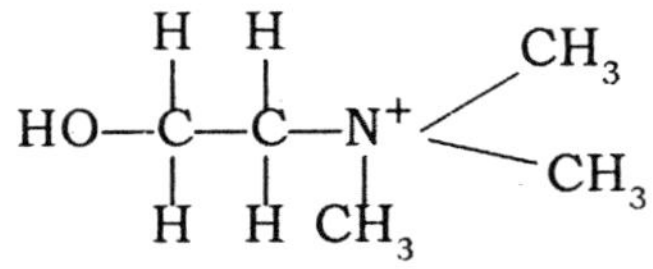

of phospholipid and a negatively charged residue of protein such as aspartyl

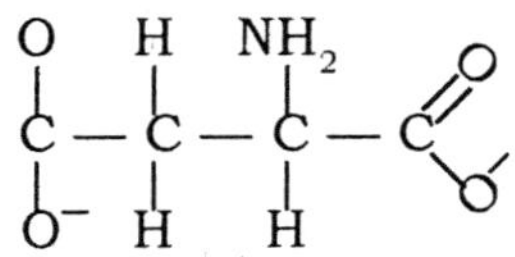

It has been reported that simultaneous and cooperative occurrence of electrostatic binding and of salt bridge is operative in bacterial cell walls.

Characteristics of different types of intermolecular forces are shown as below.

Typical bond energies for different lipid protein bonds

Bond	Energy (K cal/mole)	Dependence Distance (r)
Covalent	30-100	Maximum attraction at 1-2° A
Electrostatic	10-20	Energy $\alpha\, r^{-}$
Hydrogen bond	15-6	Maximum attraction at 2-3° A
Vanderwaals	0.5-2	Energy $\gamma\, r^{-6}$

Of these, covalent binding is the exception. Hydrogen bonding is also secondary in lipid-protein complexes, although it is indirectly important in hydrophobic bonding. Non-polar dispersion or Van der Waal forces become important when interacting groups are near. These forces are weak, and the attraction between nonpolar groups is greatly increased in aqueous media where they are pushed together by water because the water-water interaction by hydrogen bonding is much stronger than the interaction between

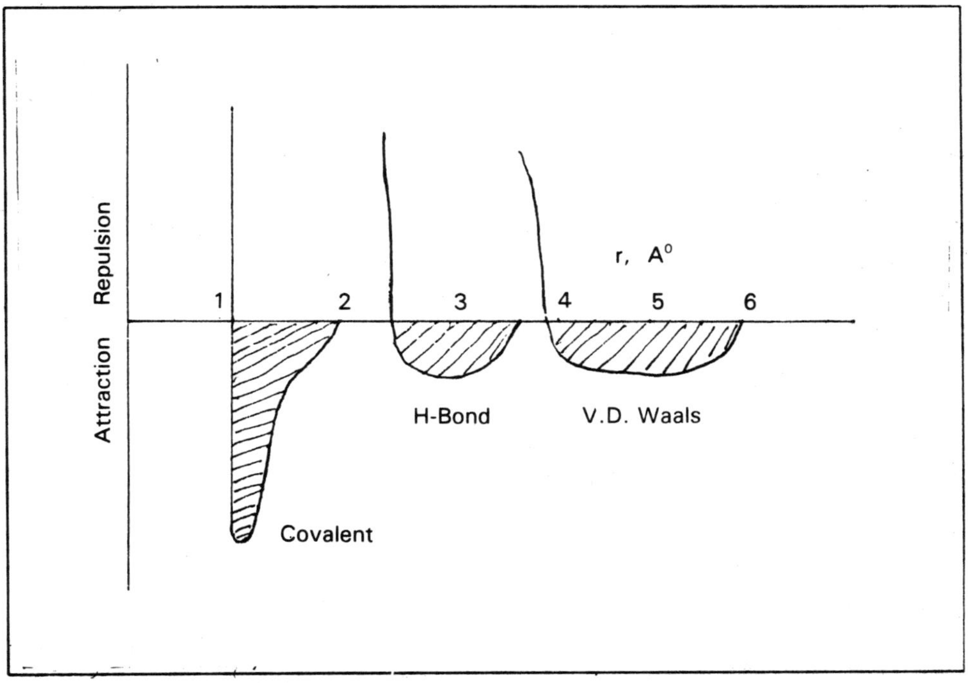

Types of intermolecular forces

water and the nonpolar groups. This phenomenon based on "hiding" nonpolar groups from water has been called "hydrophobic bonding". The total standard free energy of formation of the hydrophobic bond is larger than the energy due to Van der Waal forces alone because changes in the water structure contributes to this free energy of formation.

Interfacial Energy in Lipid-Water Systems

As a consequence of differences in attractive forces between molecules of different liquids, or differences in molecular density between phases, surface molecules possess extra energy, the free surface energy. Water surface in contract with its vapour or with air has a free surface energy (or as it is also called, surface tension) of 72 ergs/cm^2. The presence of amphipolar molecules including polar lipids and proteins lowers surface tension due to adsorption and orientation of these molecules in the surface: molecules of this type are surface active, or surfactants. Typically, this adsorption shows three phases with increasing concentration.

* Saturation and orientation in the surface.

* Region of constant surface excess (surface excess is approximately equal to surface concentration) but increasing bulk concentration. In this stage the decrease in free surface energy is proportional to the

logarithm of the surfactant concentration.

* Region of micelle formation.

The formation of micelles as well as orientation of amphipolar molecules at interfaces between water and air or between water and oil are of key importance in many colloidal phenomena, including emulsion stabilisation, foaming and solubilisation of otherwise insoluble compounds within micelles. For each type of amphipolar molecule, a critical micelle concentration range exists, a very comprehensive tabulation of these was published recently.

Micelle formation and the adsorption of amphipolar molecules at oil-water interfaces are closely related to two important phenomena.

* Biological membrane models.

* Membrane and films stabilising food emulsions.

Evolution of Concepts of Biological Membrane Structure

The formation of phospholipid bilayers is a well established phenomenon in model systems, and the phospholipid leaflet or bilayer was the earliest widely accepted model of cell membrane structure. From it grew the "unit membrane" concept. The unit membrane consists of a phospholipid bilayer with the polar groups of the phospholipids oriented outward and bound by polar interactions (probably electrostatic) to proteins in extended chain of β-configuration. Extensive work on myelin seemed to confirm this concept and it received wide acceptance in the fiftites. In the sixties, however, the concept was severely attacked by many investigators who found major contradictions between this model and new experimental findings. In particular, electron microscopic studies in a number of membranes were consistent with a globular rather than lamellar membrane. In addition, studies showed that the protein associated with the membranes is at least in part of α-helix conformation rather than in ß-configuration.

These observations led to a globular membrane model advanced primarily by Green and his co-workers at Wisconsin and resulted in a decade of often heated scientific controversy. A model was proposed.

In this model the membrane consists of repeating lipoprotein units, one layer thick in which the association of lipids with the protein is a prerequisite for membrane formation, since it prevents the protein units from forming three dimensional aggregates.

However, some very recent thought has combined the bilayer and the globular concepts.

The protein-lipid crystal proposes a double layer of globular protein subunits with phospholipid located between the globular subunits of protein in such a way that polar as well as nonpolar interactions are possible between the protein and the phospholipids. The thickness of the membrane is now determined by protein and the amount of lipid filling the spaces between the proteins can vary. The membrane is considered in constant thermal motion in the plane of membrane.

The model currently in vogue is the fluid mosaic model. In this model the phospholipid bilayer constitutes the continum of membrane forming, as it were, a “sea of liquid”. Within the bilayer, however, various protein interactions can occur. Proteins can be present as globular, α-helix containing units, or in unfolded ß-configuration. They may interact with lipids by polar or by hydrophobic bonds, and individual subunits, or their complexes may penetrate across the membrane. This system is considered to be fluid mosaic and membrane properties may change drastically with various chemical and physical influences. In particular, this model allows the protein-lipid complexes to form either hydrophilic or hydrophobic “gates” allowing transport. The concept visualises a dynamic membrane.

Lipid-Protein Interactions in Foods

Lipid protein complexes are important in properties and behaviour of many food types. The nature of the interactions involved, however, is only imperfectly understood.

Milk-fat-globule Membrane

Milk is a delicately balanced biological fluid. It is simultaneously a solution of low molecular compounds, a colloidal solution of globular proteins, colloidal dispersion of casein micelles and an oil-in-water emulsion. In fresh milk the emulsion is protected by the milk-fat-globule membrane, a lipoprotein complex which has been studied very extensively by many investigators in the concept of biological membranes.

One of the early concepts of fat globule membrane structure lays stress on a layer of high-melting triglycerides in the periphery of the fat globule which being associated with the surface layer of phospholipids, which in turn was connected to a closely oriented layer of protein.

Subsequent work has demonstrated that the membrane, including the specific enzymes associated with it, originates in the mammary gland at the time of secretion, and the membrane is similar to biological membranes. At the surface of the membrane, adsorbed lipoprotein particles are present.

Modern views centre on the following:

* The fat globule is surroundeα at the time of secretion by a lipoprotein membrane of glandular origin.

* Associated with this membrane are lipoprotein particles.

* On cooling, triglycerides crystallise first at the outermost layers of the fat, and in layers parallel to and associated with the membrane.

* As milk is stored, and treated in various ways, the membrane structure changes profoundly,

This last aspect is of most interest to the food technologists since it affects the functional properties of milk products.

Lipid-Protein Interactions in Processing of Milk Products

Membrane changes occur in various milk processing operations. Fresh, unhomogenised, cooled milk is very susceptible to creaming due to globule clustering. The clusters are apparently held together by bridges formed by a globulin-like protein which interacts with the lipoproteins of the membrane. This process is similar to flocculation of bacteria, and may involve glycoproteins as well as lipoproteins.

Though globule clusters are undesirable in milk, they are desired in ice-cream where as partial destabilisation of the fat globule membrane system causes globule clustering around the air bubbles created during aeration. Most recent work indicates that destabilisation of the membranes in ice-cream freezing actually leads to some spillage of free oil around the air-water interface, and this free oil may relate to mouthfeel of ice-cream. Ice-cream formation is a complex process in which the air-water and the oil-water interfaces apparently compete for adsorption of surface active components.

Another important process in food technology involving this competition is butter-making. In conventional batch churning, as well as in some, but not all, continuous processes, butter formation is due to a process akin to froth floatation. The flat globule concentrate at an air water interface (foam in batch churning, or film surfaces in some continuous butter makers) and the lipid proteins change rendering at least part of the surfaces more hydrophobic, i.e. positively charge, whether adsorption of parts of the membrane is a pre-requisite for such changes is not clear. The increased hydrophobity leads to spreading of oil at the air interface, supression of foam, clumping of fat into what eventually becomes the complicated water-in-oil emulsion of butter. The degree of desorption from the globule membrane during butter-making depends on pre-treatment of the cream and that the partition of heavy metals between butter and butter-milk is affected by heat denaturation of the lipoprotein complex during pasteurisation of cream.

Significant changes occur during homogenisation of milk. During homogenisation the total globule surface increases several-fold, and a lipid-casein complex forms at the surface. This complex formation involves distribution of casein micelles by homogenisation pressure, as well as the increase in globule area.

Electron microscopy has shown that casein particles and fat globules occur separately before homogenisation but form clearly discernible complexes after the process. The distortion and partial dissociation of casein micelles due to pressure is apparently a necessary factor in the complex formation. The technique of freeze-etching has been used frequently in the early 70's to obtain electron micrographs of evaporated milk, and showed that during sterilisation, the casein micelles dissociate into casein subunits of 100°A size, and these particles form a complex imbedding fat globules.

The "resurfacing" of the globules and new modes of association which exist as a result of various types of milk processing may very profoundly affect stability and properties of milk, including :

* Tendency to gel.
* Stability to metal-catalysed and photocatalysed oxidations.
* Susceptibility to enzyme-catalysed changes.

RECENT CONCEPT ON THE PHYSICO-CHEMICAL PROPERTIES OF CASEINS

Properties of Individual Caseins

Caseins are usually isolated from milk by methods involving urea or some similar denaturant in high concentration. The native conformation of these proline rich proteins may, therefore, not be fully recovered within a reasonable time after removal of the urea. However, different workers were unable to demonstrate any appreciable difference between the circular dichroism spectra of ß-casein isolated with and without exposure to urea, but the possibility of more subtle conformational differences remain, perhaps associated with special conformational states adopted during micelle assembly, or prior to phosphorylation or on binding to the colloidal calcium phosphate. Certainly, phosphorylation can induce well-defined conformational transitions in globular proteins. Studies on the isolated caseins may, therefore, not reveal the original conformation or range of conformations that allowed at least the initial stages of aggregation to occur in the Golgi apparatus.

α_{s1}-Caseins

Bovine α_{s1}-casein B, at near-neutral pH, is expected to carry a net negative charge of about -22, largely arising from the clusters of phosphoseryl and glutamyl residues between positions 45 and 75. Prolyl residues occur with a high frequency (17 out of 199), but of these, only two are found in the region containing most of the sites of phosphorylation. Earlier studies by optical rotatory dispersion (ORD), circular dichrosim (CD), and infrared spectroscopy, suggested that α_{s1}-casein has a low proportion of α-helical structure in neutral solution and more recent investigations have confirmed this finding. This is because of the presence of large number of proline residue.

The iso-electric points or pI of a protein is that pH at which it becomes less soluble or in case of casein it coagulates. The iso-electric points or pI of α', α_{s1}, K-casein and ß-casein are respectively 4.1, 4.1, 4.2 and 4.5. The building blocks of proteins that include milk proteins also are amino acids which have a salt like character and exist as Zwitterion or inner salt or dipolar ions. Since amino acids contain both acidic (-C-OH) and basic ($-NH_2$) groups within the same molecule, we may postulate an intermolecular neutralisation reaction, leading to salt formation.

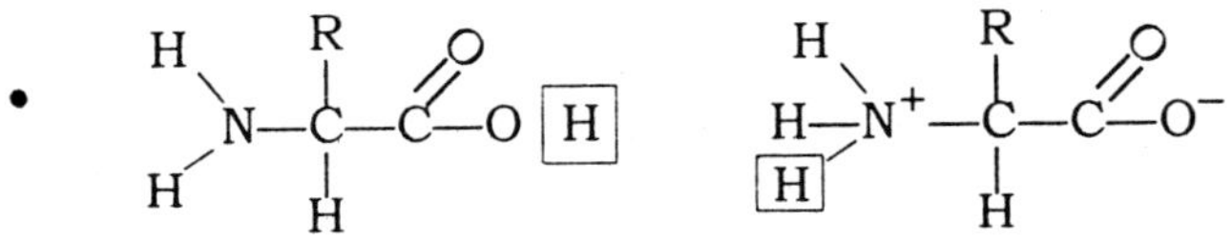

Let us take the example of alanine,

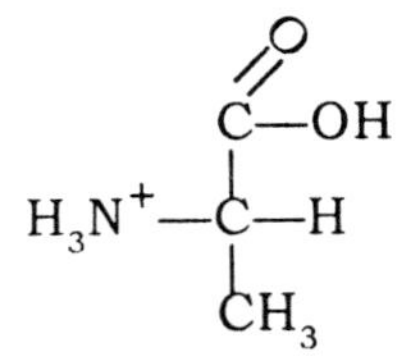

An isoelectric point (pI) its net charge is zero.

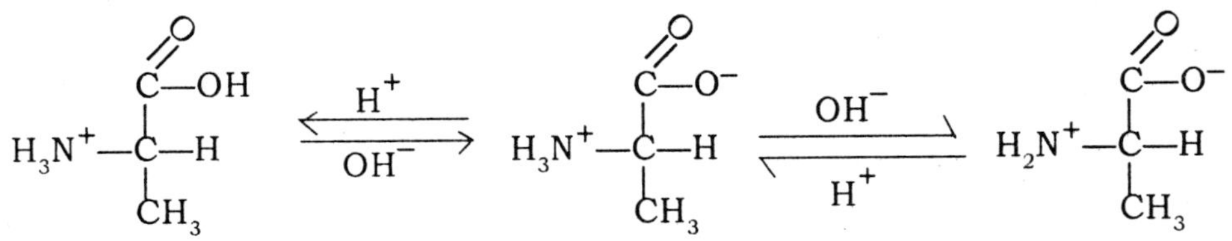

Cationic form net charge = +1 migrate to cathode	Zwitterion form at iso-electric point net charge = 0 pI = 6.0 no migration	Anionic form net charge = -1 migration to anode

All milk proteins have a nearly neutral pH and in an acid pH migrates to cathode and in a basic pH migrates to anode. Usually electrophoresis of milk proteins is done in basic pH.

The aforesaid illustration would explain why bovine α_{s1}-casein B, at near neutral pH, is expected to carry a net negative charge of about -22 which arises from the clusters of phosphoseryl and glutamyl residues.

There are reports that deconvolved and second derivative infrared spectra in the amide I region of a sample identified as α-casein. The strongest component corresponded to that expected to aise from unordered conformations. Likewise, the CD studies have revealed a low proportion of periodic structure in neutral buffers but not more than in 4.5 M guanidum hydrochloride or 6 M urea. A reduction in pH to 1.5 gave only a small increase in the calculated fractions of α helix and B strand, which contrasts with the behaviour of the somewhat similar protein phosvitin, where a significant increase in ß-strand content was observed on reducing the pH to 2. In the spectral region between 250 and 320 nm, it was observed that little change was produced by treating the α_{s1}-casein with carboxypeptidase to remove the c-terminal Trp-199 residue, indicating that Trp 164 and the adjacent Tyr-165 and Tyr-166 are in a relatively stable conformation. Using the structure predicting programme, several investigators were able to match approximately the calculated fractions of α helix and ß-strand with values obtained from the analysis of CD spectra. In the absence of satisfactory information from known protein structures it was assumed that phosphoseryl residues had zero values for the directional parameters determining the propensity to form periodic or reverse turn conformations. Nevertheless, the sequence Ser P-Ile-Ser-P-Ser P-Ser. P-Glu (residues 64-69) was found to lie at one end of an α-helical region (residues 50-69) in the middle, hydrophilic, part of the protein. Other evidence, however suggests that the phosphorylated residues are in flexible regions of the protein. The sequence of aromatic residues identified and having a stable conformation (residues 164-166) was predicted to be in a ß-strand conformation, extending from residues 164 to 168. This region of primary structure is also relatively well conserved between species, suggesting some functional role. The peptide bonds which are more sensitive to cleavage by chymosin, Phe ^{23}Phe^{24} and Phe^{24} Val^{25} lie in a region of predicted ß-strand conformation (residues 22-25), as might be expected given the preference for a ß-strand conformation around the susceptible bond in the substrates of aspartic proteinases.

α_{s2}-Casein

Bovine α_{s2}-casein A contains between 10 and 13 phosphorylated residues in three main clusters or phosphate centres, viz. 8-16, 56-61 and 129-131. It is the most hydrophilic of the Ca^{++} sensitive caseins and with ß-casein, the main substrate for the plasmin in milk. The primary structure suggests five domains, namely, an N-terminal phosphopeptide of approximately 30 residues, another acidic peptide containing the second phosphate centre in the N-terminal half of the molecule, this time about 50 residues long, followed by a predominantly hydrophobic domain of low charge density and low net charge in the middle of the chain. The third phosphate centre occurs in another acidic peptide (approximately residues 125-150) and the fifth domain is a C-terminal basic peptide, sites of preferential cleavages by plasmin are found near the boundaries of each of the domains, apart from the function between the second phosphate centre and the central domain of low charge density. Only one of the 10 prolyl residues is found in the N-terminal half of the protein and none is near a phosphate centre.

The ORD and CD spectra of α_{s2}-casein in Ca^{2+}-free phosphate buffer, pH 7.2 have been measured. The estimated content of helix is about four times as large as in α_{s1}-casein. This is due to the lower proline content of the more phosphorylated protein. There are marked differences in the primary and predicted secondary structures of α_{s2}-type caseins from different species. One notable constant feature, however, is a predicted α-helix (residues 70-83 in the cow), duplicated at residues 152-165, which is consistently found inspite of many amino acid substitutions. These α-helixes could be part of α-helix-loop-α-helix motifs, centered on major sites of phosphorylation, though other evidences indicate that the solution conformation is flexible.

ß-Casein

Bovine ß-casein A^2 carries a net charge of about -12 near neutral pH, largely arising from a hydrophilic N-terminal peptide of about 47 residues, which includes all of the phosphorylated seryl residues but only one of the 35 prolyl residues. The remaining prolyl and nearly all the hydrophobic residues are in the remainder of the molecule so that in terms of sequence and also in terms of physico chemical properties, ß-casein resembles a surfactant with a polar head and non-polar tail. This division of the ß-casein molecule into functional domains has its parallel in the exon/intron structure of the ß-casein gene. In the largely hydrophobic tails two remarkable lengths of uncharged residues occur, 51-90 and 149-168, containing respectively, 30 and 25 per cent prolyl residues.

In its hydrodynamic properties, ß-casein behaves like an expanded coil of low axial ratio, rather than as a compact globular protein. Indeed its radius of gyration calculated from the intrinsic viscosity, assuming a random coil structure, is only a little larger (5.1 nm) than the value found from low-angle X-ray scattering (4.6 nm) and is not appreciably changed in 6 M guanidium hydrochloride. However, there is a considerable amount of evidence that there are, indeed, elements of regular secondary structure in the ß-casein molecule.

Both ^{1}H and ^{31}P nuclear magnetic resonance spectroscopy show that there is a high degree of side chain mobility in the ß-casein molecule as measured by spin relaxation times. The ratio of the intensities of aliphatic and aromatic protons agrees fairly closely with that expected from the amino acid composition. Nevertheless there is a decrease in the absolute intensities in going from trifluoroacetic acid solution to 8 M urea and a further decrease in going to $^{2}H_2O$ solution, which argues that a proportion of the side chains are in fairly fixed conformational states. The backbone structure of ß-casein has been investigated by ORD and CD spectroscopy and generally interpreted in terms of α-helical, ß-strand, ß-turn, and a periodic conformations, although a polyproline II-type component has also been postulated. However, other workers have been able to simulate the CD spectrum of ß-casein very precisely using the method in which a linear combination of CD spectra of 16 proteins of known X-ray structure used, without postulating a polyproline helix. They found 7 per cent α-helix, 33 per cent ß-structure, 19 per cent ß-turns, and 42 per cent aperiodic structure. Other methods of calculation, employing different reference proteins, have given different results, notably less of the extended ß and more of the aperiodic structure. Under the solution conditions used, ß-casein is likely to be monomeric, at least from 4 to 25°C. Several workers noted a decrease in the magnitude of the CD band at 220 nm on raising the temperature to 60°C, which could be interpreted as an increase in ß-strand content at the expense of aperiodic structure, with a constant proportion of the helix. Another interesting and somewhat surprising property of ß-casein is the apparent persistence of some secondary structure in the presence of high concentration of urea or guanidium chloride.

A variety of secondary structure prediction methods has been applied to ß-casein. Regions of α-helix around residues 24, 94 and 133 and ß-strands near residues 83, 147 and 190 are widely predicted. The predicted α-helix in the N-terminal phosphopeptide region may only be stable at low pH, causing the increase in apparent helix content at pH 1.5, compared to neutrality.

^{31}P nuclear magnetic resonance studies have been reported on α_{s1} and ß-caseins, and phosphopeptides derived thereform. The spectrum of ß-casein at milk pH, and in the absence of paramagnetic ions, which can broaden considerably the ^{31}P resonances contains four peaks in the intensity ratios 1:1:2:1, accounting for all five of the phosphoseryl residues.

From the measured pK values of the resonances and their chemical shifts, individual assignments could be made, consistent with the ideal acids most strongly. Thus, the middle phosphoseryl residue (Ser P-18) of the cluster of three reasonated at the highest field strength of milk pH and had the largest pK. Its two neighbours, each with a single adjacent phosphoseryl residue formed the peak of double height. The low-field resonance was absent in the spectrum of ß-casein phosphopeptide, ß-CNA[2] (fragment 1-25), and was, therefore, due to Ser P-35 leaving Ser P-15 to be assigned to the remaining peak. The relaxation times (T_1 and T_2) of the ^{31}P nuclei were relatively long and unequal consistent with rapid segmental motions that are largely independent of the rotation and translation of the molecules as a whole. However, undoubtedly there are specific conformations of the whole ß and α_{s1}-casein proteins which the shorter peptides cannot adopt and which alter the pK values of phosphoseryl residues by bringing hydrophobic or other amino acid side chains into their vicinity K-casein.

K-Casein

The primary structure of K-casein reveals a hydrophilic C-terminal peptide containing many hydroxyamino acid residues and a predominantly hydrophobic remainder, containing, for example, all the tyrosine residues, many of which are highly conserved. Similarly, the C-terminal peptide has a higher negative charge density, particularly when glycosylated, whereas at near-neutral pH the remainder bears a small net positive charge. The chymosin-sensitive bond is at the boundary between these two domains so that cleavage brings about a profound change in the functional properties of the molecule. K-casein is unique among the caseins in having no clusters of phosphoseryl residues and in milk is linked by intermolecular disulphide bonds to form oligomers of various degrees of polymerization.

Analysis of the CD spectrum has yielded values of 14 per cent α-helix and 31 per cent ß-strand, with a possible increase in helix content observed with increase of temperature. In a more recent study, a lower fraction of α-helix was calculated but the results vary with the method of calculation. Structure prediction methods have also been applied to this protein and have given results that encourage the view that K-casein has a number of stable conformational features. Numerous workers have predicted on α-helical content of 23 per cent, with 31 per cent ß-strand and 10 per cent ß-turns. The methods used was by Chou and Fasman (1974). However, most investigators preferred the method of Lin (1974) to predict α-helix and β-strand constant because the method of Chou and Fasman (1974) was considered to over predict these elements (Lenstra, 1977). They also tested their predictions for the structure about the chymosin-sensitive bond using the later boundary analysis method of Chou and Fasman (1978), as both an extended β-structure and α-helix were considered possible conformations for this reason. The conclusion was that an extended β-structure for residues 103-108 was slightly favoured over an α-helix for residues 101-108. In another report it has been proposed that the sequence His-Pro-His-Pro-His- (residues 98-102), on the N-terminal side of chymosin-sensitive bond, and Lys-111, on its C-terminal side, are involved in electrostatic binding on either side of the active site cleft in chymosin, because synthetic substracts containing these sequences are bound more strongly and tend to be hydrolyzed more readily by chymosin.

It has also been found that in the sequence 28-79, there were three predicted lengths of extended β-structure and an helix, which could be arranged in a $\beta\alpha\beta$ configuration, forming a hydrophobic core, resistant to enzymatic digestion, as observed experimentally. In the C-terminal part of K-casein, the high charge density is likely to militate against any stable conformational states. This view was confirmed through high resolution ^{1}HNMR spectroscopic studies on monomeric K-casein and micelles of K-casein. The spectrum of the latter appeared to be a superimposition of a well-resolved component on a broadened component. A good fit to the experimental spectrum could be obtained by taking lineurdths of 25 Hz for the C-terminal residues 91 (approx.) -169 and 30 Hz for the remainder. If the side chain mobility is accompanied by flexure of the polypeptide backbone, the predicted extended β-structure around the chymosin-sensitive bond may, therefore, only be adopted on binding to the proteinase.

Binding of Calcium Ions to Caseins

Many studies have been made of the binding of Ca^{2+} to the caseins because of the undoubted importance of cation binding in the aggregations of these proteins. Binding occurs in the millimolar range of metal ion concentrations and appears to involve the phosphate moiety of the phosphoseryl residues in many of the binding sites. In this regard and also in the high contentof proplyl residues and degree of structural flexibility, the Ca^{2+} sensitive caseins resemble certain other highly phosphorylated phosphoproteins found in bones, teeth, and saliva, all are thought to be involved in binding of the surfaces of calcium phosphate structures or in controlling the stability of super saturated calcium phosphate solutions.

Addition of $CaCl_2$ to α_{s1} or β-casein in neutral solution upto a concentration of 3 mM, caused no change in the CD spectrum below 250 nm. At higher Ca^{2+} concentrations aggregation of the α-casein caused too much light scattering to make measurements possible. It has also been observed that there is no change in the CD spectrum of α_{s1}-casein in addition to $CaCl_2$ upto 5 mM and it has also been reported that no difference in the CD spectra of K-casein solution at neutral pH with and without 40 mM $CaCl_2$. There is a similar lack of effect of Ca^{2+}, on the CD spectrum of phosvitin in neutral solution, but addition of Ca^{2+} to phosphophoryn in neutral solution was reported to increase the content of β-strand structure. The absence of any profound change in the backbone conformation of the relatively flexible caseins and the low binding affinity argue against the formation of any highly specific binding sites of Ca^{2+}, such as EF hand conformation.

There can be little doubt that phosphoeryl and glutamyl (or aspartyl) residues are involved in the binding of Ca^{2+}, with a progressive increase in the contribution of carboxyl groups as more ions are found. Binding capacities of the individual caseins increase with their content of phosphorylated residues, and for a given casein, are increased by further phosphorylation and decreased by dephosphorylation. However, the most convincing evidence comes from infrared measurements. It has been shown that with α^{s1}-casein, the $-PO_3$ stretch at 98 cm^{-1} was shifted by complex formation of Ca^{2+}. Infrared measurements have also indicated the involvement of carboxyl groups.

The binding of Ca^{2+} by α_{s1}-casein also induces changes in ultraviolet absorbance, long wavelength circular dichroism, and fluorescence spectra, consistent, with a shift of chromophores to a more hydrophobic environment, but whether these changes are brought about by a conformational change in the protein or by its association is not clear. Certainly, the C-terminal region, containing the affected chromophores, is essential for the association.

Association Behaviour of Caseins

In protein interactions, it is seldom possible unambiguously to separate electrostatic effects from hydrophobic bonding and the effect of changes in the hydration of hydrophilic groups, particularly those contributing to protein charge. The interface between proteins that are bonded together sometimes show constellations of charged groups on the two surfaces that are complementary in sign, effecting a stereochemical constraint on the interaction. Furthermore, there may be hydrophobic regions that facilitate the expression of water molecules between the two surfaces as they approach. This exclusion of water can strengthen the electrostatic attraction by reducing the effective dielectric constant of the intervening medium. The synergy of electrostatic and hydrophobic factors could be essential for protein aggregation because hydration forces might otherwise be completely dominant over the electrostatic attraction between hydrophilic surfaces at small separations. All of the caseins can undergo association reactions, either with themselves or with other caseins to give a wide variety of multimers. The relatively open conformations of the caseins and the high voluminosities of the aggregates suggest that the protein-protein interface is, possibly, more likely to be defined by the isolated elements of the secondary structure than in the case of globular proteins, where tertiary structure can create the surfaces for interactions.

Association in the Absence of Calcium

α_{s1}-Casein

α_{s1}-casein associates in a series of consecutive steps rather than in the monomer-micelle manner of β- and K-casein, in a way that depends strongly on the pH and ionic strength, but the temperature has almost no effect.

Conclusion has been arrived that at pH 6.7 and 35°C, α-casein associates to form very long, warmlike chains, having a relative molecular mass per unit length of 2239 nm^{-1}. A freely formed chain of the same radius of gyration and total number of sub-units would have a segment length of about 150 nm, indicating considerable rigidity. At higher protein concentrations, there was evidence that the chains could associate laterally to create even large aggregates.

α_{s2}-Caseins

α_{s2}-Casein can undergo indefinite self association to form spherical particles. It has been observed that calculated hydrodynamic volume decreased below an ionic strength of 0.15, contrary to what might be explained by the unusual charge distribution along the amino acid sequence, with a net negative charge near the N terminus and a net positive charge near the C terminus. Increased intramolecular electrostatic attraction, for example, between lysyl and phosphoseryl side chains, could then allow a tighter packing of monomers at low ionic strength.

β-Casein

These caseins behave like simple surfactants in producing spherical soap-like micelles above a critical micelle concentration. Various findings support a monomer → n-mer equilibrium model, with n fairly large, such as the good separation between monomer and micelle peaks seen in the analytical ultracentrifuge, the agreement of the calorimetric and Van't Hoff enthalpies of association, and the existence of a sharp critical micelle concentration. The association is entropically driven and can be reversed by increasing pressure. Though the individual association steps are cooperative, a limit to growth must exist. This phenomenon is supposed to be a steric limit in the packing of head groups in a shell.

The polymerization reaction is very sensitive to temperature, ionic strength, and protein concentration, indicating a delicate balance of forces. At pH 7, ionic strength 0.2, and 21°C, the average degree of polymerisation is about 23 and the average polymer radii about 17 nm.

Several workers have reported that the β-casein micelle has a star-like structure, somewhat like a spherical soap micelle, but with stiffer arms, particularly in the middle of the micelle.

K-Casein

Soap-like micelles of reduced (SH) K-casein form above a concentration of about 5g/litre in neutral solutions of ionic strength 0.1-0.5M. Unlike the β-casein association that of K-casein is relatively insensitive to temperature, giving a constant average degree of polymerization of about 30 and a polymer diameter of 23 nm. Moreover, the ability to form micelles is not affected for reduction and alkylation of the intermolecular S-S bonds, though, for the most reproducible results it appears necessary to treat the protein with 6 M gunidium chloride and dithiothreitol, prior to dialysis against buffer, to allow slow renaturation and micelle formation.

The large specific volume of the reduced K-casein micelle can be explained by either a shell or protein with a water-filled core or by supposing, the C-terminal macropeptide of the protein extends from the surface as a random coil, trapping solvent in a "hairy layer". The ^{1}HNMR measurements are also consistent with a hairy K-casein micelle. There is a note of caution there. A quite different model has been proposed for K-casein micelles formed in a simulated milk ultrafiltrate.

Association in the Presence of Calcium Ions

α_{s1}-Casein

Horne, Parker and co-workers have described kinetic and equilibrium aspects of the reaction between α_{s1}-casein and Ca^{+2} ions. Two distinct approaches were used, one having its origin in the statistical theory of polymer condensation and the other deriving from the classical theory for the coagulation of hydrophobic collloids, both approaches have led to the proposition that net protein charge is of overriding importance in determining the nature and extent of the coagulation reactions of α_{s1}-casein.

α_{s2}-Casein

Little work has been reported on the association behaviour of α_{s2}-casein, but it is clear that precipitation by Ca^{2+} at pH 7.0 occurs at lower levels of Ca^{2+} than for the other caseins, in spite of it being the most hydrophilic casein. For example, under comparable conditions, Ca^{2+} concentrations of 2 and 8 mM are required to precipitate 80 per cent of the protein from solutions of α_{s2}- and α_{s1}-casein, respectively.

b-casein

In the Ca^{2+} induced aggregation of β-casein, a constant protein charged model was shown to be inadequate and the effects of solvation forces are more easily seen. For example, with β-casein, the number of bound Ca^{2+} ions at the critical concentration $CaCl_2$ required for precipitation increased from 4.62 to 6.75 mol/mol casein as the temperature was reduced from 40 to 20°C and 4°C, no precipitation was seen. The aggregation of this protein induced by Ca^{2+} ions, therefore, shows a marked dependence on temperature, similar to the self-association behaviour in the absence of Ca^{2+}. Unlike α_{s1}-casein, the electrostatic and solvation forces leading to association cannot be treated separately.

K-Casein

Though K-casein is described as a Ca^{2+} insensitive casein, some recent work has reported a notable different type of K-casein aggregate in a simulated milk serum compared to those proposed to be formed in Ca^{2+}-free buffers. The aggregates formed are four to five times larger in molecular weight of comparable protein concentrations. The angular dependence of neutron scattering was intermediate between that expected of a homogenous sphere and that of a statistically branded polymer. In the structural model proposed by Thurn *et al.* (1987a), substructures are linked together in a network by interactions between the outer regions of some of the polypeptide chains.

Interaction of Caseins with Calcium Phosphate

It is usually assumed that the interaction of casein with calcium phosphate is mainly mediated through the phosphorylated residues. The phosphate centres of caseins are resistant to attack and remain bound to the colloidal calcium phosphate when casein micelles are digested enzymatically. The effect of different caseins on the log time and rate of precipitation of calcium phosphate is linearly related to the content of phosphorylated residues and there are profound effects of chemical phosphorylation and dephosphorylation of caseins on the size and composition of complexes formed with calcium phosphate. Enzymatic phosphorylation of b-casein increased its binding to hydroxyapatite, particularly when the site of phosphorylation was close to the phosphate centre, indicating a possible synergy of action of phosphorylated residues close together in the sequence. Indeed, this finding may help to explain why the phosphorylated residues are predominantly clustered in the primary structure. Other types of residues may be involved in the interaction but no convincing evidence for the involvement of basic residues has yet been present though this remains a possibility meriting further investigation.

The patterns of association of the caseins induced by precipitation of calcium phosphate are quite different from those seen with Ca^{2+} alone. At a fixed total calcium concentration, the addition of increasing amounts of phosphate will at first, hardly reduce the free Ca^{2+} concentration at about neutral pH because HPO_4^{2-} ions bind only weakly to Ca^{2+}. Thus, if the initial free Ca^{2+} concentration is high enough to induce extensive aggregation of casein, the degree of association may hardly be altered by small additions of phosphate. As the concentration of phosphate is increased, precipitation of calcium phosphate occurs and at a fixed total concentration of calcium, the free Ca^{2+} concentration is high enough to induce extensive aggregation of casein, the degree of association may hardly be altered by small additions of phosphate. As the concentration of phosphate is increased, precipitation of calcium phosphate occurs and, at a fixed total concentration of calcium, the free Ca^{2+} ion concentration decreases; the formation of a casein-calcium phosphate complex can occur in which the casein acts as a protective colloid, limiting the size of the calcium phosphate particles. There can then exist calcium caseinate particles in equilibrium or in equilibrium with the casein-calcium phosphate complexes, the precise balance being sensitive to solution conditions. On further increasing the phosphate concentration, two possibilities arise, each of which could destabilize the system; first, precipitation of more calcium phosphate could exhaust the supply of stabilizing casein, leading to the precipitation of more-or-less crystalline calcium phosphate, or the free Ca^{2+} could fall to such a level that the casein dissociates from the calcium phosphate. The latter possibility must be considered because the binding of casein to the calcium phosphate in the casein micelle has been shown to depend on solution free Ca^{2+} ion concentration.

Location of K-Casein in Micelles

The high voluminosity and the appearance of micelles in electron micrographs suggest that rather than being simple uniform spheres, they are topologically more complex. An important point to be considered in relation to such particles is the meaning to be given to measured or calculated surface areas which increases with the resolution of the measurement and can include internal as well as external surface. Donnellay *et al* (1984) have drawn micelle surfaces by drawing a smooth dosed loop around the boundary of the images of micelles in electron micrographs. Effectively, an external surface is measured at a resolution below the dimension of putative submicelles. On the other hand, the surface area may be defined as the area that can make contact with a water molecule and include internal pores and cavitites as well as external surface. Therefore, the latter could be orders of magnitude larger than the external surface area defined by Donnelly *et al* (1984).

Much of the current debate on the location of K-casein is concerned with whether this component is exclusively or preferentially located on or near the external surface and the relevance of this to its size determining role. In the coat-core model (Waugh, 1971), K-casein was placed on the surface as a way of explaining the increased proportion of K-casein in smaller micelles and to account for the chymosin induced aggregation of micelles. This concept has constrained the interpretation of experiments demonstrating the inverse relation between external surface-to-value ratio and K-casein content. Other work has demonstrated that alternative models of casein micelle structure can explain the observations without requiring the K-casein to be exclusively on the external surface. For example, in the well known model of Slattery and Evard (1973), subunits of variable casein composition associate to form micelles. A geometric argument is used (Slattery, 1977) to allow the subunits to associate to an extent depending on the fraction of their surface area that is not occupied by K-casein. A proportion of the K-casein is on the internal surface of the cluster. When the external surface of a cluster of subunits is sufficiently rich in K-casein molecules, further growth is arrested. In the final stable particle, the composition of external surface achieves a constant value, irrespective of the size of the micelle.

Thus, in the model of Slattery and Evard (1973), micelle size is expected to correlate negatively with the K-casein content, especially as the fr action in the external surface becomes small compared to the total. Essentially the same argument applies if micelle growth is regarded as a branching process in which K-casein acts to terminate growth of the branches. Depending on the average functionality a proportion of the termini (hence of the K-casein) will not lie in the external surface of a crossed-linked gel. In contrast, Dalgleish *et al* (1989) ascribe the change in K-casein content with size exclusively to the change in external surface to volume ratio and reach the conclusion that all the K-casein is on the external surface.

That some of the K-casein is on the external surface of micelles is indicated by the experiments of Yoshikawa *et al* (1978), in which rapid coagulation of micelles was induced by the addition of wheat germ lectin. One argument that has been advanced in favour of the view that K-casein is exclusively on the external surface is derived from the observation that a high proportion of the K-casein in renneted milk is cleaved by the time clotting occurs. However, the chymosin molecules may be able to enter the relatively open structure of the micelles and hydrolyse the internal K-casein with equal facility. Bringe and Kinsella (1986) observed that the proportion of K-casein hydrolysed by chymosin at the clotting time decreased from 71 per cent when the micelles were suspended in 3 mM $CaCl_2$ to 26 per cent in 60 mm $CaCl_2$. It appears, therefore, that the hydrolysis of a substantial fraction of the K-casein is not required for clotting to occur and hence that the K-casein is not necessarily predominantly located on the external surface.

In principal, the proposition of K-casein on the external surface can be studied by using cross-linked or immobilised, enzymes, antibodies lectins or other specific functionalities. This approach is not without its problems. The rate of exchange of K-casein between the micelle and milk serum has not been established but may have a relaxation time of several hours, similar to that of β-casein. If this rate is of the order of the time of the experiment, or faster, the proportion apparently on the surface will be overestimated. Likewise, solubilization of micellar K-casein, e.g., by dissociation during the experiment, would lead to a similar error.

Many attempts have been made to locate K-casein in the casein micelle by electron microscopic methods, but the chief problems have been a lack of sensitivity in some of the methods used and doubts about the reliability of techniques of sample preparation in not altering micellar structure.

Electron microscopy appears to show a distribution of K-casein throughout the micelle and less certainly, a preferential location toward the pheriphery. No support is given to those experiments with immobilised reagents that appear to show that K-casein is located overwhelmingly on the external surface of micelles.

Some Aspects on the Synthesis and Secretion of Casein

Although there are many gaps in our knowledge of the synthesis and secretion of milk proteins by the lactating mammary gland, the essential features at the cellular level bear a close resemblance to the pathways identified in many other eukaryotes.

The caseins are synthesised as preproteins, with a N-terminal signal peptide in the rough endoplasmic reticulum and are able to interact with signal recognition particules and receptors in the membrane, thereby providing the topological conditions for the vectorial transfer of nascent polypeptide chains across the membrane. In the Golgi apparatus the signal peptides are excised, the caseins are phosphorylated, and (in the cow) K-casein only is glycosylated as the proteins move from the proximal to the distal regions of the Golgi apparatus vesicles, containing casein and other components of the aqueous phase of milk, bleb off the distal face of the Golgi sack and move to the apical membrane of the cell, where exocytosis occurs and the contents of the vesicles pass into the alveolar lumen. In full lactation only about 30 min elapse from synthesis to secretion. However, the precise timings, the sequence of events and their spatial location are imperfectly defined. Likewise, the pH and other ion and protein concentrations prevailing during the formation of casein micelles in secretory vesicles are completely unknown. The concentration of Ca^{2+} and H^{1} are of particular interest, together with the stage at which the calcium phosphate is formed because of the strong influence that these factors can have on the association behaviour of caseins.

Structure of Casein Genomes and mRNAs

The existence of allelic genes for all four bovine caseins has allowed classical Mendelian genetic studies to be made of their mode of inheritance. The finding is that the four types of casein are under the control of a cluster of four autosomal genes, and from the non random assortment of the alleles, a close linkage between the loci of α_{s1}-Cn (i.e., the gene coding for -Cn) α_{s1} and β-Cn and between β-Cn and K-Cn was established. Probably, there is also a close linkage of α_{s2} -Cn with the other casein genes and the hypothesis of Grausclaude (1979) is that the loci are in a sequence such as α_{s2}-Cn. β-Cn α_{s2}-Cn....K-Cn or possibly one in which the K-Cn and α_{s2}-Cn loci are interchanged. Long-range restriction analysis of bovine casein genes has confirmed that they will lie on a single chromosome in the order α_{s1}-casein .... β-cn α_{s2} -Casein......K-Cn. The loci lie within a sequence estimated to be 300 kb or less than 200 kb in length. In the cow and the rat about 90 and 150 kb, respectively of cloned sequences containing casein genes have been sequenced but no overlapping sequences have yet been identified.

Jones *et al* (1985) and Yoshimura and Oka (1989) have established the complete exon/intron structure of the rat and mouse β-casein genes, respectively, and as there is evidence of conservation of exon/intron functions in the casein gene family, this work is likely to be directly relevant to bovine β-Cn.

Jones *et al* (1985) sequenced phage clones of genomic DNA containing rat β-Cn and showed it to contain nine exons and eight intervening introns. The 5^1 non-coding region is encoded mostly by exon I_1 and partly by exon II and comparisons of the sequence with the 5^1 uncoding sequences of other casein mRNAs indicate three conserved regions that qualify as possible regulatory elements, including a putative steroid hormone binding site. Exon VIII and IX encode the C-terminal valyl residue and the 3^1 noncoding region, which includes a canonical poly (A) sequence (AATAAA). Most of the hydrophobic C-terminal region of rat β-casein is encoded in a single exon (exon VII). The ability of caseins to undergo association reactions as a part of the process of forming casein micelles and the importance of hydrophobic bonding in polymerisation suggested to Jones *et al* (1985). That exon VII, therefore, encodes a functional domain of the protein, as expected from the hypothesis of Gilberd (1978) and Blake (1978) on the significance of the exon/intron structure of eukaryotic genes. Exon II encodes the entire signal peptide sequence and the first two amino acid residues of the mature protein and also, presumably, corresponds to a functional domain of the protein. Unlike exon VII, however, the region encoding the signal peptide was found to have the most highly conserved sequence in the rate casein mRNAs. The interpretation of Jones *et al* (1985) of the exon structure in the remaining coded region of the gene is also given in terms of the functional domain hypothesis, but becomes more complicated. Exons IV and V as well as III and VI appear to be originated from an exon duplication event. They are quite similar in sequence, particularly at the 3^1 ends, both to each other and to the conserved phosphorylation sites of other casein mRNAs. This is less surprising for exons III and IV, which also encode phosphorylation sites, than for exons V and VI, which do not. Jones *et al* (1985) postulated that the latter pair of exons originated from an earlier duplication event. Jones *et al.* distinguish between a minor phosphorylation site Ser-X-Glu and a major phosphorylation site Ser-X-Glu-Glu, where X is any other amino acid. The conserved sequences of the major phosphorylation site are split between two exons, with Ser-Ser-Glu, the equivalent of a minor site of phosphorylation, at the 3^1 end of one exon and the final glutamyl residue encoded by a 5^1 GAA codon on a separate exon. A minor site is converted to a major site by just opposition, through splicing, of glutamyl codons at the 3^1 and 5^1 termini. Thus, the presence of seryl residues close to exon/intron junctions could according to the proposal of Craik *et al* (1983), facilitate their phosphorylation by membrane-bound casein kinases, because they would then be preferentially located at the protein surfaces. Jones *et al.* (1985) identify the major phosphorylation site as a functional domain of the protein but formed from one exon.

Studies of the structure of K-casein gene (Alexander *et al.*, 1988) have supported earlier findings (Jollies *et al.* 1978, Thompson *et al.* 1985) that it is not closely related to the Ca^{2+}-sensitive casein gene family and that it shows some sequence similarities to the t chain of fibrinogen, especially near exon/intron junctions in the fibrinogen gene. Alexander *et al.* (1988) observed that in spite of considerable variation in the coding sequences and the number and positions of the exons encoding the mature protein, there has been complete conservation of the positions, within the Ca^{2+}-sensitive casein genes, of the first two and the last two exons (Alexander *et al.* 1988). By these criteria there is no discernible relation between K-casein and the other caseins. The K-casein gene contains five exons coding, approximately, for (1) the 5^1 untranslated region, (2) the signal peptide, (3) the final two residues of the signal peptide together with the first nine residues of the mature proein, (4) the rest of the mature protein and a short length of 3^1 untranslated sequence. The postulated functional division of K-casein into a hydrophilic C-terminal macropeptide and a hydrophobic N-terminal para-k-casein domain does not appear to be reflected in the structure of the modern gene.

Complete cDNA coding sequences of the four bovine caseins have been reported (Stewart *et al.* 1984, 1987; Nagao *et al.*, 1984; Jimenez Flores *et al.* 1987). Like the rat α -, β-, and γ-casein mRNAs, the bovine α_{s1}-casein mRNA has a sequence in the 5^1 noncoding region that has been identified as a putative steroid hormone regulatory element. Other conserved sequences are found around the major phosphorylation sites and, particularly, in the region coding for the signal peptide. Two 24-nucleotide repeats encoding a minor phosphorylation site at their 3^1 ends were found, which, if corresponding to exons, might indicate that the major phosphorylation site is formed by a splicing event, as in the rat β-casein mRNA. In order to optimize the alignment of the bovine α_{s1}-casein with the rat α-casein mRNAs, Stewart *et al.* (1984) proposed that a number of insertion/deletion events have occurred, one of which encloses the region of the large deletion (residues 14-26) in the bovine -CN A variant. A second event largely involves insertions in the rat casein mRNA of a 10-times repeated 18-nucleotide block and coincides in position with the 24-nucleotide repeat in the bovine sequence. Thus, some parts of the sequences appear particularly prone to evolutionary change whereas others (e.g., the signal peptide and major phosphorylation site sequences) are very highly conserved.

When homologous α_{s2}-casein sequences were compared, even more insertion/deletion events were required to optimise homology (Stewart *et al.*, 1987). These features contrast with the findings on the β- and k-casein where, inspite of many amino acid substitutions, the overall architecture of the proteins appears to be conserved.

It has been recognized for many years that the casein gene family exhibits an apparently high rate of divergence, as calculated from changes in the amino acid and nucleotide sequences. The high rate of divergence in the Ca^{2+}-sensitive caseins has been taken as evidence supporting the view that these caseins generally lack secondary structural elements, which are considered unnecessary in fulfilling their role as calcium- and phosphate-transporting proteins (Stewart *et al.*, 1984, 1987). Conservation of hydrophobic binding domains still allows scope for considerable amino acid substitutions, deletions, and insertions (Hobbs and Rosen, 1982; Stewart *et al.*, 1984, 1987). To this may be addeed the consideration that the caseins are required by their biological function to form rather open, gel-like structures rather than folding into a compact state, and this does not impose a severe restraint on variation in many parts of the sequence of the mature protein.

In contrast to the homologous recognized in the 5^1 noncoding regions of all the Ca^{2+}-sensitive casein mRNAs, the corresponding region of the K-casein mRNA shows appreciable differences. There is, for example, no sequence corresponding to the putative steroid hormone regulatory element. Similarly, the length of the 3^1 noncoding region is shorter and the highly conserved signal peptide sequences of the Ca^{2+}-sensitive casein mRNAs are markedly different from that of k-casein mRNA. These observations, taken with those on the genomic structures of the caseins, support the contention that the k-casein gene ancestory differs from that of the other casein genes as was originally proposed on the basis of amino acid sequence comparisons (Gaya *et al.*, 1977; Jolles *et al.*, 1978).

Secondary Processing of Nascent Casein Chains

Just as the control and coordination of casein gene transcription and mRNA processing determine the extent of translation of individual casein proteins, control and coordination are evident at every subsequent stage, including the gross sorting of proteins into secreted and non-secreted types (Kelly, 1985) and, probably, more subtle sorting mechanisms along the secretory pathway. In many eukaryotic cells, translation and proteolytic excision of the signal peptide appear to occur concurrently. This aspect is of some importance for self-associating proteins because as long as the preprotein remains membrane bound, formation of secondary structure can occur without the complication of the association of the proteins, possibly leading to "wrong" conformational states, as has been observed in *in vitro* experiments with some oligomeric enzymes. The high proline content of the caseins suggests that they might be particularly sensitive to wrong conformational states, dependent on the sequence of folding and association steps. The formation of the correct structure of a multimeric system may require an assembly program involving some degree of cellular segregation and overall coordination of the various stages of synthesis, folding, secondary processing, and association (Jaenicke, 1982). These stages should not necessarily be regarded as either independent or successive because there is considerable scope for cooperative folding/

processing/association reactions leading to the correct product. However, at present only sketchy and indirect evidence can be gleaned of the existence of such an assembly program in the secretion of milk proteins.

Excision of Signal Peptides

The signal peptides of caseins fall into two groups, as summarised in Fig., reflecting the evolutionary development of casein genes. Thus, the Ca^{2+}-sensitive caseins all have signal peptides that are closely homologous and form one group (Fig. 7), whereas the K-casein signal peptide is longer and, among other differences of amino acid sequence, does not contain a cysteine residue (Mercier and Gaye, 1983). The presence of cysteine is a common feature of signal peptides, suggesting that a disulphide bond might be involved in their recognition and excision. Likewise, the predominance of hydrophobic amino acid residues, and one or two basic residues adjacent to the first residue, suggest that these additional structural features are significant for the function of the peptide. Also, an interspecific comparison of the sequences of casein mRNAs in the region coding for the signal peptides revealed an unusually high degree of conservation, which suggested to Hobbs and Rosen (1982) a possible additional functional role for this region of the mRNA, but this remains a speculation.

```
(a)     M K V L I L T C L V A  L A L A
            L F   F A    L   A   V
                             V
                             T
(b)     M M R N F I V V M N I L A L T L P F L A A
          M S   F L   V T                  G
```

Fig. Consensus sequences of the signal peptides of (a) the three Ca^{2+}-sensitive caseins and (b) the k-caseins, taken from sequences for the cow, sheep, man, guinea pig, rat, mouse, and rabbit.

Phosphorylation: Phosphorylation of caseins occurs as newly synthesised proteins move along the secretoryy pathway and interact with at least two different membrane-bound kinases (Bingham, 1987). Mercier and Gaye (183) have summarised the evidence that phosphorylation is a post-translational event which occurs in the Golgi apparatus and this finding has been confirmed by recent work. Thus, the pulse-chase experiments of Boulton *et al.* (1984), in the lactating guinea pig mammary gland, support the idea of a precursor-product relationship between (^{35}S) methionine-labelled preproteins and phosphorylated forms : no phosphorylated early or intermediate forms of the caseins were identified, whereas excision of the signal peptide could be demonstrated 10-20 min before the incorporation of ^{32}P into newly synthesised protein. Secretion occurred 10-20 min later, as demonstrated by the appearance, then, of fully phosphorylated caseins in the incubation medium. Immunocytochemical studies at the light level have shown that casein kinase and galactosyltransferase have a similar subcellular distribution in lactating mammary gland cells (Moore *et al.*, 1985).

The mammary gland casein kinases are highly specialised and specific enzymes. Apart from the caseins, a number of other examples are known where protein phosphorylation occurs in an ordered sequence of steps, catalysed by different kinases (Fiol *et al.*, 1990). Membrane-bound, serine-specific, cyclic AMP-indpendent kinases requiring a divalent cation for maximum activity have been found in the rat, cow, and mouse (Leiderman *et al.*, 1985). The polyamines, spermine and spermidine are able to influence the selective phosphorylation of caseins at physiological concentrations. Of the two cyclic nucleotide-indpendent kinases isolated from lactating mouse mammary gland, the activity of only one was stimulated by polyamines, particularly at relatively high (>10 mM) concentrations of Mg^{2+} (Leiderman *et al.*, 1985). An interesting feature of mammary gland casein kinase activity is that the optimum Mg^{2+} concentration is about 10 mM, but this can be replaced by Ca^{2+} to give conditions liekly to cause the substrate to aggregate extensively. Phosphorylation and micelle formation driven by a divalent cation concentration in the millimolar range could, therefore, be concurrent processes.

Brooks and Landt (1984) and Brooks (1987) isolated a calmodulin-dependent kinase from rat and bovine mammary acini which selectively phosphorylated k-casein rather than dephosphorylated α_{s1}-casein or β-casein.

In all the species that have been investigated, casein phosphorus is in the form of phosphomonoesters of seryl and, to a much lesser extent, thereonyl residues. The primary recognition site is invariably of the form Ser/Thr-X-Y, where Y is either a glutamyl or occasionally an aspartyl residue and X-may be any amino acid. Phosphorylation can then proceed in the direction of the N-terminus, provided a suitable hydroxyamino acid residue is available in the sequence at position n and an acidic residue (Glu/SerP/Asp) is in the n + 2 position. Nevertheless, not all possible sites are phosphorylated completely, or even partly. Mercier (1981) has constructed a hierarchy of effectiveness of the primary phosphorylation sites:

Ser-X-Glu>Thr-X-Glu>Ser-X-Asp>Thr-X-Asp

If X is basic or particularly bulky, it may reduce the extent of phosphorylation: the best example is provided by the α_{s2}-casein C variant in which Ile replaces Thr-130 to reduce the extent of phosphorylation of Ser-129 and Ser-131. Although the triplet recognition site is necessary for phosphorylation by mammary Golgi kinases, not all sites are actually phosphorylated, so there may be a further topological requirement, such as surface location for the site or even a β-turn conformation. Some evidence for this has come from studies on the kinetics of phosphorylation of synthetic model substrates for mammay gland and other casein kinases (Meggio *et al.*, 1989).

Glycosylation : Only the k-casein component of bovine casein is glycosylated, by the O-glycosyl rather than the more complex N-glycosylation mechanism, probably in the Golgi apparatus. Evidence that the three enzymes involved in glycosylation of k-casein are all located in the Golgi apparatus of the lactating mammary gland was provided by Takeuchi *et al.* (1984). The activities of N-acetylgalactosaminyl transferase, galactosyltransferase, and sialyltransferase were all concentrated more than 10 fold, compared to cell homogenates, in a fraction enriched in membranes from the Golgi apparatus. No such concentration of activity was found in fractions enriched in Golgi vesicle membranes or other subcellular components. The relative order of phosphorylation and glycosylation of caseins in the Golgi apparatus is not known with any certainty, but there is some indirect evidence to indicate that phosphorylation precedes glycosylation. Takeuchi *et al.* (1984) found that dephosphorylation of k-casein A 1-P gave a substrate that was glycosylated at about the same rate as the native substrate. In contrast, phosphorylation of dephosphorylated k-casein containing three carbohydrate side chains was reduced compared to dephosphorylated, unglycosylated k-casein. There is inhibition of ATP and Mg^{2+} when the extravesicular free Ca^{2+} concentration is submicromolar (West, 1981; Neville *et al.*, 1981). The dependence of lactose synthase and Golgi kinase activity on free Ca^{2+} concentration has been used to estimate the free ion concentration within the vesciles; in rat mammary Golgi vesicles, studies of the rate of casein phosphorylation indicated the concentration to be above 0.1 mM (West and Clegg, 1981), whereas in mouse mammary Golgi vesicles a value of about 2-4 mM was indicated (Neville and Staiert, 1983). These estimates are of an order similar to values for the free Ca^{2+} concentrations in the aqueous phase of rat and cow milks, respectively (Holt and Jenness, 1984; Holt *et al.*, 1981; Geerts *et al.*, 1983), possibly indicating that a similar concentration is sustained throughout the assembly of the micelles.

It has been suggested that the accumulation of calcium within Golgi vesicles may be intimately linked to the mechanism of lactose synthesis in the complex, coordinated process of milk secretion (Holt, 1983). The lactose synthase complex (EC 2.4.1.22) comprises two subunits, galactosyltransferase and α-lactalbumin, and is located on the luminal surface of Golgi membranes. Galactosyltransferase has a high-affinity site for Mn^{2+} and a low-affinity divalent metal ion-binding site that is half saturated at about 1 mM free Ca^{2+}, so lactose synthesis could be initiated and modulated by the free Ca^{2+} concentration. The enzyme substrates, glucose and UDP-galactose, enter the Golgi lumen, presumably through membrane carriers, and form UDP and lactose neither of which can permeate the Golgi membrane, but the former is hydrolysed by a membrane-brand Ca^{2+} dependent nucleoside diphosphatase to produce UMP and phosphate, both of which can equilibrate with the cytosol (Kuhn and White, 1975; Navratnam *et al.*, 1986). It appears reasonable to suppose, however, that part of the phosphate generated within the lumen by lactose synthesis is incorporated in the casein micelle by calcium phosphate (Holt, 1983); A further consequence of lactose synthesis is the dilution of Golgi vesicle contents by osmotic swelling. It has long been recognised that lactose synthesis regulates the volume of the aqueous secretion (Taylor and Husband, 1922) and Linzelle and Peaker (1971) suggested that this occurs at the level of the Golgi vesicle. In effect, proteins and, specifically, caseins present in the Golgi dictoyosome are diluted during transit to the alveolar lumen so that casein micelle formation occurs at some casein concentration intermediate between the high concentration present in the dictyosome and that in the aqueous phase of milk (Holt, 1983).

Uncontrolled precipitation of calcium phosphate would present a hazard to the functioning of any cell. The binding of the calcium phosphate produced in Golgi vesicles to the casein phosphoproteins offers a means of controlling the precipitation process and preventing the formation of relatively insoluble crystalline forms of calcium phosphate. In effect, limiting the growth of calcium phosphate to the size of the micellar microgranules prevents a highly cooperative first order phase transition. Hold and Sawyer (1988 a,b) suggested that this may be one of the biological functions of casein and Holt and van Kemenade (1989) have discussed the more general questions of the physiochemical properties to be expected of phosphoproteins involved in the control of calcification - in particular, the advantages of a flexible solution conformation is being able bind rapidly to calcium phosphate nuclei, forming spontaneously from a supersaturated biological fluid. This speculation may go in some way to explain why the caseins have suchh an open and flexible conformation.

It is aruged here that the formation of casein micelles is a highly controlled process, producing a macromolecular complex with a specific structure and function. This point needs to be stressed because of a long-standing view of caseins as “random coil-type proteins” which associate to produce a largely random coil complex having only a nutritional function.

REFERENCES

Alexander, J., Stewart, A.F. Mackinlay, A.G., Kapelinskaya, T.V., Tkach, M., and Gorodetsky, S. (1988). *Biochemical Journal* **178**:395-401.

Andrews, A.L., Atkinson, D., Evans, M.T.A., Fincer, E.G., Green, J.P., Phillips, M.C. and Robertson, R.N. (1979). *Biopolymers* **18**:1105-1121.

Berry, G.P. and Creamer, L.K. (1975). *Biochemistry* **14**:3542-3545.

Blake, C.C.F. (1978). *Nature* (London) **273**:267.

Bloomfield, V.A. and Mead, R.J. Jr (1974). *Journal Dairy Science* **58**:592-601.

Bonsing, J. and Mackinlay, A.G. (1987). *Journal Dairy Research* **54**:447-461.

Boulton, A.P., Pascall, J.C. and Craig, R.K. (1984). *Biochemical Journal* **222**:502-510.

Bringe, N.A., and Kinsella, J.E. (1986). *Journal of Dairy Research* **53**:371-379.

Buchheim, W., and Schmidt, D.G. (1979). *Journal of Dairy Research* **46**:277-280.

Byler, D.M. and Farrell, H.M. Jr. (1989). *Journal of Dairy Science* **72**:1719-1723.

Byler, D.M. and Susi, H. (1986). *Biopolymers* **25**:469-487.

Carroll, R.J. and Farrell, H.M. Jr. (1983). *Journal Dairy Science* **66**:679-686.

Caspar, D.L.D., and Klug, A. (1962). Cold Spring Harbor Symposuin *Quantitative Biology* **27**:1-24.

Chaplin, B., and Green, M.L. (1981). Netherlands *Milk Dairy Journal* **35**:377-380.

Chou, P.Y. and Fasman, G.D. (1974). *Biochemistry* **13**:222-242.

Chou, P.Y. and Fasman, G.D. (1978). *Advances in Enzymology* **47**:45-148.

Cookson, D.J., Levine, B.A., Williams, R.J.P., Jontell, M., Linde, A., and de Bernard, B. (1980). *Europeon Journal of Biochemistry* **110**:273-278.

Craik, C.S., Rutter, W.J. and Fletterick, R. (1983). *Science* **220**: 1125-1129.

Creamer, L.K., Berry, G.P. and Mills, O.E. (1977). NewZealand Journal of *Dairy Science and Technology* **12**:58-66.

Creamer, L.K., Richardson, T., and Parry, D.A.D. (1981). *Archives of Biochemistry and Biophysics* **211**: 689-696.

Dalgleish, D.G. (1973). *Europeon Journal of Biochemistry* **40**:375-380.

Dalgleish, D.G. (1979). *Journal of Dairy Science* **46**: 653-661.

Dalgleish, D.G. (1980a). *Biophysical Chemistry* **11**:147-155.

Dalgleish, D.G. (1980b). *Journal of Dairy Research* **47**: 231-235.

Dalgleish, D.G. (1987). In "Cheese: Chemistry, Physics and Microbiology" (P.F. Fox, ed.)Vol. I, pp. 63-96. Elsevier Applied Science London.

Dalgleish, D.G. (1988). *Journal Dairy Research* **55**:521-528.

Dalgleish, D.G. and Holt, C. (1988). *Journal of Colloidal Interface Science* **123**:80-84.

Dalgleish, D.G. and Law, A.J.R. (1988). *Journal of Dairy Research* **55**:529-538.

Dalgleish, D.G. and Parker, T.G. (1979). *Journal of Dairy Research* **46**:259-263.

Dalgleish, D.G. and Parker, T.G. (1980). *Journal of Dairy Research* **47**:113-122.

Dalgleish, D.G., Horne, D.S. and Law, A.J.R. (1989). *Biochemistry Biophysics Acta* **991**:383-387.

Davies, Dr. R. (1990). *Annual Review of Biophysical Chemistry* **19**: 189-215.

Dickson, I.R. and Perkins, D.J. (1971). *Biochemical Journal* **124**:235-240.

Donnelly, W.J., McNeill, G.P., Buchheim, W. and McGann, T.C.A. (1984). *Biochemical and Biophysical Acta* **789**:136-143.

Farrell, H.M., Jr., and Thompson, M.P. (1988). In "Calcium Binding Proteins" (M.P. Thompson, ed.) 2, pp. 31-51. CRC Press, Boca Raton, Florida.

Ferretti, L., Leone, P., and Sgaramella, V. (1990). *Nucleic Acids Research* **18**:6829-6833.

Fosser, R.C. (1979). *American Journal of Physiology* **236**:C286-C292.

Garnier, J. (1966). *Journal of Molecular Biology* **19**:586-590.

Garnier, J., Osguthorpe, D.J. and Robson, B. (1978). *Journal of Molecular Biology* **120**:97-120.

Gaye, P., Gautron, J.P., Mercier, J.C. and Haze, G. (1977). *Biochemical and Biophysical Research Communication* **79**:903-911.

Geerts, J.P., Bekhof, J.J. and Scherjon, J.W. (1983). *Netherlands Milk Dairy Journal* **37**:197-211.

Gilbert, W. (1978). *Nature* (London) **271**:501.

Graham, E.R.B., Malcolm, G.N. and McKenzie, H.A. (1984). *International Journal of Biological Macromolecules* **6**:155-161.

Green, M.L., Hobbs, D.G., Morant, S.V., and Hill, V.A. (1978). *Journal of Dairy Research* **45**:413-422.

Crosclaude, F. (1979). *Proceedings of International Confederation of Animal Blood Groups, Biochem. Polymorphisms 16th Leningard,* 1978 Vol. 1, pp. 54-92.

Haga, M., Yamauchi, K., and Aoyagi, S. (1983). *Agricultural and Biological Chemistry* **47**:1467-1471.

Heth, A.A. and Swaisgood, H.E. (1982). *Journal of Dairy Science* **65**:2047-2054.

Hobbs, A.A. and Rosen, J.M. (1982). *Nucleic Acids Research* **10**:8079-8098.

Holt, C. (1983). *Journal of Theoretical Biology* **101**:247-261.

Holt, C. and Jenness, R. (1984). *Comparative Biochemistry and Physiology* **77A**:275-282.

Holt, C. and Sawyer, L. (1988a). *Protein Engineering* **2**:251-259.

Holt, C. and Sawyer, L. (1988b). *Protein Engineering* **3**:273.

Holt, C., Muir, D.D. and Sweetsur, A.W.M. (1978b). *Journal of Dairy Research* **45**:47-52.

Holt, C., Dalgleish, D.G. and Jenness, R. (1981). *Annals of Biochemistry* **113**:154-163.

Holt, C., Hasnain, S.S. and Hukins, D.W.L. (1982). *Biochemical and Biophysical Acta* **719**:299-303.

Holt, C., Davies, D.T. and Law, A.J.R. (1986). *Journal of Dairy Research* **53**:557-572.

Holt, C., van Kemenade, M.J.J.M., Harries, J.E., Nelson, L.S., Jr., Bailey, R.T., Hukins, D.W.L., Hasnain, S.S. and de Bruyn, P.L. (1988). *Journal of Crystal Growth* **92**:239-252.

Holt, C., van Kemenade, M.J.J.M., Nelson, L.S., Jr. Sawyer, L., Harries, J.E., Bailey, R.T. and Hukins, D.W.L. (1989a). *Journal of Dairy Reseach* **56**:411-416.

Holt, C. van Kemenade, M.J.J.M., Nelson, L.S., Jr. Harries, J.E., Bailey, R.T., Hukins, D.W.L., Hasnain, S.S. and de Bruyn, P.L. (1989b). *Materials Research Bulletin* **23**:55-62.

Horisberger, M. and Vauthey, M. (1984). *Histochemistry* **80:**9-12.

Horisberger, M. and Vonlanthen, M. (1980). *Journal of Dairy Research* **47**:185-191.

Horne, D.S. (1982). *Journal of Dairy Research* **49**:107-118.

Horne, D.S. (1983). *International Journal of Biological Macromolecules* **5**:296-300.

Horne, D.S. (1984a). *Journal of Colloid Interface Science* **98**:537-548.

Horne, D.S. (1984b). *Biopolymers* **23**:989-993.

Horne, D.S. (1986). *Journal of Colloid Interface Science* **111**:250-260.

Horne, D.S. (1987). *Faraday Discussions of the Chemical Society (Land)* **83**:259-270.

Horne, D.S. (1989). *Journal of Physiology* **22**:1257-1265.

Horne, D.S. and Dalgleish, D.G. (1980). *Internationl Journal of Biological Macromolecules* **2**:154-160.

Horne, D.S. and Davidson, C.M. (1986). *Colloid Polymis Science* **264**:727-734.

Horne, D.S. and Davidson, C.M. (1990). *Milchwissenschaft* **45**:712-715.

Horne, D.S. and Moir, P.D. (1984). *International Journal of Biological Macromolecules* **6**:316-320.

Horne, D.S. and Parker, T.G. (1981a). *Journal of Dairy Research* **48**:273-284.

Horne, D.S., Parker, T.G. and Dalgleish, D.G. (1989). In "Food Colloids" (R.D. Bee, P. Richmond and J. Mingins, eds.), No. 75, pp. 400-406. Spec. Publ. R. Soc. Chem., Cambridge.

Humphrey, R.S. and Jolley, K.W. (1982). *Biochimical and Biophysiacl Acta* **708**:294-299.

Imade, T., Sato, Y. and Noguchi, H. (1977). *Agricultural and Biological Chemistry* **41**: 2131-2137.

Israelachvili, J. and Pashley. R. (1982). *Nature* (London) **300**:341-342.

Jaenicke, R. (1982). *Biophysics of Structure and Mechanism* **8**:2310256.

Jimenez-Flores, R., Kang, Y.C. and Richardson, T. (1987). *Biochemical and Biophysical Research Communication* **142**:617-621.

Jolles, J., Allais, C. and Jolles, P. (1970). *Helvetica Chimical Acta* **53**:1918-1926.

Jolles, P., Loucheaux-Lefebvre, M.H. and Henschen, A. (1978). *Journal of Molecular Evolution* **11**:271-277.

Jones, W.K., Yu-Lee, L.Y., Clift, S.M., Brown, T.L., and Rosen, J.M. (1985). *Journal of Biological Chemistry* **260**:7042-7050.

Kajiwara, K., Niki, R., Urakawa, H., Hiragi, Y., Donkai, N. and Nagura, M. (1988).*Biochemical and Biophysical Acta* **955**:128-134.

Kaminogawa, S., Shimizu, M. and Yamauchi, K. (1983). *Proceedings of World Conference on Animal Production* 5th, Vol. 2, pp. 681-682.

Keenan, T.W. and Dylewski, D.P. (1985). *Journal of Dairy Science* **68**:1025-1040.

Kelly, R.B. (1985). *Science* **230**:25-32.

Kuhn, N.J. and White, A. (1975). *Biochemical Journal* **148**:77-84.

Lee, S.L., Veis, A. and Glonek, T. (1977). *Biochemistry* **16**:2971-2979.

Lenstra, J.A. (1977). *Biochemical and Biophysical Acta.* **491**:333-338.

Levitt, M. and Chothia, C. (1976). *Nature* (London) **261**:552-558.

Linzell, J.L., and Peaker, M. (1971). *Journal of Physiology* London **216**:683-700.

Loucheaux-Lefebvre, M.H., Aubert, J.P. and Jolles, P. (1978). *Biophysics Journal* **23**:323-336.

Mehaia, M.A. and Cheryan, M. (1983a). *Journal of Dairy Science* **66**:390-395.

Mehaia, M.A. and Cheryan, M. (1983b). *Journal of Dairy Science* **66**:2474-2481.

Mercier, J.C. and Gaye, P. (1983). In "*Biochemistry of Lactation*" (T.B. Mepham, ed.) pp. 177-227. Elsevier, Amsterdam.

Nagao, M., Maki, R., Sasaki, R., and Chiba, H. (1984). *Agricultural of Biology Chemistry* **48**:1663-1667.

Navaratnjam, N., Virk, S.S., Ward, S. and Kuhn, N.J. (1986). *Biochemical Journal* **239**:423-433.

Neville, M.C. and Staiertg, P.A. (1983). *Journal of Cellular Biology* **97**:442a.

Neville, M.C., Selker, F., Semple, K. and Watters, C.D. (1981). *Journal of Membrance Biology* **61**:97-105.

Ono, T. and Obata, T. (1989). *Journal of Dairy Research* **56**:453-461.

Ono, T., Kaminogawa, S., Odagiri, S. and Yamauchi, K. (1976). *Agricultural and Biological Chemistry* **40**:1717-1723.

Ono, T., Yahagi, M. and Odagiri, S. (1980). *Agricultural and Biological Chemistry* **44**:1499-1503.

Ono, T., Yada, R., Yutani, K. and Nakai, S. (1987). *Biochemical and Biophysical Acta.* **911**:318-325.

Parker, T.G. and Dalgleish, D.G. (1977a). *Journal of Dairy Research* **44**:79-84.

Parker, T.G. and Dalgleish, D.G. (1977b). *Biopolymers* **16**:2533-2547.

Parker, T.G. and Dalgleish, D.G. (1981). *Journal of Dairy Research* **48**:71-76.

Parry, R.M., Jr., and Carroll, R.J. (1969). *Biochemical and Biophysical Acta* **194**:138-150.

Parsegtan, V.A. (1982). *Advances in Colloid Interface Science* **16**:49-56.

Payens, T.A.J. and Vreeman, H.J. (1982). In "*Solution Behaviour of Surfactants*" (K.L. Mittal and E.J. Fendler, eds.), pp. 543-571. Plenum, New York.

Provencher, S.W. and Glockner, J. (1981). *Biochemistry* **20**:33-37.

Raap, J., Kerling, K.E.T., Vreeman, H.J. and Visser, S. (1983). *Archives of Biochemistry and Biophysics* **221**: 117-124.

Rau, D.C., Lee, B. and Parsegian, V.A. (1984). *Proceedings of the National Academy of Sciences* USA, **81**:2621-2625.

Rollema, H.S., Vreeman, H.J. and Brinkhuis, J.A. (1984). In *"Proceedings of the 22nd Congress Ampere* (K.A. Muller, R. Kind, and J. Ros. eds.) pp. 494-495. Ampere Committee, Zurich.

Rollema, H.S., Brinkhuis, J.A. and Vreeman, H.J. (1988). *Netherlands Milk and Dairy Journal* **42**:233-248.

Schmidt, D.G. (1980). *Netherlands Milk and Dairy Journal* **34**:42-64.

Schmidt, D.G. (1982). In *"Developments in Dairy Chemistry"* (P.F. Fox ed.), pp. 61-86, *Applied Scientific Publication Barking,* UK.

Schmidt, D.G., and Both. P. (1982). *Milchwissenschaft* **37**:336-337.

Schmidt, D.G. and Buchheim, W. (1976). *Netherlands Milk and Dairy Journal* **30**:17-28.

Schmidt, D.G. and Koops, J. (1977). *Netherlands Milk and Dairy Journal* **31**:343-351.

Schmidt, D.G. and Poll, J.K. (1989). *Netherlands Milk and Dairy Journal* **43**:53-62.

Schmidt, D.G., van der Spek, C.A., Buchheim, W. and Hinz, A. (1974). *Milchwissenschaft* **29**:455-459.

Schmidt, D.G., Koops, J. and Westerbeek, D. (1977). *Netherlands Dairy Journal* **33**:40-48.

Slattery, C.W. (1975). *Biophysical Chemistry* **3**:83-89.

Slattery, C.W. (1977). *Biophysical Chemistry* **6**:59-64.

Slattery, C.W. and Evard, R. (1973). *Biochemical and Biophysical Acta* **317**:529-538.

Sleigh, R.W., Mackinlay, A.G. and Pope, J.M. (1983). *Biochemical and Biophysical Acta* **742**:175-183.

Snoeren, T.H.M., van Markwijk, B. and van Montfort, R. (1980). *Biochemical and Biophysical Acta* **622**:268-276.

Stewart, A.F., Willis, I.M. and Mackinlay, A.G. (1984). *Nucleic Acids Research* **12**:3895-3907.

Stewart, A.F., Bonsing, J., Beattie, C.W., Shah, F., Willis, I.M. and Mackinlay, A.G. (1987). *Molecular Biology Evolution* **4**:231-241.

Swaisgood, H.E. (1982). In "*Developments in Dairy Chemistry*" (P.F. Fox, ed.), pp. 1-59, Elsevier Appl. Sci., Barking, U.K.

Tai, M. and Kegeles, G. (1984). *Biophysical Chemistry* **20**:81-87.

Takase, K., Niki, R., and Arima, S. (1980). *Biochemical and Biophysical Acta* **622**:1-8.

Taylor, J.E. and Husband, A.D. (1922). *Journal of Agricultural Sciences* **12**:111-124.

Thompson, M.D., Dave, J.R. and Nakhasi, H.L. (1985). *DNA* **4**:263-271.

Threadgill, D.W., and Womack, J.E. (1990). *Nucleic Acids Res.* **18**:6935-6942.

Thurn, A., Burchard, W., and Niki, R. (1987a). *Colloid and Polymer Science* **265**:653-666.

Thurn, A., Burchard, W., and Niki, R. (1987b). *Colloid and Polymer Science* **265**:897-902.

Van Kemenade, M.J.J.M. (1988). Ph.D. Thesis, University of Utrecht, The Netherlands.

Vreeman, H.J., Brinkhuis, J.A. and van der Spek, C.A. (1981). *Biophysical Chemistry* **14**:185-193.

Walstra,, P. (1990). *Journal of Dairy Science* **73**:1965-1979.

Walter, P., Gilmore, R., and Blobel, G. (1984). *Cell* (Cambridge, Mass.), **38**:5-8.

Waugh, D.F. (1971). In *"Milk Proteins"* (H.A. Mackenizie, ed.), Vol. 2, pp. 3-85, Academic Press, New York.

Waugh, D.F., Slattery, C.W. and Creamer, L.K. (1971). *Biochemistry* **10**:817-823.

West, D.W. and Clegg, R.A. (1981). *Biochemical Society Transactions* **9**:468-469.

Yoshikawa, M., Sasaki, R., and Chiba, H. (1981). *Agricultural and Biological Chemistry* **45**:909-914.

APPENDIX–I

A list of Research Laboratories associated with Dairy Science particularly Dairy Processing

Dairy Chemistry Division,
National Dairy Research Institute,
Karnal–132 001 (Haryana)

Dairy Technology Division,
National Dairy Research Institute,
Karnal–132 001 (Haryana)

Dairy Microbiology Division
National Dairy Research Institute,
Karnal–132 001 (Haryana)

Dairy Chemistry Section,
Southern Regional Station of
National Dairy Research Institute
Adugodi Post Office,
Bangalore–560 030 (Karnataka)

Dairy Microbiology Section,
Southern Regional Station of
National Dairy Research Institute,
Adugodi Post Office,
Bangalore–560 030 (Karnataka)

Dairy Technology Section,
Southern Regional Station of
National Dairy Research Institute,
Adugodi Post Office,
Bangalore–560 030 (Karnataka)

Dairy Chemistry Department,
S.M.C. College of Dairy Science,
Gujarat Agricultural University,
Anand Campus,
Anand–388 110

Dairy Microbiology Department,
S.M.C. College of Dairy Science,
Gujarat Agricultural University,
Anand Campus,
Anand–388 110

Dairy Technology Department,
S.M.C. College of Dairy Science,
Gujarat Agricultural University,
Anand Campus,
Anand–388 110

Department of Dairy Chemistry,
Bidhan Chandra Krishi Viswavidyalaya,
Mohanpur, Nadia, West Bengal–741 252

Department of Dairy Bacteriology,
Bidhan Chandra Krishi Viswavidyalaya,
Mohanpur, Nadia, West Bengal–741 252

Department of Dairy Chemistry,
Sanjay Gandhi Institute of Dairy Technology,
Dhelwan, Lohiyanagar,
Patna–800 020 (Bihar)

Department of Dairy Science,
AES College,
Hingoli–431 513 (Maharashtra)

Department of Dairy Science,
Madras Veterinary College,
Madras–7

Department of Dairy Chemistry,
College of Dairy Science,
Rajasthan Agricultural University,
Udaipur (Rajasthan)

Department of Dairy Chemistry,
Dairy Science College
U.A.S. Hebbal,
Bangalore-560 024 (Karnataka)

Department of Dairy Technology,
Dairy Science College,
U.A.S. Hebbal,
Bangalore-560 024 (Karnataka)

Dairy Microbiology Department
Dairy Science College,
U.A.S. Hebbal,
Bangalore-560 024 (Karnataka)

Department of Dairy Science,
College of Veterinary Science,
Tirupati Andhra Pradesh Agricultural University
Tirupati-517 502
Dist. Chittor, Andhra Pradesh

Department of Dairy Science,
Jawahar Bharati,
Kavali-524 202
Dist. Nellore, Andhra Pradesh

National Dairy Development Board,
Anand-388 001 (Gujarat)

Dairy Technology Department,
Allahabad Agricultural Institute,
Allahabad-211 007

Department of Animal Husbandry and Dairying,
Marathwada Agricultural University,
Parbhani-431 402 (Uttar Pradesh)

Department of Dairy Science,
Veterinary College,
Mannuthy-680 651
Trichur (Kerala)

Department of Animal Husbandry and Dairying,
R.B.S. College,
Bichpuri,
Agra-283 105 (Uttar Pradesh)

Department of Food Science and Technology,
G.B. Pant University of Agriculture and Technology,
Pantnagar-263 145, Nainital

Department of Animal Science,
G.B. Pant University of Agriculture and Technology,
Pantnagar-263 145, Nainital

Department of Animal Husbandry and Dairying,
C.S. Azad University of Agriculture and Technology,
Kanpur-208 002

Military Farms School and Research Centre,
Meerut Cantt-250 001 (Uttar Pradesh)

Department of Dairy Science,
Andhra Pradesh Agricultural University,
Rajendranagar,
Hyderabad-500 030

Department of Animal Husbandry and Dairying,
Punjab Rao Krishi Vidyapeeth,
Akola-444 104

Department of Dairy Science,
College of Agriculture,
Jobner–303 329 (Rajasthan)

Department of Chemistry,
Loyola College,
19, Pushpa Nagar 1st,
Madras–600 034

Institute of Dairy Technology (COAS)
Kamareddy–503 111 (Andhra Pradesh)

Department of Dairy Science,
Madras Veterinary College,
Madras–600 007

Department of Food Science and Technology,
Punjab Agricultural University,
Ludhiana–141 004 (Punjab)

Department of Animal Science,
Punjab Agricultural University,
Ludhiana–141 004 (Punjab)

Food Science Division,
Central Food Technology Research Institute
Mysore–570 013

Cooperative Dairy Training Institute,
C-383, Indira Nagar,
Lucknow (Uttar Pradesh)

Agricultural College,
Vellayani,
Trivandrum–695 582

Birsa Agricultural University,
Ranchi–834 007

Department of Animal Science,
North-Eastern Hill University
College of Agriculture,
Medziphema-797 106

Department of Animal Husbandry and Dairying,
Amar Singh College,
Lakhaoti-245 407

APPENDIX–II

Standards of various milk and milk products in India
P.F.A. standards for different classes and designation of milk

Class of milk	Designation	Locality	Minimum milk fat	Per cent solid-not-fat
Buffalo milk	Raw, pasteurised, boiled, flavoured and sterilised	Haryana	6.0	9.0
Cow milk	Raw, pasteurised, boiled, flavoured and sterilised	Haryana	4.0	8.5
Goat or sheep milk	Raw, pasteurised, boiled, flavoured and sterilised	Haryana	3.5	9.0
Standardised milk	Raw, pasteurised, boiled, flavoured and sterilised	All India	4.5	8.5
Recombined milk	Raw, pasteurised, boiled, flavoured and sterilised	All India	3.0	8.5
Toned milk	Raw, pasteurised, boiled, flavoured and sterilised	All India	3.0	8.5
Double toned milk	Raw, pasteurised, boiled, flavoured and sterilised	All India	1.5	9.0
Skimmed milk	Raw, pasteurised, boiled, flavoured and sterilised	All India	not more than 0.5%	8.7

Standards for different classes and designations of milk as per P.F.A. (1976)

Class of milk	Designation	Locality	Milk Fat per cent minimum	S.N.F. per cent minimum
Buffalo milk	Raw, pasteurised, boiled, flavoured and sterilised	Assam, Bihar Delhi, Haryana, Gujarat, Maharashta, Punjab, Uttar Pradesh West Bengal.	6.0	9.0
		Rest of the country	5.0	9.0
Cow milk	Raw, pasteurised, boiled, flavoured and sterilised	Chandigarh, Haryana, Punjab,	4.0	8.5
		Orissa	3.0	9.0
		Rest of the country	3.5	8.5
Goat or sheep milk	Raw, pasteurised, boiled, flavoured and sterilised	Chandigarh, Haryana, Kerala, Madhya Pradesh, Uttar Pradesh, Punjab	3.5	9.0
		Rest of the country	3.0	9.0
Standardised milk	Raw, pasteurised, boiled, flavoured and sterilised	All India	4.5	8.5
Recombined milk	Raw, pasteurised, boiled flavoured and sterilised	All India	3.0	8.5
Toned milk	Raw, pasterurised, boiled, flavoured and sterilised	All India	3.5	8.0
Double toned milk	Raw, pasterurised, boiled flavoured and sterilised	All India	1.5	9.0
Skimmed milk	Raw, pasterurised, boiled, flavoured and sterilised	All India	0.5	8.7

Note: When milk is sold without labelling the class, the standard prescribed for buffalo milk shall apply.

Standards of Various Milk Products as per PFA

1.	Cream	Milk fat 25 per cent (minimum)
2.	*Dahi* or curd	*Dahi* shall have the same minimum percentage of milk fat and milk solids-not-fat as the milk from which it is prepared.
3.	Paneer/*Channa*	Moisture 70.0 per cent (minimum) Fat 50.0 per cent of the dry matter (minimum)
4.	Skim milk Paneer/*Chhana*	Moisture 70.0 per cent (maximum) Fat 13.0 per cent of the dry matter (maximum)
5.	Cheese (Hard)	Moisture 43.0 per cent (Maximum Fat 42.0 per cent of the dry matter (minimum) Stabiliser/emulsifier/NISIN 0.1 per cent.
6.	Processed cheese	Moisture 47.0 per cent (maximum) Fat 40.0 per cent of the dry matter (minimum) Salt 3.0 per cent (maximum) Stabiliser/emulsifier-4.0 per cent (Minimum) or sorbic acid-0.1 or Nisin 0.1 per cent.
7.	Ice cream	Fat-10.0 per cent (minimum), 8.0 per cent (minimum) if containing fruits of nuts. Protein-3.5 per cent (minimum) T.S.-36 per cent (minimum) Stabilizer/emulsifier -0.5 per cent (Maximum)
8.	Milk ices/Dollies	Fat 2.0 per cent (maximum) Protein 3.5 per cent (minimum) T.S. 20 per cent (minimum) Stabilizer/emulisifer-0.5 per cent (Maximum)
9.	Condensed milk unsweetened evaporated milk	Fat 8.0 per cent (minimum) Total milk solids-26 per cent (minimum)
10.	Condensed milk sweetened	Fat-9.0 per cent (minimum) Total milk solids-31.0 per cent (minimum) Cane Sugar-40.0% (minimum)
11.	Condensed skimmed milk unsweetened (evaporated skimmed milk)	Total milk solids-20.0 per cent (minimum) MIlk fat-not more than 0.5 per cent by weight

	Product	Standard
12.	Condensed skimmed milk sweetened	Total milk solids (26.0 per cent min.) Cane sugar 40.0 per cent (minimum) Milk fat-Not more than 0.5 per cent by weight
13.	Whole milk powder	Fat 26.0 per cent Acidity 1.2 per cent (L.A.) maximum solubility 98.5 per cent (Spray) 85 per cent (Roller) S. Index 2.0 (ml) Spray 15.0 (ml) Roller Moisture 5.0 per cent (maximum) B.H.A. - 0.01 per cent, coliform 90/g, SPC - 50,000 (g) Sodium citrate 0.3 per cent
14.	Skimmed milk powder	Milk fat 1.5 per cent (maximum) Moisture 5.0 per cent (maximum) Acidity 1.5 per cent (maximum) Sodium citrate 0.3 per cent Maximum SPC - 50,000/g E. col. Coliform - 90/g
15.	Partially skimmed milk powder	Moisture - 5.0 per cent (maximum) Fat - It should be between 1.6 to 24.0 per cent, Acidity 1.5 per cent (L.A.) maximum, Sodium citrate - 0.3 per cent BHA - 0.01%

The minimum solubility index of the product

	Roller dried	Spray dried
Solubility per cent	85.0	98.5
Solubility index	15.0 ml	2.0 ml

	Product	Standard
16.	*Khoa*	Moisture 28.0 per cent (maximum) fat 20.0 per cent (minimum)
17.	Table Butter	Fat - 80.0 per cent (minimum) Salt 3.0 per cent (maximum) Curd 1.5 per cent (minimum) Diacetyl - 4.0 PPM (maximum)
18.	*Deshi* Butter (Cooking)	Fat - 76.0 per cent (minimum)
19.	*Ghee*	Butyro-refractometer reading 40.0 to 43.0 reading at 40°C Minimum Reichert value - 28.0 per cent of free fatty acids as 3.0, oleic acid (maximum limit), per cent of moisture (maximum limit 0.5 per cent)

20.	Infant Milk Food	Moisture - 5.0 per cent (maximum) Fat - 18.0 per cent (maximum) - 28.0 per cent (maximum) Total carbohydrate - 35.0 per cent (Min.) Milk proteins - 20.0 per cent (minimum) Ash - 8.5 per cent maximum SPC - 50,000/gm SI - 2.0 ml (spray dried) 15.0 ml (roller dried)
21.	Soy-whey Food	Fat - 12 per cent Protein - 23 per cent Carbohydrates - 50.5 per cent Ash - 5.0 per cent Moisture - 1.5 per cent (Fortification with vitamins and iron)
22.	Coffee complete	Instant coffee - 7.2 per cent SNF - 20.0 per cent Sugar - 39.0 per cent Water - 25.0 per cent Total solids - 75.0 per cent Milk fat - 8.5 per cent
23.	Tea complete	Fat - 13.0 per cent SNF - 31.5 per cent Sugar - 39.5 per cent Tea solids - 14.0 per cent Moisture - 2.0 per cent

AGMARK STANDARDS FOR GHEE

	Test	All-India	Regional Winter	Regional Summer
1.	Baudouin		-Negative-	
2.	Phytosterol Acetate		-Negative-	
3.	B.R. reading (40°C)	40.0-43.0	41.4-44.0	42.50
4.	R.M. Value	Not less than 28.0	Not less than 23.0	Not less than 21.0
5.	P. value	1.0-2.0	0.5-1.2	0.5-1.0
6.	Moisture (%)	-Not more than 0.3-		
7.	Free fatty acid (% oleic)			
	a) Special grade Agmark Red label	-Not more than 1.4-		
	b) General grade Agmark green label	-Not more than 2.5-		

SUBJECT INDEX